AF902705

A Brief History of Transportation in China

Huapu Lu
Editor

A Brief History of Transportation in China

人民交通出版社
China Communications Press

Springer

Editor
Huapu Lu
Institute of Transportation Engineering
Tsinghua University
Beijing, China

Map Content Approval Number: GS(2025) 3696

ISBN 978-981-97-9850-6 ISBN 978-981-97-9851-3 (eBook)
https://doi.org/10.1007/978-981-97-9851-3

Jointly published with China Communications Press Co., Ltd
The print edition is not for sale in China (Mainland). Customers from China (Mainland) please order the print book from: China Communications Press Co., Ltd

© China Communications Press Co., Ltd 2025
This work is subject to copyright. All rights are solely and exclusively licensed by the Publisher, whether the whole or part of the material is concerned, specifically the rights of reprinting, reuse of illustrations, recitation, broadcasting, reproduction on microfilms or in any other physical way, and transmission or information storage and retrieval, electronic adaptation, computer software, or by similar or dissimilar methodology now known or hereafter developed.
The use of general descriptive names, registered names, trademarks, service marks, etc. in this publication does not imply, even in the absence of a specific statement, that such names are exempt from the relevant protective laws and regulations and therefore free for general use.
The publishers, the authors, and the editors are safe to assume that the advice and information in this book are believed to be true and accurate at the date of publication. Neither the publishers nor the authors or the editors give a warranty, express or implied, with respect to the material contained herein or for any errors or omissions that may have been made. The publishers remain neutral with regard to jurisdictional claims in published maps and institutional affiliations.

This Springer imprint is published by the registered company Springer Nature Singapore Pte Ltd.
The registered company address is: 152 Beach Road, #21-01/04 Gateway East, Singapore 189721, Singapore

If disposing of this product, please recycle the paper.

Foreword

Professor Huapu Lu is the author of *A Brief History of Transportation in China*, and it was a great honor for me to receive an invitation to write the Foreword.

Professor Huapu Lu and I have been colleagues for many years, and we have also developed a strong friendship. He has dedicated his life to teaching and researching transportation. Not only does he make great impact on transportation research on the global stage he has also conducted in-depth research and made significant contributions to China's transportation industry. His extensive knowledge and numerous publications make him the ideal candidate to author this book.

As widely recognized, transportation is the bedrock of human civilization and the lifeblood of a nation's development. China's rich and enduring civilization has given rise to a unique transportation system. The integration of territories, political cohesion, economic development, cultural interchange, and even the strengthening of national defense all rely on the indispensable role played by transportation.

Since the time of the Xia and Shang dynasties, China has been accumulating a wealth of experience in the field of transportation. The saying, "History is the narration of the past to instruct future generations", underscores the fact that history is the best teacher. Comprehensive works on the history of transportation in China, however, have been relatively scarce. Eighty-five years ago, historian Mr. Bai Shouyi authored *A History of Chinese Transportation*, which detailed China's transportation development up until the 1930s. He undoubtedly served as a pioneer in the study of Chinese transportation history. Nonetheless, due to the limitations of the era, the content in the book concerning modern transportation appears somewhat lacking. In the past few decades, various government agencies have started to place greater emphasis on the study of China's transportation history, resulting in the publication of numerous works. However, the issue of fragmented writing remains conspicuous. Different sectors such as roads, water transportation, railways, aviation, etc. have operated independently, each documenting its own developmental history, which has made it challenging for readers to gain a comprehensive understanding of the overall development of transportation in China.

Although titled A Brief History, Prof. Huapu Lu's book *A Brief History of Transportation in China*, is anything but brief in content. It delves into the historical evolution of transportation in China over thousands of years and illustrates the journey of modern transportation development in the country. Notably, Prof. Huapu Lu's comprehensive approach deserves praise. To present a holistic view of China's transportation development, he united experts in the history of various transportation modes to collaborate on this work. This not only harnessed the collective wisdom of the team but also endowed this book with a higher degree of comprehensiveness, accuracy, and authority.

This book follows the thread of transportation development along a continuous timeline of historical dynasties, with a primary focus on presenting the factual accounts of transportation development, key events, and their evolving characteristics. It paints a fundamental picture of the origins, formation, and development of transportation. Simultaneously, the author delves into the exploration of the mutual relationship between transportation and Chinese culture, elucidating how transportation is integrated with politics, economics, society, and international exchanges. This not only reflects the supporting role of transportation but also highlights its transformative effects. Additionally, the author has, based on ancient texts and historical sources, created road network maps from different eras and thoughtfully selected photographs of significant transportation landmarks to facilitate readers' intuitive comprehension.

What's even more commendable is that the book's comprehensive and in-depth discussion of the trajectory and accomplishments in the development of transportation in modern China. It covers the remarkable progress of the new transportation sector. The author highlights innovative achievements such as China's bridges, roads, vehicles, and ports that have left the world astounded. Drawing from this, the author offers a compelling conclusion: transportation is the vanguard of progress, an essential element for national development, and the cornerstone of a strong nation. It will undoubtedly play a critical role in the great rejuvenation of the Chinese nation.

A Brief History of Transportation in China is a splendid creation with a well-structured narrative, abundant visual content, and easy-to-understand language. The publication of this book not only helps readers to comprehensively grasp the main facets of China's transportation development but also fosters the spirit of self-improvement among those in the transportation industry, thereby bolstering the efforts to build a powerful transportation nation.

July 2022

Fu Zhihuan
The President of the Chinese Academy
of Engineering
Beijing, China

Preface

Upon the gracious invitation from the China Communications Press, I enthusiastically agreed to take up the task of writing *A Brief History of Transportation in China*. This project has been a cherished aspiration of mine for many years.

Where there are people, there is transportation. As the saying goes, "There were no roads in the world; people walked and made them." Therefore, in narrating a concise history of transportation, it is essential to harken back to the experiences of those who walked. The remarkable Chinese nation, from its very inception, embarked on a journey to explore the universe's order, ponder the philosophy of life, seek harmony in all things, and yearn for a better life. This journey was rooted in their innate qualities of diligence, kindness, tolerance, resilience, and wisdom. They held themselves to a high standard, warmly welcomed visitors from distant lands, finding immense joy in the company of friends, advocated for tolerance and comprehension, pursued noble aspirations, and respected the laws of natural development. Throughout history, the Chinese nation has been the architects of a better life and the catalysts for harmonious coexistence among all living things. This book's exploration of China's transportation history is deeply intertwined with this natural backdrop and cultural context.

In fact, whether transportation, urban planning, or the entirety of human society, they are all an intricate whole. Various aspects and numerous factors are interconnected and interdependent, making them inseparable. Therefore, when contemplating the framework for this book, I think it is imperative to adopt a comprehensive perspective that spans geography, politics, economics, society, culture, and technology. By integrating this with the broader historical and cultural context, I aimed to craft a concise and easily comprehensible history of transportation. This book will not only offer a clear exposition of fundamental historical facts and the evolution of transportation methods but will also unveil the internal and external conditions influencing transportation development, along with the philosophical and cultural connotations it carries.

Chinese transportation experts, through their unwavering diligence, wisdom, dedication, and loyalty, have forged marvels that have become world-renowned, leaving an indelible imprint on history and an invaluable legacy. These accomplishments encompass the road networks that solidified national unity, the canals that drove the nation's economic flourishing, the ancient Silk Road stretching from the distant past to the future, China's high-speed rail system marked by the indelible imprints of diligence and wisdom, the extensive road networks to support the economy, the epic narratives of maritime transportation, the rapid advances in civil aviation and space transport, the extensive pipeline infrastructure that spans the vast expanse of China, the robust urban transportation systems that underpin the healthy growth of cities, and a diverse array of transportation masterpieces and historical spectacles. From the ancient Tea-Horse Road to the hanging roadways, from the virtue of benevolence to the concept of a shared human destiny, the Chinese nation, with its outstanding qualities of kindness, maturity, and resilience, upholds the noble ideals of a global community and shared prosperity. In the narrative of transportation development, they continue to inscribe new chapters.

The historical evolution of transportation is a continuous narrative, therefore, this book underscores four distinctive features in its composition. Firstly, the book revolves around the development of transportation as its central theme, sketching the fundamental fabric of Chinese society along a chronological timeline that spans dynasties, with a focus on recounting pivotal events that propelled transportation advancements. Secondly, the book ensures historical accuracy. The authors have meticulously consulted authoritative historical works, engaged in multifaceted analysis, and judiciously referenced ancient texts. For contemporary historical content, the book has invited numerous authoritative experts from various domains within the transportation sector to provide oversight and, in some cases, direct contributions to specific sections, ensuring the book's precision and credibility. Thirdly, the book delves into the interplay and impact of transportation on Chinese culture. It seamlessly integrates the essence of traditional Chinese culture and philosophical thought into the transportation narrative, elucidating the intricate connections between transportation and culture, particularly in relation to the concept of a shared human destiny and the "Belt and Road" initiative. Fourthly, the book is richly illustrated. While narrating the historical development of road networks and other aspects, the authors have created road network maps for different historical eras based on ancient texts and historical records. These illustrations include annotations that denote the correspondence between ancient and contemporary place names, facilitating readers' intuitive comprehension and cross-reference.

This book, from its inception in planning and conceptualization, to its framework design and content creation, has received significant attention from the Ministry of Transport and generous support from numerous experts in the field of transportation. The China Communications Press has provided continuous guidance throughout the entire process. Editor-in-Chief Han Min elucidated the book's background, purpose, and essential requirements. Director He Liang and Editor Qu Wencong have not only answered queries and provided clarification throughout the writing process but have also been actively involved in editing and improving the content of each chapter, essentially becoming contributors to the book. Academician Fu Zhihuan, Academician Huang Weihe from the Chinese Academy of Engineering, and Chairman Weng Mengyong from the China Highway and Transportation Society have respectively recommended authors for the sections on railways, pipeline transportation, and highways. Vice Chairman Guan Jizhen from the China Intelligent Transportation Association organized a special expert discussion meeting for the section on intelligent transportation and provided a reference outline. Several experts in intelligent transportation contributed short articles on cutting-edge developments. Deputy Director Peng Xiao from the Urban Transportation and Rail Transit Research Center at the Transport Research Institute of the Ministry of Transport contributed a historical review of the Ministry of Transport's urban public transportation development. Director Li Jing of the Space Division at the China Transport and Communication Information Center authored the section on Beidou applications in the transportation industry. Chairman Zhang Huahui of the Energy Industry and Trade Group in Nanan City, Fujian Province, and Chairman Yi Bing of the Chongqing Transport Hub Company provided comprehensive support and enthusiastic assistance for the in-depth research on China's maritime trade center during the Song and Yuan Dynasties and the Transit-Oriented Development (TOD) case in Shapingba, Chongqing. Due to space limitations, it's not feasible to list all the scholars and experts who have offered genuine help and enthusiastic support. I would like to sincerely thank all the experts, colleagues, and friends for their invaluable assistance.

This book is the result of the painstaking efforts of dedicated experts and represents a collective endeavor. The principal authors are: Huapu Lu, Director of the Institute of Transportation Research at Tsinghua University; Zhu Jian'an, Head of the Public Security Ministry's Road Traffic Safety Research Center Education Office; Wang Jifeng, Director of the Comprehensive Transportation Research Institute at the Traffic Branch of the China Academy of Urban Planning and Design; Liu Changjian, Head of the Transportation Economics Division at the Water Transport Institute of the Planning and Research Institute of the Ministry of Transport; Xu Deqian, Vice Director of the China Highway Society Office and Deputy Secretary-General of the Committee on Traffic History and Culture; Liu Zhongmin, Director of the Archives and Historiography Center at China National Railway Group Limited; Zhang Zhiming, Deputy Chief Engineer of China Communications Construction Company Ltd.; Guan Zhongyuan, President of the Journal of Oil and Gas Storage and Transportation at the National Pipeline Network Group Research Institute; Zhang Yu, Executive Editor of "*Urban Transportation*" magazine; Prof. Zhang Tangxian from National Taiwan University; Ye Zhenxiang, Senior Engineer at the Institute of Transportation Research at Tsinghua University; Bao Weigang, Deputy Director of the Technology Center at China Communications Construction Company Ltd.; Wang Jing, Associate Professor in the Department of Urban and Rural Planning at the School of Architecture and Urban Planning at Beijing University of Civil Engineering and Architecture; Li Ziming, Deputy Head of the Historiography Office at the Archives and Historiography Center of China National Railway Group Limited; Zhang Yongbo, Senior Engineer at the Institute of Transportation Research at Tsinghua University; Lu Yang, Chairman of Beijing PwC Technology Co., Ltd.; Yu Zhongtao, Deputy Director of Design II (Inland Waterways) at China Communications Waterway Planning and Design Institute Ltd.; Wang Tianshi, Doctoral candidate at the Institute of Transportation Research at Tsinghua University; Bai Zhuotong,

Doctoral candidate at the Institute of Transportation Research at Tsinghua University; Liu Ruoyang, Master's candidate at the Institute of Transportation Research at Tsinghua University; Li Xiaoming, Senior Engineer at the Shaanxi Highway and Tunnel Museum.

The principal authors, from the day they accepted the invitation just before the 2022 Spring Festival, committed themselves to the creation of this book with the aim of producing a refined work that embodies comprehensive historical facts and serves as a historical resource for guidance. They forfeited the precious time they would have spent celebrating the Spring Festival with their families and enjoying leisure vacations. Instead, they immersed themselves in deep contemplation, diligent research, meticulous refinement, and repeated consideration, demonstrating a profound sense of responsibility and a strong mission.

This book aims to integrate transportation with economic, social, and cultural aspects and to link the evolution of transportation with the development of political, economic, technological, and other elements. Our aspiration is for this book to become a reference for researchers and educators in the field of transportation history, a trusted guide for transportation researchers and technologists, and a historical repository for management personnel in the transportation industry. Furthermore, we hope it will be widely accessible to the general public, allowing readers to gain a comprehensive understanding of the historical panorama of China's transportation and to deeply comprehend the relationship between transportation development and the nation's prosperity and the rejuvenation of its people.

Given the authors' limitations in knowledge and the broad scope of the content, this book surely has its shortcomings. We sincerely invite readers to offer criticism and correction!

Beijing, China
June 2022

Prof. Huapu Lu

Contents

1 A Rich Legacy: The Birth of Chinese Transportation 1

2 Ancient Transportation: From the Yinxu Carriage to the Emperor Carriage 19

3 Unification of the Four Seas: Establishment of the Qin Empire and Qin Chi Road 49

4 Postal Service System: An Everlasting Communication System 67

5 Ancient Path of Qinling Mountains: Rise of the Great Han Empire Under the Advancing Secretly by an Unknown Path 91

6 Millennium Canal: Prosperity Forged by Water Transport of Grain 107

7 Crossing Business and Travel: Ancient Tea Horse Road and Horse Caravan Transportation 143

8 Urban Roads: Inheritance and Development of Traditional Chinese Urban Planning Wisdom 165

9 Silk Road: Common Wealth and Great Unity 197

10 Maritime Area: The Rise of Maritime Transport and Economic Development 219

11 Driving Progress Through Transportation: The Role of China's Highway Development in Economic Prosperity 259

12 High-Speed Railway Racing: A Hundred Years of Dramatic Changes in China's Railways 291

13 Through Mountain and Cross Sea: The Outstanding Chinese Bridge and Channel 321

14 Flight Dreams: From Ancient Aspirations to Contemporary Realization 367

15 Urban Transport: Supporting and Leading the Way for Rapid Urbanisation and Motorisation in China 391

16 Transportation Towards the Globe: Great Upsurge of Ports and Water Transportation 419

17 Pipeline Transportation: Energy Arteries Spanning Thousands of Miles 461

18 Diverse: Wonders of Transportation Creation Throughout the Ages 485

19 Transport in the Treasured Island: The Roller-Coaster Development 507

20 Leap of Civilization: The Road to Sustainable Urban Transportation 531

21 Intelligent Empowerment: Enabling People to Enjoy Mobility, Facilitating Goods to Flow Freely 559

22 Glory Decade: Filling Gaps, Creating Miracles, Transport Development Reaches New Heights 589

23 Carrying on the Past and Opening up the Future: Comprehensive Transportation Enters the Era of Integrated Development 637

Chapter 1
A Rich Legacy: The Birth of Chinese Transportation

Transportation encompasses the movement of people and goods through space. It co-evolved with human society and stands as a fundamental element and driving force for human survival and progress. The history of transportation is an integral part of human development.

Tracing back along the historical trajectory, we embark on a journey into the early stages of Chinese civilization. This chapter serves as both the prologue of this book and a meticulous exploration of the living environment, the developmental context, and the historical threads of transportation in ancient China. In terms of the timeline, this chapter covers the widest historical span, spanning approximately 2500 years, from the mythical times of the Three Sovereigns and Five Emperors to the period before the unification by the Qin dynasty.

Factual evidence indicates that the evolution of transportation is inseparable from the development of politics, economics, culture, and society. In different historical eras, the primary driving forces and characteristics of transportation development vary. The development of transportation is intricately connected to the core challenges faced by humanity during specific epochs and the level of advancement in productive forces. During the Xia, Shang, and Zhou periods, the Chinese nation grappled with fundamental tasks such as ensuring survival, fostering ethnic cohesion, establishing societal rituals, and initiating cultural development. As this process unfolded, the Chinese nation continued to assimilate and evolve, drawing from diverse regions and ethnicities, ultimately giving rise to a resilient and cohesive Chinese civilization. Continuous integration and development defined this era.

© China Communications Press Co., Ltd 2025

H. Lu (ed.), *A Brief History of Transportation in China*,
https://doi.org/10.1007/978-981-97-9851-3_1

The Origin of the Term "Transportation" ("交通" (Jiāotōng) in Chinese)

In the passage from "*Zhuangzi—Tian Zifang*" (《庄子·田子方》), it states, "Profound yin is serene, profound yang is radiant; serenity originates from the heavens, radiance emerges from the earth. The interplay of these two elements leads to harmony and the birth of all things." ("至阴肃肃,至阳赫赫;肃肃出乎天,赫赫发乎地。两者交通成和而物生焉") [1]. In this context, "交通" signifies a connection or resonance, an interaction between these elements.

In "*Guanzi—Dudi*" (《管子·度地》), it is mentioned, "As the mountains and rivers dry up, the weather descends, and the earth's energy rises, all things are interlinked." ("山川涸落,天气下,地气上,万物交通") [2]. In this context, "交通" conveys the idea of mutual interconnectedness and interflow.

In the "*History of the Former Han Dynasty—Biography of Jiang Chong*" (《汉书·江充传》), it is noted, "They conspired with the influential and cunning individuals in different commanderies and territories, launching illicit attacks and looting, while the authorities couldn't curb their actions." ("交通郡国豪猾,攻剽为奸,吏不能禁") [3]. In this context, "交通" signifies collusion and collaboration for nefarious purposes.

In Tao Yuanming's "*Record of the Peach Blossom Spring*" (《桃花源记》) from the Jin Dynasty, he described, "Paths and roads crisscrossed and connected." ("阡陌交通") [4]. In this context, "交通" conveys the notion of interconnectedness and mutual accessibility.

During the Tang Dynasty, Liu Zongyuan, in his "*Response to Wei Zhongli's Discussion on the Teaching of the Dao*" (《答韦中立论师道书》), he wrote "expanding lateral communication." ("旁推交通") [5]. In this context, "交通" signifies interaction and communication.

In Volume Ten of the "*The Unauthorized biography of Han's Poems*" (《韩诗外传》), Han Yu wrote: "Maintaining connections with virtuous individuals throughout one's lifetime ensures a life without difficulties." ("与士交通终身无患难") [6]. In this context, "交通" signifies social interaction and communication.

From the examples mentioned above, we can discern that the term "交通" originally carried a wealth of meanings and found its application in diverse contexts. Just as in the "*Book of Changes*", which proclaims, "Heaven and Earth intersect, and all things communicate" ("天地交而万物通") [7], where "交" signifies coming together and "通" implies mutual connection. This same interpretation is found in Kang Youwei's "*Book of Great Unity*" (《大同书》) [8]. As Hu Shi notes in his "*Introduction to the National Language and Grammar*", "In later times, land transportation witnessed the advent of human-drawn carts, horse-drawn carriages, trains, and trams, while water transportation saw the introduction of steamboats. Human interaction became more convenient, reliable, and rapid." ("后来陆地交通有了人力车、马车;火车、电车;水路交通有汽船,人类交通更方便了,更稳当了,更快捷了。") [9]. It is thus evident that the term "交通" gradually evolved to encompass all forms of transportation and the postal and telecommunications industries in the modern and contemporary eras.

Profound Impact of China's Natural Geographical features on Transportation Development

The customs, traditions, economy, society, and cultural progress of a nation are profoundly interconnected with the natural geographical environment it occupies. Understanding the natural environment that shapes the existence and evolution of the Chinese nation is crucial for a comprehensive understanding of Chinese transportation.

Within China's expansive territory of 9.6 million square kilometers, one can witness majestic plateaus, towering mountain ranges, rolling hills, vast plains, and basins encircled by mountain ranges. China presents the rich diversity of the five primary landforms commonly found on Earth. The overall topographical feature of China is marked by higher elevations in the west and lower elevations in the east, leading to the sweeping macroscopic pattern of major rivers flowing eastward across the vast expanse of China, organized into three distinct terraces. This not only offers a wide range of natural landscapes for human activities and the development of agriculture and industry but also provides a natural setting and abundant resources for the harmonious integration of diverse civilizations on their distinctive paths of development [10].

China's terrain exhibits a west-to-east elevation gradient, with higher elevations in the west and lower ones in the east, sloping from the inland regions toward the ocean. This geophysical characteristic finds its origins in vast tectonic movements that unfolded from 80 to 38 million years ago. The northward movement of the Indian Ocean plate, coupled with its swift collision with the Eurasian plate, released an immense amount of energy, leading to an extraordinary surface uplift. This gave rise to the Earth's highest and most recently formed plateau—the Qinghai-Xizang Plateau, constituting the first tier of China's topographical features, as depicted in Fig. 1.1.

As the Qinghai-Xizang Plateau came into existence, the immense forces of this grand collision also began to propagate outward, subjecting regions with pre-existing elevations to compression. This continuous pressure led to the further uplift of the terrain, gradually forming the Loess Plateau, the Yungui Plateau, and the Inner Mongolia Plateau. Thus, the second tier of China's topography was born. At this point, China's landscape exhibited three distinct tiers: the first tier, with an average elevation exceeding 4,000 m, is the Qinghai-Xizang Plateau; the second tier comprises the Inner Mongolia Plateau, Loess Plateau, Yungui Plateau, Tarim Basin, Junggar Basin, and Sichuan Basin, with elevations ranging from 1,000 to 2,000 m; and the third tier, with an average elevation below 500 m, encompasses the Northeast Plain, North China Plain, and the middle and lower reaches of the Yangtze River Plain, bordered by the undulating hills of Northeast China, Shandong, and Southeast China. The boundary between the first and second tiers is marked by the towering Kunlun Mountains, Altai Mountains, Qilian Mountains, and Hengduan Mountains, while the second and third tiers are separated by the continuous expanse of the

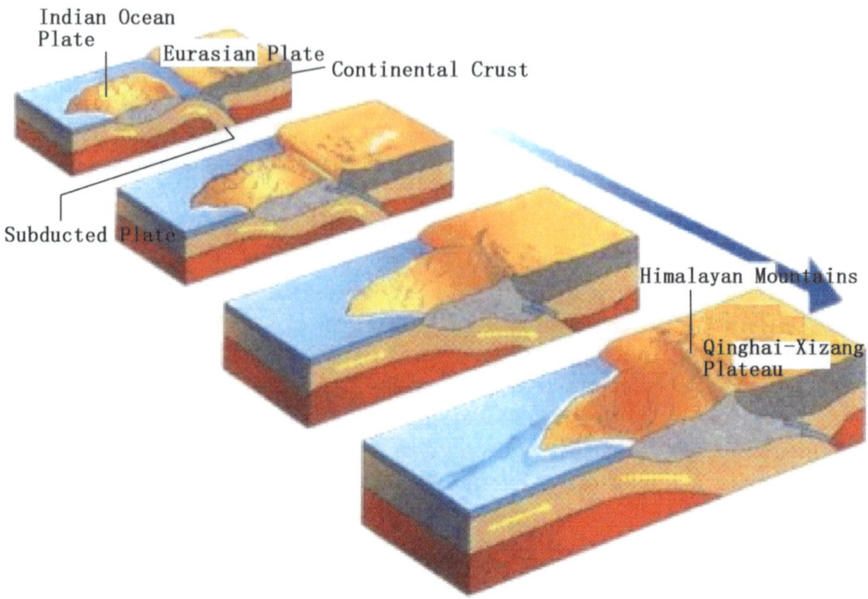

Fig. 1.1 Schematic diagram of the formation of the Qinghai-Xizang Plateau

Greater Khingan Mountains, Taihang Mountains, Wushan Mountains, and Xuefeng Mountains, as depicted in Fig. 1.2.

China's unique three-tiered topography has presented significant challenges in the construction of transportation infrastructure, primarily in two key aspects. The first is the challenge of traversing these tiers. Creating smooth transitions for the east–west transportation corridors to smoothly cross from the third tier, with elevations below 500 m, into the second tier, with elevations ranging from 1,000 to 2,000 m, and even efficiently accessing the first tier, with elevations from 3,000 to 4,000 m, has long been a goal pursued by builders. The second challenge is the expansion within these tiers. Within the second tier, the north–south transportation arteries must navigate their way from the Inner Mongolia Plateau and the Loess Plateau, crossing the Qinling Mountains to enter the Sichuan Basin and the Yungui Plateau. Mountains cover 33% of China's total land area, and each region is marked by towering peaks and a complex network of valleys. Even within the third tier, transportation construction from north to south must overcome the challenges posed by the hills of Northeast China, Shandong, and Southeast China. The Chinese people, overcoming these formidable challenges, have achieved remarkable feats and miracles in the history of China's transportation development.

The 32nd parallel north stands as one of the most distinguished latitudinal lines in China's three-tiered topographical terraces. A closer look at the topographic profile along this latitude vividly reveals its unique structure: the first tier terrace sculpted by the Qinghai-Xizang Plateau, the second tier terrace shaped by the Sichuan Basin, and the third tier terrace formed by the middle and lower reaches of the Yangtze

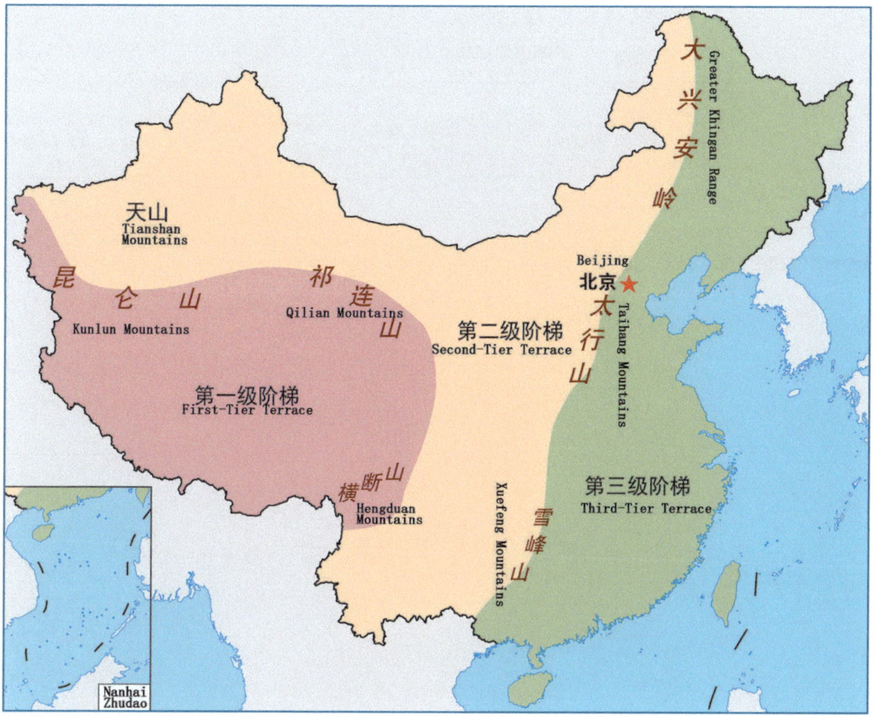

Fig. 1.2 Three-tier topography of China

River Plain, as depicted in Fig. 1.3. This geographical configuration has given rise to the remarkable pattern of China's major rivers flowing from west to east, as beautifully illustrated in Fig. 1.4. The Yangtze and Yellow Rivers, crossing these three terraces, have not only nurtured countless civilizations but have also become emblematic of their indomitable spirit. The abundance of river systems serves a dual purpose. On one hand, it provides a robust foundation for inland river transport. Even in the era of well-developed railway and road networks, waterway freight, leveraging the inherent advantages of golden waterways such as the Yangtze, Huai, and Pearl Rivers, still constitutes over 15% of China's total cargo transport volume. This underscores the irreplaceable role of waterway freight in China's transportation landscape. On the other hand, it symbolizes the intricate challenge and source of national pride in constructing road and railway bridges spanning numerous rivers and water bodies. Today, China boasts a staggering 800,000 road bridges, with more than 150 of them gracefully spanning the main channel of the Yangtze River. What once seemed insurmountable has now transformed into a tapestry of interconnected pathways.

The distinct characteristics of China's three-tiered terraces contribute to the remarkable diversity of its geographical landscapes. However, the formation of these terraces has more far-reaching consequences than just shaping the land; it

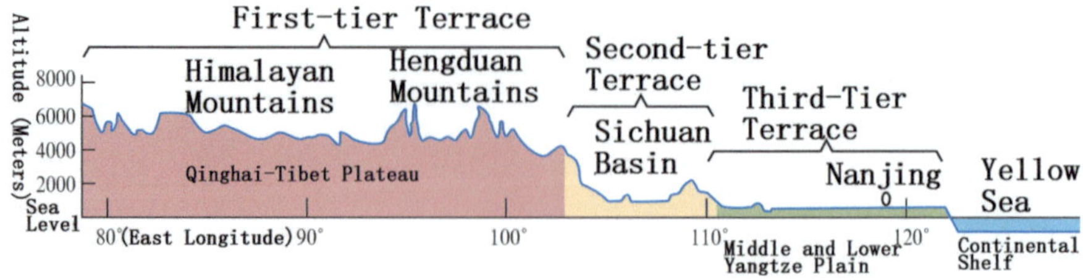

Fig. 1.3 Schematic cross-section of China's topography at north latitude 32

Fig. 1.4 Flow direction of rivers originating from the Qinghai-Xizang Plateau

also has a profound impact on atmospheric circulation and climate. In the absence of geographical features, Earth's rotation alone would establish a regular flow of the atmosphere, resulting in a planetary wind system. This system controls regions along the 30th parallel north, subjecting them to subtropical high-pressure systems. In this scenario, air descends from the upper atmosphere to the surface, making it challenging for descending moist air to condense and generate rainfall, ultimately leading to increasingly arid climates in these areas. Consequently, globally, the 30th parallel north is often associated with arid deserts, particularly the contiguous belt spanning from North Africa to West Asia. Without the Qinghai-Xizang Plateau, even China's Jiangnan region would have succumbed to arid desertification.

The Qinghai-Xizang Plateau region receives a higher influx of solar radiation compared to the plains. This surplus of solar energy warms the atmosphere above the plateau's surface, causing the heated air to rise and leading to a drop in surface air pressure. This reduction in air pressure draws in airflows from the surrounding regions into the Qinghai-Xizang Plateau. Consequently, both the South Asian monsoon from the Indian Ocean and the East Asian monsoon from the Pacific Ocean are influenced by and converge upon the Qinghai-Xizang Plateau. Notably, the East Asian monsoon, originating from the vast expanse of the Pacific Ocean, continuously brings in copious moisture into the heartland of Jiangnan in China. This monsoon system effectively counteracts the influence of the planetary wind system over the middle and lower reaches of the Yangtze River region, preventing it from becoming a desert along the 30th parallel north. This intricate interaction has played a pivotal role in shaping the fertile "Land of Fish and Rice" and the misty, rain-soaked landscapes of Jiangnan that we admire today.

However, the Qinghai-Xizang Plateau also serves as a barrier, impeding the northward movement of moisture-laden South Asian monsoon winds from the Indian Ocean. Instead, most of this moisture is redirected to the southern slopes of the Himalayan Mountains. This diversion leaves China's northwest region even drier, contributing to the natural environmental pattern of arid deserts in the northwest. The formidable westerly jet stream is similarly compelled to circumvent the Qinghai-Xizang Plateau, traveling eastward along the northwestern foothills. As it progresses, it carries with a substantial volume of sand and dust particles, which gradually settle in areas west of the Taihang Mountains and north of the Qinling Mountains, eventually forming the present-day Loess Plateau. Over the course of extensive periods, this process has given rise to China's three major natural regions: the temperate and humid eastern monsoon region, the windy and arid northwest region, and the world's "Third Pole," the frigid Qinghai-Xizang Plateau, as portrayed in Fig. 1.5. Influenced by distinct natural regions and diverse topographies, vast China has also forged a rich and diverse ecosystem, boasting 36,000 plant species out of a global total of 225,000, which accounts for 16% of the world's total plant species. Furthermore, China is home to more than 10% of the world's terrestrial vertebrate species. The emergence of the Qinghai-Xizang Plateau forms the bedrock of China's current topography, precipitation, climate, vegetation, and more.

One could argue that the distinctive three-tiered terraces of China's topography have carved out unique features in the natural climate distribution and hydrological

Fig. 1.5 China's three major natural zones

characteristics of the Chinese land. These features have acted as a bulwark against the potential desertification of vast regions in China. The three major natural regions, the diverse topographic landscapes, and the wealth of biological resources have laid the groundwork for the development and prosperity of the Chinese nation. China's illustrious 5,000 years of civilization have unfolded in harmony with the rivers flowing down these terraces and have thrived thanks to the continuous inflow of moisture-laden monsoons, giving rise to an enduring agricultural culture. The Chinese people, driven by dreams and aspirations, have embarked on a long journey from ancient times, relying on the Earth's bestowed natural endowments.

Likewise, China's transportation endeavors, in their grand historical construction, have faithfully followed the unique geographical features of the nation. From overcoming the challenges posed by the three-tiered terraces to bridging the divides over numerous rivers, traversing the deserts and seas, and enduring the harsh environments of frozen terrains, this journey culminates in the current path of safeguarding the abundant natural resources and embracing environmental conservation. This developmental process stands as a testament to the strenuous journeys of China's great transportation builders who started from scratch, challenged themselves, and, most significantly, exhibited immense courage and created wonders on a monumental scale.

The Fushan Alliance: Unification of Tribes Under a Common Symbol ("合符釜山")

The birth of the Chinese nation is an inescapable consequence of regional amalgamation and the blending of diverse ethnic groups. In ancient times, humans struggled against the formidable forces of nature, and even achieving the basics like sustenance and warmth was a formidable challenge. Hence, there existed an innate impetus for amalgamation and progress. As different ethnic groups intermingled and the early ancestors of the Chinese continued to fuse and develop, the Chinese nation gradually progressed from a period of living in small communities to forming tribes and eventually evolving into the early structures of nationhood. The event of the Hefu State represents a crucial milestone in this historical journey.

The "*Records of the Grand Historian: Annals of the Five Emperors*" (《史记·五帝本纪》) documents that the Huangdi "clashed with the Yandi on the plains of Banquan in three significant battles until his objectives were realized." ("与炎帝战于阪泉之野……三战,然后得其志") Subsequently, he " made the unification at Fushan and established the capital at the foot of Zhuolu" ("合符釜山,而邑于涿鹿之阿") [11].

Prior to the "Unification at Fushan," the Huangdi first vanquished the Yandi in the "Battle of Banquan." Following this victory, he reached a peace agreement with the Yandi at the North He Mountain in Fushan. Later, the Huangdi successfully captured and defeated Chiyou in the "Battle of Zhuolu," integrating Chiyou's followers. This transformative event led to a profound integration, merging the Huangdi's tribal coalition with the Yandi's, and the tribal coalitions of the Eastern Yi peoples. Consequently, this marked the creation of a unified multi-ethnic Huaxia nation [12].

The term "合符釜山" refers to the alliance assembly held in Fusan. "诸侯咸尊轩辕为天子" means that all the lords recognized Xuanyuan as the Emperor, signifying the collective acknowledgment of Huangdi as the leader of all [11]. This marked the establishment of the foundation of Chinese civilization, a pivotal moment in the unification and strengthening of the Chinese nation. Following this, Huangdi transitioned from being a tribal alliance leader to the leader of the Huaxia people. He guided the early Chinese people away from their nomadic lifestyles, ushering in the era of agriculture. His inventions, such as sericulture, the wheeled cart, and writing, as well as his contributions in music, medicine, mathematics, and more, laid the groundwork for Chinese civilization. Huangdi is widely recognized as the common ancestor of the various Chinese ethnic groups and is celebrated as the human progenitor of Chinese culture.

Dayu's Water Management and the Road System of the Xia Dynasty

Throughout the annals of history, water has served as the wellspring of human existence and the cornerstone of human progress. Yet, it is the menace of relentless flooding that loomed as one of the gravest perils confronting early human civilizations. The management of water resources became the chief dilemma during the era of the Five Emperors, particularly with Emperor Shun's reign, marked by this imperative struggle for survival.

As documented in the "*Records of the Grand Historian—Annals of the Five Emperors*", during Emperor Yao's reign, the turbulent floodwaters inundated the landscape, leaving the common people in deep despair [11]. In the era of Emperor Shun, he began the process of dividing the realm into twelve provinces and took measures to clean up and manage the rivers.

In the "*Records of the Grand Historian—Annals of the Zhou Dynasty*," there is a significant passage by Tai Shi Bo Yang Fu that emphasizes the vital role of water management: When water sources are obstructed, it spells the inevitable demise of a nation. The smooth flow of water within the soil is necessary for the prosperity of the populace. When these vital water channels within the soil are obstructed, it leads to the people's impoverishment, and the nation's survival is at stake. Historical precedents are evident; the desiccation of the Yi and Luo rivers contributed to the fall of the Xia Dynasty, and the depletion of the Yellow River led to the collapse of the Shang Dynasty. Now the virtue of the Zhou Dynasty resembles the twilight years of the Xia and Shang, with its water sources blocked. Once obstructed, they are doomed to run dry. The choice of a capital city must involve proximity to mountains and rivers. The disintegration of mountains and the drying up of water sources serve as unmistakable harbingers of a nation's decline, as the depletion of water sources will inevitably provoke the collapse of mountains [11].

Dayu, the founder of the Xia Dynasty in ancient China, ascended to the throne during a time when the land was divided into tribal alliances. Emperor Shun, who was a paramount leader in one of these alliances, designated Gun, who was the father of Dayu, as the person to be responsible for flood control. Gun's approach to managing floods was obstructive, but it became clear that attempting to block the torrents of water was an insurmountable task. After more than a decade without success, Gun was held accountable for his failure to address the water-related crises. Following his father's footsteps, Dayu assumed the mantle of flood control and learned valuable lessons from Gun's failure. He adopted a strategy that combined both obstruction and clearance, a tactic that yielded success. One of his most notable historical contributions was the "Way of the Nine Rivers" which involved the management of nine rivers, including the Wei River, the Yellow River, the Yangtze River, the Huai River, the Yan River, and the Loess River, among others. This initiative aimed to mitigate the grave water-related threats to the survival of the Huaxia people. To implement the "Way of the Nine Rivers" Dayu initiated a comprehensive system of engineering, encompassing the projects of the "Dredging of the Nine Rivers", the "Embankment of Nine Marshes", and the "Survey of Nine Mountains" [11].

The concept of the "Dredging of the Nine Rivers" involved the development of nine key transportation routes, setting the stage for extensive flood control measures. It serves as a remarkable example of "prioritizing transportation." Dayu initiated the establishment of these nine transportation routes, with a collective length of approximately 8,000 km, as illustrated in Fig. 1.6.

The first route was the "Hutai Passage". It stretched from the northeastern slopes of Hukou Mountain, traversed the Leishou Mountain, and continued to Mount Taiyue. Hukou Mountain was situated in Jixian County, Linfen, Shanxi; Leishou Mountain, also known as Zhongtiao Mountain, was in Puxian County, Linfen, Shanxi. Mount Taiyue spanned three cities in Shanxi: Jinzhong, Changzhi, and Linfen, with its highest peak, Huoshan, found in Huozhou City, Linfen. This mountain was considered one of China's "Nine Great Mountains" in ancient times. Historical records indicate that during Emperor Yao's reign, the capital city of Pingyang was established in the Yaodu District of Linfen, Shanxi. In the time of Emperor Shun, the capital city of Puban was in Puzhou Town, Yongji City, Yuncheng, Shanxi. In 2007, archaeological findings at the "Taosi Site" confirmed the existence of the ancient city of Pingyang, celebrated as "China's First Prehistoric City." This underscores that the

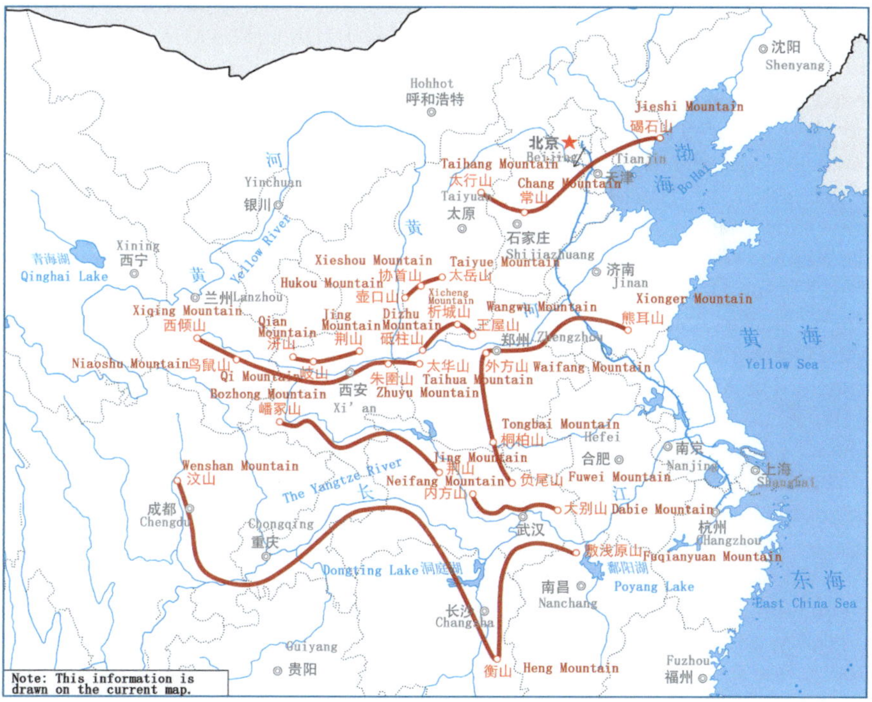

Fig. 1.6 Schematic diagram of the 9 transportation routes established by Dayu

Hutai Passage served as a pivotal transportation artery in the thriving regions of that era and played a crucial role in supporting the transportation needs of the Yellow River Upper Reaches Management Project, spanning approximately 200 km in total.

The second is the "Qianjing Passage". It stretches eastward from Qianshan, crosses Qishan, and reaches Jingshan. Qianshan is situated in Qianyang County, Baoji, Shaanxi, marking the source of the Qian River. Qishan lies in Baoji's Qishan County and is recognized as the birthplace of the later Zhou Dynasty. Jingshan can be found in Fuping County, Weinan, Shaanxi. The Qianjing Passage plays a pivotal role in supporting the transportation needs of the Wei River Management Project and covers approximately 400 km.

The third is the "Diwang Passage". It leads eastward from Dizhushan, traverses Xichengshan, and extends to Wangwushan. Dizhushan is situated in Shan County, Sanmenxia, Henan. Xichengshan is in Henghe Town, Yangcheng County, Jincheng, Shanxi. Wangwushan can be found in Jiyuan, Henan. The Diwang Passage serves as a crucial transportation support line for the Yellow River Middle Reaches Management Project, spanning roughly 300 km.

The fourth is the "Taijieshi Passage". This passage heads eastward from the Taihang Mountains, crosses Changshan, and arrives at Jieshishan. The Taihang Mountains are located at the confluence of four provinces and municipalities: Shanxi, Hebei, Henan, and Beijing. The principal peak, Wutai Mountain, is situated in Wutai County, Xinzhou, Shanxi. Changshan is in Zhengding County, Shijiazhuang, Hebei. Jieshishan has two locations, one in Wudi County, Binzhou, Shandong, and the other in Changli County, Qinhuangdao, Hebei. Regardless of whether one considers the Shandong or Hebei Jieshishan, the Taijieshi Passage covers approximately 700 km and plays a crucial role in supporting the transportation needs of the Yellow River Lower Reaches and the Lao River Management Projects.

The fifth is the "Xitai Passage." It extends eastward from the Xiqing Mountains, crosses the Niaoshu Mountain and Zhuyu Mountain, and arrives at Mount Taihua. The Xiqing Mountains are situated in the southeastern part of Qinghai, forming part of the Kunlun Mountain Range. Niaoshu Mountain can be found in Weiyuan County, Dingxi, Gansu. Zhuyu Mountain is in Gangu County, Tianshui, Gansu, belonging to the Qinling Mountain Range. Mount Taihua is situated in Huayin City, Weinan, Shaanxi. The Xitai Passage plays a crucial role in supporting the transportation needs of the Weak Water and Yang Water Management Project, covering an extensive distance of approximately 1400 km.

The sixth is the "Xiongfu Passage." It heads southwest from Xionger Mountain, crosses the Waifang Mountain and Tongbai Mountain, and reaches Fwei Mountain. Xionger Mountain is in Zaozhuang, Shandong. Waifang Mountain, also known as Song Mountain, is situated in Dengfeng City, Zhengzhou, Henan. Tongbai Mountain can be found in Tongbai County, Nanyang, Henan. Fwei Mountain is in Hengshan, Anlu City, Xiaogan, Hubei. The Xiongfu Passage serves as a crucial transportation support line for the Yanshui, Huai River, and the upper and middle reaches of the Yangtze River Management Projects, with a total length of approximately 1500 km.

The seventh is the "Bojing Passage." It extends northeast from Bozhong Mountain and reaches Jingshan. Bozhong Mountain is in Mian County, Hanzhong, Shaanxi, serving as the source of the Han River. Jingshan can be found in Fuping County, Weinan, Shaanxi. The Bojing Passage is of paramount importance in supporting the transportation needs of the Heishui River Management Project, spanning a total distance of approximately 500 km.

The eighth is the "Neida Passage." It heads eastward from the Neifang Mountain and reaches the Dabie Mountains. The Neifang Mountain is in Zhongxiang City, Jingmen, Hubei. The Dabie Mountains are situated between Hubei, Henan, and Anhui, with one of its main peaks, Tian Tangzhai, found in Luotian County, Huanggang, Hubei. The Neida Passage plays a pivotal role in supporting the transportation needs of the upper and middle reaches of the Yangtze River Management Project and spans approximately 500 km in total.

The nineth is the "Wenfu Pass". It extends from Wenshan in the east, crosses Hengshan, and reaches Fujianyuanshan. Wenshan is traditionally believed to be the starting point of Dayu's management of the Yangtze River, located in Wenchuan County, Aba, Sichuan. Hengshan is situated in Hengyang, Hunan. Fujianyuanshan is Mount Lu in Jiujiang, Jiangxi. The Wenfu Pass serves as the transportation support line for the Heishui and the upper and middle reaches of the Yangtze River management projects, covering approximately 2500 km.

The phrase "Embankment of Nine Marshes " signifies the construction of nine flood control embankments, while "Survey of Nine Mountains" means the assessment of nine major mountains to design routes for diverting floodwaters. To maintain water infrastructure and facilitate management, Dayu went on to "Opening of Nine Provinces".

The notion of "Opening of Nine Provinces" involved the delineation of nine water management responsibility areas, which later evolved into administrative regions [13]. According to the official "*Chronology of the Xia, Shang, and Zhou Dynasties*", formally published by the Dynasty Dating Project, it is estimated that Dayu was born around 2133 BC.

In Dayu's endeavors to manage water, he used carts to traverse dry roads and boats to navigate water routes. For muddy and slippery terrain, he employed sleds, and in mountainous areas, he wore clogs with teeth for better grip. His constant companions were tools for measuring levelness, a measuring line, compasses, and squares for drawing maps. He worked year-round according to the appropriate seasons. Ultimately, he divided the land into nine provinces, established thoroughfares across these provinces, constructed embankments to control the waters of nine marshes and lakes, and surveyed the mountain ranges of the nine provinces. At the same time, he instructed Boyi to distribute rice seeds and taught the people to cultivate in areas with excess moisture. He directed Houji to provide food to the people during times of scarcity. In regions with food shortages, he facilitated the transfer of surplus food from more abundant areas, ensuring equitable food distribution among the various lords. Dayu also established a tribute system based on the resources of each region and assessed the topography of mountains and rivers to determine if it's convenient for the lords to pay their tributes [11].

According to "*Records of the Grand Historian—Annals of Xia*" (《史记·夏本纪》), Dayu was the great-grandson of Changyi, the second son of the Huangdi, making him the great-great-grandson of the Huangdi [3]. Changyi was granted a fief in Ruoshui (now known as the Yalong River) and "married a woman from the Shushan clan." Legend has it that Dayu was born in what is now Beichuan County, Mianyang City, Sichuan. Due to his busy public duties, Dayu did not marry until he was 30 years old. Just four days after his marriage, he hurriedly left home to dedicate himself to managing water-related issues for 13 years, even passing his family's doorstep thrice without entering. In 2060 BC, Dayu passed away due to illness in Shanyin (modern-day Shaoxing, Zhejiang) at the age of 73 and was buried at the northern foot of Mount Kuaiji. In 1959 BC, Dayu's fifth-generation descendant, Shaokang, enfeoffed his illegitimate son Wuyu in Kuaiji, establishing the ancient Yue State and safeguarding Dayu's tomb. This lineage has been passed down for 145 generations. Dayu's tomb is in Yuling Village, Yuecheng District, Shaoxing City, Zhejiang, occupying only 1 mu (approximately 666.67 square meters). It is simple yet dignified. Yuling Village has 150 households with the surname Si, who are believed to be the descendants of Wuyu. In 210 BC, the First Emperor of Qin, Qin Shi Huang, visited Mount Kuaiji to pay his respects at Dayu's tomb, initiating an elaborate national memorial ceremony in honor of Dayu. Subsequently, throughout dynasties and generations, this practice of sending special envoys for tribute continued.

In "*Han Feizi—Five Vermin*" (《韩非子·五蠹》), it is acknowledged that both water management and road construction are formidable undertakings, and Dayu deserves recognition as a trailblazer in the field of flood control and road building [14].

The successful water management efforts of Dayu had a significant impact on society and the economy. It led to the unification of the nine provinces, allowing people to reside within the country's borders. Roads were constructed across the nine mountain ranges, water sources were cleared in the nine major rivers, and embankments were built around the nine large lakes. Princes from all corners of the realm could now convene in the capital for gatherings and audiences. The management of the six storehouses—metal, wood, water, fire, earth, and grain—was handled efficiently. Land quality in different regions was categorized, facilitating the proper assessment of taxes based on these categories. Tax rates were determined according to three distinct soil grades. Princes were also enfeoffed within the nine provinces, receiving land and family names. It was emphasized that virtue should be given the highest priority, and adherence to the various measures should not be compromised [11].

During Emperor Shun's reign, Dayu made the most significant contributions. He opened nine major mountain ranges, managed nine lakes and marshes, dredged nine rivers, and defined the boundaries of the nine provinces. Regions from all over came to pay their tributes as required. "*Records of the Grand Historian*" (《史记》) states, "The land stretched for five thousand miles, reaching even the remote areas. To the south, he pacified Jiaozhi, to the north, he expanded. In the west, he handled the Rong, Xizhi, Qusou, Di, and Qiang. In the northern mountains, he dealt with the Rong, Fa, and Xishen, while to the east, he managed the Chang and the Niaoyi. Within the four seas, all recognized the achievements of Emperor Shun." [11].

Dayu's establishment of the nine routes not only facilitated extensive flood control projects but also laid a solid foundation for the governance and economic growth of the early Xia dynasty. When Dayu first assumed the throne of Xia dynasty, the capital remained in Pingyang (now Yaodu District, Linfen, Shanxi). Subsequently, the capital was relocated to Anyi (now Xiaxian, Yuncheng, Shanxi) and Jinyang (now Yongji, Yuncheng, Shanxi). Later, the capital was moved to Yangcheng (now Gaocheng Town, Dengfeng, Henan) and Yangdi (now Hongchang Town, Yuzhou, Xuchang, Henan). Yangcheng and Yangdi served as the historic heartlands of Dayu's tribe. The move from Shanxi to Henan was a response to the need for political stability.

After Dayu's son, Xia Qi, took control of the government, he relocated the capital to the "Yellow Terrace Hill". When his son Taikang ascended the throne, he moved the capital to Zhenxun (now Zhai Town, Yanshi District, Luoyang, Henan). However, due to Taikang's excessive indulgence in luxury and neglect of state affairs, the leader of the Eastern Barbarians, Houyi, seized the opportunity to seize power. He made Tai Kang's brother, Zhongkang, a puppet ruler, a period known as the "Loss of the State by Taikang". According to expert research, the route of Houyi's march to seize power was from Diqiu (now Hualong District, Puyang, Henan) westward to Zhenxun, covering approximately 400 km. After Zhongkang served as a puppet for 13 years, he died of illness, and his son Dixiang succeeded to the throne. Shortly thereafter, Houyi simply took over and directly declared himself king, an event known as the "Houyi's Usurpation of Xia". Eight years later, Houyi's general, Hanzhu, followed a similar pattern, leading a rebellion and conspiring with Houyi's personal guards to kill Houyi and his family, once again seizing power. After 32 years of a "no-king period," Shaokang, a posthumous son of Emperor Dixiang, rose in rebellion, regained power, and restored the rule of the Xia dynasty. He became the sixth king of the Xia dynasty, an event known as the "Revival by Shaokang". According to expert research, Shaokang's route to regain the country was from Shang (now Gaoxin Town, Suiyang District, Shangqiu, Henan) heading north to Diqiu, first capturing areas such as Guo (now Laizhou, Shandong), Yi (now Decheng District, Dezhou, Shandong), Lixia (now Licheng District, Jinan, Shandong), and other places, dismantling Hanzhu's base, and then marching west to capture Zhenxun. These large-scale migrations were all facilitated by an advanced road system.

Social and Cultural Development and Road Systems of the Shang and Zhou Dynasties

The Shang and Zhou dynasties witnessed the initial thriving of Chinese culture. In the middle of the Shang dynasty, oracle bone script and bronze inscriptions made their appearance. The wars between the Shang dynasty and its neighboring regions facilitated the amalgamation of various ethnic groups and expanded the territorial boundaries. During the Western Zhou period, the state was effectively governed, social etiquettes were well-established, the populace grew more affluent, and culture experienced unprecedented development, leading to the emergence of diverse schools of thought. The system of tribute from different regions during this time suggests that China's road infrastructure had seen considerable development in terms of scale and quality [11].

The Western Zhou dynasty established the "Zhou Roads". In "*the Book of Songs—Xiao Ya*" (《诗经·小雅》), it is recorded: "The Zhou Roads is like a whetstone, its straightness like an arrow." [15]. The "Zhou Roads" was the primary transportation axis that connected the Western Zhou's capital, Haojing (now Doumen Town, Chang'an District, Xi'an, Shaanxi), with the Eastern capital, Luoyi (now the old city of Luoyang, Henan). It extended eastward along the south bank of the Wei River, passing through Liyi (now Lintong District, Xi'an, Shaanxi), Zhengyi (now Huaxian, Weinan, Shaanxi), and then reached Taoyuan Pass (now Hanguan Town, Lingbao City, Sanmenxia, Henan), running along the south bank of the Yellow River until it reached Luoyi (now the old city of Luoyang, Henan). The total distance was approximately 500 km. Later, the Western Zhou dynasty extended the "Zhou Roads" further east, passing through Gongyi (now Gongyi City, Zhengzhou, Henan), Dongguo (now Xingyang City, Zhengzhou, Henan), Guancheng (now Guancheng District, Zhengzhou, Henan), Laoqiu (now Chenliu Town, Kaifeng, Henan), and reaching Songyi (now Suiyang District, Shangqiu, Henan). Then, it divided into two routes heading northeast: the eastern route started from Songyi, moving northeast through Pengcheng (now Tongshan District, Xuzhou, Jiangsu), Tancheng (now Tancheng County, Linyi, Shandong), Qiyi (now Yishui County, Linyi, Shandong), Juxian (now Dongguan Town, Juxian County, Rizhao, Shandong), Yingqiu (now Yingqiu Town, Changle County, Weifang, Shandong), and then southward, passing through Lixia (now Licheng District, Jinan, Shandong), until it reached Qufu (now Qufu City, Jining, Shandong). The western route started from Songyi, proceeding directly north to Qufu. Meanwhile, between Pengcheng and Qufu, as well as between Tancheng and Qufu, direct connecting routes were also constructed [12]. The "Zhou Roads" mainly consisted of a central road, which radiated to the surrounding vassal states on the plains, forming "Zhou Roads extension lines." This network connected the Western Zhou dynasty with significant vassal states situated in the east, such as Qi, Lu, Wei, Song, Cao, and others. Its political, military, and economic significance was considerable.

The "Zhou Roads" and its extension lines served not only as the central transportation main artery with Chengzhou and Luoyi as its focal points but also as a political, economic, and military lifeline incorporated into the national governance system. It became a crucial facility for the Western Zhou dynasty to strengthen its rule, acting as a national defense transportation route. For over 2,000 years, from the Western Zhou period to the Tang and Song dynasties, China consistently used the Xi'an-Luoyang-Kaifeng route as its primary road axis. Even during the Yuan, Ming, and Qing dynasties, this road axis remained a vital transportation route in northern China. In the "*Shang Shu—Zhou Shu—Zi Cai*" (《尚书·周书·梓材》), Luoyi, the Eastern capital of the Western Zhou dynasty, is referred to as the heart of China [13]. The earliest written record of this route is found in the inscriptions on the He's urn unearthed in Shaanxi. Therefore, the historical significance of the "Zhou Roads" is profound, representing a milestone in ancient Chinese road transportation history.

In the waning years of the Western Zhou dynasty, inept rule led to social turmoil and popular uprisings. The well-known story of "Teasing the Dukes with Beacon" unfolded during this period. Ultimately, King You of Zhou was assassinated by the Quanrong and Shenhou, leading the nobles to crown the previously deposed Crown Prince Yijiu as the King, known as King Ping of Zhou. King Ping of Zhou relocated to Luoyi (modern-day Luoyang, Henan), marking the beginning of the Eastern Zhou period, which can be divided into the " the Spring and Autumn " and "Warring States" periods. These times witnessed frequent warfare, rapid economic growth, and an unprecedented demand for transportation, further shaping the road network into a well-established system. As depicted in Fig. 1.7, the road system during the Spring and Autumn period achieved a considerable scale. Without a doubt, this development laid a solid foundation and provided essential support for the formation and growth of a centralized nation.

Simultaneously, the well-developed road transportation system further stimulated the deep integration and unified development of cultures among the various ethnic groups in Huaxia. It also provided fertile ground for the emergence of the Hundred

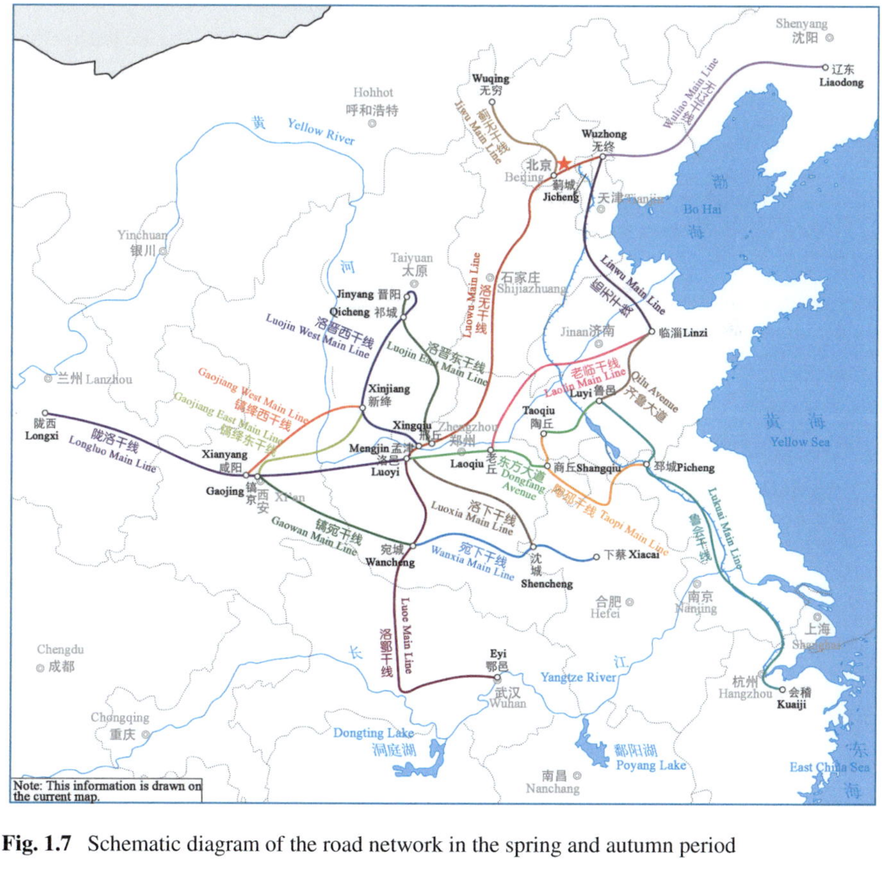

Fig. 1.7 Schematic diagram of the road network in the spring and autumn period

Schools of Thought during the Spring and Autumn and Warring States periods, fostering the formation of various philosophical schools and the rapid development of early philosophy, laying the foundation for Chinese philosophy and culture. In a certain sense, the intellectual and cultural diversity arising from the Hundred Schools of Thought, as well as the achievements of classical philosophy, can be regarded as the "genetic repository" of Chinese culture. Subsequent developments and refinement of various philosophical ideas and traditional culture in China can almost all be traced back to the rich heritage of classical philosophy from the Hundred Schools of Thought, including Confucianism, Taoism, Legalism, Mohism, the School of Names, and Yin-Yang philosophy.

References

1. *Zhuangzi*. Shanghai: Shanghai Lexicographical Publishing House, 2003.
2. *Guanzi*. Commentary by Fang Xuanling, Supplementary Notes by Liu Xiaoyi, Edited and Annotated by Shanghai Ancient Books Publishing House, 2015.
3. Ban Gu. *Han Shu*. Beijing: Zhonghua Book Company, 1997.
4. Tao Yuanming. *The Complete Works of Tao Yuanming*. Annotated by Tao Shu, Edited by Gong Bin. Beijing: Zhonghua Book Company, 2015.
5. Liu Zongyuan. *The Collected Works of Liu Hedian*. Shanghai: Shanghai Ancient Books Publishing House, 2008.
6. Han Ying. *The Annotations of the Unauthorized biography of Han's Poems*. Beijing: Zhonghua Book Company, 2020.
7. *Zhou Yi*. Shanghai: Shanghai Ancient Books Publishing House, 1987.
8. Kang Youwei. *Book of Great Unity*. Shanghai: Shanghai Ancient Books Publishing House, 2019.
9. Hu Shi. *Collected Works of Hu Shi*. Beijing: Peking University Press, 2020.
10. Star Planet Institute, China Qinghai-Xizang Plateau Research Association. *This Is China*. Beijing: CITIC Publishing Group, 2019.
11. Sima Qian. *Records of the Grand Historian*. Beijing: Zhonghua Book Company, 1959.
12. Qin Guoqiang. *A History of Chinese Transportation*. Shanghai: Fudan University Press, 2012.
13. *Shang Shu*. Translated and Annotated by Wang Shishun and Wang Cuiye. Beijing: Zhonghua Book Company, 2012.
14. *Han Feizi*. Translated and Annotated by Gao Huaping, Wang Qizhou, and Zhang Sanxi. Beijing: Zhonghua Book Company, 2015.
15. *The Book of Songs*. Translated and Annotated by Wang Xiumei. Beijing: Zhonghua Book Company, 2015.

Chapter 2
Ancient Transportation: From the Yinxu Carriage to the Emperor Carriage

The birth of transportation was an important symbol of human wisdom and civilization development. From a flat boat to a wheel, relying on various means of transportation, the scope of human exploration and cognition of the world has been expanded, the frequency of interaction has increased significantly, and the development of transportation has contributed to the progress of world civilization and the fusion of cultures. At the same time, with the continuous development of human society, transportation has also gradually evolved from simply serving transportation, giving rise to functions such as transporting goods and engaging in warfare, while also being endowed with richer meaning of rituals and ceremonies. Today, the transportation has become an innovative driving force for economic development, and "transportation+" has become the most promising and spatial direction in economic development. Reviewing the origin and development of ancient transportation in our country not only enables us to understand the splendid and ancient history and culture of our country, as well as the great and extraordinary national wisdom, but also inspires us to think about the future and inspire us to innovate.

Origin of Transportation Vehicles in China

The cart was the most common form of transportation vehicle. In ancient China, the car was all animal powered, with the main sources of power being cows and horses, with mostly wooden components, equipped with wheels and axles for people to ride on or to transport goods. Regarding the invention of the Chinese vehicle, there were three key figures, namely, Yellow Emperor, Chengdu and Xizhong. At present, based on archaeological evidence and historical records, it was possible that the Yellow Emperor invented circular wheel hubs and created the prototype of vehicles; Chengdu and Xizhong utilized the traction of livestock to advance vehicle technology and invented truly widely applicable vehicles.

© China Communications Press Co., Ltd 2025

H. Lu (ed.), *A Brief History of Transportation in China*,
https://doi.org/10.1007/978-981-97-9851-3_2

Vehicle Made by Yellow Emperor

Yellow Emperor (Fig. 2.1), the original surname was Gongsun, later changed his surname to Ji and the name was Xuanyuan. It was said that he became the leader of the Ancient Chinese Tribal Alliance in the first year of Yellow Emperor (2698 BC) [1, 2]. According to the records of "Biographies of Five Emperors • Shiji", the Yellow Emperor knew astronomy and geography, calculated the calendar, astronomy, yin-yang and five elements, and used the ten heavenly stems and twelve earthly branches to form the a cycle of sixty years, which was used as the twelve zodiac animals to record the time, this system has been used to this day. He also invented writing, music, sericulture, medicine, and arithmetic, taught the people to sow grains, implemented the Land System, and established the earliest agricultural production system. It was probably the field work that inspired Yellow Emperor to make vehicles [3].

According to the seventh volume of the "Lu Shi" written by the historian Luo Mi of the Southern Song Dynasty, "Xuanyuan began to make chariots on the north side of Kongsang, inheriting the knowledge of his predecessors and opening up his wisdom. When he saw the windblown grass, he began to make chariots, using wood as wheels, jade as axles, horizontal wood as shafts, and straight wood as front crossbars to worship the highest deity in heaven, and so was known as the Xuan Yuan" [5]. That was to say, when the Yellow Emperor was working in the fields, he was inspired

Fig. 2.1 Yellow Emperor [4]

by the rotation of the wind canopy and invented the vehicle. Another theory was that when the Yellow Emperor was working in the field, a sudden strong wind blew away his hat. Because the brim of the hat was circular, it fell to the ground and kept rolling forward, which inspired the Yellow Emperor to create the first cart in history. Cang Jie proposed to the Yellow Emperor to name it "Cart", also known as "Xuan Yuan". In fact, the "Xuan" in the title of Yellow Emperor refers to the horizontal wooden bar that pulls vehicles, while the "Yuan" was the straight wooden bar that connects vehicles. The literary figure of the Eastern Han Dynasty, Wang Yi, annotated "Xuan Yuan was unattainable" in Qu Yuan's "Chu Ci·Yuan You" as "Xuan Yuan," which was the title of the Yellow Emperor. The Yellow Emperor began to make clothes and carts, and people all over the world called him the "Xuan Yuan", [6] indicating the relationship between the Yellow Emperor and carts".

In addition, Cui Bao, the Grand Tutor of the Western Jin Dynasty, recorded in "Gu Jin Zhu" that "Yellow Emperor and Chi You were at war in the wilderness of Zhuolu. Chi You created a thick fog, and the soldiers lost their direction. Therefore, Yellow Emperor created a guide cart to indicate the direction, and ultimately captured Chi You, who ascended to the throne as a result." [7]. Liu Shu, a historian of the Northern Song Dynasty, wrote in "Tong Jian Wai Ji" that "Chi You created a dense fog that caused the army to lose its direction, so Xuan Yuan was used as a guide cart to indicate the direction." [8]. Both Gao Cheng, a historian of the Southern Song Dynasty, recorded in "Shi Wu Yuan Ji", and Ma Su, a historian of the Qing Dynasty, recorded in "Yi Shi" that "Empyrean Fairy made a guide cart for the Yellow Emperor and placed it in front of the queue, with the cart on right." [9, 10]. Taiping Yulan quoted Eastern Jin astronomer Yu Xi in "Zhi Lin" that "Yellow Emperor and Chi You engaged in battle in the wilderness of Zhuolu. Chi You cast spells and created thick fog, causing the army to lose its direction. Yellow Emperor ordered Feng Hou to make a guide cart based on the Big Dipper Star to distinguish direction and ultimately captured Chi You." [11]. Yellow Emperor ordered Feng Hou (a descendant of Fu Xi, rumored to be the prime minister of the Yellow Emperor) to use mechanical force to create the guide cart and distinguish direction in the war, winning the victory. Scholars have specifically described the guide cart made during the reign of the Yellow Emperor, which used natural magnets to make a wooden arm to indicate direction, and the rotating mechanism of the wooden arm was installed on the wooden car [12], as shown in Fig. 2.2. The Ji Li drum cart was a vehicle used to record the distance traveled by the vehicle. According to records, the Ji Li drum cart has two layers, with a bell on the upper layer and a drum on the lower layer, there were wooden robots holding wooden hammers, and the cart beats the drum once every mile and the bell once every ten miles, as shown in Fig. 2.3.

From the above historical data, it can be seen that the immortal Empyrean Fairy created a guide cart for the Yellow Emperor, helping the Yellow Emperor defeat Chi You and to unify China, which was obviously a myth full of romanticism. From the perspective of social productivity development, even if cart was invented during the Yellow Emperor period, it was only the structural style of unicycles, driven by human labor, and was not exquisite and practical guide cart. There were two possibilities for the internal structure of the guide cart: utilizing the directionality of the gear system

Fig. 2.2 Guide cart [13]

Fig. 2.3 Ji Li drum cart [14]

or the directional polarity of the magnet. According to archaeological results, gears appeared in China between the Warring States period and the Western Han dynasty. Therefore, the Yellow Emperor fought against Chi You with the use of the guide cart and the Ji Li drum cart, it was more likely that historians advanced the more advanced vehicles invented by later generations through romantic imagination for thousands of years to express their reverence for the Yellow Emperor.

Vehicle Made by Xi Zhong

In the "Si Kao Ji" written by historian Qiao Zhou during the Three Kingdoms period, it was recorded that "the Yellow Emperor began to make cart to carry heavy loads, making places with long distances close at hand. In the Shao Hao era, people began to drive oxen; in the Xia Yu era, Xi Zhong began to drive horses." [15]. This means that after the invention of the rickshaw during the Yellow Emperor period, by the Shao Hao era (the eldest son of the Yellow Emperor), ox-powered chariots had already been introduced. During the Xia Yu era, Xi Zhong, who was responsible for managing vehicles, further created a more efficient horse drawn carriage. Xi Zhong, surnamed Ren, was from the Xue State (now Tengzhou, Shandong). He was a craftsman during the Xia Dynasty and was revered by later generations as the originator of the vehicle. There were records in "Shi Ben·Zuo Pian" and "Lv Shi Chun Qiu·Wu Gong Lan" that "Wang Hai began to tame cows." [16, 17], while in "Mo Zi ·Fei Ru", "Shi Ben·Zuo Pian", "Lv Shi Chun Qiu·Jun Shou Pian", and "Huai Nan Zi·Xiu Wu Pian", there were four words, namely "Xi Zhong made the cart" [17–19]. The "Shuo Wen Jie Zi" mentions that "the vehicle was invented by Xi Zhong of the Xia Dynasty." [20], clearly indicating that Xi Zhong of the Xia Dynasty was the ancestor of cart making (Fig. 2.4), and here the cart should mean carriage. Mr. Li Xueqin, the chief scientist of the Xia, Shang, and Zhou Dynasty Chronology Project, once said that the "carriage" made by Xi Zhong should be a carriage with certain technical standards and significant innovations.

Fig. 2.4 Statue of Xi Zhong, the ancestor of the cart [21]

Xi Zhong was the official in charge of carts during the Xia Dynasty. The theory that Xi Zhong made carts indicates that the real carts or carriages in ancient China were invented and utilized during the Xia Dynasty. The records of the "Shi Ji·Xia Ben Ji" also tells the story of the Da Yu's tamed the flood, stating that "Da Yu traveled by cart". If Sima Qian's statement was reliable, then the creation of Chinese carriages was completed at least during the Yao Dynasty and Shun Dynasty and was fully utilized.

Based on current historical records, the Yellow Emperor invented the simple rickshaw over 4,700 years ago. Until more than 700 years later in the Xia Dynasty, oxcarts and horse-drawn carriages appeared, which were driven by animal power. After research, the China Pre-Qin History Society, the China Culture Promotion Association, the China Society of Automotive Engineers, and the China Association of Automobile Manufacturers have unanimously recognized that Xicun Village in Zaozhuang, Shandong was the residence of Xi Zhong over 4,000 years ago, and also the birthplace of the earliest carriage in China (Fig. 2.5). Therefore, we can believe that Xi Zhong was not the original creator of carts. His contribution should have been the combination of carts and horses, and he played an important role in the improvement of vehicle technology and the establishment of cart system.

Fig. 2.5 Xi Zhong's hometown in Xicun Village, Taozhuang Town, Xuecheng District, Zaozhuang, Shandong Province [22]

Birth of Vessels

An important feature of China's natural environment was that there were many rivers and lakes. The Yangtze River basin and the Yellow River basin were the cradle of the birth of Chinese civilization. In order to better reproduce and maintain life, ancient people relied on observation and experience to set up camps in areas with sufficient water sources, live along the water, and fully utilize water resources in times of extremely low productivity.

The birth of vessels was much earlier than that of vehicles, and there were abundant archaeological artifacts. According to current archaeological discoveries, the invention of Chinese vessels can be traced back to the middle and early Neolithic period, about 3000 years earlier than the Yellow Emperor era. In November 2002, a canoe dating back approximately 8000 years was unearthed at the archaeological site of the Kuahuqiao Neolithic Ruins in Hangzhou, Zhejiang Province, as shown in Fig. 2.6. This was the only canoe from the Neolithic period unearthed in China, and it was also the oldest transportation vehicle in China with tangible evidence. It can be seen from this that the inventors of Chinese vessels were not ancient people during or after the Yellow Emperor's era, but may have been late Neolithic Homo sapiens who lived about 5000 years earlier than the Yellow Emperor.

The wooden oars used for rowing, which was also called "oar" by ancient people, should be a product of the mature rowing technology. Eight wooden oars were unearthed from the Hemudu Neolithic Ruins in Yuyao, Zhejiang Province (approximately 7000 years ago), which were made of raw wood and resemble those of later generations, but with slightly smaller shapes. The handle and blades of the wooden oars were processed from a whole section of wood, with meticulous workmanship.

Fig. 2.6 Canoes unearthed from the Kuahuqiao Ruins [23]

The blades were flat willow shaped and gradually thinner from top to bottom, the length of handle was about 35 cm, the width was 14 cm, and the thickness was 1.5 cm. There was a long circular concave hole on one side of the upper end of the handle, which was not much different from the wooden paddle used today. There were even exquisite linear patterns engraved, as shown in Fig. 2.7.

In addition, an ancient oar about 5000 years old was also discovered at the Qianshanyang Neolithic Ruins in Huzhou, Zhejiang. It was made of green oak wood and had a long oar, a wing length of about 86 cm, and a wing width of about 19 cm. It was much larger than the wooden oar unearthed at the Hemudu Ruins. This was likely a large oar that needs to be tied to the foot of the oar, rather than the hand oar commonly used in canoes, as shown in Fig. 2.8.

In addition, among the physical clues tracing the origin of Chinese boats, a complete pottery boat was unearthed at the Hemudu Ruins (Fig. 2.9), and a simulated boat pottery was also unearthed at the Guojia Village Ruins in Dalian, Liaoning

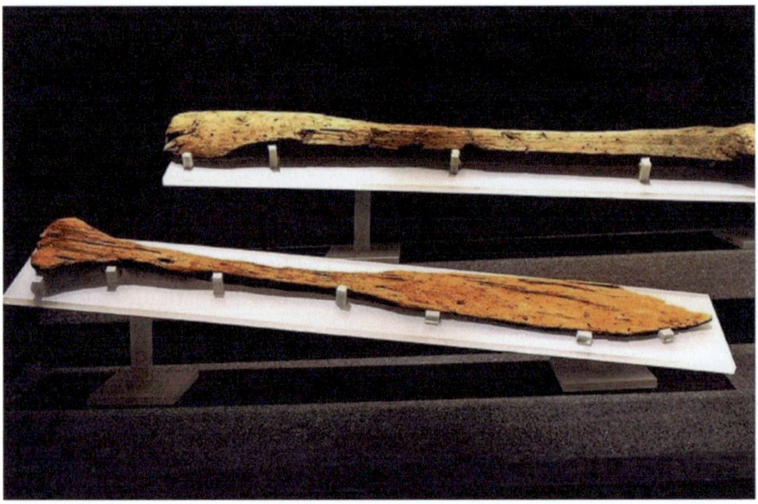

Fig. 2.7 Wooden oars unearthed from the Hemudu Ruins [24]

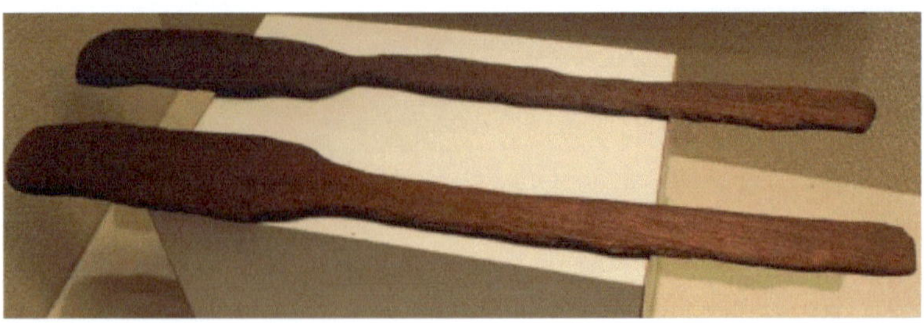

Fig. 2.8 Ancient oars unearthed from the Qianshanyang Ruins [25]

Fig. 2.9 Pottery boats unearthed from the Hemudu Ruins [26]

Fig. 2.10 Pottery boats unearthed from the Guojia Village Ruins [27]

(Fig. 2.10). These two pottery boats have realistic images and can be seen as models of canoes, which was of great significance for exploring the forms of early Chinese canoes.

There were also many legends in ancient books about the origin of Chinese boats, but the people referred to as boats were not the same. The "Shi Ben·Zuo Pian" and "Shuo Wen Jie Zi" record that the boat was invented by two ministers of the Yellow Emperor, Gong Gu and Huo Di. In the "Zhou Yi·Xi Ci Xia", it was believed that the Yellow Emperor himself "hollowed out wood to make a boat and cut it into oars" [28]. The "Classic of Mountains and Rivers·Hai Nei Jing", states that Xi Zhong's father, Pan Yu, began making boats [29]. In "Mo Zi·Fei Rong Xia", it was believed that Qiao Chui, a brother of the same generation in Pan Yu, was the one who made a boat [18]. In "Lu Shi Chun Qiu·Wu Gong Lan", it was believed that Yu Xu was the one who made the boat [17]. Although the characters who started making boats in the various literature legends mentioned above were different, they all date back to the period from the Yellow Emperor to the Xia Dynasty.

It was not difficult to see that ancient Chinese transportation vehicles were nurtured in the Neolithic period of primitive society. The earliest transportation vehicles began with driftwood, rafts, and bamboo rafts on water, and gradually developed into canoes. Compared to this, land transportation vehicles appeared relatively late. The birth of water transportation vehicles during the Neolithic period in China may have inspired and promoted the gradual formation of land transportation vehicles. Although no remains of vehicles have been discovered in the culture of the Neolithic period, based on the level of chariot making technology discovered in archaeological studies during the Shang Dynasty and Zhou Dynasty, the Xia Dynasty may not necessarily be the earliest period of vehicle invention. With the continuous deepening of archaeological and research work, the invention of vehicles may be traced back to even earlier periods.

Earliest Excavated Vehicles in the Yinxu

Although there were numerous records of cart making in historical documents and legends, it was unfortunate that no vehicle remains before the Shang Dynasty have been found in archaeology so far. The rut relics have been discovered in the Xiadu Ruins in Erlitou, Yanshi, Henan Province (Fig. 2.11). These two ruts were located between the early road soil on the south side of the main road in the palace area, roughly parallel in shape, the overall length of the ruts was more than 5 m, and continue to extend east–west. From the shape of the ruts, rut groove show a groove, the distance between the two ruts was about 1 m. According to the research, these two ruts were formed about 3700 years ago. This was to explore the evolution of China's ancient vehicle development of important information.

Since the Yin Xu in Anyang in 1928, the earliest vehicle specimens in China have been unearthed in the chariot and horse pits of the Shang Dynasty. Currently, there were six Shang Dynasty chariot and horse pits (Fig. 2.12) and rut relics displayed in the Yin Xu. Each pit buries one carriage, with five pits burying two horses along with the carriage, and four pits burying one person each [31]. The carriage has a beautiful appearance, a sturdy structure, a lightweight body, fast operation, and a balanced center of gravity, showcasing the exquisite design of early Chinese animal-drawn carts.

Combined with the unearthed objects and the shape of the character "Che" in the Shang Dynasty oracle bone inscriptions (Fig. 2.13), it can be seen that the Shang Dynasty carriage was a curved single shaft, single shaft, two wheels, a horizontal rectangular carriage, a rear door, a straight or curved balance, with some copper accessories and a wooden main body, powered by two horses.

The carriage structure of the Shang Dynasty did not differ due to the different uses of the vehicles, only one form can be seen, which was used as a tank for warfare and as a carriage for carrying people. The carriage of the Shang Dynasty was a symbol of power and status, which could only be used by nobles and nobles, and was used for burial after death. The carriage shape in the Shang Dynasty laid the foundation for the single shaft carriage in China and had a huge impact on the carriage shape in the later Zhou Dynasty.

The carriage of the Shang Dynasty was mainly composed of several major parts, including a transverse beam, a shaft, a carriage, wheels, and axles (Fig. 2.14). Transverse beam was mostly a long round rod or a straight transverse beam, and there

Fig. 2.11 Rut Relics on the Erlitou Ruins [30]

were also curved transverse beams with upturned ends. The head end of the shaft was generally curved upwards, with a length of 2.6 ~ 2.9 m. The carriage was spacious, with a general depth of about 1 m, and the horizontal width of 1.3–1.5 m. The wheel diameter was 1.2–1.5 m, and the rail gauge was 2.1 ~ 2.4 m. The majority of wheel spokes were 18, and the length of the axle was generally about 3 m. The carriages of the Shang Dynasty were cleverly designed to organically pull horses and bodies together [32].

Technological Evolution from Single Shaft Carriage to Double Shaft Carriage

After King Wu of Zhou Dynasty conquered the Shang Dynasty in 1046 BC, he accepted the suggestion of Duke of Zhou to build Luoyi (now Luoyang, Henan), excavate roads, manufacture vehicles, and develop transportation. On the basis of the single shaft carriage in the Shang Dynasty, the Western Zhou Dynasty further refined the craftsmanship of carriages and carried out significant innovations in vehicles. In

Fig. 2.12 Chariot and Horse Pit in Yin Xu [33]

Fig. 2.13 Oracle bone inscription of carriage

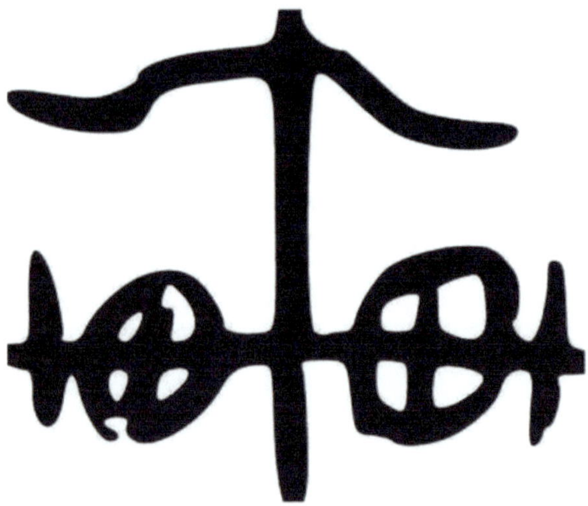

the "Zhou Li·Kao Gong Ji", it was recorded that "making a carriage was the most common if it was necessary to gather multiple trades to make a single object" [35]. From the archaeological excavations, it can be seen that there were two major changes in the carriage of the Western Zhou Dynasty.

A change was reflected in the components, carriage decorations, and horse decorations, such as copper Luanling, carriage transverse beam decorations, and carriage

Fig. 2.14 Carriage structure of Shang Dynasty [34]

hubs, which were clearly more exquisite and exquisite than those of the Shang Dynasty, and were more magnificent in appearance. The Western Zhou carriage Luanling (Fig. 2.15), which was currently housed in the Shaanxi History Museum and unearthed in Fufeng, Shaanxi in 1999, was a decoration on carriage during the Western Zhou period, similar to later bells, and it was popular for several hundred years until the Warring States period. The word "imperial carriage" used by ancient people to refer to the carriage of emperors originated from this object. The shape and structure of this Luanling may seem simple, but in reality, it required high requirements for bronze casting technology and manufacturing processes. Its exterior was a thin layer of enclosed spherical bronze frame, with a carved hollow surface, the internal was wrapped a small stone ball. During the carriage movement, the stone ball collided with the bronze frame and emits a crisp bell, which was not only beautiful and fashionable, but also pleasant to listen to. It can also serve as a safety reminder for nearby pedestrians.

In addition, the bronze carriage transverse beam decorations of the Western Zhou Dynasty (Fig. 2.16) currently collected in the Luoyang Museum were installed at both ends of the carriage transverse beam to protect it and have a certain carriage transverse beam effect, were also a symbol of personal wealth and identity. In addition, there was a well preserved bronze carriage hub from the Western Zhou Dynasty, collected in the Luoyang Museum, this was the most important bronze component on the carriage of the Western Zhou Dynasty, which not only has high manufacturing difficulty but also needs to bear the weight of the carriage, while being lightweight and flexible. The carriage hub determined the overall quality and performance of the carriage.

Fig. 2.15 Green carriage Luanling in Western Zhou Dynasty [36]

Fig. 2.16 Bronze carriage transverse beam decoration of the Western Zhou Dynasty [37]

This bronze wheel hub of the Western Zhou Dynasty (Fig. 2.17) was cylindrical in shape, divided into three layers from the inside to the outside. The inside was a hollow truncated cone, which was convenient for fitting the axle. Its working principle and appearance were very similar to modern car wheels.

Fig. 2.17 Bronze carriage hubs of the Western Zhou Dynasty [37]

The second change was from one horse to many horses. In the "Zhou Li", the carriage, together with the Ding and Gui, symbolized status and power, became important burial objects. In the "Li Ji·Wang Du Ji", it is recorded that "the emperor used six horses for driving, the feudal princes used four horses, the officials used three horses, the scholars used two horses, and the commoners could only use one." [38]. However, in the use of words, the carriage driving two horses was called "Pian", driving three horses was called "Can", driving four horses was called "Si", and driving six horses was called "Liu Fei". Regarding the content of using carriage to represent identity and status will be described in detail later.

The Qin bronze carriages and horses unearthed in June 1978 on the west side of the Qin Mausoleum in Lintong District, Xi'an, Shaanxi Province, were now housed in the Emperor Qin Shi Huang of Qin Mausoleum Museum. These are the largest, most complex, and most complete ancient carriages and horses in Chinese archaeological history. A group of two Qin bronze carriages and horses were unearthed and have been broken into over 3000 pieces, after 8 years of careful repair, they were finally formed. Two bronze chariots and horses, one was a standing chariot and the other was a seat chariot, both of which were ancient single shaft double wheeled chariots, and were made to half the size of a real human chariot and horse. The copper chariot and horse were cast entirely in bronze, with gold and silver ornaments weighing over 14 kg. More than 3500 components were assembled through various processes such as casting, inlay, welding, sub mother buckle connection, and hinge connection. The No. 1 chariot (Fig. 2.18) has a total length of 2.25 m, a height of 1.52 m, a weight of 1.06 tons, a shaft length of 1.83 m, a carriage width of 74 cm, and a depth of 49 cm. There were railings on the front, left, and right sides of the carriage, with a

Fig. 2.18 Names of No. 1 Qin copper chariot and horse and its parts [39, 40]

crossbar at the front for passengers to support, and a chariot door at the back. There was a cross shaped umbrella seat inside the carriage, on which a long handled copper umbrella was inserted. The bronze imperial figurine stood under the umbrella, and the carriage was equipped with weapons such as copper crossbows, copper arrows, and copper shields. The biggest difference between No. 2 chariot (Fig. 2.19) and No. 1 chariot was the carriage: Firstly, No. 1 chariot was convertible, while No. 2 chariot was enclosed; Secondly, the No. 1 chariot belonged to a standing passenger, while the No. 2 chariot plane had a convex shape for sitting. The No. 2 chariot was divided into two rooms in the front and back. The front room was a small area where the emperor sits, with a painted pottery figure on knees. The back room was where the owner sits, with a floor plan that is approximately square, width of 78 cm and depth of 88 cm. There was an oval shaped canopy on the upper part of the carriage, covering the front and back rooms. The front room of the carriage also has four horizontal beams at the bottom of the chariot, with railings standing on the left, right, and front sides.

Qin bronze chariots and horses were all driven by four horses. The two horses in the middle used to drive the shaft are called "Fu Horse", while the two horses on both sides used to pull the chariot are called "Can Horse". Fu Horse drives the shaft at a relatively fixed distance, while Can Horse is relatively free. The excavation of copper chariots and horses fully showcases the technology of ancient Chinese horse drawn carriages in front of people. In order to ensure that the four horses could be placed in their respective positions and not crowded with each other, there were side drives for driving horses installed on the outer chest of the two horses. In addition,

Fig. 2.19 Names of No. 2 Qin copper chariot and horse and its parts [39–41]

Can Horse and Fu Horse each had a rope. The four horses could use the rope to pull the front part of the rope of the two horses and tie it to the yoke, then go down the shaft, merge it into one strand at the intersection of the shaft and the bottom of the carriage, and then extend it backwards to tie it to the axle. Secondly, the front end of the rope for the two horses was in the shape of a loop, which was wrapped around the horse's chest. The back end passes through a copper ring fixed at the front of the carriage bottom, and then extended parallel to the bottom of the carriage and was tied to the rear of the carriage bottom. The power ropes of two horses in uniform were tied to the middle of the axle, while the power ropes of two horses were tied to the bottom of the carriage respectively. The distance between the force points and the two wheels was equal, and each relatively independent part was pulled. The three force points formed an isosceles triangle, and the forced on the two wheels and the carriage were equal.

And controlling the horse is achieved through the rope in the hand of the rider, also known as the "bridle rope". The wooden pole used to drive livestock in front of the carriage connected the two horses in the middle, and their internal bridle function could be ignored, so they were tied in front of the carriage. The remaining six bridle ropes were held in the hands of the rider respectively. The bridle rope on the right side of the horse was held in the right hand, and the bridle rope on the left side of the horse was held in the left hand. When turning right, pull the bridle rope on the right hand, and when turning left, pull the bridle rope on the left hand. Control was simple.

Entering the Han Dynasty, double shafted chariots began to become popular. Before Emperor Wu of Han, there was a period when single and double shafted chariots coexisted; By the middle and late period of the Western Han Dynasty, the double shafted chariots began to gradually become popular; After the Eastern Han Dynasty, the double shafted chariot basically replaced the single shafted chariot. Archaeological discoveries of stone, brick, and tomb murals from the late Western

Fig. 2.20 Double shafted chariot [42]

Han Dynasty and Eastern Han Dynasty had a large number of double shafted chariots, confirming this process of change. From archaeological data, the appearance of the double shafted chariots could be traced back to the late Warring States period at least: double shafted chariots with one horse were found in the No.1 chariot and horse pit of the Ma'anzhong Late Warring States Period Tomb in Huaiyang, Henan and the Qin Tomb in Qin'an, Gansu (Fig. 2.20); The lacquer pot unearthed from the Chu Tomb in Changsha, Hunan also depicted a horse drawn carriage. These could all be seen as the starting point of the double shafted chariot. The structure of a double shafted chariot was basically the same as that of a single barreled carriage, except for two shafts. At the beginning, the carriage shaft was still in an upward curved form, to prevent breakage, two wooden poles were often added between the middle of the carriage shaft and the yoke to reinforce it. Later, it gradually evolved into a straight form. The emergence of the double shafted carriage broke through the limitation that a single shafted carriage could only be driven by two horses, making it possible for a single horse to pull a carriage, and thus ushering in a new stage of development for ancient Chinese chariots from the four horse high carriage to the single horse light carriage. As for the popularity of the double shafted chariot in the Han Dynasty, one theory is that the cavalry was the main military force at that time. Emperor Wu of Han requisitioned a large number of horses to resist the Xiongnu, while the horses used for daily transportation were strictly controlled. People found it difficult to use the Four Horses to pull the Chariot, so they had to rely on transforming the shape of the Chariot to seek a solution. The small horse carriage with two legs driving one horse may have emerged to meet this need and quickly became popular.

Rise and Fall of Carriages and Ox Carts

Generally speaking, the driving force of ancient vehicles was mainly horses and cows. At the beginning of the Records of the "Shi Ji·Ping Zhun Shu", it was written: "Even the emperor did not have four horses of the same color to drive, and some generals could only ride on ox carts." [3]. Before the rule of Emperor Han Wen and Emperor Han Jing in the early Han Dynasty, the country's finances were scarce, and even four carriages of the same color that met the etiquette system could not be raised from below the emperor, as a result, the generals and prime ministers could only ride on ox carts. But at that time, the ox cart was only a temporary replacement, and its status was obviously not as noble as the carriage. After the rule of Emperor Han Wen and Emperor Han Jing, the economy of the Han Dynasty greatly improved, and carriages reappeared in the martial arts world, therefore, from the early Western Han Dynasty to the middle of the Eastern Han Dynasty, ox carts were mostly used for carrying goods, and people would be despised by the esteemed. According to the Records of the "Si Ji·Wu Zong Shi Jia", "Later on, some of the poor among the feudal lords could only ride on ox carts." [3]. However, when riding on carriages, the requirements for etiquette also increased day by day. For example, the nobles and nobles who rode on carriages were subject to many rules such as "appearance and behavior when riding" and "appearance and behavior when standing on carriages", and must always maintain a gentleman's demeanor, rather than being arbitrary. In the late Eastern Han Dynasty, this situation underwent a disruptive change. According to the "Hou Han Shu·Liu Kuan Zhuan", "Liu Kuan once went on a trip and a person lost a cow. He went to Liu Kuan's car to claim it. Liu Kuan didn't say anything and got off the car and walked home. After a while, the person who recognized the cow found the cow to return Liu Kuan, kowtowed and apologized, saying," I feel ashamed to have offended an elder like you, no matter how you punish me." Liu Kuan said, "Things are similar, and mistakes are also allowed. Fortunately, you worked hard to bring them back. Why apologize?" [43]. Liu Kuan was a famous official of the Eastern Han royal family, indicating that ox carts were also vehicles used by high-ranking officials and nobles during the late Eastern Han Dynasty.

After the Wei Dynasty and Jin Dynasty, the trend of riding ox carts became even more prevalent. The ox cart moved slowly and steadily, and the carriage was spacious and tall. If it was slightly modified, a shed and a curtain could be installed on the carriage, and a few seats could be laid inside the carriage for easy sitting and sleeping. This was the most suitable for the privileged and wandering aristocrats. So, since the Wei Dynasty and Jin Dynasty, ox carts have gradually gained the favor of aristocratic families. Riding ox carts is no longer a lowly thing, but has become a fashionable trend. Especially after the Yi Guan Nan Du of the Eastern Jin Dynasty, there were more cattle and fewer horses on the Jiangzuo River, which was also one of the reasons for the rise of ox carts. According to the "Jin Shu·Yu Fu Zhi", "At the end of Emperor Wu of Han's promotion of gratitude, the feudal lords were weak, and the poor even rode on ox carts. Later, they saw some nobles. Since the spiritual offering, the emperor and scholars thought they would ride on ox carts frequently, and the highest authority went out to the court to mourn and ride on them." [44]. According to records, Emperor Yuan of Jin also liked to ride on ox carts, and his ministers imitated him. The important minister Wang Dao, as a prime minister, also rode on short legged calves, leaving behind a historical joke of "a chariot pulled by a calf, a dust brush made from a cinnabar tail. It refers to the chariot used by high-ranking officials and nobles, with exquisite and magnificent dust brushes". The increase in the value of ox cars is also reflected in the system of riding sedans during travel. According to the ritual system at that time, officials of different levels used different colors and materials for car covers and different body decorations. The "Jin Shu" also records a type of mica chariot at that time: "Decorate a calf chariot with mica, officials are not allowed to ride it, in order to give the ears of princes." [44]. Although the image of the mica chariot is no longer verifiable, we can imagine the shining decoration of the ox chariot from this.

During the Northern Dynasty and Southern Dynasty, ox carts became increasingly popular. The imperial carriage that the Emperor of Northern Wei rode on during his travels was meant to "drive twelve cows" [45], indicating that the use of ox carts in the Northern Dynasty was even more prosperous than in the Northern Dynasty and Southern Dynasty. The Northern Dynasty was like this, and the Southern Dynasty was not inferior. It was precisely because the noble families were greedy for comfort and indulged in pleasure that various high-end ox carts rapidly developed, resulting in the almost disappearance of Han dynasty carriages. In the tombs of the Wei Dynasty, Jin Dynasty, Northern Dynasty and Southern Dynasty, using pottery ox carts for burial was a nationwide revolutionary burial custom. The pottery ox cart of the Western Jin Dynasty (Fig. 2.21) unearthed in Jiyuan, Henan Province, is made of mud and gray pottery, has an arched roof, with an open front and a square door at the back, the wheels have 12 spokes, the ox horns were bent forward, the eyes were round, the limbs were stable, and the cow was fat and strong. The image of the cow was lifelike, as if it was ready to take its owner on a trip.

Until the Sui Dynasty and early Tang Dynasty, the trend of high-ranking officials and nobles riding ox carts was still very prevalent [53]. Until Emperor Taizong of the Tang Dynasty, who advocated for the rule of horses and even carved six horses in Zhaoling, the number of carriages gradually increased. At this point, the system of officials riding ox carts has been completely abandoned. After the Song Dynasty, ox carts became a specialized tool for freight transportation and agriculture.

Fig. 2.21 Pottery Ox carts from western Jin Dynasty Tombs [46]

Other Functions of Vehicles: Ritual and Warfare

The Book of Yi Li records the number of horses used by people of different noble levels when riding in carriages, representing the symbolic role of the rituals of ancient transportation. For example, "the emperor's carriage was driven by six horses, the feudal lords used five horses, the nobles used four horses, the officials used three horses, the scholars used two horses, and the commoners used one horse". A large number of Eastern Zhou tombs and chariot and horse pits were discovered during archaeological drilling at the Eastern Zhou Wangcheng Site in Luoyang, Henan Province. One of the largest pits, north–south length of 42 m, east–west width of 7.2 m, depth of 1.8 m, the pit buried 26 cars, horses 68, vehicles were placed in the north–south longitudinal double rows, the head of the car, the head of the horse are to the south.. Among the vehicles, there are 1 carriage that can handle 6 horses, 8 carriages that can handle 4 horses, and 17 carriages that can handle 2 horses. The carriage carrying six horses was placed in the position of the second one from north to south on the west train. The car has a square shape, with clear marks on the carriage, axle, wheels, and axle. Placed 6 complete horse skeletons on both sides of the carriage shaft in front of the carriage: 3 on the east side of the carriage shaft, with the head south and the tail north, lying on the side facing east; 3 on the west of the car shaft, with the head south and the tail north, lying on the side facing west, arranged neatly and orderly. Although it has undergone nearly 3000 years of baptism, the shaft, body components, and horse bones are well preserved, with a grand scale, which clearly confirms the scene recorded in ancient literature that "the emperor rode a carriage driven by six horses" (Fig. 2.22).

This archaeological achievement provides strong factual basis for the controversy among Eastern Han Confucian scholars over whether the Zhou Emperor rode six horses or four horses.

According to the records in the "Spring and Autumn Annals of Gongyang" and "Yi Li · Wangdu Ji", the scholar Xu Shen of the Eastern Han Dynasty believed that in the ancient system, "the emperor's carriage was driven by six horses". According to the records in the Book of "Mao Shi" that "from the emperor to the officials,

Fig. 2.22 Excavated site of "emperor using six horses to pull his carriage"

they all rode carriages driven by four horses, while scholars rode carriages driven by two horses" [47], and the records in the Book of Zhou Li that "riding in a cart driven by four horses," the scholar Zheng Xuan of Eastern Han Dynasty believed that "the carriage in which the Son of Heaven rode was driven by four horses". Zheng Xuan refuted Xu Shen's the Book of "Refuting the Differences in the Five Classics" by saying, "If the Emperor of Zhou did indeed drive six horses, why didn't the article "Xiao Ren" mention using six horses as the counting unit? Why didn't the feudal lords offer six horses in the article "Gu Ming?". "Today, the emperor drives six horses, which is naturally the etiquette of the Han Dynasty and different from ancient times." [48]. When Liu Zhao, a Liang scholar from the Southern Dynasty, annotated the "Yu Fu Zhi" in the Book of Later Han, he summarized the arguments and evidence of this debate, saying: "Now emperors ride carriages driven by six horses." It is a system that began in the Han Dynasty and is different from ancient systems [49]. Since then, scholars from the Tang Dynasty and Song Dynasty have been controversial about this for more than 2000 years, until a conclusion is reached below, finally uncovering the historical mystery of whether the emperor drove six horses or four horses in ancient times.

In addition to the number of horses, the ceremonial system also originated from the Western Han Dynasty, which allocated the texture of components (such as gold, silver, copper, etc.) for different levels of passengers, the patterns of chariot and horse decorations (such as dragons, phoenixes, tigers, leopards, etc.), the size and materials used for chariot covers (such as cloth, silk, etc.), and the shape and materials used for chariot canopies. In addition the main carriages used by nobles and officials of all sizes, there are also regulations on the number of accompanying carriages and riding officials. In short, the etiquette level characteristics of vehicles are becoming increasingly prominent, which to some extent also affects the etiquette culture of modern racing teams.

In addition, vehicles were commonly used in warfare in ancient times and were one of the main equipment for combat armies. The development history of ancient chariots in China is also long-standing, with a history of using ancient chariots for warfare dating back to the Xia Dynasty and Shang Dynasty. There is a detailed record of the chariot shapes during the Shang and Zhou dynasties in the Book of "Zhou Li·Kao Gong Ji". In 1936, a chariot of the Shang Dynasty was unearthed for the first time in the chariot and horse pit of Yin Ruins in Anyang, Henan Province. It was made of a single shaft, two wheels, a long hub, and a rectangular vehicle with short transverse width and vertical length. The rear end of the shaft was pressed between the carriage and the axle. After the end of the shaft was slightly exposed from the carriage, the front end of the shaft was transversely placed with a car scale, and two yokes were attached to the scale to drive the horse. The hub of a tank is generally much longer than that of a civilian vehicle, with a wheel diameter of about 1.3–1.4 m and 18–24 spokes. The width of the carriage is generally 1.3–1.6 m and the depth is 0.8–1 m. Due to the large wheel diameter, the carriage is wide and the depth is short, and it is a single shaft, which can further increase stability and protect the side from enemy vehicles approaching.

The most glorious era of ancient Chinese chariots should have been during the Spring and Autumn period and the Warring States period, when chariots had become the main combat force of major powers. The number of chariots has even become the standard for measuring the strength of a great country, and the allusions of "a country with a thousand chariots" and "a powerful state with ten thousand chariots" have been around since the Spring and Autumn and Warring States periods. Although collectively referred to as "ancient chariots," the roles, shapes, and performance standards of chariots in different periods of Chinese history also vary slightly. According to the military work "Methods of the Sima" of the Spring and Autumn period, "the chariots of the Xia Dynasty were called hook chariots, mainly focusing on their balance performance; the chariots of the Shang Dynasty were called Yin chariots, focusing on speed; and the chariots in the Zhou Dynasty were also known as Yuan Rong, paying more attention to overall quality" [50]. The power of ancient Chinese chariots lies not in their individual combat performance, but in their ability to form formations for combat. The ancient chariot groups formed in groups can easily break through enemy infantry formations.

The chariots of the Shang Dynasty were arranged with two wheels and a single shaft, with the shaft arranged between two or four horses. The number of horses used to tow the chariots was mainly determined by the width and weight of the chariot's body. There would be three armored soldiers on the car: one person is responsible for long-range shooting using bows and arrows, known as the "the leader of a military unit", with a status similar to the commander of a modern tank;

Another warrior, wielding a spear, was responsible for close range combat, known as the "a person accompanying a rider in an ancient carriage"; The last warrior is located in the middle and front position of the vehicle, responsible for driving, known

Fig. 2.23 Schematic diagram of Shang Dynasty warfare chariots [51]

as "imperial hand" (Fig. 2.23). In practical use, if it is a group operation that requires the formation of a large formation of tanks, all infantry will be deployed to the rear of the tank group to avoid the tanks rolling over our own infantry; If it is a small battle formation, each tank is usually equipped with 15 infantry soldiers to clean up the enemy forces that have been dispersed by the tanks. This type of tank form and combat organization had a good effect in practical use and was used until the Spring and Autumn Period and the Warring States Period.

The Spring and Autumn Period and the Warring States Period were the heyday of the development of ancient Chinese chariots. Most of the warring areas had relatively flat terrain, providing a foundation for large-scale chariot warfare. At that time, the division of ancient tanks was also more detailed, and according to different uses, offensive tanks and transport tanks were further divided. Compared with the Xia Dynasty and Shang Dynasty, although there was not much difference in the overall shape of tanks during the Spring and Autumn and Warring States periods, a large amount of metal was used in the body materials. Especially with the maturity of bronze smelting technology, most soldiers began to use more lethal bronze weapons for combat. In order to better protect the safety of armored vehicles, during the Spring and Autumn and Warring States periods, the guardrails on the ancient chariots began to be raised, and leather and bronze armor were laid to further enhance defense performance. Therefore, the overall weight of the chariot was relatively large, and the wheels and shaft structure were correspondingly strengthened. Of course, ancient chariots at that time were not only used for offense, but their tall body was also a good barrier for defense. In the Battle of Maling between Qi and Wei in the 27th year of King Xian of Zhou (342 BC), Qi's strategist Sun Bin used a large number of tanks to block the Wei army's route, surrounded them in groups, and annihilated them all.

During the Eastern Han and Three Kingdoms periods, ancient chariots had already declined. Due to the expansion of combat areas, tanks were unable to demonstrate their advantages in mountainous and hilly areas, so their position began to be gradually replaced by more flexible cavalry. However, in a few flat terrain areas such as plains, in order to resist the impact of cavalry, tanks continued to survive on the battlefield in the form of defensive vehicle formations by further strengthening their defense capabilities. Among them, the most famous was the defensive "Wugang Vehicle" developed by Shu Prime Minister Zhuge Liang to resist the cavalry of Cao Wei.

After the Three Kingdoms period, ancient Chinese chariots did not immediately disappear, but gradually retreated to the rear. In addition to the transport type ancient chariots used for transporting materials, most of the later ancient chariots were mainly used for siege, such as the ancient Chinese cloud ladder chariots, throwing chariots, and collision chariots, all of which were derivatives of ancient chariots. Even during the Ming Dynasty and Qing Dynasty, China still had records of using tanks. The rise of firearms in the Ming Dynasty led to the heavy weight of multi barrel firearms, which were difficult to transport and use solely by human labor. Therefore, people invented the "Seven-Star Musketeer Wagon", which actually installed seven musketeers on the car to drive forward and stopped shooting when needed. This type of tank had the strongest combat performance in the development of ancient Chinese tanks. Ming Dynasty general Qi Jiguang even developed more powerful "a one-sided compartment war chariot" (Fig. 2.24), but due to their overall bulkiness, they were not widely promoted and used. In the Qing Dynasty, with people's further emphasis on cavalry and archery technology, the development of ancient chariots officially came to an end.

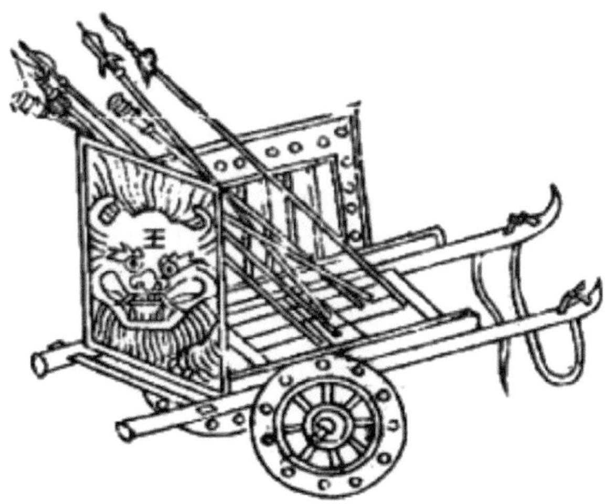

Fig. 2.24 "One-sided compartment war chariot" in Ming Dynasty [52]

References

1. 徐水釜山文化研究会.炎黄之始釜山考[M].东莞:九鼎传播有限公司, 2009. Xushui Fushan Cultural Research Association, *The exploration of the origins of the Yan and Huang Emperors conducted by the Fushan Association*, Dongguan: Jiuding Communication Co., Ltd., 2009.
2. 柏杨.中国人史纲[M].北京:人民文学出版社, 2011. Bai Yang, *Outlines of Chinese history*, Beijing: People's Literature Publishing House, 2011.
3. 司马迁.史记[M].北京:中华书局, 1959. Sima Qian. *Records of the Grand Historian.* Beijing: Zhonghua Book Company, 1959.
4. 趣历史.黄帝——传说中的中原各族的共同祖先[EB/OL]. [2022-05-16].http://www.qulishi.com/renwu/huangdi/.Interesting History: Huang Emperor - The Legendary Common Ancestor of All Ethnic Groups in the Central Plains[EB/OL]. [2022-05-16]. http://www.qulishi.com/renwu/huangdi/.
5. 罗泌.路史[M]//王云五.丛书集成初编.北京:商务印书馆, 1936. Luo Mi, *Lu Shi (A Comprehensive History)*, Wang Yunwu. Series of Early Collected Works. Beijing: The Commercial Press, 1936.
6. 王逸.楚辞章句[M].黄灵庚,点校.上海:上海古籍出版社, 2017.Wang Yi. *Chu Ci Zhangju (Commentaries on the Elegies of Chu).* Annotated and Collated by Huang Linggeng. Shanghai: Shanghai Ancient Books Publishing House, 2017.
7. 崔豹.《古今注》校笺[M].牟华林,校笺.北京:线装书局, 2015. Cui Bao, *Annotations and Corrections for Gu Jin Zhu (Comments and Notes on Ancient and Modern Times).* Edited and Annotated by Mou Hualin. Beijing: Xianzhuang Book Bureau, 2015.
8. 刘恕.通鉴外纪[M].北京:中央编译出版社, 2015. Liu Shu. *Tong Jian Wai Ji (Additional Comments of History as a Mirror*). Beijing: Central Compilation and Translation Press, 2015.
9. 高承.事物纪原[M].北京:中华书局, 1989. Gao Cheng. *Shi Wu Ji Yuan (The history of Origins of Things).* Beijing: Zhonghua Book Company, 1989.
10. 马骕.绎史[M].上海:上海古籍出版社, 1993. Ma Su. *Yi Shi (A Comprehensive Interpretation of Pre-Qin History).* Shanghai: Shanghai Ancient Books Publishing House, 1993.
11. 李昉,等.太平御览[M].上海:上海古籍出版社, 2008. Li Fang, et al. *Tai Ping Yu Lan (Encyclopedia covering various fields such as politics, economy, culture, history, geography, science and technology, etc.).* Shanghai: Shanghai Ancient Books Publishing House, 2008.
12. 曲辰.轩辕黄帝史迹之谜[M].北京:中国社会科学出版社, 1992. Qu Chen. *The Mystery of the Historical Traces of Huang Emperor (Xuan Yuan).* Beijing: China Social Sciences Press, 1992.
13. 百度百科.指南车[EB/OL]. [2022-05-15]. https://baike.baidu.com/item/%E6%8C%87%E5%8D%97%E8%BD%A6?sefr=cr.Baidu Baike: *Guide Vehicle (a reference to an ancient Chinese vehicle designed to maintain a fixed direction, used for navigation in ancient times).* [EB/OL]. [2022-05-15]. https://baike.baidu.com/item/%E6%8C%87%E5%8D%97%E8%BD%A6?sefr=cr.
14. 盛世文玩.国家宝藏·一张从明代来的世界古地图[EB/OL]. (2021-04-13) [2022-05-15]. https://www.sohu.com/a/460428058_182897. Sheng Shi Wen Wan. *National Treasure: An Ancient World Map from the Ming Dynasty.* [EB/OL]. (2021-04-13) [2022-05-15]. https://www.sohu.com/a/460428058_182897.
15. 谯周.古史考[M].扬州:广陵古籍刻印社, 1984. Qiao Zhou, *Gui Shi Kao (Ancient History Research of China).* Yangzhou: Guangling Ancient Books Typography and Printing Society, 1984.
16. 世本八种[M].秦嘉谟,等,辑.北京:国家图书馆出版社, 2008.*Shi Ben Ba Zhong (Research on the History of the Pre-Qin Period).* Compiled by Qin Jiamo, et al. Beijing: National Library of China Press, 2008.
17. 吕不韦.吕氏春秋[M].高诱,注.毕沅,校.徐小蛮,标点.上海:上海古籍出版社, 2014. *Lü Buwei. Lü's Almanac*. Annotated by Gao You, Collated by Bi Yuan, Punctuated by Xu Xiaoman. Shanghai: Shanghai Ancient Books Publishing House, 2014.
18. 墨子[M].毕沅,校注.吴旭民, 点校.上海:上海古籍出版社, 2014. *Mo Zi*. Collated and Annotated by Bi Yuan, Punctuated and Checked by Wu Xumin. Shanghai: Shanghai Ancient Books Publishing House, 2014.

19. 刘安.淮南子[M].许慎,注.陈广忠, 点校.上海:上海古籍出版社, 2016. Liu An, *Huai Nan Zi*. Collated and Annotated by Bi Yuan, Punctuated and Checked by Wu Xumin. Shanghai: Shanghai Ancient Books Publishing House, 2014.
20. 许慎.说文解字[M].杭州:浙江古籍出版社, 2016. Xu Shen. *Shuowen Jiezi (Explaining Simple and Analyzing Compound Characters)*. Hangzhou: Zhejiang Ancient Books Publishing House, 2016.
21. 九派新闻.走进车祖奚仲故里奚村 感受省级传统村落魅力[EB/OL]. (2022-02-07) [2022-05-16]. https://baijiahao.baidu.com/s?id=1724104817047803334&wfr=spider&for=pc.Jiupai News. *Visit the hometown of Xi Zhong, the legendary inventor of the cart, and feel the charm of the provincial-level traditional village*. [EB/OL]. (2022-02-07) [2022-05-16]. https://baijiahao.baidu.com/s?id=1724104817047803334&wfr=spider&for=pc.
22. 董龙江.美篇[EB/OL]. [2022-05-16]. https://www.meipian.cn/1wb0pwai. Dong Longjiang. *Beautiful Articles* [EB/OL]. [2022-05-16]. https://www.meipian.cn/1wb0pwai.
23. 史上浙一天.8000年前浙江人造了世界最早独木舟[EB/OL]. [2022-05-05]. https://zj.zjol.com.cn/news/488071.html. *On this day in history, Zhejiang. 8,000 years ago, the people of Zhejiang province built the world's earliest canoe*. [EB/OL]. [2022-05-05]. https://zj.zjol.com.cn/news/488071.htm.
24. 余姚新闻网.中国乡愁丨最初的家园——河姆渡[EB/OL]. [2022-05-06]. http://www.yynews.com.cn/system/2020/09/06/012147007.shtml. *Yuyao News Network. Chinese Nostalgia|The Original Homeland - Hemudu*. [EB/OL]. [2022-05-06]. http://www.yynews.com.cn/system/2020/09/06/012147007.shtml.
25. 上海交通大学董浩云航运博物馆.水田畈遗址出土的木桨[EB/OL]. [2022-05-06]. http://www.maritimemuseum.sjtu.edu.cn/index.php?m=content&c=index&a=show&catid=55&id=316. Dong Haoyun Shipping Museum, Shanghai Jiao Tong University. *Wooden paddles unearthed from the Shuitianfan site*. [EB/OL]. [2022-05-06]. http://www.maritimemuseum.sjtu.edu.cn/index.php?m=content&c=index&a=show&catid=55&id=316.
26. 中华古玩网.陶舟形器[EB/OL]. [2022-05-14]. http://www.gucn.com/Info_MuseumCurio_Show.asp?Id=1657. China Antique Network. *Pottery Boat-shaped Vessel*. [EB/OL]. [2022-05-14] http://www.gucn.com/Info_MuseumCurio_Show.asp?Id=1657.
27. 巴比葫芦.略谈大连地区的远古捕鱼活动[EB/OL]. (2017-03-29) [2022-05-14]. https://www.sohu.com/a/131006599_665821. Babi Hulu. *A Brief Discussion on Ancient Fishing Activities in Dalian Area* [EB/OL]. (2017-03-29) [2022-05-14]. https://www.sohu.com/a/131006599_665821.
28. 周易[M].杨天才,译注.北京:中华书局, 2022. *I Ching (The Book of Changes)*. Translated and Annotated by Yang Tiancai. Beijing: Zhonghua Book Company, 2022.
29. 山海经[M].周明初,校注.杭州:浙江古籍出版社, 2011.*Shan Hai Jing (Classic of Mountains and Seas)*. Edited and Annotated by Zhou Mingchu. Hangzhou: Zhejiang Ancient Books Publishing House, 2011.
30. 人民网.二里头——来这里可以触摸最早的中国[EB/OL]. [2022-05-14]. http://www.qiyuandi.cn/portal.php?mod=view&aid=632. People's Network. *Erlitou - Here You Can Touch the Earliest China*. [EB/OL]. [2022-05-14]. http://www.qiyuandi.cn/portal.php?mod=view&aid=632.
31. 殷墟博物馆.殷墟博物馆[EB/OL]. [2022-05-14]. http://www.ayyx.com.Yin Ruins Museum. *Introduction to Yin Ruins Museum*. [EB/OL]. [2022-05-14]. http://www.ayyx.com.
32. 郑若葵.中国史话:交通工具史话[M].北京:社会科学文献出版社, 2012. Zheng Ruokui. *Chinese History: Chronicles of Transportation Tools* . Beijing: Social Sciences Academic Press, 2012.
33. 殷墟博物馆.中国最早的车马遗迹——殷墟车马坑[EB/OL]. (2007-11–26) [2022-05-14]. http://www.ayyx.com/News/Article/124. Yin Ruins Museum. *The Earliest Chariot and Horse Remains in China - the Chariot and Horse Pit of Yin Ruins*. [EB/OL]. (2007-11–26) [2022-05-14]. http://www.ayyx.com/News/Article/124.
34. 360个人图书馆.殷苑拾珍第四期:商代马车[EB/OL]. (2019-03-18) [2022-05-14]. http://www.360doc.com/content/19/0318/09/91189_822371482.shtml. 360 Personal Library. *Yin Yuan Treasures Collection, Issue 4: Chariot of the Shang Dynasty* [EB/OL]. (2019-03-18) [2022-05-14]. http://www.360doc.com/content/19/0318/09/91189_822371482.shtml.

35. 周礼[M].陈戍国,点校.长沙:岳麓书社, 1989. *The Rites of Zhou*. Edited and collated by Chen Shuguo. Changsha: Yuelu Publishing House, 1989.
36. 新浪时尚.涨知识 藏在博物馆里的生僻字原来这么读[EB/OL]. (2022-03-15) [2022-05-14]. http://fashion.sina.com.cn/style/man/2022-03-15/1602/doc-imcwiwss6157273.shtml. Sina.com, Fashion Column. *Expand Your Knowledge. How to Pronounce Those Rare Characters Hiding in Museums*. [EB/OL]. (2022-03-15) [2022-05-14]. http://fashion.sina.com.cn/style/man/2022-03-15/1602/doc-imcwiwss6157273.shtml.
37. 琅琊神话.这3件青铜文物证明西周马车的先进工艺,原来中国古人已造车4千年[EB/OL]. (2018-03-12) [2022-05-14]. https://baijiahao.baidu.com/s?id=1594721063022157205&wfr=spider&for=pc. Langya Mythology. *These three bronze artifacts serve as evidence of the advanced craftsmanship of the Western Zhou Dynasty chariots, revealing that the ancient Chinese people have been building vehicles for 4,000 years*. [EB/OL]. (2018-03-12) [2022-05-14]. https://baijiahao.baidu.com/s?id=1594721063022157205&wfr=spider&for=pc.
38. 礼记[M].陈澔,注.金晓东, 点校.上海:上海古籍出版社, 2016. *The Book of Rites*. Annotated by Chen Hao and Jin Xiaodong. Shanghai: Shanghai Ancient Books Publishing House, 2016.
39. 百度百科.秦铜车马[EB/OL]. [2022-05-16]. https://baike.baidu.com/pic/%E7%A7%A6%E9%93%9C%E8%BD%A6%E9%A9%AC/10684517/0/8b13632762d0f7033304dd1f06fa513d2797c5bb?fr=lemma&ct=single#aid=0&pic=8b13632762d0f7033304dd1f06fa513d2797c5bb. Baidu Baike. *Qin Terracotta Chariots and Horses*. [EB/OL]. [2022-05-16]. https://baike.baidu.com/pic/%E7%A7%A6%E9%93%9C%E8%BD%A6%E9%A9%AC/10684517/0/8b13632762d0f7033304dd1f06fa513d2797c5bb?fr=lemma&ct=single#aid=0&pic=8b13632762d0f7033304dd1f06fa513d2797c5bb.
40. 小草.西安游(10)秦始皇兵马俑[EB/OL]. (2019-12–09) [2022-05-16]. http://www.meipian.cn/2kj3kiu0. Xiao Cao. *Xi'an Travel Album (10): Terracotta Army of Qin Shihuang*. [EB/OL]. (2019-12–09) [2022-05-16]. http://www.meipian.cn/2kj3kiu0.
41. 古城弥勒佛.秦陵彩绘铜车马[EB/OL]. (2010-11-28) [2022-05-16]. http://bbs.photofans.cn/blog-238242-39612.html. Ancient City Maitreya Buddha|Qin Mausoleum Painted Bronze Chariots and Horses [EB/OL]. (2010-11-28) [2022-05-16]. http://bbs.photofans.cn/blog-238242-39612.html.
42. 西北旅媒研究院.一眼千年,去这些博物馆欣赏镇馆之宝[EB/OL]. [2022-04-15]. https://www.sohu.com/a/279962015_429223?_f=index_pagerecom_7. Northwest Tourism Media Research Institute. *A Glance at a Thousand Years, Visit These Museums to Appreciate Their Most Prized Possessions*. [EB/OL]. [2022-04-15]. https://www.sohu.com/a/279962015_429223?_f=index_pagerecom_7.
43. 范晔.后汉书[M].李贤,等,注.北京:中华书局, 1997. Fan Ye. *History of the Later Han Dynasty*. Annotated by Li Xian, et al. Beijing: Zhonghua Book Company, 1997.
44. 房玄龄,等.晋书[M].北京:中华书局, 1997. Fang Xuanling, et al. *History of the Jin Dynasty*. Beijing: Zhonghua Book Company, 1997.
45. 魏收.魏书[M].北京:中华书局, 1997. Wei Shou. *History of the Wei Dynasty*. Beijing: Zhonghua Book Company, 1997.
46. 历史烟霭.以牛车为尊:魏晋南北朝时期,牛车为何成为了“豪车”[EB/OL]. (2020-03-08) [2022-05-16]. https://baijiahao.baidu.com/s?id=1660577186732203911&wfr=spider&for=pc. Historical Haze. *The Prestige of Ox-Drawn Carriages: Why Did Ox-Drawn Carriages Become "Luxury Vehicles" During the Wei, Jin, Southern, and Northern Dynasties?* [EB/OL]. (2020-03-08) [2022-05-16]. https://baijiahao.baidu.com/s?id=1660577186732203911&wfr=spider&for=pc.
47. 毛亨.毛诗注疏[M].郑玄,笺.孔颖达,疏.陆德明,音释.朱杰人,整理.上海:上海古籍出版社, 2013. Mao Heng. *Mao's Annotations on the Book of Songs,* Explanations by Kong Yingda, Phonetic Annotations by Lu Deming, Compiled by Zhu Jieren. Shanghai: Shanghai Ancient Books Publishing House, 2013.
48. 郑玄.驳五经异义[M].扬州:江苏广陵古籍刻印社, 1984. Zheng Xuan. *Criticism of The Annotations on the Five Classics*. Yangzhou: Jiangsu Guangling Ancient Books Carving and Printing Press, 1984.

49. 司马彪.续汉书[M].刘昭,注.北京:中华书局, 1965. Sima Biao. *Continuation of the History of the Han Dynasty*. Annotated by Liu Zhao. Beijing: Zhonghua Book Company.
50. 吴子·司马法[M].陈曦,译注.北京:中华书局, 2018. Wu Zi. *Master Ssu-ma's Military Methods*. Translated and Annotated by Chen Xi. Beijing: Zhonghua Book Company, 2018.
51. 闲叙春秋.探一探牧野之战到底有多壮观,纣王七十万大军为何遭此惨败[EB/OL]. (2019-07-31) [2022-05-16]. https://baijiahao.baidu.com/s?id=1640466980024169094&wfr=spider&for=pc. A Leisurely Discussion on the Spring and Autumn Period. *Exploring the Magnificence of the Battle of Muye and Why King Zhou's 700,000 Troops Suffered Such a Catastrophic Defeat.* [EB/OL]. (2019-07-31) [2022-05-16]. https://baijiahao.baidu.com/s?id=1640466980024169094&wfr=spider&for=pc.
52. 二桂.戚继光的装甲部队"车营", 到底取得了多大的战果[EB/OL]. (2019-04-10) [2022-05-16]. https://zhuanlan.zhihu.com/p/61936818. Er Gui. *What Were the Actual Achievements of Qi Jiguang's Armored Troops, the 'Cheying'?* [EB/OL]. (2019-04-10) [2022-05-16]. https://zhuanlan.zhihu.com/p/61936818.
53. 魏徵.隋书[M].北京:中华书局, 2020. Wei Zheng. *History of the Sui Dynasty*. Beijing: Zhonghua Book Company, 2020.

Chapter 3
Unification of the Four Seas: Establishment of the Qin Empire and Qin Chi Road

In the 26th year of the Emperor Qin Shi Huang's reign (221 BC), the Qin Dynasty was established, and China officially entered the era of "comprehensive unification" of a centralized dynasty. The term "comprehensive unification" first appeared in the "Spring and Autumn Annals of the Gongyang", which reads: "Why do you say 'in the first month of the Zhou calendar'? This is because it is necessary to mark that the whole country is using a unified calendar" [1]. Emperor Qin Shi Huang pioneered the establishment of a unified national system in the fields of politics, economy, and cultural thought, and took a series of measures to strengthen centralization, playing an important role in consolidating national unity and rapidly developing economy and culture. Therefore, "comprehensive unification" refers not only to regional unification, but also to the integration of national political, economic, and ideological and cultural systems. The unification of the six states by Qin put an end to the chaotic situation of feudal separatism since the Spring and Autumn and Warring States periods, and opened up a new era for the establishment and development of China, a unified multi-ethnic country. In the following 2000 years of history, although there were also brief periods of division, unity and concentration have always been the mainstream of historical development, which is closely related to the institutional system created by the Qin Dynasty.

Looking back at this period of history, while sorting out historical facts with transportation as the main line, we are also pondering: in the process of Qin's completion of the great cause of unification, what transportation advantages did this regime located on the outskirts of the Central Plains rely on to ultimately win? What transportation measures were taken to consolidate the unified state power after reunification? What impact did these measures have on future generations? This chapter attempts to reveal the relationship between transportation, politics, and economy by answering these questions.

© China Communications Press Co., Ltd 2025

H. Lu (ed.), *A Brief History of Transportation in China*,
https://doi.org/10.1007/978-981-97-9851-3_3

Road Transportation Before Qin's Unification

Struggle for Power in the Spring and Autumn Period and the Warring States Period

The Spring and Autumn Period refers to the historical period from 770 BC when King Ping of Zhou moved eastward to Luoyi until the death of King Jing of Zhou in 476 BC. "Spring and Autumn" is named after the Annals of the State of Lu, "Spring and Autumn". According to legend, this book was revised by Confucius and recorded the history from the year of Lu Yin (722 BC) to the 14th year of Duke Ai of Lu (481 BC). The Spring and Autumn period was a period of great division in Chinese history. The political and military strength of the Zhou royal family is declining day by day, and the major feudal states are becoming stronger. The great drama of competing for hegemony is repeatedly staged, with the "Five Hegemons of the Spring and Autumn Period" taking turns to appear. Sima Zhen's "Records of the Grand Historian" states: "With the call to suppress the unjust monarch, the Zhou royal family is revered as the co ruler of the world, and power is presided over by the Five Hegemons. The Five Hegemons refer to Duke Huan of Qi, Duke Wen of Jin, Duke Mu of Qin, Duke Xiang of Song, and King Zhuang of Chu." [2]. However, "Xunzi's" Wang Ba" points out: "Therefore, Duke Huan of Qi, Duke Wen of Jin, King Zhuang of Chu, King Helu of Wu, and King Goujian of Yue all come from remote countries, but they are influential and expand to the Central Plains region for no other reason, just because they can uphold integrity. This is what is said to be because they uphold integrity. The situation of claiming hegemony over the world." [3]. As the monarch of the Western powerhouse Qin, Duke Mu of Qin was also regarded as one of the hegemons of the Spring and Autumn period.

The Warring States period refers to the period from 475 to 221 BC, which was the later period of the Eastern Zhou Dynasty until the unification of the Qin Dynasty. Due to the continuous warfare among various countries, it was later known as the "Warring States". The name "Warring States" comes from the "Strategies of the Warring States" compiled by Liu Xiang of the Western Han Dynasty. During the Warring States period, various countries no longer held the banner of "respecting the Zhou", and the seven major powers confronted each other, each annexing small neighboring countries. Wars continued one after another, presenting a situation of great chaos in the world. The geographical situation of the Seven Heroes of the Warring States period is: the State of Chu is in the south, the State of Zhao is in the north, the State of Yan is in the northeast, the State of Qin is in the west, the State of Qi is in the east, the State of Han and the State of Wei are in the center [4].

Among them, the State of Chu had a vast territory and a large population, and its development was greatly hindered by the formation of the State of Qi to resist the State of Qin. So, the State of Qin sent Zhang Yi to the State of Chu and advised the State of Chu to leave the State of Qi and follow the State of Qin, with a price of 600 li from the land of Shang Yu. King Huai of Chu was greedy for cheap goods, so he broke up with the State of Qi. When the State of Chu sent people to important places,

Qin Prime Minister Zhang Yi cunningly argued that "Leyi and the King of Yan agreed to meet at the city gate, about six miles apart.". King Huai of Chu launched an attack on the State of Qin and suffered a great defeat before returning. The State of Chu was weak and isolated, so the State of Qin marched eastward into the Central Plains.

Road Transportation During the Spring and Autumn Period and the Warring States Period

During the Spring and Autumn period, various countries spared no effort in opening up roads in their respective capitals in order to expand their strength. They competed for dominance on the premise of convenient transportation, improving regional road traffic. Due to the fact that the Zhou royal family was still a nominal monarch at that time, the locations of wars and alliances among various countries were not far from the location of the Zhou royal family in Chengzhou (now Luoyang, Henan). Therefore, road transportation centered around Chengzhou, connecting the States of Song, Lu, and Qi to the east, the west to the States of Qin, Long, connecting the States of Chu, Wu, and Yue to the south, and north to the States of Jin and Yan [5].

To the Warring States period, the Seven Heroes established their own separatist regimes, and in order to seize more land, population, and rent, the annexation war became increasingly fierce, and the scale of the war continued to expand. Meanwhile, due to the use of ironware, the promotion of cattle farming, and the further development of agriculture and handicrafts, commerce has gradually flourished. With the needs of political, military, and economic development, the range of road traffic has become wider. It extends from Inner Mongolia, Zhangjiakou, and Chengde in Hebei Province to the north, from Yunnan and northern Guangxi in the south, to southeastern Gansu and most of Sichuan in the west, and connects to present-day Xinjiang, as well as Korea peninsula and the coast of the East China Sea in the east (Fig. 3.1) [5].

Transportation Advantages of the State of Qin

The State of Qin is located in the northwest, within the four fortresses surrounded by mountains and rivers. In order to accept the positive influence of Central Plains culture, it is necessary to persist in moving eastward in order to take the initiative in communication and expand national strength. The fortress where Qin expanded eastward was Hangu Pass (now Sanmenxia, Henan), and the passage connecting Qin Capital and Hangu Pass was called "the road to Hangu Pass", covering a distance of about 300 km. The Hangu Pass and the road to Hangu Pass had particularly important strategic significance for the State of Qin, and the joint efforts of other countries to

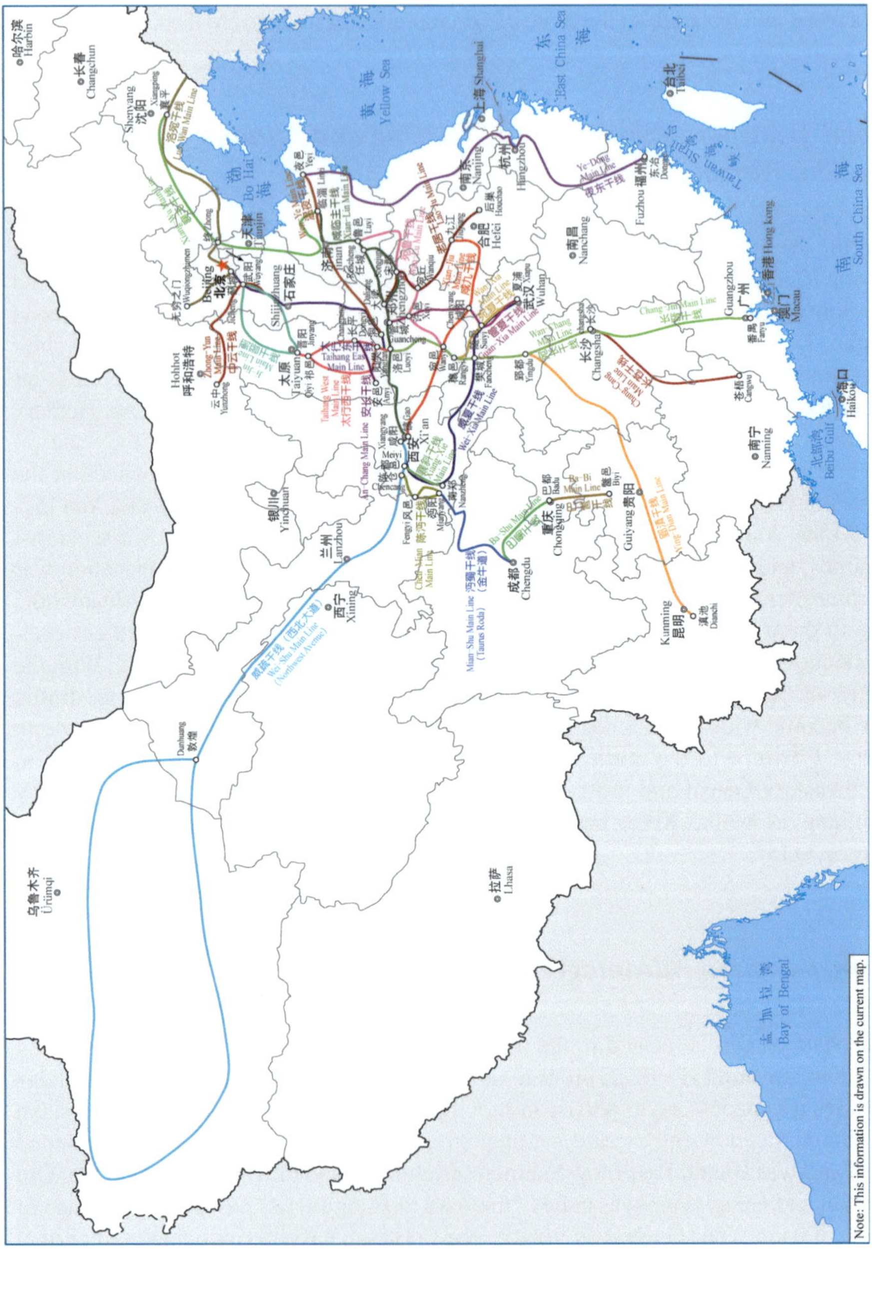

Fig. 3.1 Schematic diagram of main roads during the Warring States period

conquer Qin were all blocked at Hangu Pass. The Qin army galloped along the Hangu Road, ensuring the efficiency of military operations.

The waterway located in the western part of the State of Qin is another important transportation artery and closely related to the development of the State of Qin itself. In the 11th century BC, the 25th grandson of Boyi, Zhongxie, was settled in Xichui (now Longnan, Gansu) by the Shang Dynasty. After the fall of the Shang Dynasty, some of the Ying clan were sent here. Afterwards, the Qin-Ying people began to operate in the local area, first settling their capital in Qianyi (now Longxian County, Baoji, Shaanxi), and then moving eastward along the Qianhe River to Yongcheng (now Fengxiang County, Baoji, Shaanxi). In order to resist the invasion of the Western Rong, Qin Xianggong began to fight against it continuously until Qin Mugong "dominated the Western Rong region", and the Qianhe River Valley Road was the lifeline of the Qin state backed by Guanzhong. The year-round battles with the Western Rong objectively enhanced the military strength of the State of Qin.

The management of the Shu Road by the Qin people has also attracted worldwide attention. The construction of the plank road leading to Bashu by the Qin people significantly improved the traffic conditions of the Bashan Road in the Qinling Mountains. The Qin army took control of Bashu along the plank road, acquiring resources in this area, changing the strength comparison with Eastern powers, and forming a strategic situation of flanking the main enemy state of Chu, which is of great significance for achieving reunification in the end [6].

Another advantage of transportation development in the State of Qin is the advancement of vehicle manufacturing technology. In the early days of China, vehicles were all single barreled, requiring the carriage of two or four livestock, while double barreled carriages only required the carriage of one livestock, which was greatly simplified and easier to handle. The emergence of the double barreled carriage was a significant progress in the history of transportation, and its widespread application in the State of Qin also added advantages in transportation.

Due to the relatively complete transportation conditions, advanced transportation technology, and strong transportation forces, the Qin army was able to adapt to the difficult conditions of long-term expeditions, constantly receiving sufficient personnel and supplies, thus having a strong military force that never tires out in a hundred battles. In addition, the development of transportation in the State of Qin was conducive to communication between the Qin people and various parties, absorbing the positive factors of other regional cultures, and making Qin culture more open and enterprising. Therefore, "the land knows no bounds, and its people share a common world" and "the people are prosperous and prosperous because of this, the country is prosperous and strong" (Li Si's "Letter of Advising and Chasing Guests") [7]. The advantage of transportation construction by the Qin people was one of the important reasons why the State of Qin was able to adapt to the historical trend and advance eastward year after year, finally completing the great cause of unification during the reign of Emperor Qin Shi Huang [6].

Transportation in Other Countries [8]

The transportation of the State of Qi is centered around Linzi, with two parts: land and water. The land transportation lines of the State of Qi, known as the "Five Capitals" of the State of Qi, were centered around Linzi, Lingqiu (now Gaotang County, Liaocheng, Shandong), Guyi (now Donga Town, Pingyin County, Jinan, Shandong), Jimo (now Pingdu, Shandong), and Pinglu (now Wenshang County, Jining, Shandong), forming a road network connecting more than 70 cities in the State of Qi. The water transportation line of the State of Qi mainly utilizes the Jishui River, which leads westward through the Yellow River and Wei River to reach the capital city of Wei State, Daliang (now Kaifeng, Henan), the capital city of Eastern Zhou Dynasty, Luoyi (now the old urban area of Luoyang, Henan), and the capital city of the State of Qin, Xianyang (now Weicheng District, Xianyang, Shaanxi). Southward through Huanggou and Hangou, it leads to the Yangtze River Basin. In addition, the state of Qi also has seaports such as Jieshi, Zhifu, and Langya. Going south leads to the States of Wu and Yue, and going north leads to Liaodong and the Korean Peninsula.

In the early Spring and Autumn period, the State of Yan had weak national strength, with its ruling areas including present-day Beijing and the central and northern parts of Hebei. It was almost destroyed due to the invasion of the Shanrong army, but was ultimately preserved with the military assistance of the State of Qi, which respected the Zhou royal family and repelled the invasion of ethnic minorities. In the 75th year of King Nan of Zhou (300 BC), Yan general Qin Kai led his army to launch a northern expedition, expanding the territory controlled by the State of Yan and constructing the Great Wall. Subsequently, he established five commanderies, including Shanggu, Yuyang, Youbeiping, Liaoxi, and Liaodong, opened up roads and transportation, and strengthened his national strength, finally ranking among the "Seven Heroes of the Warring States period".

The State of Zhao moved its capital to Handan in the 16th year of King An (386 BC). Handan is located in the south of Hebei, bordering the Taihang Mountains in the west and the North China Plain in the east. It is known as the "famous capital of the world". At the time of the State of Zhao's capital relocation, Handan was already an important transportation hub. In order to ensure the connection with the west of the Taihang Mountains, the State of Zhao cleared the "Gankou Xing", "Jingxing", "Feihu Xing", and "Puyin Xing", and set up passes such as the Infinity Gate (now Zhangbei County, Zhangjiakou, Hebei), Juzhu Pass (now Dai County, Xinzhou, Shanxi), Hongshang Pass (now Tang County, Baoding, Hebei), Gaoque Pass (now Urad Back Banner, Bayannur, Inner Mongolia), and Jingxing Pass (now Jingxing Mining Area, Shijiazhuang, Hebei). In the ninth year of King Nan of Zhou (306 BC), King Wuling of Zhao passed the "Hu Fu Cavalry Shooting" reform and began to form cavalry to counter the invasion of the Xiongnu. He pursued the victory to Jiuyuan (now Jiuyuan District, Baotou, Inner Mongolia) and immediately opened up the "Lingjiu Avenue" (from Lingqiu to Jiuyuan). This plan involved a strategic detour from the north towards the capital of Qin, Xianyang, forcing Qin to open up the "Xianjiu Avenue" to confront it, ultimately forming the embryonic form of the later "Qinzhidao Road".

The transportation of the State of Wei includes land and waterways. The road transportation centers around the ancient capital Anyi (now Xia County, Yuncheng, Shanxi) and the capital Daliang (now Kaifeng, Henan), forming three major roads from Anyi to Daliang, from Xinjiang to Jinyang, and from Daliang to Handan. King Wei Hui excavated "the Hong Canal" centered around Daliang to connect the Yellow River and the Yingshui River, strengthening the connection between the Central Plains and various regions in Jiangnan, making Daliang a hub for water and land transportation in the Central Plains region.

The focus of the State of Han's territory is in the central and western parts of Henan Province, close to the so-called "China" city of Luoyang. However, with the decline of the Zhou royal family and the rise of various countries, the central position of Luoyi gradually lost. Compared to the States of Zhao and Wei, the State of Han lacks geographical advantages. The State of Zhao established the capital of Handan and controlled the north–south transportation arteries; The State of Wei established its capital in Daliang and controlled the east–west transportation arteries. However, Yangzhai, the capital of the State of Han (now Yuzhou City, Xuchang, Henan Province), had no room for development. It was bordered to the south by the outpost of the State of Chu, to the north by the State of Zhao, to the east by the State of Wei, and to the west by Luoyang, facing the threat of the State of Qin. Therefore, the situation was relatively poor and it was the first country among the six states to be conquered by the State of Qin.

The State of Chu was once the largest country in terms of area during the Spring and Autumn and Warring States periods, with its influence expanding from present-day Hubei and Hunan to Guangdong, Guangxi, Jiangxi, Zhejiang, Fujian, Jiangsu, as well as eastern Sichuan, most of Anhui, and southern Henan. In order to consolidate its territory, the State of Chu successively opened three communication arteries from the capital city of Ying (now Jingzhou District, Jingzhou City, Hubei Province) to the capital city of Zheng (now Xinzheng City, Zhengzhou, Henan Province), from the capital city of Ying to the capital city of Song (now Suiyang District, Shangqiu, Henan Province), and from Yingdu to Dianchi (now Kunming City, Yunnan Province). Although the State of Chu first arrived in the Sichuan Yunnan region, the State of Qin used the "the Jinniu Road" to quickly break through the State of Shu and the State of Ba, and occupied central Guizhou, cutting off the transportation routes of the State of Chu army.

Unification of the Qin Empire and the Establishment of the Centralization System

The State of Qin Conquered Six States and Unified the World

In the late Warring States period, the State of Qin's military strength was unprecedentedly strong, accelerating the pace of unifying China. In the 51st year of King Zhaoxiang of Qin (256 BC), he launched an attack on the ruler of Western Zhou. In the same year, King Nan of Zhou died without a queen, and the sacrificial offerings were cut off. The nine tripods were relocated to the State of Qin. In the first year of Qin Zhuangxiang (249 BC), King Zhuangxiang destroyed the Eastern Zhou Dynasty, which had its capital in Gong. As a result, the Zhou royal family died before the six states. In the tenth year of Emperor Qin Shi Huang (237 BC), Ying Zheng took power and appointed Wei Liao and Li Si, intensifying the planning of the action plan to destroy the six states. In just over 10 years, the six states were annihilated, completing an unprecedented cause of reunification. The situation map of the State of Qin's annihilation of the six states is shown in Fig. 3.2.

The State of Qin eliminated the six states and established a unified multi-ethnic country for the first time in Chinese history, ending the chaotic situation of the Warring

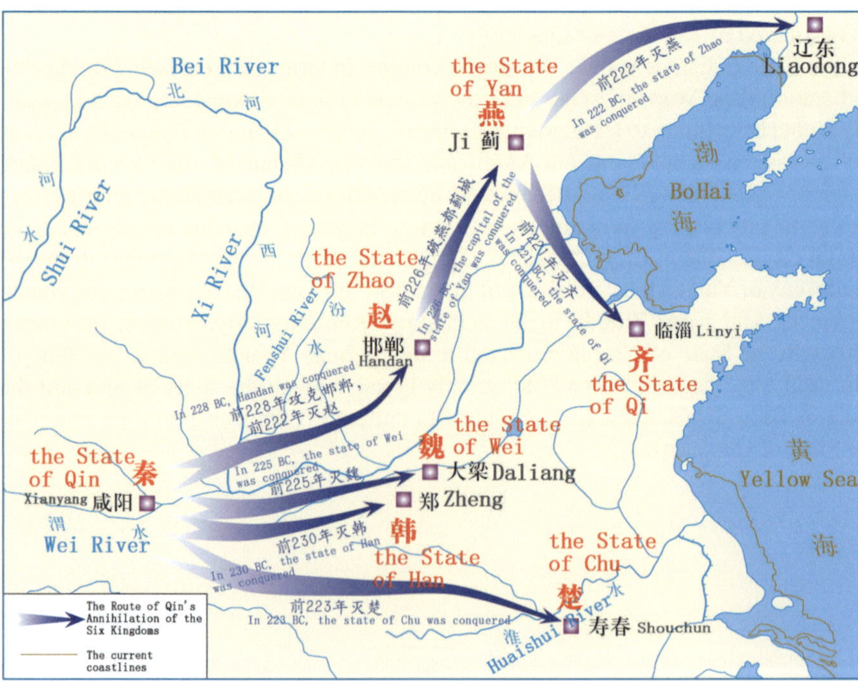

Fig. 3.2 Situation of the state of Qin's annihilation of the six states

States period. The State of Qin was able to unify China due to its enormous economic, political, and military advantages, which can be described as an unstoppable trend.

Establishment of Centralized System of Power

The Qin Empire had a vast territory, with Xianyang as its capital. Its territory extended to the sea in the east, Longxi in the west, Lingnan in the south, and Hetao, Yinshan, and Liaodong in the north. In order to consolidate his ruling position and demonstrate his supreme authority, Emperor Qin Shi Huang took a series of measures and established a centralized system and system.

Firstly, Emperor Qin Shi Huang followed Li Si's suggestion to abolish the system of enfeoffment of feudal lords and fully implement the county system. The state was divided into 36 districts, and later it was gradually added to more than 40 districts. These districts are completely controlled by the central government and the emperor, and are local administrative units under the jurisdiction of the central government. Local administrative institutions are divided into two levels: district and county. The district is equipped with a governor, a captain, and a supervisor (Inspector of the Imperial Court). The district governor was in charge of his punishment, the district lieutenant assisted the county governor, and handled military affairs under the supervision of the district supervisor. County, with orders for households above 10,000, and chiefs for households below 10,000. County Magistrates and Magistrates have officials such as Cheng, Wei, and other subordinates. The main officials of districts and counties are appointed and dismissed by the central government.

Secondly, Emperor Qin Shi Huang adjusted and expanded the official system of the State of Qin, and established a new set of government institutions that met the needs of a unified country. This was known as the "the Three Councillors of State and the Nine Ministers". "Three Councillors" refers to the Prime Minister, Grand Commandant, and Imperial Censor. The Prime Minister has two members, left and right, who are in charge of political affairs. The Grand Commandant is in charge of military affairs and is not frequently appointed. The Imperial Censor is responsible for assisting the Prime Minister, holding the position of Secretary of the Imperial Library, and supervising all officials. Three Councillors are not under the jurisdiction of each other, but directly under the emperor, which facilitates the concentration of imperial power. "Nine Ministers" refers to Fengchang, Tingwei, Zhisu Neishi, Dianke, Langzhongling, Shaofu, Weiwei, Taipu, Zongzheng. Their respective responsibilities are: Tingwei, in charge of judicial affairs; Managing the internal history of the millet and overseeing national finance and taxation; Fengchang, in charge of the ancestral temple's sacrificial rituals; Dianke, handling domestic affairs and foreign relations of various ethnic minorities; Langzhongling, in charge of the imperial attendants and guards; Shaofu is in charge of the income from mountains, seas, land, and resources exclusively for the needs of the royal family, as well as the government's handicrafts; Guard, in charge of palace guards; Tai Fu, in charge of palace chariots and horses; Zong Zheng is in charge of the affairs of the emperor's clan. But whether it is the Three Councillors or the Nine Ministers, they are appointed, dismissed, and transferred by the emperor, and are not hereditary.

Thirdly, the Qin Empire improved the registered residence land system and military system to ensure the stability of the imperial rule. In terms of the registered residence system, as early as the Shang Yang Reform Period, it was stipulated that both men and women should be listed in the registered residence after birth and expelled after death. Later, after improvement, the registered residence gradually added age, Fu registration, land and other items, which became the certificate of people's identity. In terms of military affairs, the State of Qin army is an unprecedentedly powerful military force. The military mobilization is based on the copper tiger talisman. The tiger talisman is split into half, the right half is controlled by the emperor, and the left half is in the hands of the leader. Only when the left and right talismans match can the military be mobilized. This is an important system to ensure that military power is in the hands of the emperor.

Finally, Emperor Qin Shi Huang established a complex ritual and a grand ceremony of enfeoffment that was suitable for the emperor's position, instilling in his subjects the concept of sacred imperial power as the ideological foundation of the centralized system. Emperor Qin Shi Huang built many palaces near Xianyang and built the magnificent Afang Palace south of the Wei River. The layout of the palace is inspired by the Purple Micro Palace in the sky, symbolizing the unity of the world. Emperor Qin Shi Huang also pre built a tomb on Mount Li, in which mercury was used to represent the rivers, oceans, and streams. The tomb was infused with information, with astronomy above and geography below, demonstrating the supreme power of the emperor and immortality.

Unification of Weights and Measures for Text Transportation and Its Historical Contributions

Unified Text: "Unified Writing System"

Before the unification of the State of Qin, the "abnormal writing" brought many inconveniences to the implementation of political orders and economic and cultural exchanges. Therefore, in the year of unifying the whole country, Emperor Qin Shi Huang ordered the use of simplified Qin Xiaozhuan as the standard script, abolishing the large seal script since the Western Zhou Dynasty and the ancient script and other variant scripts commonly used in the Eastern Six States. Led by Prime Minister Li Si and Zhao Gao, a small seal script was written, which was stipulated as a compulsory textbook for students. Those who entered the government needed to pass the exam before they could serve as junior officials.

Due to the strengthening of royal power since the Warring States period, government affairs have been complex, and the number of official documents has greatly increased. Although it was initially simplified, the small seal script still found it difficult to write and did not meet the needs. Therefore, in practice, a more simplified font, known as "the official script of the Qin Dynasty," emerged. By the Han Dynasty, clerical script was very popular, which is the predecessor of the commonly used regular script today.

China has a vast territory and a large population, with significant differences in dialects among different regions. Having this unified writing system as a written language is of great significance for strengthening regional connections and consolidating national unity.

Unified Transportation: "All Wagons Have the Same Track Width"

In ancient times, dirt roads would form two ruts with the same width as the wheels after repeated rolling. During long-distance transportation, keeping the wheels on the ruts can ensure smooth walking and significantly reduce animal power consumption and axle wear. Before the unification of the State of Qin, the size of carriages varied in different regions, so the ruts were also wide and narrow, making it inconvenient for carriages to pass.

After unification, the Qin Dynasty formulated the law of "all wagons have the same track width", which stipulated that the distance between the two wheels of a vehicle should be changed to six feet, making the distance between the wheels the same. This allowed roads across the country to be compressed into lanes of equal width within a few years. This not only reduced the cost of transporting goods and passengers, but also facilitated the rapid delivery of imperial military supplies to any county in the country. "all wagons have the same track width" became one of the important strategic measures for the unification of the Qin Dynasty.

Unified Currency and Weights and Measures

Emperor Qin Shi Huang abolished the different shapes, weights, and sizes of currency in various countries during the Warring States period, and instead used gold as the upper currency, with Yi (20 liang) as the unit of measurement; Using the circular square hole copper coin from the old line of the State of Qin as the lower coin, it is said to be half a liang and weighs as much as it is written.

Emperor Qin Shi Huang used the measurement standard instrument established by Shang Yang to unify the national measurement system. In order to ensure the national tax revenue, it is required to implement unified measures such as Dou, Tong, Quan, Heng, Zhang and Chi throughout the country. Among them, 1 standard Chi is about 0.23 m today, and 1 standard Sheng is about 0.2 L today.

The unity of currency and measurement provides convenient conditions for economic development, has a positive effect on the unity of tax and salary systems, and is conducive to the development of commodity economy and national unity.

Formation and Impact of the National Road Network

The transportation industry experienced unprecedented development in the Qin Dynasty, especially the construction of Qin Imperial Road, which basically laid the basic framework of China's land transportation network. This was a significant contribution of the Qin Dynasty to future generations.

Composition of the National Road Network

According to the Annals of Emperor Qin Shi Huang in the Records of the Grand Historian, build Imperial Road [9] in the 27th year of Emperor Qin Shi Huang's reign (220 BC). The construction of Imperial Road was based on the transportation of various countries during the Warring States period, and after renovation and communication, the roads of various countries were incorporated into the national transportation system, forming a general pattern of the national land transportation network [6]. This is the most distinctive achievement of the Qin Dynasty's transportation construction. This project took about 10 years to connect important cities across the country, especially the capitals of various countries in Shandong. The multiple tours of Emperor Qin Shi Huang and Emperor Qin II utilized this transportation network to reach various parts of the country.

The main transportation network of the Qin Dynasty included various forms such as Imperial Road, Straight Road, and Gallery Road. Among them, the Imperial Road mostly utilized the original roads of the Warring States period, with a focus on connecting and repairing them; Straight Road specifically refers to the newly built South North Passage from Yunyang to Jiuyuan after the unification of the Qin Dynasty; Gallery Road was a road from Shaanxi to Sichuan that was built during the Warring States period to overcome the barriers of the Qinling Mountains. Most of it was built on steep cliffs in the valleys, and was later incorporated into the road network through drilling holes and wooden frames.

The overall structure of the road network consists of 6 radial trunk lines centered around Xianyang and 2 peripheral trunk lines.

The six radial trunk lines are:

Northwest direction: It is called the Longxi North Tunnel, from Xianyang to the west, is the eastern section of the Silk Road. Emperor Qin Shi Huang's first tour is traveling westward through this road, the Annals of Emperor Qin Shi Huang in the Records of the Grand Historian: "In the 27th year, Emperor Qin Shi Huang's trip first crossed Longshan to Longxi, then went north to patrol the northern prefecture (now Qingyang area), and finally returned by the Huizhong Road on the east side of Longshan (also known as Liupan Mountain), which happened to take a detour around Longshan." [9].

Due north direction: named "Straight Road", it starts from Yunyang and leads directly to Hetao Wuyuan. According to the Annals of Emperor Qin Shi Huang in the Records of the Grand Historian, "In the 35th year, roads were constructed from Jiuyuan all the way to Yunyang, relying on the mountain terrain to dig stones, fill ditches and valleys, and continue until they reach Straight Road." [9]. According to the Annals of the Six Kingdoms in the Records of the Grand Historian, "in the 35th year of Emperor Qin Shi Huang, "the road was built along the Straight Road, reaching Jiuyuan, and then leading to Ganquan." [9]. This road leads directly to the front line of the resistance against the Xiongnu, Jiuyuan County. It is a major north–south passage that reflects the military and administrative efficiency of the Qin Empire. It can not only timely transmit the military situation on the front line to Xianyang, but also facilitate the emperor's inspection of the front line. Its role is very significant. The places that Qin Zhi Dao passes through are mostly hills, ravines, and deserts, and even by current standards, the construction difficulty is still very high. However, during the construction of Mengtian, the road was leveled and filled with ditches, ensuring maximum straightness. Local relics can still be seen in Inner Mongolia, with a width of nearly 60 m (Fig. 3.3).

Northeast direction: It is called Handan Guangyang Road, passing through Hedong and Shangdang, or heading north from Henei to Handan, Guangyang, and Youbeiping, leading to Yan and Zhao. This avenue already held an important position during the Warring States period. Famous historian Shi Nianhai once pointed out: "There is a main road on the eastern side of the Taihang Mountains, parallel to the Taihang Mountains, running north–south. The capital cities of Zhao, Handan, and Yan, Jidu, are on this transportation line." [11].

Due east direction: named Sanchuan Donghai Road, starting from Xianyang, passing through the northern foothills of Huashan, passing through Hangu Pass, passing through "Sanchuan", heading east, and connecting to Linzi County to the north at Daliang, and then to Sishui County and Jiujiang County to the south, finally

Fig. 3.3 Restored Qin Zhi Dao in Daqin Zhidao cultural tourism Scenic Area, Ordos City [10]

ending in boundless ocean. It connects two cities, Chang'an and Luoyang, as well as multiple cities in North and East China, making it the transportation artery with the largest transportation volume during the Qin Dynasty and Han Dynasty.

Southeastern direction: Named Nanyang Nanjun Road, starting from Xianyang, going out of Lantian, crossing Wuguan, passing through Danyang and Nanyang, and reaching to Jingzhou. After crossing Yunmengze, it may head south to Jiuyi, Guilin, and Panyu along the Xiangjiang River, or west to Qiandian along the Yuanjiang River.

Southwest direction: named Hanzhong Bashu Road, it is an ancient plank road that runs from Chencang to Nanzheng and then through Jinniu Road to reach Shuzhong. The roads in the Qinling Mountains are treacherous and the construction is extremely arduous. After being opened, they have undergone extensive repairs and improvements, forming several routes that cross the Qinling Mountains south, such as the Old Road, Baoxie Road, Tangluo Road, and Ziwu Road. Some of the remains of the boardwalk along certain routes still exist today (Fig. 3.4).

The two main lines on the periphery are:

Beibian Road: Along the Great Wall, it crosses the border from Jiuyuan to Daijun, passes through Youbeiping to Jieshi Mountain, and extends to Liaodong. This main line originated from the construction of the Great Wall, and due to the needs of construction and defense, a cross east–west transportation avenue appeared along the Great Wall. The construction of the Great Wall required hundreds of thousands of laborers, and heavy troops were gathered along the route for years to guard and defend. They once marched northward from various counties in the north to attack the Xiongnu. Therefore, Beibian Road has the conditions for organizing construction, mobilizing troops, and transporting military supplies.

Fig. 3.4 Baoxie ancient plank road ruins [12]

Binghai Road: also known as Binhai Avenue, runs along the eastern coast, from Langya to Pengcheng, heading south across the river and up to Kuaiji Mountain, and heading north around the Bohai Sea to Liaodong County. The route is speculated based on the records of Emperor Qin Shi Huang and Qin II's patrols. According to the Annals of Emperor Qin Shi Huang in the Records of the Grand Historian, after unifying the country, Emperor Qin Shi Huang went on five expeditions, four of which went to the seaside, often traveling alongside the sea. Emperor Qin II toured counties and counties, once "went north to the Jieshi Mountain, and then along the coast southwards to Kuaiji prefecture", and then "returning to Liaodong". This avenue intersects with the Sanchuan Donghai Road and the Guangyang Road in Handan, connecting the prosperous land of Qi and Chu with other regions to mobilize various materials and directly support the central government materially.

Shape and Regulations of Qin Imperial Road

According to the "Book of Han", Jia Shan, a native of the Western Han Dynasty, said about the shape of the Imperial Road: "The width of the road is fifty steps, and a tree is planted every three zhangs. The rampart is made very strong, and metal cones are used to compact it. Green pine trees are planted on both sides of the rampart. In order to make the Imperial Road so magnificent, future generations cannot even insert a slope or a foot." [13].

The width of the road is fifty steps, equivalent to about 70 m today. Archaeologists have discovered a north–south ancient road relic east of Donglong Village in Yaodian Town, Xianyang, Shaanxi. The width of the road is about 50 m and built on raw soil, with Han Dynasty cultural layers on both sides. According to research, this road is a transportation artery that connects Xianyang Palace and Afang Palace from north to south, and should be included in the road transportation system. In addition, the No.1 Avenue discovered near the Qin Xianyang Palace is located 220 m north of the north wall, and the road surface has been damaged. The current widest point is 54.4 m, which the width is generally 40–50 m. The middle of the road surface is 10–15 cm higher than both sides, forming a fish ridge shape. The north and south sides of the main road are covered with mud, resembling a drainage ditch on the road surface. From the analysis of form and scale, this road may be related to the raceway. The remains of the Straight Road built during the reign of Emperor Qin Shi Huang have been found in many places in Shaanxi, such as Chunhua, Xunyi, Huangling, Fuxian, and Ganquan, with the width of the road often ranging from 50 to 60 m [14].

There are two explanations for planting a tree every three zhangs. One explanation is that the road surface with the width of the three zhangs in the center of the road is exclusively for the emperor, and others are not allowed to enter, planting trees as the boundary. Wang Xianqian pointed out in the "Supplementary Annotations to the Book of Han": "Wang Xianshen explained that the width of the three zhangs area is the central area of the palace, and only the emperor can pass through it. Trees are planted as boundaries." [15]. Another explanation is that roadside trees are planted every three zhangs along the road. Yang Shuda mentioned in the “Observation of the Book of Han": “Planting a tree every three zhangs means planting a tree every three zhangs on both sides of the road.” [16].

“Building the rampart very strong” refers to the construction of the road foundation, which aims to be solid. Through multiple layers of ramming, the entire road surface is raised above the ground, and wide and gentle slopes are formed on both sides. From the roadbed section of the Fangjiahe Straight Road ruins in Ganquanqiao Town, Shaanxi Province, clear compacted layers can also be seen. The remains of the Fangjiahe Qin Straight Road are said to have “rammed several flat square partition walls on the outer side of the extensively filled roadbed, and filled with soil to form a slope or road surface” [17]. This architectural form can also be seen in the remains of the Straight Road in Huagoukou section of Fuxian County, Shaanxi Province, or can be used as an explanation for “building the rampart very strong".

"Also use metal cones to compact it", use metal cones to build it to make it stable.

Plant green pine on both sides of the rampart, and the trees planted on both sides of the Imperial Road are green pine. Some studies suggest that in addition to pine, there are also trees that are easy to propagate through cuttings as roadside trees, such as cypress, catalpa, locust, cypress, sandalwood, elm, etc.

"In order to make the Imperial Road so magnificent that future generations cannot even insert a single foot on a slope," it is said that the selection of road routes should strive for straightness, slow down the slope, and expand the radius of the curve in order to improve traffic speed. The selection of roads and routes is exceptionally cautious, and local officials cannot make unauthorized changes. In order to reduce the number of channels and bridges, it is better to change the direction of the original channels. The fact that many sections of the Imperial Road have become timeless passages indicates that the rationality of the Imperial Road design and route selection can withstand the test of history.

In order to ensure the efficient transmission of military, political and other information, a pavilion is built every ten miles on the road as a management post for section security, a rest station for pedestrians on the road, and a junction for postal personnel.

There is also a strict management regulation on the Imperial Road, which prohibits unauthorized passage in the Imperial Road, reflecting a sense of hierarchy. At that time, the Imperial Road was a multi lane road with a divided road surface and divided into three parts. The “Three Zhangs in Central” was known as the “Heavenly Road", and authorized aristocratic bureaucrats could bypass it. Those who traveled along the Imperial Road without an imperial order were severely punished.

Influence of Qin Imperial Road

The national transportation trunk network, with the Imperial Road as the backbone, connected the various feudal states that were separated during the Warring States period. It not only improved military and administrative efficiency, but also provided convenience for trade, commerce, and cultural dissemination. This played a positive role in the development of politics, economy, military, and culture at that time, and provided facilities for the unification of centralized countries.

The construction of Qin Imperial Road can be seen as the starting point for national highway network planning and construction, basically laying the foundation for the road system structure of subsequent dynasties and even affecting the overall pattern of China's national trunk highway network more than 2000 years later. This cannot be said to be a great miracle.

References

1. *Commentary and Annotation on the Gongyang Commentary on the Spring and Autumn Annals.* Compiled by He Xiu, Interpreted by Xie Fang, Annotated by Xu Yan. Shanghai: Shanghai Ancient Books Publishing House, 1990.
2. Sima Zhen. *The Index of Shi Ji.* Xi'an: Shaanxi Normal University Press, 2018.
3. Xun Kuang. *Xun Zi.* Shanghai: Shanghai Ancient Books Publishing House, 2010.
4. Bai Zhide. *Twenty Lectures on Bai Shouyi's Historiography: Ancient Times - Xia, Shang, Zhou, Spring and Autumn, and Warring States.* Beijing: China Friendship Publishing Corporation, 2013.
5. The Chinese Highway Transportation History Compilation Committee. History of Road Transportation in Ancient China. Beijing: China Communications Press, 1994.
6. Wang Zijin. *History of Qin and Han Dynasties' Transportation.* Beijing: Social Sciences Academic Press, 2020.
7. Xiao Tong. *Wen Xuan (Selections from the Ancient Literature of China).* Shanghai: Shanghai Ancient Books Publishing House, 1986.
8. Qin Guoqiang. *China's Transportation History.* Shanghai: Fudan University Press, 2012.
9. Sima Qian. *Shi Ji.* Beijing: Zhonghua Book Company, 1959.
10. Qishu interpret Shi Ji. *Detailed Account of Zhang Liang's Escape Plan After Failed Attempt to Kill Qin Shi Huang.* [EB/OL]. (2018-01-17) [Accessed on 2022-04-26]. https://baike.baidu.com/tashuo/browse/content?id=ab5d18cefc34d485fa13666b&fr=vipping.
11. Shi Nianha. *Collection of Essays on Rivers and Mountains.* Beijing: The Joint Publishing Company Ltd., 1963.
12. Zhang Yongtao and Chen Hao. *Many Ancient Sites Along the Baoxie Road.* [EB/OL]. (2019-08-15) [Accessed on 2022-04-26]. https://www.chinahighway.com/article/65303.html.
13. Ban Gu. *History of the Han Dynasty.* Beijing: Zhonghua Book Company, 1997.
14. Wang Zijin. *Characteristics of Ancient Chinese Transportation Systems: Focusing on Cultural Relics and Materials from the Qin and Han Dynasties* [J]. Social Sciences, 2009(7):9.
15. Ban Gu. *Supplementary Annotations to the History of the Han Dynasty, Annotated by Wang Xianqian.* Beijing: Bibliographic Literature Publishing House, 1995.
16. Yang Shuda. Insights into the History of the Han Dynasty. Beijing: The Commercial Press, 2017.
17. Sun Wenbo. Research Collection on the Qin Zhidao. Xi'an: Shaanxi Normal University Press, 2018.

Chapter 4
Postal Service System: An Everlasting Communication System

The postal service system in ancient times in our country has a long history. In ancient times, postal service were a horse delivery network with stations as the main body and a foot delivery network with express delivery shops as the main body. According to "Shuo Wen Jie Zi", "post service" mainly refers to rest stations on the border, while "post station" is a place used to deliver messages or goods, usually equipped with horses for quick movement [1]. From the definition of "Shuo Wen Jie Zi", postal service are the institution for transmitting letters in border areas, postal service and station are the carriage for transmitting official documents. Postal service are the official information transmission system established in China based on transportation arteries and stations, and the information transmitted includes both political and military information as well as frontline collection. The deployment and arrangement of national military and political affairs, the collection and consolidation of intelligence in various regions, and the management dynamics at the grassroots level all needed to be transmitted through postal service in the absence of electronic information technology in ancient times, making their status extremely important. The accessibility of postal service played a crucial role in stabilizing the political power of various dynasties and profoundly influenced the historical process of ancient China.

Origin and Development of Postal Service System

Legend has it that Yao and Shun both set up "wooden plaques for people to express their opinions" (Fig. 4.1) to observe gains and losses, and to accept criticism and advice. The "Annals of Emperor Xiaowen in the Records of the Grand Historian" records: "In ancient times, when governing the country, the court set up "a banner for promoting virtuous deeds" to write good deeds and wooden plaques for criticism of the court's politics, which were used to smooth the way of governing the country

© China Communications Press Co., Ltd 2025

H. Lu (ed.), *A Brief History of Transportation in China*,
https://doi.org/10.1007/978-981-97-9851-3_4

Fig. 4.1 Wooden Plaques model for people to express their opinions of Yao Temple in Shanxi Province [2]

and attract admonitors." [3]. The wooden sign that makes people raise their opinions is a crossbar crossed on the column head. People can not only write their opinions on managers on it, but also set it as a signpost at each intersection. Later, it gradually evolved into ornamental column. This is the earliest record in Chinese written records of expressing opinions and transmitting information upwards, and can be considered as the embryonic form of communication postal service.

By the time of Shun, the "wooden plaque for people to express their opinions" had formed a political system. In the "Annals of the Five Emperors" of the Records of the Grand Historian, it is mentioned that Shun appointed officials specifically responsible for receiving information, in order to "understand the ears and eyes of all directions" [3]. These officials traveled day and night to various places to listen to the opinions of the people and convey Shun's intentions to everyone. This kind of acceptance system was the earliest organized communication activity for exchanging opinions from top to bottom.

During the Xia Dynasty, people's communication activities became more diverse and complex. The Book of Shangshu · Xiashu, introduces a way for the government to announce orders: in March each year, a proclamation officer is selected and appointed to hold a wooden tablet on various transportation arteries to announce orders [4]. This is the earliest way for our country to issue national documents. The oracle bone inscriptions of the Shang Dynasty also often recorded information exchange, with the most famous being the divination of Wu Ding. Due to his wife Fu Hao being a famous female general (Fig. 4.2), Wu Ding often divined and asked "letter from Fu Hao" or "no letter from Fu Hao" during her expedition. This "letter" means information, which means asking if there is a messenger coming. The Shang Dynasty established many strongholds and rest areas on the main road, which was the prototype of post stations. These strongholds were originally known as "Die" and were defensive fortifications built with wooden fence walls, with one located 50 miles away; Later on, it developed into a "Ci" and could temporarily reside; Later, "Ji" was established, which was a roadside inn specifically designed for the Shang kings and nobles, and is considered the earliest post station in China [5].

During the Shang Dynasty, there was no system for segmented delivery of information, and message orders were usually transmitted by one person to the end, so the work of messengers was both arduous and dangerous. There is a oracle bone record from the reign of King Wu Ding of the Shang Dynasty, where an elderly messenger

Fig. 4.2 Fu Hao Tomb in Yin Ruins

walked for 26 days and 600 li on the road, but died before reaching his destination. There is also an oracle bone inscription from the reign of King Zugeng of the Shang Dynasty, which records that a messenger set off from dusk and traveled for 48 days to reach his destination, covering a total of 1200 li [5]. It can be seen from this that the postal service of the Shang Dynasty had already reached a high level and held a leading position in the world at that time.

Although the communication methods during the Xia Dynasty and Shang Dynasty were relatively primitive, they laid the foundation for information transmission in China. By the Western Zhou Dynasty, China's postal system gradually developed rapidly and had formed a relatively complete postal service system, with different methods of document transmission also having different names. The post roads of the Western Zhou Dynasty gradually began to provide service facilities, setting up various rest stations called Lu, Wei, and Shi. According to the Book of Rites of Zhou, there was a house located in 10 li, where there was food and drink; there is a place to stay in 30 li, called Wei; A Shi is set up in 50 li, and there is a waiting hall in the Shi to receive messenger guests [28]. These locations are similar to the service areas of highways nowadays, allowing messengers traveling on the road to no longer have to sleep and eat outdoors, but to have better food and rest along the way.

In addition, the biggest feature of communication in the Western Zhou Dynasty was the use of beacon fires in military affairs. Once an emergency occurs, the guards emit smoke by lighting firewood and wolf droppings in the beacon tower, allowing the outside world and the guards of the next beacon tower to immediately notice, and the next beacon tower will also be lit. Information is transmitted through beacon towers, and the message will soon reach the capital city (Fig. 4.3). Similarly, if a major event occurs in the capital, the news will spread from the opposite direction to the border. The beacon system was a very effective communication system in ancient China, which was adopted and used from the Western Zhou Dynasty to the Han Dynasty.

Fig. 4.3 Schematic diagram of the great wall beacon fire transmission information [6]

Fig. 4.4 Du Hu Fu [7]

In addition to booklets, ancient people also used certain objects to convey information. The bronze national treasure Du Hufu (Fig. 4.4) unearthed in 1973 is the treasure of the Shaanxi History Museum. It was a kind of voucher used by the State of Qin during the Warring States period to convey orders to subordinate departments, recruit soldiers, and be used for various affairs. The talisman reads: "The right half of the talisman for dispatching troops is in the hands of the monarch, and the left half is in the hands of the general. Anyone who wants to mobilize an army, put on armor, or use more than fifty soldiers must verify the ruler's talisman before he can act." The meaning is that this talisman is held half by the king and half by the generals of Dudi, and if more than 50 soldiers are needed, the two talismans must be combined before action can be taken.

Milestone in Changing Postal Services by Pedestrian to Postal Services by Horse Riding During the Qin Dynasty and Han Dynasty

After the unification of China by Qin Dynasty, a series of laws and systems were established to ensure timely delivery of official documents and letters within a larger territory. The Book of "Xing Shu Lv" unearthed in Hubei was a legal provision of the Qin Dynasty regarding the transmission of documents, which divided documents into two categories: urgent documents and ordinary documents. Urgent documents, including imperial edicts, must be immediately conveyed without delay; ordinary documents are also required to be sent out on the same day, without backlog, and violators shall be dealt with according to the law [8].

In order to ensure that no secrets were leaked during the delivery of letters and that letter slips were not forged, the Qin Dynasty also made several legal provisions. For example, seal mud was generally used at the knot of the letter slip, and a seal was affixed to prevent any leaks during the delivery process; Different documents are written in different languages. Simplified books are written in either large or small seal script, symbols are passed on using engraved symbols, seals are sealed in Miao seal script, banner script is written in bird script, and official documents are written in clerical script [5]. These regulations effectively prevent the forgery of documents. In addition, the Qin Dynasty also stipulated that the time of origin and receipt of documents should be recorded. The above regulations all indicate that the postal communication system of the Qin Dynasty was more standardized. The Qin Dynasty consolidated the centralization system through this institutionalized and standardized communication system. The court can continuously receive reports from various regions, timely understand border defense and civilian dynamics, and take decisive military measures. The developed Qin Imperial Road, along with the scattered post stations, buildings, and military facilities along the way, provides a basic guarantee for the postal service system.

In the Han Dynasty, there were significant changes in the postal service system, with the most significant changing postal services by pedestrian to postal services by horse riding. Li Xian's note in the Book of the Later Han Dynasty quoted the "Compendium of Customs and Traditions" by Ying Shao, the chief magistrate of Mount Taishan in the Eastern Han Dynasty, which said: "The postal service was changed into a post station. The post station was set according to the distance of the road." [9]. The "postal services by horse riding", also known as the post station or transmission station, refers to a large post station equipped with horses and specially used to deliver emergency official documents. The core of changing postal services by pedestrian to postal services by horse riding is to transform the original vehicle or person transmission method into a single person horse riding transmission method, and to stipulate that "one postal service by pedestrian should be set up every five li and one postal service pavilion should be set up every ten li", "one post station should be set up every thirty li". Postal service by pedestrian and postal service pavilion is a transportation and rest station for short distance messengers, also known as a post station, which refers to the facilities for long-distance transmission of letters and documents. It was used to transport urgent and important documents of the Han Dynasty. The advantage of post houses lies in their fast transmission, usually dominated by light vehicles and fast horses. The post station will prepare carriages and horses in advance, ready for use by the traveling post envoys at any time.

The Xuanquanzhi Ruins in Dunhuang County of the Han Dynasty is the oldest, largest, and most well preserved ancient postal reception institution discovered by archaeology to date. It is the first site on the eastern section of the Silk Road from the Western Regions to the Central Plains. It was a famous reception and transfer station for people and mail between Anxi (now Jiuquan, Gansu) and Dunhuang during the Han Dynasty and Tang Dynasty. It is 56 km from Anxi in the east and 64 km from Dunhuang in the west, with a total area of approximately 22,500 square meters. The Xuanquanzhi Ruins is a square small castle with its door facing east and surrounded by tall courtyard walls, with a side length of 50 m. In the southwest corner, there is a corner tower protruding from the dock body (Fig. 4.5). Over 23,000 inscribed bamboo slips, silk books, paper documents, and various relics from the Western and Eastern Han dynasties were unearthed from the Xuanquanzhi. These bamboo slips contain a large number of various books related to the functions of post offices, which fully reflect the functions of post station in the Han Dynasty. According to the textual analysis of bamboo slips, there are 47 officials and soldiers in the Xuanquanzhi, with

Fig. 4.5 Xuanquanzhi Ruins [11]

around 40 horses and 15 carriages passed on. If necessary, civilian carriages and horses were also requisitioned for transporting personnel or supplies [10].

The post station system in the Han Dynasty was very strict, with documents to be registered and recorded. Postal workers also had to wear eye-catching special decorations, with red turbans on their heads, red sleeves on their arms, and red-and-white pouches on their backs. In terms of speed, a horse can travel 300–400 li a day, a cart can travel about 70 li, and a foot can travel 40–50 li a day.

The most famous cultural relic among the artworks showcasing ancient post is the brick portrait of the "Painting of a Postal Courier" from the Wei Dynasty and Jin Dynasty (Fig. 4.6). It is the treasure of the Gansu Museum and also the "image ambassador" of China Post. The length of brick is 35 cm and the width is 17 cm, with a beige background and black outline. The painter captured a moment in the process of ancient post delivery, using form to depict the spirit. With just a few strokes, he vividly and accurately reproduced the historical scene of post envoys holding documents and galloping horses on the northwest border of the Silk Road over 1600 years ago, using a combination of realism and freehand brushstrokes.

Fig. 4.6 Portrait bricks of the "painting of a postal courier" [12]

Postal Service in the Tang Dynasty

During the Sui Dynasty and Tang Dynasty, the postal service industry further developed and the road network became more perfect. Since the Southern and Northern Dynasties, the responsibilities of the postal service system have continued to increase. It is responsible for transmitting national official documents, letters, and emergency military intelligence, as well as handling various affairs such as picking up and dropping off officials, maintaining public order, chasing criminals, rescuing disaster areas, and escorting prisoners. During the Sui Dynasty and Tang Dynasty, post roads and stations were spread throughout the country. According to the Book of the "Tang Liu Dian" during the reign of Emperor Xuanzong of Tang Dynasty, there were a total of 260 water post stations and 1297 land post stations in China. There were more than 20,000 people engaged in postal work, including 17,000 post drivers, [13] which even exceeded the number of people engaged in postal work during the Republic of China period. The largest post station in the Tang Dynasty was called Dutingyi, which was a post station in the capital city of Luoyang during the Wu Dynasty and Zhou Dynasty. According to the famous geographer Song Minqiu of the Song Dynasty, in the Book of the "Henan Zhi", there was a "Zuojingwuwei and Dutingyi" [14] in the Qinghua Fang area of Luoyang city during the Tang Dynasty. This was the first Li Fang on the north side of Xuanrenmenwai Street, approximately located in the area north of the western section of Dongdajie Street in the old urban area of Luoyang today.

The land post stations in the Tang Dynasty were divided into six classes: first class post stations were equipped with 20 post drivers, second class post stations were equipped with 15 post drivers, third class and below were gradually reduced, and sixth class post stations were equipped with 2–3 post drivers. And the water post stations is divided into three classes: the busy water post station is equipped with 12 post drivers, the idle water post station is equipped with 9 post drivers, and the more idle water post station is equipped with 6 post drivers. In addition to the post station, Tang Dynasty also built a post road with extending in all directions with Chang'an as the core. Liu Zongyuan detailed seven important radial post roads centered around the capital city of Chang'an in his Book of the "Record of Guan Yi Shi Bi": the first is the northwest post road from Chang'an to the Western Regions, passing through Jingzhou, Huizhou, Lanzhou, Shanzhou, Liangzhou, Guazhou, and Shazhou to reach the Anxi Protectorate; The second is a post road from Chang'an to the southwest, passing through Xingyuan, Lizhou, Jianzhou, Chengdu, Pengzhou, and Qiongzhou to reach the present-day Sichuan Tibet region; The third is a post road from Chang'an to Lingnan, which runs directly from Chang'an through Xiangzhou, Ezhou, Hongzhou, Jizhou, and Qianzhou to Guangzhou; The fourth is a post road from Chang'an to the southeast coast, passing through Luoyang, Bianzhou, Sizhou, Yangzhou, Suzhou, Hangzhou, Yuezhou, Quzhou, and directly reaching Quanzhou, Fujian; The fifth is a post road from Chang'an to the northern grassland area, from Chang'an to Tongzhou, then passing through Hezhong Prefecture, Jinzhou, Daizhou, and Shuozhou, directly reaching the Northern Chanyu Protectorate; The sixth is the post road from Chang'an to the northeast region of Shandong; The seventh is from Chang'an to Jingzhou, Kuizhou, Zhongzhou, and other areas in Yunnan, Guizhou, Sichuan [15].

The postal efficiency of the Tang Dynasty was very high, ensuring the implementation of various central system throughout the country. During the Tang Dynasty, fast horses could run 180 li a day, and they could be increased to 300 or even 500 li in case of emergency. When An Lushan rebelled, Fanyang was 3000 li away from the Huaqing Palace. Emperor Xuanzong of the Tang Dynasty received intelligence six days later. In order to ensure the smooth flow of post roads, the Tang Dynasty government also set up full-time postal system officials. According to the records in the Book of the "Tang Liu Dian", the Ministry of War had a department called the Department of Chariots and Horses, which was specifically responsible for the management of the country's royal carriages and the postal and courier system. The Department of Chariots and Horses also managed horse affairs, facilitating the unified use of horses in postal system. There were also personnel specialized in managing postal system in local areas. Under the governors of various provinces, there were four postal inspectors who specialized in managing postal system. In each state, the postal system was divided and managed by the state soldiers, Cao Sibing, and the county magistrate concurrently managed postal affairs [13]. This comprehensive postal system managed over 20,000 postal officials, postal workers, and a total of approximately 50,000 li of postal routes across the country. In order to ensure the normal operation of postal system and avoid interference along the way, the Tang Dynasty also set up defense troops at each station. During the reign of Emperor

Zong of the Tang Dynasty, 300 post guards were stationed every two post stations on both sides of the canal from Luoyang to the Huai River. According to the regulations of the Tang Dynasty, postal system institutions across the country had different postal properties to ensure the normal expenses of postal system. These post houses include post houses, post fields, post horses, post boats, and related postal system tools, daily office supplies, and food and accommodation needs for the building. During the Tang Dynasty, land post stations were equipped with post horses, while water post stations were equipped with post ships. Each upper level post station of land post station was equipped with about 60 horses, medium level post stations were equipped with 40 horses, and lower level post stations were equipped with 10 horses. The water post station was equipped with relay ships, ranging from 4 to 1. In addition, the government also provided fixed funding subsidies to each post station every year. A complete postal system management system and sufficient postal funds ensured the normal operation of the Tang Dynasty postal system for a considerable period of time.

Such a developed post station played a significant role in the political operation and economic and cultural prosperity of the Tang Dynasty. The first is the military role. During the An Lushan Rebellion, the post station was able to report the news to Emperor Xuanzong of Tang as soon as possible, providing sufficient time for early defense. Cen Shen wrote in his poem "Sending a Letter to Judge Yuwen on the Way to Cross the Longshan Mountains for the First Time" that "the post station followed the post station, and the post station made the horse ride at a gallop" [16]. He vividly portrayed the speed of the post cavalry. The second is the role in politics. According to incomplete statistics, over 500,000 official documents were sent from the central government to local areas during the Tang Dynasty, all of which were transmitted through postal system. The emperor collected intelligence from various regions through post stations and issued orders to them. Post stations became a link for communication between the central and local governments, ensuring that administrative orders from the central government were implemented in border areas to continuously strengthen centralization and prevent local rebellions and corruption of officials. Thirdly, it plays an economic role. Taxes and tribute from various regions are handed over to the court through post stations, ensuring the daily needs and financial revenue of the court. Merchants from various regions conduct commercial transactions through post roads, which support the national economic market.

Postal System in the Song Dynasty

Entering the Song Dynasty, the postal system began to shift from national administrative management to militarized management. Firstly, from the perspective of the central management system, the Ministry of War specifically manages the regulations, personnel allocation, and horse delivery equipment of postal system. At the same time, the Shumiyuan (Commission of Military Affairs) also participated in management, responsible for the distribution of post horses and the issuance of trust tokens for post delivery. Two institutions mutually constrained each other. This indicates that the influence of postal services was increasing, and in order to prevent authoritarianism, it was necessary to establish a constraint mechanism. Secondly, the Northern Song Dynasty implemented the method of using soldiers instead of the common people as postal personnel, completely managing the transmission of letters according to military organization. The delivery of mail and documents had been completely separated from the inn where former officials stayed in terms of functions. The post station has evolved into a simple government guesthouse, serving as a place for officials and envoys to stop and rest midway; The institutions responsible for delivering official government documents and personal letters were known as "Di" (mentioned in Shen Kuo's "Dream Pool Essays"). "Di" was further categorized into "Jijiao Di" (for urgent delivery), "Ma Di" (for horse-drawn delivery), and "Bu Di" (for foot delivery). There are courier runners, also known as postal soldiers, who are the main personnel responsible for delivering official documents.

Delivery by foot was used for the transmission of general documents, which was carried out by personnel relay walking. This type of transmission also bore the heavy task of transporting official goods at a slower speed.

Horse-drawn delivery service was used for transmitting urgent documents, usually not official items, and it was fast. However, because most of the horses responsible for this task were selected by the military as the remaining old, weak, and sick horses, it was also not possible to transmit the most urgent documents at the fastest speed.

Express delivery service was the title of the fast military postal system in the Song Dynasty. It was gradually established and developed in the special environment of the Northern Song Dynasty's battles with Liao, Jin, and Western Xia, and was initially only set up in key areas and routes throughout the country. According to the "Yongle Canon" of the Ming Dynasty, in the second year of the Jingde reign of Emperor Zhenzong of Song (1005 AD), "the soldiers of Jijiaopu on both sides of Hebei Province were ordered not to accept document tasks from other places except for delivering documents to the Zhending General Administration and Xiongzhou." [17]. Their function was to transmit documents at a speed of 400 li per day while shutting them down to reach the capital Bianliang. During the reign of Emperor Shenzong of the Song Dynasty, there was also a gold character plaque for express delivery service. A gold character plaque is a type of communication plaque, the length over a cun, made of wood and engraved with the words "words used by the emperor are not allowed to enter the store". The transfer speed of the gold character plaque's express delivery service is faster than that of the general express delivery service [18]. The characteristic of this communication method is that it is sent directly from the imperial court, without going through the Shumiyuan or provincial departments, reducing unnecessary handover procedures and preserving confidentiality. It is transmitted by relay along the way, non-stop day and night, and does not enter the delivery shop for handover, reducing the stay time at the delivery shop. When transmitting, they passed by like a flash of lightning, and those who saw them all instinctively moved out of their way [19]. The most famous golden plaque in Chinese history was the urgent foot delivery in the 11th year of Shaoxing in the Southern Song Dynasty (1141). Qin Hui, under the name of Emperor Gaozong of Song, issued 12 gold character plaques in a day to convey orders, ordering Yue Fei, who was fighting on the front line of resistance against Jin, to retreat.

Inspection Tour and Postal Roads Between Yuan Dynasty's Two Capitals

During the Yuan Dynasty, the postal system developed even more. In order to adapt to the rule of the vast territory, the Yuan government carried out active reforms in the postal system and greatly expanded the range of postal roads (Fig. 4.7). As early as the time of Genghis Khan, many new postal stations were established in the western region. Genghis Khan's son, Vogtai, and grandson, Batu, pushed the Mongol postal roads all the way to Europe.

Fig. 4.7 Main postal roads of the Yuan Dynasty

After Emperor Kublai Khan unified the Central Plains, he established a strict "jamuci" system on the vast territory. The so-called "jamuci" is a transliteration of Mongolian post transmission. The postal roads of the Yuan Dynasty were divided into three types: one was called Tieligan Dao, which is the cart road; The second was called Mulian Dao, also known as horse road; The third was called Nalian Dao, also known as the small road [5]. From a regional perspective, the Tieligan Dao and Mulian Dao are mostly used as postal stations between Lingbei Shangdu and Dadu; Nalian Dao was used for military affairs in the northwest, with most of the stations located in present-day Gansu Province. The "jamuci" system is a systematic and strict postal system, including the management regulations of post stations, the responsibilities of post officials, post station equipment, and the tax collection system for station households. During the reign of Emperor Kublai Khan of the Yuan Dynasty, the "Jamuci Regulations" were formulated, which were the basic management regulations related to post stations at that time. They standardized the organization and leadership of post stations, the management of horses, the supply of food and drink at post stations, the acceptance of horses and the restraint of station officials, the inspection of symbols, the management of pastures, the supervision of envoys, and timely promotion. During the Yuan Dynasty, each post station was equipped with post officials and appointed leaders, responsible for supplying good horses, inspecting post officer certificates, and counting post station equipment. The specific regulations for the management of these postal stations better ensured the development of postal system in the Yuan Dynasty.

The postal roads of the Yuan Dynasty were more developed than those of the previous dynasties. According to the "Geographical Records of the Yuan Dynasty", post stations were scattered throughout the country at that time, and the postal roads were bustling with people coming and going [20]. Italian traveler Marco Polo vividly described the prosperity of postal services during the reign of Emperor Kublai Khan: "From the city of Khan Bali, there are roads leading to various provinces in all directions. On each road, that is to say, according to the location of the town, there are postal stations every 40 or 50 km, with hotels built to accommodate passing business travelers. These buildings are magnificent, with decorated rooms, silk curtains and door curtains hanging for the use of high-ranking officials and nobles. Even if the princes stay in such postal services, they will not lose their dignity." Because all the necessary items can be obtained from nearby towns and fortresses, and the court also regularly supplies certain post stations." [21]. According to travelogues, each post station always has over 400 horses for the use of the Khan's messengers. The couriers deliver urgent documents and can run 320 km per day. They all carry a sign with a falcon painted on them, as a symbol of speeding.

In the first year of the Zhongtong era (1260), Kublai Khan, the Great Khan, was located north of the Jinlian River and named this place Kaiping Prefecture, which was later renamed Shangdu. Shangdu was an important political center of the Yuan Dynasty and also an important hub for northern post stations. In 1272, Kublai Khan moved his capital to Dadu and decided to hold a meeting of various dynasties in Shangdu every summer. In the autumn, he returned to Dadu, officially establishing the patrolling system of two capital in the Yuan Dynasty. The schematic diagram of its route is shown in Fig. 4.8.

At that time, there were four post roads leading from Dadu to Shangdu, of which three passed through the present-day Zhangjiakou area. These post roads are well-equipped, with complete post station construction, a continuous flow of vehicles and horses, pedestrians facing each other, and transportation thriving for a while.

The first one is the Bolao Post Road. Bolao is in Mongolian, which means West Road. Before the establishment of the Yuan Dynasty, this road was the route from the mainland to the northern desert. This post road starts from Juyongguan in Dadu, enters Longqingzhou, passes through Yulin Station, Leijia Station, Jimingshan Station, exits Deshengkou, crosses Huling Mountain, passes Fuzhou, turns east to Chahannaoer Palace, connects with Wangyun Road, and heads north to Shangdu, with a total length

Fig. 4.8 Schematic diagram of the patrolling routes of the two capitals during the Yuan Dynasty

of about 700 km. Before Kublai Khan established the Yuan Dynasty, this road was the "Zheng Roan", and all emperors, envoys, and officials followed this road. In the third year of the Zhongtong reign (1262), Wangyun Yilu was changed to Yilu Zhenglu. But every autumn when the Emperor of the Yuan Dynasty returned from Shangdu to Dadu, he often took this road.

The second one is Wangyun Post Road, also known as Wangyun Road. After ascending to the throne in Kaiping, Kublai Khan decided to take a path from Yanqing to Wangyun and establish the Haiqing Station. He also stipulated that in the future, except for urgent military affairs, envoys and officials with the Haiqing card should enter Wangyun Station. If there was no urgent business Haiqing card, they cannot leave Wangyun and were only allowed to enter the Grand Station. This was also the famous Haiqing Post Road in history, which was the shortest path and only allows messengers with important official documents and military orders to use it. Wangyunyi Road starts from Dadu, passes through Changping Xindianyi, exits Juyongguan, passes through Longqingzhou Jinshan Station, Yulin Station, Diao'e Station, Chicheng Station, Longmen Station, Dushikou Station, Ming'an Station, Chahannaoer Xinggong, Huanzhou Station, and ends at Shangdu Kaiping, with a total length of about 600 km.

The third one is Heigudong Road, also known as Nian Road, which is exclusively used by emperors to patrol every year and is prohibited for ordinary people to walk. Because it passes through Heigu, it is called Heigudong Road. There are two Nian roads, one for each round trip. From Dadu to Shangdu, take the Dong Road, and from Shangdu to Dadu, take the Xi Road. This is known as "returning from east to west". This road has 18 Nabo, which are the Emperor's mobile tent. The first Nabo out of Dadu is Dagou, passing through Huanghoudian, Zaojiao, Longhutai, Bangchuidian, Guanshan, Shaling, Niuquntou, then to Zhenggudian, Nihe'er, Nanpo, and finally to Shangdu.

The fourth one is Gubeikou Road, also known as Dong Road due to its easternmost location. This road does not pass through the Zhangjiakou area and was exclusively used by inspectors and the military. It starts from Dadu, passes through Shunyi, Tanzhou, Gubeikou, and Xuanxingzhou, and continues northwest along the Luan River, passing through Dongliangting to Shangdu.

Main Post Road of the Ming Dynasty

In the first year of Hongwu (1368), Ming Taizu Zhu Yuanzhang established the Ming Dynasty and ordered the rectification and restoration of postal stations throughout the country, vigorously promoting the construction of postal facilities in border areas. Except for the southwestern Yunnan Guizhou region, Zhu Yuanzhang opened a large number of post roads and set up post stations in the northeastern, northern, and northwestern border areas of China, greatly strengthening the connection between the central and border areas. In the early Ming Dynasty, a series of military strongholds were successively established in the area from Yalu River to Jiayuguan, known as the "Jiu Jiang". Post stations were set up along more than 8000 li of the Jiu Jiang. Zhu Di, the founder of Ming Dynasty, also established the Nurgan Regional Military Commission and 45 post stations in the Heilongjiang and Wusuli River basins, effectively managing the administrative affairs of this border area.

In order to strengthen its rule over the northeastern border, the Ming Dynasty established numerous post stations in the Liaodong region. With Liaoyang as the center, there are four post roads radiating outward: a southwest route to Shanhaiguan, with 17 postal stations; a northbound route to Kaiyuan, with 5 postal stations; and the last southeast route to Jiuliancheng, with 7 postal stations. Each postal station was equipped with post soldiers, sedan bearers, boat carriages, horses and donkeys, etc. In order to operate the Nurgan area, the Ming Dynasty government not only set up post stations in the area, but also set up shipyards in Jilin and opened water transportation post roads. Starting from Liaoyang, they traveled upstream along the Liao River to Kaiyuan, passed through Luyi to reach the Songhua River, and then took post boats from the Songhua River to Nurgan. This water land intermodal transportation post road has a total of 3,000 officers and soldiers, providing over 20,000 stones of military food annually. It passes through Hupi Postal Station, Shenyang Postal Station, Yinzhou Postal Station, Kaiyuan Postal Station, etc. It is a channel for the Ming Dynasty government to connect with the vast Northeast region. The main postal routes of the Ming Dynasty are shown in Fig. 4.9.

In addition, the Ming Dynasty strengthened the construction of post roads in Xizang. Xizang was still called Tubo in the Yuan Dynasty, and the Yuan Dynasty government set up postal stations there. According to the Book of the Jing Shi Da Dian, there were 28 large postal stations and 7 or 8 small postal stations in Xizang in the Yuan Dynasty. At that time, religious leaders and local officials in Xizang often came and went on the post roads with books, post coupons and official documents [22]. In order to strengthen ties with Xizang, the Ming Dynasty government has repeatedly ordered the restoration and construction of post roads and postal stations connecting Uszang (the name of Xizang in the Ming Dynasty) to the mainland. In the fifth year of the Yongle reign (1407), the Ming Dynasty government ordered Shanhua Wang, Hujiao Wang, Zanpu Wang, and the Imperial Preceptor of Uszang to work together to restore the postal station and open up the post road of Uszang in Yazhou. This is another new post road from the mainland to Uszang, following the Ganzang Post Road. This post road has convenient transportation and a very abundant supply of clothing and food along the way, which not only ensures the smooth flow of imperial edicts, decrees, and documents from the Ming Dynasty government in the Uszang region, but also facilitates economic and cultural exchanges between Uszang and the mainland.

The Ming Dynasty also established post stations for foreign economic exchanges in the southeastern coastal areas. Ming Chengzu established Huaiyuan Postal Station in Guangdong, Laiyuan Postal Station in Fujian, and Anyuan Postal Station in Zhejiang, specifically responsible for receiving foreign envoys and merchants. According to the Ming Dynasty's legal treatise "Ming Hui Dian", "There were postal stations located from the capital to various parts of the country. In the capital, they were called Huitongguan, while in other places, they were called Shuima Postal

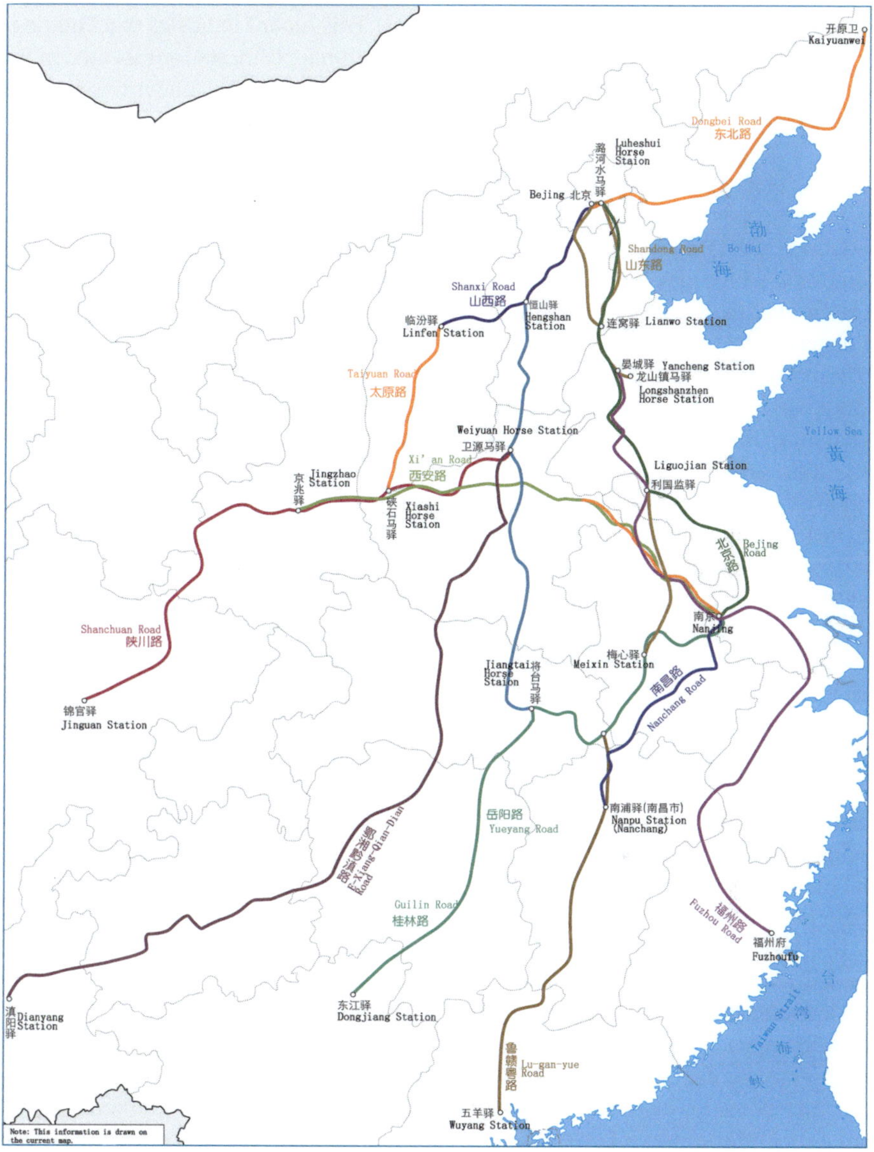

Fig. 4.9 Schematic diagram of post road main lines in the Ming Dynasty. *Note* Self-drawn by the Author

Station and Official Transportation Office." [23] This record indicates that Huitongguan in the Ming Dynasty was the main hub of national postal stations located in the capital, Beijing at that time. The Huitongguan has two functions, one is to serve as a postal station for transmitting letters, and the other is to serve as a national level guesthouse. This place can provide food and accommodation for foreign envoys, royal officials, and senior officials. Sometimes, the government also holds state banquets here to entertain tribute personnel from neighboring countries.

During the Ming Dynasty, some local post offices were responsible for receiving domestic and foreign guests. In the book "Renovation of Wuhu County Post House" written by Liu Xuan, a student of the Ming Dynasty, it is mentioned that Wuhu is a must pass place for overseas Southeast countries to go to Beijing, and it is "transportation hub of the world, so post house is set up." Foreign envoys from Southeast countries must first temporarily stay at the Wuhu Postal Station before "changing official boats" to go north [24]. Suzhou is also a major transportation hub city in the southeast. In the Ming Dynasty, the Gusu Postal Station was extremely exquisite, with pavilions, halls, buildings, and platforms built. There were 20 large character antithetical couplets in the stone pillars of its main gate: "When guests arrive, we cook tea for them, and although we are travelers, we have the right to entertain them as hosts. At night, we hang up lanterns, the light of which mingles with the moonlight, and the post pavilion in the distance is also faintly visible in this light, reflecting the Xujiang River." This couplet reflects the service scene of the Ming Dynasty Post Station in China. Along the northern post roads of the Ming Dynasty, there were also many magnificent buildings with a high level of garden art. During the Ming Dynasty, Yongnian District in Handan, Hebei Province was known as Guangping Prefecture. During the Wanli period, the local governor built a "Lotus Pond Resort" building in the east of the prefecture to receive guests from various regions. According to records, there were 5 main halls, 3 ear rooms, a water pool, and many floating bridges, painting pavilions, and pavilions. The name was elegant, like a private garden.

In the Ming Dynasty, just like in the Yuan Dynasty, ordinary documents were handed over to walking post shops, while important and urgent documents were handed over to horse post for processing. Official Transportation Office is a transportation agency specialized in transporting military supplies and tribute goods, in addition to regular transport and post. This type of agency is divided into two types: land transportation and water transportation. The establishment of Official Transportation Office represented a major advancement in transportation during the Ming Dynasty, providing specialized institutions for the transportation of goods. Official Transportation Office is responsible for specialized officials, with one ambassador and one deputy envoy each, and another centurion. The transportation task of the land post is undertaken by military soldiers, while the waterway post is managed by local shipowners. This type of delivery basically adopts a fixed point, route, and relay method. This specialized delivery service effectively organizes land transportation as well as sea and river transportation.

Postal Delivering System and Department of Postal Delivering System in the Qing Dynasty

Before the Qing Dynasty, although "post and postal stations" were often referred to collectively, in reality, "post" and "postal stations" were two different types of organizational structures with different functions. Since the Han and Tang dynasties, post has been a communication organization responsible for delivering official documents, also known as delivery; However, the postal station is actually only responsible for providing various transportation and communication tools, and also has the nature of a guesthouse. During the Qing Dynasty, these two types of organizations were integrated, and postal stations went from indirectly serving communication envoys to directly handling communication affairs. This essentially simplifies the communication system and greatly improves work efficiency. The management of postal stations in the Qing Dynasty was under the jurisdiction of the Ministry of War, with a dedicated Vehicle and Driver Department and seven appointed officials in charge of national postal stations. At the same time, two specialized institutions were set up near the Donghua Gate of the Imperial Palace, jointly managed by the two ministers of Manchu and Han, to manage the communication between the capital and various post offices. There was a horse house under it, which was dedicated to managing post drivers and horses. Establish a reporting office to receive and dispatch official documents and military intelligence.

According to records, communication during the Qing Dynasty reached the fastest speed in history. In the past, a maximum of 500 li was transmitted in a day and night, and during the Qing Dynasty, the fastest transmission of official documents by horse delivery could reach 800 li in a day and night. During the Kangxi, the rebellion of Wu Sangui was quelled, and military intelligence was transmitted from the southwest to the capital, covering a distance of over 5000 li. Later, when Shi Lang was sent to recapture Taiwan, he reported victory from Fujian to the capital, covering a distance of over 4800 li, fast horse communication can be reached in just 1 week.

Postal stations in the Qing Dynasty were more common than those in the Ming Dynasty, especially in some remote county-level areas where "county postal station" was newly established. This type of postal service conveys information between counties, making up for the shortcomings of trunk stations. County postal station was not a formal post station, but it was equipped with communication horses called "post horses", which actually serve as a postal station. Due to the addition of county postal station, the postal service throughout the Qing Dynasty was greatly improved compared to before. By the time of Guangxu, there were 139 formal and informal postal stations, including county postal stations, only in Shandong province.

After the establishment of the Grand Council in the Yongzheng Dynasty, the postal service industry in the Qing Dynasty had new developments. The Grand Council can directly issue the emperor's oracles or imperial edicts downwards. Sometimes, these imperial edicts can be handed over directly from the Grand Council to the Ministry of War in feudal China's the department of victorious news without going through the handling of the the Outer Court and the Inner Cabinet, and sent to the postal station for downward delivery. These important documents often have the words "Immediately post by fast horse" written on them, indicating that they are urgent documents. Some require a tight time limit, directly stating "urgent with six hundred li", and even some require "urgent transmission of 800 li by post", that is, a request for delivery at a speed of 600 li or 800 li per day. This approach is both confidential and efficient.

During battles with border warlords and local forces, emperors such as Kangxi, Yongzheng, and Qianlong had to wait for front military reports every night. If the reports did not arrive for a day or two, they would be restless and unable to sleep at night. At that time, it was 10,000 li away from Beijing to the front line of the Junggar War, and it took more than a month for documents to be exchanged. Qianlong often had to anticipate the situation several months later before making a decision. It can be affirmed that if there were no developed postal facilities at that time, the emperors of the Qing Dynasty could hardly have timely understood the military situation ahead, made timely decisions, and effectively controlled the battlefield ahead.

After the mid Qing Dynasty, the imperial system collapsed, and with the development of the old style postal service, there were naturally many irreparable drawbacks. Especially the laboring masses living under the imperial post office were in dire straits. Qin Songling, a poet of the Qing Dynasty, described the miserable life of the postmen in his poem of "Dian Fu Xing", "After running and working hard, the face turned dark, and the body and mind were exhausted to the point of emaciated appearance and spirit. The deep water submerged the ankles and covered them with mud, and the clothes were worn out and the skin on the shoulders was exposed. The pain in the heart was too hard to speak out, want to tell but first choked up." [25]. In such poor conditions of service stagehands, naturally, there can be no incentive to work, have fled. In the northeast of Wuqing District, Hebei Province, there is a Hexi postal station located at the main waterway thoroughfare of Jingdong, which has always been the throat of transportation throughout the dynasties. In the early Qing Dynasty, the postal service here was very developed, with 152 servants and 33 post horses. But by the late Qing Dynasty during the Guangxu period, the post affairs were in a slump, leaving only 30 servants and 24 post horses. At that time, a lower level official who also managed post affairs wrote the Hexi Post Affair Diary. In his writing, the Hexi Post Affair had become a dilapidated organization: dilapidated houses, stables about to collapse, old and weak sick horses, and hungry and cold stagehands made up a bleak picture. In Heshi Post Affair, there were also illegal officials who kept on extorting money, which caused delay of documents and telegrams, blockage of communication, and postal service could not be carried out normally. Such a post station, of course, could not fulfill the important responsibility of postal communication, but became a social wart and had no need to exist.

Experienced innovators have long proposed to abolish the outdated traditional post stations mentioned above. Feng Guifen, a famous philosopher in the late Qing Dynasty, wrote a special article titled "Cai Yi Zhan Yi", which deeply exposed the malpractices of the Qing Dynasty government's postal stations. The article pointed out that a certain Taoist official used 500 miles of courier service for the purpose of sending greetings for festivals and birthdays, greatly wasting human and material resources. Feng Guifen exclaimed loudly, "The state has filled the private pockets of officials with limited national treasury funds, while also conveying insignificant letters. What is the use of such a post station?" He believed that it should be decided to cancel the post station and establish modern postal services. This is not only beneficial for officials, but also for the people. It can not only save national expenses, but also borrow millions of surplus postal revenue [29]. Under the pressure of the situation, the Qing Dynasty government began to establish a new type of postal service in the 22nd year of the Guangxu reign (1896), gradually replacing post stations. In 1913, after the Xinhai Revolution, the Beiyang government finally announced the revocation of all post stations [30]. From then on, the human horse postal system in ancient China officially came to an end, and a new chapter was opened in the postal reform that adapted to the rapid development of productivity brought about by the industrial revolution.

Leapfrog Development of Contemporary Postal Services

At the beginning of the founding of the People's Republic of China, in order to quickly unify the national postal service and restore postal communication, the central government quickly established the Ministry of Posts and Telecommunications, and in 1950, the Postal Administration was also established. At that time, the postal industry was in great need of development, with incomplete networks and outdated production equipment. The total length of postal routes in China was only 706,000 km, and there were only 26,000 postal service bureaus, most of which were concentrated in large and medium-sized cities. Counties, townships, and rural areas were still blank areas that postal services could not cover. For a long period of time thereafter, the services that postal branches were able to handle mainly focused on letters, parcels, remittances, and newspaper distribution. In 2005, the State Council issued the "Postal System Reform Plan". Two years later, according to the postal system reform concept of "one separation, two reforms, and four improvements", the State Postal Administration and China Post Group Corporation unveiled their licenses, and the separation of government and enterprises was basically completed. In 2008, the "large department system" reform was launched, and the postal industry added new impetus to the "big transportation". Postal and express delivery enterprises also took advantage of the development of the comprehensive transportation system [26].

Since the 18th National Congress of the Communist Party of China, the postal industry has continued to leverage the advantages of postal system reform, taking innovation as the primary driving force for development. Driven by innovation, the postal industry is making people's lives more convenient and making enterprise development more dynamic. In 1949, the total amount of postal services in China was only CNY160 million. By the end of 2021, the total amount of postal services in China had reached CNY1264.2 billion, an increase of over 7,900 times over the past 70 years [31]. The postal system has become a strategic infrastructure of the country and an organizational system of society. As an important component of modern service industry, the postal industry plays an increasingly important foundational role in serving the country's economic and social development and improving people's livelihoods, as a modern leading industry that promotes the transformation of circulation methods and consumption upgrading.

In addition to the official China Post, private express delivery companies in China have also emerged like mushrooms after rain with the development of e-commerce. In 2009, the Postal Law of the People's Republic of China was revised to confirm the legal status of private express delivery at the national level. At present, the coverage rate of rural express delivery outlets in China has exceeded 98%, with an average annual service user number exceeding 100 billion person-time. In 2020, China's express delivery business volume completed 83.336 billion pieces, ranking first in the world for seven consecutive years, accounting for more than 60% of the global express delivery business volume [27]. Behind these historical coordinates is the steady progress of China's postal industry reform and the acceleration of industry development. At the same time, an industrial ecosystem that optimizes the development environment and enhances governance capabilities is being improved: devices such as information and communication, transportation tools, and handheld terminals have become powerful tools for improving production and service efficiency; The injection of capital, talent, management and other elements has accelerated the pace of building modern enterprises; Deeply linked with industries such as e-commerce and manufacturing, and effectively connected with the comprehensive transportation system, the express delivery industry's industrial chain is constantly expanding. The Action Plan for Building a Postal Power proposes to basically build a postal power by 2035, achieve global network access, intelligent facilities and equipment, intensive development methods, diversified service supply, and basically achieve modernization of industry governance system and governance capacity. From ancient relay stations to the next day delivery by plane, China's postal and express delivery industry has entered a new era of 2.0, moving forward with high quality development.

References

1. Xu Shen. *Shuowen Jiezi (Explaining Simple and Analyzing Compound Characters)*. Hangzhou: Zhejiang Ancient Books Publishing House, 2016.
2. A Chapter on Aesthetics. The 20th Pilgrimage to Pray in Temples - Yao Temple [EB/OL]. (2019-08-03) [2022-04-24]. https://www.meipian.cn/2ak9y47s.
3. Sima Qian. *Shi Ji*. Beijing: Zhonghua Book Company, 1959.
4. Shang Shu (Book of Documents). Translated and Annotated by Wang Shishun and Wang Cuiye. Beijing: Zhonghua Book Company, 2012.
5. Zang Rong. Ancient Chinese Post Stations and Postal Services. Beijing: China International Broadcasting Press, 2009.
6. Longmen Fugue. What Was the Real Function of the Great Wall in Ancient Times? [EB/OL]. (2018-09-25) [2022-03-15]. https://iask.sina.com.cn/d-b/5ba9dbc00cf2c394227e93d0.html.
7. China Social Sciences Net. Treasure of the Museum | Du Tiger Token [EB/OL]. (2022-01-19) [2022-04-12]. https://baijiahao.baidu.com/s?id=1722344470415560050&wfr=spider&for=pc.

8. Zheng Ruokui. Illustrated Dictionary of Ancient Chinese Transportation. Kunming: Yunnan People's Publishing House, 2007.
9. Fan Ye. History of the Later Han Dynasty. Annotated by Li Xian et al. Beijing: Zhonghua Book Company, 1997.
10. Gansu Provincial Museum. The Han Dynasty Postal Service on the Silk Road [EB/OL]. (2020-06-11) [2022-04-05]. https://baijiahao.baidu.com/s?id=1669179759177711754&wfr=spider&for=pc.
11. Beijing Evening News. Listen to the Call of the Desert, Take a Virtual Tour of Dynamic Dunhuang! [EB/OL]. (2021-10-08) [2022-03-15]. http://finance.sina.com.cn/jjxw/2021-10-08/doc-iktzqtyu0201707.shtml.
12. China Gansu News. Appreciation of Gansu Cultural Relics ⑨ · Courier Map Mural Brick [EB/OL]. (2022-06-10) [2022-06-15]. https://baijiahao.baidu.com/s?id=1735241932348509461&wfr=spider&for=pc.
13. Li Linfu et al. Tang Code. Edited by Chen Zhongfu. Beijing: Zhonghua Book Company, 1992.
14. Song Minqiu. Local Chronicles of Henan. Chengdu: Sichuan University Press, 2007.
15. Liu Zongyuan. Collected Works of Liu Hedong. Shanghai: Shanghai Ancient Books Publishing House, 2008.
16. Cen Shen. Annotated Poems of Cen Jiazhou. Beijing: Zhonghua Book Company, 2004.
17. Xie Jin et al. Yongle Canon. Beijing: Bibliographical Literature Publishing House, 1983.
18. Li Tao. Continuation of the Comprehensive Annals of the History of the Five Dynasties. Edited by Huang Yizhou et al. Shanghai: Shanghai Ancient Books Publishing House, 1986.
19. Shen Kuo. Dream Pool Essays. Shanghai: Shanghai Bookstore Publishing House, 2003.
20. Song Lian et al. History of the Yuan Dynasty. Beijing: Zhonghua Book Company, 1976.
21. Marco Polo. The Travels of Marco Polo. Translated by Liang Shengzhi. Beijing: China Literature and History Publishing House, 2008.
22. Zhao Shiyan. Collected and Edited Jingshi Dadian (Comprehensive Treatise on Statecraft). Edited by Zhou Shaochuan, Wei Xuntian, and Xie Hui. Beijing: Zhonghua Book Company, 2020.
23. Shen Shixing et al. Ming Huidian (Code of Ming Dynasty). Beijing: Zhonghua Book Company, 1989.
24. Integrated Local Chronicles of China: Anhui Prefecture and County Local Chronicles Album. Nanjing: Jiangsu Ancient Books Publishing House, 1998.
25. Zhang Yingchang. Collected Poems of the Qing Dynasty. Beijing: Zhonghua Book Company, 1960.
26. China Transportation News Network. Black Horse Adds Wings to Reach All Directions – A Summary of Achievements in the Postal Industry during the 70th Anniversary of the Founding of New China [EB/OL]. (2019-10-15) [2022-04-24]. https://news.china.com/zw/news/13000776/20191015/37222546_1.html.
27. China Business Journal. A Brief History of China's Express Delivery Industry: The Path to Becoming a 'Powerful Postal Country' from Large to Strong [EB/OL]. (2021-07-03) [2022-04-24]. https://baijiahao.baidu.com/s?id=1704227129094639537&wfr=spider&for=pc.
28. Zhou Li (Rites of Zhou). Edited by Chen Shuguo. Changsha: Yuelu Bookstore, 1989.
29. Liu Jinzao. Continued Documentation and Research of the Qing Dynasty. Hangzhou: Zhejiang Ancient Books Publishing House, 1988.
30. Zhang Youcai. Establishment and Development of Post Stations in Hezhou during the Ming and Qing Dynasties [EB/OL]. [2022-04-05]. http://www.lxzbwg.com/Article/Content?ItemID=67b7e936-06b4-44af-8267-5d0d57ffdb11.
31. State Post Bureau of China. State Post Bureau Announces the Operation of the Postal Industry in 2021 [EB/OL]. (2022-01-14) [2022-04-24]. https://m.thepaper.cn/baijiahao_16293264.

Chapter 5
Ancient Path of Qinling Mountains: Rise of the Great Han Empire Under the Advancing Secretly by an Unknown Path

At the end of the Qin Dynasty, the extreme extravagance and excessive taxation of the rulers led to widespread complaints among the people, and the strict laws of the Qin Dynasty left the people with nowhere to go. As a result, the Daze Township Uprising led by Chen Sheng and Wu Guang finally broke out, marking the first peasant uprising in Chinese history. Among many uprising armies, although Liu Bang led his army to attack Xianyang first, he was forced by the powerful Xiang Yu to withdraw from Guanzhong and led his troops to retreat to Hanzhong. To confuse and paralyze Xiang Yu, Liu Bang burned down all the plank roads along the Qinling Mountains leading from Hanzhong to Guanzhong during his retreat, expressing his determination not to return to Guanzhong again. Liu Bang, who had regrouped, dispatched Han Xin to fight back to Guanzhong and compete for the world. Before embarking on the expedition, Han Xin dispatched troops to repair the burnt boardwalk and put on a posture of advancing from the original route. Upon hearing the news, Xiang Yu's defenders in Guanzhong closely monitored the repair of these plank roads and dispatched the main force to closely guard against various checkpoints along this route. Han Xin drew the attention of the enemy on the front line of the plank road while sending a large army to bypass Chen Cang and launch a sudden attack, defeating Zhang Han in one fell swoop, pacifying the San Qin, and taking a decisive step towards Liu Bang's unification of the Central Plains. This is the story of "openly building plank roads and secretly passing the Chen Cang", which also showcases the historical glory of the Ancient Path of Qinling Mountains to the world.

© China Communications Press Co., Ltd 2025

H. Lu (ed.), *A Brief History of Transportation in China*,
https://doi.org/10.1007/978-981-97-9851-3_5

Ancient Path of Qinling Mountains and Empire of Huaxia

"For about 48,000 years since then, Qin and Shu have been obstructed by the Qinling Mountains and have never communicated back and forth." [1]. Li Bai described the towering and steep Qinling Mountains in his book "Shu Dao Nan". A Qinling Mountains, with the length of about 3000 li, isolates the communication between Qin and Shu. The Qinling Mountains is one of the important mountain ranges in China, which together with the Huaihe River form the boundary between the north and south of China. The geographical range of the broad Qinling Mountains is shown in Fig. 5.1. The Qinling Mountains distinguish the geographical climate, natural environment, historical culture, and social life of northern and southern China, blocking the cold air from moving south and also blocking a large amount of water vapor from moving north. It is the location where the zero temperature isotherm and the 800 mm precipitation line pass through in January, so the northern and southern slopes of the Qinling Mountains present a completely different ecological landscape. To the north of the Qinling Mountains is the Yellow River Basin water system, and to the south is the Yangtze River Basin water system, which to some extent creates the differences between the north and south of China.

The Qinling Mountains have nurtured two prosperous areas, the Chengdu Plain and the Guanzhong Plain, giving birth to the ancient capital of Chang'an. For 3000 years, many dynasties such as Zhou, Qin, Han, Sui, and Tang have established their

Fig. 5.1 Schematic diagram of the geographical range of the generalized Qinling Mountain

capitals here. The Western Zhou royal family, together with various tribes, defeated King Zhou of Shang Dynasty and settled at the foot of the Lishan Mountains in the Qinling Mountains, marking the beginning of the 2000 year era of the Great Qinling Mountains in China. Throughout history, dynasties have centered around the Chang'an and Luoyang lines of the Great Qinling Mountains. Before the eastward migration of the Zhou royal family, the Western Zhou Dynasty was backed by the Qinling Mountains and ruled the land of China with Zhou rituals and feudal systems. After the Zhou royal family moved eastward, the Qin people entered Guanzhong and arrived at the fertile land at the foot of Lishan Mountain.

The ancestors of the Qin people were former officials of the Shang Dynasty. After the Zhou Dynasty conquered the Shang Dynasty, they were forced to move westward to the Tianshui River Valley in Gansu Province, northwest of the Qinling Mountain, to raise horses for the Zhou royal family and resist the Western Rong forces. Tianshui, Gansu was the birthplace of the Qin people. In the first year of King Ping (770 BC), due to his meritorious service in escorting King Ping of Zhou, Duke Xiang of Qin was appointed as a vassal state, and Qin Yiyue became a vassal state. After the decline of the Zhou royal family, the Qin people moved to Guanzhong and seized the important road of Hangu Pass, which led eastward. In the eighth year of King Xian's reign (361 BC), after Qin Xiaogong ascended to the throne, he highly valued Shang Yang and implemented reforms, establishing a new capital in Xianyang. During the reign of King Huiwen of Qin, the Qin army crossed the Qinling Mountain, annexed Bashu, and captured Hanzhong in the state of Chu. At this point, the west of the Yellow River and the Bashan Mountain of the Qinling Mountain have all returned to the State of Qin. During the reign of King Zhaoxiang of Qin, the State of Qin invaded the south and north, expanding its territory to the middle and lower reaches of the Jianghan River. In the 26th year of Emperor Qin Shi Huang's reign (221 BC), King Ying Zheng of Qin finally unified the six states and established the first unified centralized state in Chinese history.

In the chaotic era of the late Qin Dynasty, Liu Bang led the vanguard troops of the uprising army to bypass the easily defended and difficult to attack Hangu Pass, and advanced in a detour from the southwest of the Qinling Mountain, taking the lead in entering Guanzhong and annihilating the Qin Dynasty. Previously, the nominal leader of the rebel army, King Huai of Chu, and the generals agreed that the first to enter Guanzhong would be the King of Guanzhong. Liu Bang surrendered to Xiang Yu, who held a heavy army, and gave up Guanzhong, using Hanzhong and Bashu as his strongholds to become the King of Han. Later, he once again sent troops to compete with Xiang Yu for the world, starting from the plank roads along the Qinling Mountain.

Between the deep valleys of the Qinling Mountain, a group of ancient roads composed of multiple steep passages connect Bashu, Qiangzang, Ganqing, Guanzhong and other places. The Ancient Path of Qinling Mountain now appears to be some inconspicuous mountain paths, but over a history of 3000 years, it has connected the Qinling Mountain with the Yellow River and Yangtze River basins in the most primitive way (Fig. 5.2).

Fig. 5.2 Chencangdao historical site [2]

After the decline of the Han Empire, the Ancient Path of Qinling Mountains once again became a passage for north–south warfare. Zhuge Liang's claim of leaving the Qishan Mountain six times is actually leaving the Qinling Mountains. Zhuge Liang launched a northern expedition against Cao Wei and advanced upstream along the Western Han River in the Qinling Mountains to capture Longxi. He intended to advance further towards Guanzhong, but unfortunately died before his victory. Later, when Cao Wei conquered Shu, he also advanced through the Ancient Path of Qinling Mountains. General Deng Ai of the State of Wei wrapped himself in a blanket and descended from the Yinping Path by climbing trees and scrambling along cliffs. He led his troops in person to raid Chengdu, forcing Liu Shan, the last emperor of the Shu Han Kingdom, to surrender and leave the city.

In the heyday of the Tang Dynasty, the great master Li Shimin wrote in his poem "Looking at the Zhongnan Mountain": "This mountain peak is layered and overlooks the Wei River. The emerald green mountains rise into the clouds, as if piercing into the sky. When the red sun rises from the mountain range, it seems to support the mountain peak; and when the emerald green rocks are shrouded in mist, the mist seems to be stored in the crevices of the rocks." [1]. In his later years, the poet Wang Wei lived in seclusion at the Zhongnan Mountain in the Qinling Mountains and wrote a famous landscape poem: "Along the way to the end of the water, sit down and enjoy the beautiful scenery that comes when the clouds rise. By chance, he meets the mountain." The old man in the forest was chatting and laughing with him, forgetting the time to go home [3]. Yang Guifei loved to eat lychee, is from the south, the transportation through the Qinling Mountains in the Ziwu Road all the way to Chang'an. Han Yu was exiled to Chaozhou and missed his family on his long journey. When he looked back at Chang'an in the Qinling Mountains, he exclaimed, "Where is the hometown of Qinling Mountains shrouded in dark clouds? Even the horses in Lan Guan refused to move forward despite the heavy snow" [4]. After the Tang Dynasty, with the gradual southward shift of China's economic center, the economic status of the Yellow River Basin was gradually surpassed by the south, and the political status of the Guanzhong region began to decline. The imperial capital never returned to Chang'an or the Qinling Mountains.

Winning Secret Hidden in the Ancient Path of Qinling Mountains

The Chu Han War is extremely famous in the history of world wars, and the entire campaign embodies many military wisdom. In 209 BC, the uncles and nephews of Xiang Liang and Xiang Yu from Chu, following Chen Sheng and Wu Guang, launched the Kuaiji Uprising. Liu Bang, the former head of the Sishui Pavilion, also rose up in response in Peixian and surrendered to Xiang Liang. At that time, the leaders of each tribe proclaimed themselves as kings, and it was difficult for the uprising to gather strength. Therefore, Xiang Liang adopted Fan Zeng's suggestion and supported Xiong Xin, the grandson of the former King Huai of Chu, as king, still known as the "King Huai of Chu". In order to boost his morale, King Huai of Chu agreed with his generals that "the one who takes the lead in Guanzhong will be the king there" [5], so Liu Bang led his troops to the west with the intention of entering Guanzhong first. Usually, the way to enter Guanzhong from the Central Plains is through Luoyang along the Yellow River to the west, crossing the Hangu Pass to enter Guanzhong, i.e. "Tongguan Avenue", but Liu Bang's army suffered a great defeat in Luoyang, and it was difficult for them to enter the same route. After strategic analysis, Liu Bang led his team to choose to detour through the Wuguan Road in search of distance. This strategic adjustment is very necessary. Wuguan Road is an ancient road that connects the Guanzhong Plain with the Nanyang Basin in Henan Province. It passes through the Funiu Mountains and Qinling Mountains, and enters the Guanzhong Plain through Shangluo, which is a steep terrain. The Qin army in Guanzhong focused their defense on Tongguan Avenue, while their defense on the front line of Wuguan Road was weak. Although choosing the Wuguan Road is dangerous and far away, it can avoid direct confrontation with the main forces of the Qin army, and also allow other uprising forces and defending forces to consume each other on the Tongguan Road. As expected, Liu Bang entered through Wuguan Road and defeated the Qin army in Lantian, and forced Prince Ying of Qin to surrender in the fall of 207 BC, thus ending the rule of Qin Dynasty. Xiang Yu, on the other hand, led his troops to enter Guanzhong via Tongguan Road, nearly two months after Liu Bang. The routes of Liu Bang and Xiang Yu into the Guanzhong are shown in Fig. 5.3.

After pacifying Guanzhong, Xiang Yu submitted a letter to King Huai of Chu requesting the title of meritorious minister. King Huai replied, "as promised", that is to say, he recognized the status of Liu Bang as the king. Xiang Yu was furious, and with the advantage of his military power, he made himself a feudal lord and set his capital at Pengcheng (now Xuzhou). Xiang Yu designated Ba, Shu, Hanzhong and other places and appointed Liu Bang as the King of Han; he also divided Guanzhong into three parts and appointed Zhang Han as the King of Yong, Sima Xin as the King of Sai and Dong Fei as the King of Zhai, collectively known as the "Three Qin". Xiang Yu's attempt to control the Guanzhong Plain through the power of the three kings of Guanzhong, and to lock Liu Bang in the vast Qin Mountains for an eternity, made him unable to turn around.

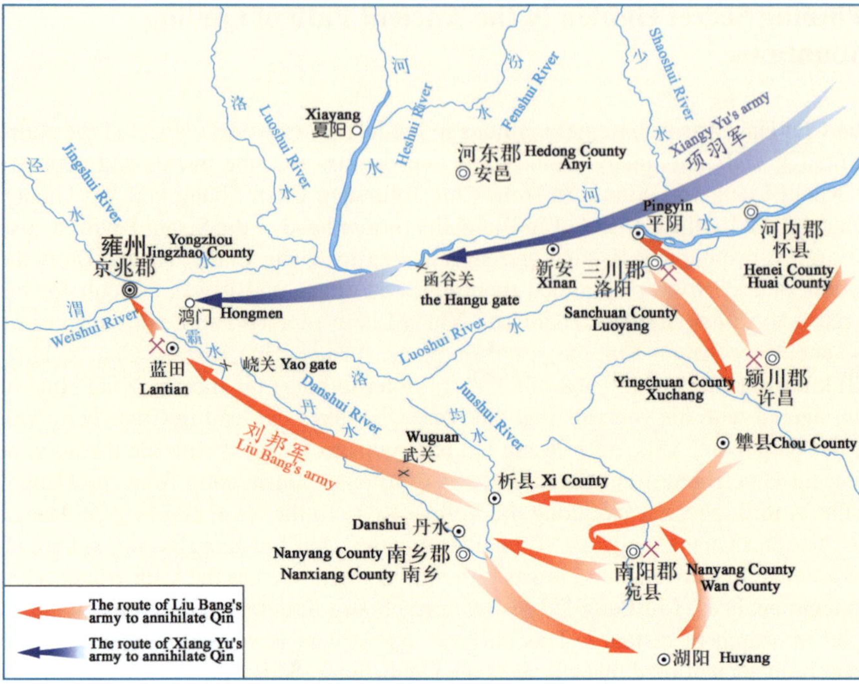

Fig. 5.3 Route for Liu Bang and Xiang Yu to enter the pass and Extinguish Qin

At first, Liu Bang was dissatisfied and planned to attack Xiang Yu, but the strategist Xiao He advised him to occupy Hanzhong. So Liu Bang adopted Xiao He's suggestion and accepted the title of King Han, and went to Hanzhong to settle in April 206 BC. The Eastern Han cliff carving "Ode to the Stone Gate" reads: "Liu Bang, the Gaozu Emperor, accepted the Mandate of Heaven and rose up in Hanzhong. He took the road of the Ziwu Valley, and marched out from the Scattered Pass into the land of Qin." [6]. From this, it can be inferred that Liu Bang chose the Ziwu Road to enter Hanzhong, that is, from Xianyang to the south through Ningshan to Shiquan, and then to the west through Yangxian to Hanzhong (there is also a saying that he chose the Baoxie Road to enter Hanzhong, that is, from Meixian in Baoji to the south through Taibai County and Liuba County to Hanzhong)". The schematic diagram of the Ancient Path of Qinling Mountains from Hanzhong to Guanzhong is shown in Figs. 5.4. On the way to Hanzhong, Liu Bang ordered Zhang Liang to burn down the boardwalk passing through the Ziwu Road. On the one hand, it was to prevent the feudal lords from sneaking attacks, and on the other hand, it was to numb Xiang Yu and show him that he had no intention of making a comeback. Xiang Yu really fell into the trap, and his troops returned to Pengcheng and ordered the main force to move to Shandong to suppress the rebellion of the State of Qi.

Four months after Liu Bang entered Hanzhong (August 206 BC), he dispatched troops to attack Zhanghan in Guanzhong. To further confuse his opponents, Liu

Fig. 5.4 Ancient path of Qinling Mountains from Hanzhong to Guanzhong

Bang listened to Han Xin's plan and arranged for his soldiers to rebuild the burnt plank road on the Ziwu Road (or Baoxie Road), so as to induce Zhang Han to send heavy troops to guard the above passes, but he himself quietly chose the westernmost Chencang Road to enter Guanzhong, which caught Zhang Han by surprise and made him retreat, and finally besieged Zhang Han and forced him to commit suicide. In the same month, Sima Xin, the king of Sai, and Dong Fei, the king of Zhai, surrendered. By then, all the three Qin lands belonged to Liu Bang. This is the allusion to the saying "to build a road in the open, but to do it in the dark". After occupying the rich land of Ba Shu and San Qin, Liu Bang used it as a strategic base to launch the 4-year-long Chu-Han War, and eventually forced Xiang Yu to cut his own throat in Wujiang River, winning the world and settling down in Chang'an, establishing the greatness of the Han Dynasty. The road network of Han Dynasty was further developed, and the main lines of transportation are shown in Fig. 5.5.

At the beginning of the war between Chu and Han, there was a significant difference in power between the two sides. If the two armies were to confront each other head-on on the plain, Liu Bang had little chance of winning. However, the vast Ancient Paths of Qinling Mountains provided excellent conditions for Tibetan soldiers to win in the mountains and become the "secret code" for winning the world. From abandoning Tongguan Avenue and choosing Wuguan Road for the first time to enter Guanzhong, to burning down the Ziwu Road (or Baoxie Road) boardwalk to avoid Chu's sharpness, then rebuilding the boardwalk to confuse opponents, and finally defeating the enemy with one strike from Chencang Road, the ancient roads in the Qinling Mountains were fully valued and applied by Liu Bang in military activities, thus opening the prelude to the prosperity of a multi-ethnic unified country. It

Fig. 5.5 Schematic diagram of road traffic arteries in the Han Dynasty

can be said that without the military strategic support provided by the Ancient Path of Qinling Mountains, perhaps there would not have been the rise of the Great Han Empire.

King Liu Bang of Han utilized the Ancient Path of Qinling Mountains to conquer the world, and the development of these ancient paths began in an earlier period. In the more than 3000 li stretch of the Qinling Mountains, thousands of descendants of China have built a large number of ancient roads in the Qinling Mountains at different

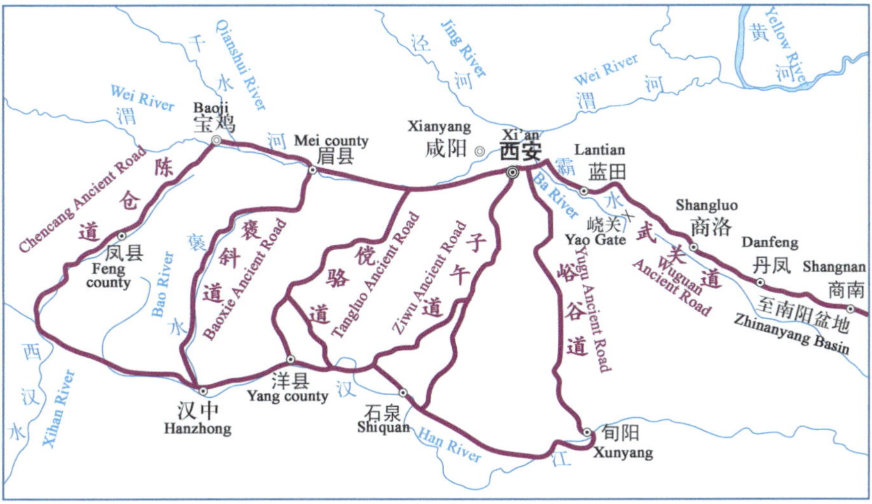

Fig. 5.6 Six ancient paths of Qinling Mountains

times in order to travel further. Among them, the most famous six are Chencang Road, Baoxie Road, Tangluo Road, Ziwu Road, Yugu Road, and Wuguan Road (Fig. 5.6), which are important channels for connecting Guanzhong and Hanzhong.

The history of the Ancient Paths of Qinling Mountains is very long. In ancient times, the ancestors of the Yongliang region (now Gansu, formerly part of the Yongliang and Liangzhou regions) used grass as a path for hunting, crossing the mountains and creating many north–south paths. The wars between tribes further expanded these paths. According to the "Wu Di Ben Ji" in the Records of the Grand Historian, "When there was disobedience in the world, the Yellow Emperor went forth to conquer, and left after pacifying a place, splitting the mountains and opening up paths along the way, and never residing anywhere in peace." [5]. During the Western Zhou Dynasty, the Guwuguan Road, the Ziwu Road, the Baoxie Road, and the Chencang Road were successively constructed. The Western Zhou Dynasty was divided into scattered states on the south bank of the Wei River in present-day Baoji. During the reign of Emperor Qianlong of the Qing Dynasty, the San Shi plate unearthed here was a bronze artifact from the early Western Zhou Dynasty, with the inscription "Zhou Dao" on it. According to historian Wang Guowei's research, this Zhou Road is the Chen Cang Road. However, the emergence of the Qinling Mountains may have occurred earlier than the Western Zhou Dynasty. The oracle bone inscriptions unearthed from Zhouyuan bear the inscription of King Wen of Zhou, who "attacked Shu". To the south of Zhouyuan is the Xiegu Kou, which is the starting point of Baoxie Road. Legend has it that when Zhou conquered the Shang Dynasty, ethnic groups from Bashu participated in the war, and they probably used the ancient paths to cross the Qinling Mountains to fight in the Central Plains.

In ancient times, the Qinling Mountains were considered the biggest obstacle in the world, with high mountains and dense forests, but many roads were built between them. According to statistics, in addition to the six ancient paths mentioned above, there are also over 50 small paths hidden in the valleys on the north slope. The Huayang Guozhi·Shu Zhi believes that the Baoxie Road was opened during the era of the Three Sovereigns and Five Emperors [7], while the Qing Dynasty's Gu Zuyu's "Notes on Reading Historical Records" states that "the Baoxie Road was built at the time of Xia Yu" [8]. It can be confirmed that the opening of the various roads in the Qinling Mountains was earlier than that of the Western Zhou Dynasty, which led to the later King You of Zhou taking the Baoxie Road to attack the Baoguo and obtaining the Baosi. Based on the unearthed cultural relics and related book descriptions, it can be inferred that Chencang Road, Baoxie Road, and Wuguan Road were opened the earliest, possibly before the Western Zhou Dynasty; Yugu Road was opened during the Spring and Autumn period; The Ziwu Road was opened during the Qin Dynasty; The Tangluo Road was opened during the period from the Eastern Han Dynasty to the Three Kingdoms.

Chencang Road

Chencang Road, Baoxie Road, Tangluo Road, and Ziwu Road all belong to the Shu Road leading to present-day Sichuan. Chencang Road, also known as the Gudao Road or Jialing Road, is named after the ancient Chencang County. This ancient path starts from Chencang in Baoji, extends southwest to the Qinling Mountains, then exits Dashan Pass and runs along the upper reaches of the Jialing River to the valley all the way to Fengxian. It then passes through present-day Liangdang and Huixian to present-day Lueyang, connects with the Jushui Road, and reaches the Hanzhong area. The total length of the road is estimated to be about 1190 km according to records from the Tang and Song dynasties, and about 1260 km according to records from the mid Qing Dynasty.

The most famous fortress on Chencang Road is Da Shan Guan. Da Shan Guan is located in the southwest Dashan Ridge of Baoji City, controlling the north–south throat of Chencang Road. It is the only fortress in the southwest of Guanzhong and also one of the famous Guanzhong Four Passes. The wars on Chencang Road mostly revolve around Da Shan Guan. According to historical records, there were at least 70 wars that occurred at Da Shan Guan. According to the "Du Shi Fang Yu Ji Yao", Da Shan Guan "holds onto the main thoroughfare of north–south transportation. If the north does not occupy it, it will be impossible to open roads to Liangzhou and Yizhou; if the south does not occupy it, it will not be able to occupy the Guanzhong area" [8]. The strategic situation of Da San Guan was so important that it has naturally become a battleground for military strategists throughout history. Liu Bang passed through this pass after Chen Cangdao returned to the Three Qin in Hanzhong; Yan Cen of the Eastern Han Dynasty led troops into Sanguan to Chencang. Cao Cao attacked Zhang Lu from Chencang through the bulk of the pass, Zhuge Liang out of the bulk of the pass around the Chencang, Wan Yan Zongbi (Jin Wuzhu) for the opening into the channel of Shu and the Southern Song Dynasty general Wu Jie repeatedly tugged at this. The occurrence of these wars mentioned above all demonstrates the military importance of Da San Guan. Da San Guan has been a popular tourist destination for literati, high-ranking officials, and ordinary people from ancient times to the present day. Wang Bo, Wang Wei, Cen Can, Du Fu, Li Shangyin, and Su Dongpo had all left poems about the Da San Guan. And Lu You wrote the most poems about Da San Guan, which had the greatest influence. For example, "I remember beating the Jin army at Gua Zhou Du, rushing towards the battleships on a snowy night; crossing the warhorses in the autumn wind, and galloping freely, the victories of recapturing Da San Guan were frequent." "One day, when the year was coming to an end, I raised my whip and arrived at Da San Guan." "Da San Guan towered high into the clouds, overlooking the enemy's camp; the clear Wei River was like a jade belt, displaying the military's formidable lineup." "The enemy still occupied favorable terrain, and the attacking army had not yet captured Da San Guan, and the situation remained." [9].

Baoxie Road

Baoxie Road is the first plank road recorded in historical records that crossed the Qinling Mountains, and it is also the main passage between Guanzhong and Hanzhong in ancient times. It is named after its northern entrance in Xiegu, Meixian County, and its southern exit in Baogu, Hanzhong. According to historical records, Baoxie Road has a total length of about 250 km and is the most convenient passage among ancient roads that crossed the Qinling Mountains and led to the land of Bashu (Fig. 5.7). The most famous one on the Baoxie Road is the Shimen Shisan Ping. Shimen is a tunnel located at the southern end of Baoxie Road, opened during the Yongping period of the Eastern Han Dynasty. Since the Han and Wei dynasties, a large number of cliff carvings with inscriptions and records have been carved on the cliffs on both sides of the Baohe River. In the early 1970s, Shimen was submerged due to the construction of water conservancy projects, and the most representative 13 stone carvings were relocated to the Hanzhong Museum for preservation, collectively known as the "himen Shisan Ping", with the most famous being Cao Cao's two characters "Gunxue".

Tangluo Road

Tangluo Road, also known as Dangluo Road, runs south from Xi'an, turns west through Huxian, reaches Zhouzhi, and then travels southwest for 15 km before finally reaching Hanzhong. Among the four main roads leading from Guanzhong to Hanzhong, Tangluo Road has the shortest distance but is the most difficult to pass. Among the various ancient roads recorded in history that crossed the Qinling Mountains, Tangluo Road is closest to the main peak of the Qinling Mountains, Taibai

Fig. 5.7 Baoxie Road ruins park [10]

Mountain, known as the "Huaxia Dragon Vein". According to records from the Tang and Song dynasties, the total length of Tangluo Road is estimated to be about 375 km, while according to records from the Qing Dynasty, it is about 425 km. Tangluo Road passes through multiple large mountains and excavation is very difficult. During the Three Kingdoms period, Zhuge Liang once set up defenses at dangerous roads, while Jiang Wei repeatedly attacked Wei through this route. In addition, it is said that during the An Lushan Rebellion, the death of Yang Guifei at Mawei was replaced by someone. The real Yang Guifei quietly descended from Mawei Slope, entered Tangluo Road, crossed Huangbaiyuan, and reached Hanzhong directly. She then entered the Yangtze River along the Han River to Yangzhou, changed her name and surname to Taizhen, and later crossed eastward to Japan. History and legend are difficult to distinguish, but the strategic role of Tangluo Road can still be seen.

Ziwu Road

Ziwu Road was opened in the late Qin and early Han dynasties. It was a major transportation route from Chang'an to the Ba and Shu areas in central Han during the Han and Tang dynasties. It had important political and economic value and was once a national post road. Ziwu Road travels south from Xi'an, passes through the Ziwu Valley, and finally reaches Hanzhong. According to the records in the Tang Dynasty's "Tong Dian", its length is estimated to be about 420.5 km, and according to the records in the mid Qing Dynasty, it is about 477.5 km. In ancient times, the Ziwu Road had important military value. In the late Eastern Han Dynasty, Cao Cao, Ma

Chao, and Han Sui fought a great war in Guanzhong, and the people of Guanzhong fled to Hanzhong for refuge through the Ziwu Road. After the An Lushan Rebellion, the Tang Dynasty strengthened the defense of the southern valley, and Ziwu Pass became an important garrison.

Yugu Road

Yugu Road is actually composed of three valley roads, namely Xigu Road, Yigu Road, and Kugu Road, all named after the valley at the northern foot of the Qinling Mountains, which formed relatively late. Yugu Road was a road from Chang'an to Ankang in ancient times. It crossed the Qinling Mountains from Chang'an and reached Ankang through the northern boundaries of Zhashui, Zhen'an, and Xunyang. The total length was about 350 km and was mainly used for north–south goods exchange, trade, and other purposes.

Wuguan Road

Wuguan Road, also known as Shangshan Road, was opened in the late Shang and early Zhou dynasties. It is connected by the Bashui River Valley on the north side of the Qinling Mountains and the Danshui River Valley on the south side. It was an important transportation route from ancient Chang'an to many southeastern regions and also a passage for Liu Bang to enter Guanzhong. The natural scenery along Shangshan Road was very beautiful. Tang Dynasty poet Wen Tingyun once wrote the famous poem "Shang Shan Zao Xing": "Waking up at dawn, the bells of the carriages and horses are already vibrating; traveling all the way, I can't help but be saddened by the thought of my hometown. The rooster's voice is loud and clear, the thatched grass store is bathed in the afterglow of the dawn moon; the footprints are faint, the wooden boardwalk is covered with the frost of early spring. The withered mistletoe leaves fall all over the wild road of the barren mountain; the pale white flowers of Hovenia, brightly open in the mud wall of the post station. Thus I recalled the beautiful scene of last night's dream of Duling; a flock of wild geese was frolicking in the round and winding pond." [11]. The poem paints a beautiful landscape picture for posterity.

Miracle of Cliff-Side Plank Path

Qinling Mountains and Bashan Mountains, two huge barriers, run in an east–west direction and stretch for thousands of li, blocking the southward view of the Guanzhong and Central Plains regimes and the journey south to Bashu. Even in the

Tang Dynasty, Li Bai lamented that "for about 48,000 years since then, Qin and Shu have been obstructed by the Qinling Mountains and have never communicated back and forth." "Shudao Road is too difficult to walk, it's almost impossible to reach the blue sky" [12]. In order to solve traffic problems, people adapted to local conditions, drilled holes, erected wooden stakes, built bridges and pavilions on cliffs, paved wooden boards, and opened up "special" roads for pedestrians and vehicles to pass through. Due to the influence of terrain and construction level, the Ancient Path of Qinling Mountains all advance along river valleys. In areas with relatively flat terrain and wide riverbanks, the roads are gentle and flat; The road sections with steep terrain and crossing mountains are often steep roads or steps made up of stones, extending forward along the mountain.

The plank path is often located in barren mountains and steep cliffs, with high mountains and dangerous waters, thorns everywhere, no human habitation, and the presence of wild animals. Moreover, it is often adjacent to deep rivers and streams. The construction environment of the plank path is extremely dangerous, with simple construction tools and great difficulty in construction. The construction process is extremely difficult. In Li Bai's poem "Shu Dao Nan", it is written that "the five strong men of the State of Shu were crushed to death due to the collapse of the mountains and the earth, and only then did the two places have a ladder and plank road that began to connect." [12]. This is the story of the "five strong men sacrificed their lives to hack through the mountain pass.

The construction of the plank path is extremely difficult, more than in the cliff hole buried pile method, in the cliff wall cut into 30 cm square, 50 cm deep hole, divided into the upper, in the lower three rows, are inserted into the stake; then in the upper row of stakes on the shelter, in the middle of the row of stakes on the pavement into the road, the lower row of stakes on the wood for the frame (Fig. 5.8). In order to ensure safety, people often install railings on the side of the plank path near the river and at bends to prevent people, horses, and vehicles from accidentally falling into the river. In order to prevent the falling rocks and rocks above the cliff from injuring people and animals, people have built a roof on the plank path in that area. The plank path with a ceiling and railings looks like a long string of aerial corridors and pavilions from a distance, hence it is called the "Pavilion Road". Due to the bridge passing across the water and the plank path running alongside the water, the plank path is called the "Bridge Pavilion".

Tangluo Road is the most difficult and shortest road to cross the Qinling Mountains. Due to the fact that this road passes through many river canyon areas, there are nearly a hundred plank paths built on the cliffs along the way. During its heyday, there were over 90,000 mountain path with trestles and galleries from Chang'an to Hanzhong, each measuring a total of 270,000 m.

When the Baoxie Plank Path was built to Jitouguan on the Baohe River in Hanzhong, it encountered unprecedented resistance—a rocky mountain. The terrain of Bao Gu Kou is very complex and dangerous, and the common techniques of drilling holes, burying piles, and building a plank path on the cliff are ineffective here. Therefore, the Baoxie Plank Path can only be repaired up to Baogukou. During the Yongping period of the Eastern Han Dynasty, Emperor Ming issued an edict to

Fig. 5.8 Schematic diagram of Plank Path construction [13]

repair the Baoxie Plank Path in order to further communicate with Qin and Shu, and convened skilled craftsmen from all over the country to discuss.

Finally, the physical method of "burning with fire and washing with water" was adopted to excavate a tunnel through the mountain with the length of 15 m, the height of 4 m and the width of 4 m, known as the "Stone Gate" in history (Fig. 5.9). This not only makes the passage of the Baoxie Plank Path smoother, but also achieves the integration of Baoxie Road and Jinniu Road. This move pioneered the use of manual tunneling in the world, creating the first mountain crossing tunnel in world transportation history where people and vehicles run parallel.

Fig. 5.9 Stone Gate Baoxie Plank Path [13]

The plank path is soaked in the blood and sweat of the working people, and also contains amazing ancient wisdom and technology. The plank path is a huge civil works project that predates the Great Wall. Its function is to connect, symbolize communication, extension, and development, and it has transportation, military, and political strategic functions. Its status is not inferior to that of the Great Wall and the Grand Canal of Beijing and Hangzhou.

After experiencing wind and rain erosion, war, and the baptism of time, the ancient plank path has long been dilapidated and abandoned, with few human traces, leaving only broken walls and ruins. Nowadays, the Beijing Kunming Expressway, Baomao Expressway, and Xicheng High Speed Railway, which span across the Qinling Mountains and connect Sichuan and Shaanxi, have all been completed and opened to traffic. The road connecting the north and south foothills of the Qinling Mountains has been compressed to within a few hours, forming a new three-dimensional transportation network along with air routes.

References

1. Peng Dingqiu, et al. The Complete Collection of Tang Poetry. Beijing: Zhonghua Book Company, 1985.
2. SOHU. The Past and Present of the Qinling Ancient Road [EB/OL]. (2019-07-10) [2022-04-25]. https://www.sohu.com/a/325884309_120047347.
3. Wang Wei. The Collected Works of Wang Mojie. Beijing: China Bookstore, 2013.
4. Han Yu. Annotated Collection of Han Changli's Works by Five Hundred Scholars. Annotated by Wei Zhongju. Beijing: Zhonghua Book Company, 2019.
5. Sima Qian. *Shi Ji*. Beijing: Zhonghua Book Company, 1959.
6. Ouyang Xiu. The Collected Works of Ouyang Xiu. Shanghai: Shanghai Bookstore, 1989.
7. Chang Qu. Records of Huayang Prefecture. Jinan: Qilu Book Press, 2010.
8. Gu Zuyu. Interpretation and Summary of Historical Geographical Treatises. Shanghai: Shanghai Bookstore Publishing House, 1998.
9. Lu You. Annotated Jiannan Poetry Collection. Shanghai: Shanghai Ancient Books Publishing House, 1985.
10. Beijiang Xianyun Yehe. A Sunny Spring Day in Hanzhong: The Shimen Plank Road (Baoxie Road) [EB/OL]. (2015-04-27) [2022-03-26]. https://you.ctrip.com/travels/shanxi100057/2336793.html.
11. Wen Tingyun. The Collected Works of Wen Tingyun. Beijing: Central Compilation and Translation Press, 2015.
12. Li Bai. Collected Poetry of Li Bai. Changsha: Yuelu Book Society, 1987.
13. Qinli Technology. The Ninth Wonder of the World - Ancient Plank Roads, Rivaling the Great Wall and the Grand Canal [EB/OL]. (2021-02-03) [2022-03-26]. https://www.sohu.com/a/448440907_120633302.

Chapter 6
Millennium Canal: Prosperity Forged by Water Transport of Grain

The canal system in China is a world miracle created by China and an inevitable choice determined by China's natural geographical characteristics. In the long years dominated by water transportation, the construction of a north–south water transportation corridor was not only a necessity for social and economic development at that time, but also a great feat of ancient humanity. Today, China's canal system has fully connected the Haihe River, Huaihe River, Yellow River, Yangtze River and the Pearl River water systems, forming the world's largest, busiest and still playing an important role in the artificial canal system, which is an unparalleled world miracle.

Canal is to communicate with the region or rivers, lakes and seas and artificially excavated waterways, is the human conquest, the use of nature's major innovation. In addition to the basic functions of shipping, the canal also serves functions such as irrigation, flood diversion, drainage, and water supply. As a long-standing water transportation mode, it has played a role in promoting economic prosperity, cultural communication, and urban development, effectively maintaining national unity and stability.

The emergence of canals originated from the practical needs of ancient transportation. The terrain of our country is characterized by a high elevation in the west and a low elevation in the east. The main rivers flow into the sea from west to east, and there is a lack of north–south water transportation channels. In order to connect various water systems, ancient Chinese ancestors began to excavate canals since the Spring and Autumn period, continuously improving the water transportation system.

Pre-Qin Period: Beginning of the Canal

The earliest historical materials related to canal excavation in China appeared during the Spring and Autumn period. At that time, various vassal states, for their own development and communication with the outside world, dug canals in the controlled

© China Communications Press Co., Ltd 2025

H. Lu (ed.), *A Brief History of Transportation in China*,
https://doi.org/10.1007/978-981-97-9851-3_6

areas to communicate with natural water bodies. The earliest country to excavate canals was the State of Chu. During the reign of King Zhuang of Chu (613 BC–591 BC), Sun Shu'ao stimulated the Jushui River on the banks of the Yunmengze River in the Jianghan Plain to create the Yunmengze Pool. He excavated the Yunmengtong Canal, which was known as the Western Canal of Chu, with the Han River Basin having smooth waterways and vast plains in the Yunmengze area [1]. During the reign of King Ling of Chu (540 BC–529 BC), he also opened a canal from Zhanghuatai (now north of Jianli, Hubei) to facilitate the transportation of water. During the reign of King Zhao of Chu (516–489 BC), Wu Zixu led the Wu army to attack the State of Chu and dredge this transportation route to enter, hence the name "Zi Xu Du" [2]. Wu Zixu was still opening canals within the territory of the State of Wu, but these canals were not long and had limited effectiveness [3, 4].

King Fuchai of Wu (reigned from 495 to 473 BC) built Hancheng (now Yangzhou, Jiangsu) and opened a ditch on the side, diverting the river water to the north between the two lakes of Wuguang Lake (now Shaobo Lake) and Luyang Lake (now Gaoyou South, Jiangsu), and placing it in Fanliang Lake (now Gaoyou Lake), then flowed northeast to the two lakes of Bozhi and Sheyang (now Baoying East, Jiangsu), to Jiaye Lake, and then northwest to the ancient end of Huai'an to enter the Huai River, connecting the grain road, also known as the Chudongfang Canal, also known as Hangou, which is the predecessor of the Zhongli Canal in the present-day Beijing Hangzhou Grand Canal. Hangou connects the two major water systems of the Yangtze River and Huaihe River, and is the first water conservancy project in China to cross and connect different water system in the north and south through a canal, with a huge impact.

In 484 BC, King Fuchai of Wu opened a deep ditch in the northeast of Dingtao, Shandong Province to divert the water from Heze to the southeast and enter the Sishui River. As its water source comes from Heze, it is called the Heshui River, which is the earliest canal in the Yellow River Basin and a follow-up project of the Hangou River. Its ancient path is equivalent to the Wanfu River in the southwest of Shandong Province and the north of Jinxiang. The excavation of the Heshui River enabled the previously disconnected rivers of the Yangtze River, Huaihe River, Yellow River and Jishui River to connect, becoming the main inland waterway for east–west exchanges in the Central Plains region [5].

The ancient Jiangnan River is one of the canals built by the State of Wu in the the Taihu Lake area during the Helu and Fuchai periods. Water flows out of the Pingmen Gate in Wudu (now the North Gate of Suzhou, Jiangsu); traveling northwest, passing through Chaohu Lake (now Caohu Lake), passing through Taibodu, passing through Meiting (now Meicun, Southeast Wuxi, Jiangsu), entering Yanghu Lake (now Furong Lake), leaving Yupu (now Jiangyin Ligang, Jiangsu), entering the Yangtze River and arriving at Guangling (now Yangzhou, Jiangsu). To the south of the Wudu, Haining opened the "Bai Chi Du" to the ancient Qiantang River to transport grain and fodder. The ancient Jiangnan River and Baishi Dushi were the predecessors of the Jiangnan Canal [6].

In order to improve the conditions of transportation, King Goujian of Yue made a treatment of the natural river and excavated the Shanyin waterway. The waterway started from Dongguomen in Shaoxing, passed through Liantang Village (now Dongguan Street, Shangyu District, Shaoxing, Zhejiang), passed through Cao'e River, Yongjiang River, and Yaojiang River, and entered the East China Sea, becoming the predecessor of the later Zhejiang East Canal [7].

In the ninth year of King Hui of Wei (361 BC), the state of Wei opened Dagou in the north of Yuanyang County, Henan Province, and diverted the Yellow River to flow south into Putianze (Guda Lake, located in the west of Zhongmou, Henan Province). It then diverted water from Putianze to the north of the capital city of Daliang (now Kaifeng, Henan Province), bypassing the city to the south, passing through Tongxu and Taikang, and injecting sand water. It then went south to Chen (now Huaiyang, Henan Province) and southeast into the Huaihe River tributary Yingshui, forming Honggou. The excavation of the Honggou connected several rivers between the Yellow River and the Huai River, including Jishui, Pushui, Sishui, Heshui, Suishui, Yingshui, and Rushui, forming a water transportation network in the Huanghuai Plain with artificial canals as the main stream and natural rivers as branches (Fig. 6.1). Honggou water transportation network not only promoted commercial exchanges, but also helped improve agricultural irrigation conditions along the way, becoming an important factor for the State of Wei's prosperity and strength, and entering the "Seven Heroes of Warring States". The formation of the Honggou River system is a summary of the initial period of the Pre Qin Canal, and also foreshadows the arrival of a new era.

Fig. 6.1 Plan sketch of Honggou water system

In addition, the world-famous Dujiangyan Irrigation Project was built in the last year of Emperor Zhaoxiang of the Qin Dynasty (about 276.251 BC). It is a large-scale comprehensive water conservancy project organized and built by Li Bing and his son, the prefect of Shu County, on the basis of the Jianpeng built by the legendary King Bieling of the ancient Shu State. It consists of Feng Shui Yu Zui, Fei Sha Yan, Bao Ping Kou and other parts. In addition to flood control and irrigation functions, Dujiangyan Irrigation Project also has comprehensive functions such as navigation aid. The construction of Dujiangyan Irrigation Project has laid a rich water source foundation for the development of Chengdu Plain into a land of abundance.

The canals of the pre Qin period mostly relied on the favorable conditions of natural rivers and lakes in the plain areas at that time, which were densely distributed and convenient for manual excavation and connection. They were often scattered in layout, lacking system, and had few water diversion and backwater facilities. The engineering was not yet perfect, and the shipping capacity was weak. However, they had a profound impact on the development of canals in later generations.

During the pre Qin period, river and canal dredging, maintenance and management were under the jurisdiction of water conservancy, and there were no dedicated departments or officials; the water transport of grain was only temporary and seasonal, and there are no dedicated departments or dedicated officials. Si Kong was the highest administrative officer in charge of soil and water engineering in the central government during this period. The Book of Shang Shu records that "Boyu was appointed as the Si Kong to manage soil and water". During the Western Zhou Dynasty, there were three main administrative officials in the central government, one of which was the "Si Gong", also known as the "Si Kong" [8]. According to "Kao Gong Ji", water conservancy engineering is an important part under the leadership of Si Kong [9].

The canals in the pre Qin period mainly maintained national unity and social stability through the function of their water transport of grain, creating favorable conditions for socio-economic development, and indirectly promoting the rise and development of commodity economy, social and cultural exchange, and regional development. Under the influence of the above effects, the canal facilitated the expansion of territory and laid the foundation for the establishment of a unified empire. The successive opening of many canals has achieved the connection of north–south waterways, further facilitating north–south exchanges and promoting the social and economic development of the north and south. During the Spring and Autumn period and the Warring States period, the excavation of numerous canals directly promoted the development of shipbuilding and ship repair industries, as well as indirectly promoted the development of industries such as iron smelting, textile, salt making, and painting. It also brought irrigation and flood discharge benefits to agriculture, promoting its development. The geographical masterpiece "Yu Gong", written during the Warring States period, was named "Da Yu Governed Flood". It records the local customs, local products, and tribute taxes, indicating that the scope, direction, and transportation capacity of land and water transportation had initially formed a scale and had already been recognized by people at that time [10].

Spiritual Heritage of Da Yu Governed Flood [4]

Da Yu Governed Flood is one of the famous myths and legends in ancient China, which is recorded in many documents. The Yuwang Temple (Fig. 6.2) located at the Yuhui Village site in the western suburbs of Bengbu City, Anhui Province, has a history of thousands of years. Numerous literati and eunuchs throughout history, such as Di Renjie, Liu Zongyuan, Su Shi, came here to visit and pay their respects, leaving behind a large number of poems and inscriptions.

Although Da Yu Governed Flood did not directly open canals, it was the first complete record in Chinese history of the great feat of Chinese people opening up water conservancy and fighting against floods, and played an important role in the development of Chinese civilization. The spirit of Da Yu Governed Flood is not only one of the origins of the Chinese national spirit, but also an important symbol of the Chinese national spirit. Liang Qichao said: the achievements of the Da Yu laid the foundation for the material unity of China; The virtues of Da Yu serve as the foundation for the spiritual unity of China. The valuable exploration and experience of Da Yu Governed Flood directly laid the foundation for the excavation of canals in later generations.

The spirit of Da Yu has the following points. Firstly, ethnic supremacy: The starting point of water management is to protect more tribes and their lives, farmland, property, mountains and forests, and actively develop irrigation and land reclamation to promote agricultural production. He was the promoter and leader of the early great integration of the Chinese nation. Secondly, selflessness and selflessness: He works

Fig. 6.2 Yuwang temple ruins

hard personally, holds tools in his hands, disregards the small family, "passes through the door three times but does not enter", and fights with the people through the wind and rain, fighting against the flood, reflecting the noble quality of giving up the small family and caring for everyone. Thirdly, innovation and exploration: carefully summarizing the experience and lessons of predecessors, studying the laws and methods of water control, creating and inventing measurement tools, and proposing dredging, flood discharge, and water control strategies for dredging and guiding rivers. Fourthly, honesty and simplicity: simple clothing and formal ceremonial attire, simple food and exquisite offerings, living in low rise houses and dedicated to water management.

Li Bing and his son inherited and carried forward the spirit of the Da Yu, comprehensively rectifying the water conservancy in Shu County (western Sichuan), and laying the foundation for the basic pattern of water conservancy in western Sichuan. For over 2000 years, their work has played a huge role in flood control and irrigation, making the Chengdu Plain a "land of abundance". They are loyal to their duties and work hard: repairing water conservancy and developing agriculture; Dredging waterways, building roads and bridges to facilitate transportation; Opening salt wells, developing industry and commerce, achieving brilliant political achievements. They respect nature and pursue science: mastering hydrogeology, adapting to local conditions, and proposing reasonable water management plans. They inherit and innovate, focusing on the overall situation: comprehensively considering flood control, irrigation, water transportation and other functions, and promoting policies from a larger range of water systems.

Qin and Han Dynasties: Canal in Its Early Stages of Prosperity

During the process of unifying China, the State of Qin reorganized and modified the canals left over from the pre Qin era, and excavated the Lingnan Canal to connect the Lingnan region. In the 37th year of Emperor Qin Shi Huang's reign (210 BC), Emperor Qin Shi Huang dispatched 3000 prisoners from Dantu (now Dantu District, Zhenjiang, Jiangsu) to Qu'a (now Danyang, Jiangsu), in accordance with the hilly terrain, to excavate a new river channel called Dantu Waterway, which connected to the ancient Jiangnan River opened by King Fuchai of Wu in the southeast and reached Kuaiji County (now Suzhou, Jiangsu), laying the basic direction of the Jiangnan Canal in the Sui Dynasty [4, 7, 11].

Lingqu is located in Xing'an County, Guilin, Guangxi. It was excavated in the 33rd year of Emperor Qin Shi Huang (214 BC) and has been in operation for over 2200 years. It connected the Xiangjiang River (the Yangtze River system) and the Lijiang River (the the Pearl River system). It was a water transport hub connecting Huguang in the north and Guangdong in the south. It also plays an irrigation role. It is a famous water conservancy project in ancient China, and one of the oldest canals in the world. It plays an important role in the development of Lingnan. From then on, the four major water systems of China, including the Yellow River, Huai River, Yangtze River, and Pearl River, were connected by canals. Ships in the Yellow River basin could sail directly to the Lingnan region by waterway, which was a major development of China's water transportation (Fig. 6.3).

The main canal engineering in the Han Dynasty included the excavation of a canal with a total length of over 300 li in the sixth year of the Yuanguang reign of the Western Han Dynasty (129 BC), which was the predecessor of the Guangtong Canal in the Sui and Tang dynasties, transporting over four million stones of grain annually [12]; In the 24th year of the Jianwu reign of the Eastern Han Dynasty (48 AD), the Yangqu Canal was excavated to solve the water supply in Luoyang and connect the Yellow River and Luoshui; In the 12th year of Yongping in the Eastern Han Dynasty (69 AD), Wang Jing was dispatched to dredge the Bian Canal, which lasted for over a thousand li, forming a backbone waterway for maintaining the water transport of grain between the Yellow River and Huaihe River [7]. The main canal engineering during the Cao Wei period included: in the 9th year of Jian'an (204 AD), Cao Cao dredged the Baigou, in the 18th year of Jian'an (213 AD), the Licao Canal was excavated [5], and in the 11th year of Jian'an (206 AD), the Pinglu Canal and Quanzhou Canal were excavated. At the turn of the Jin Dynasty, the Zhejiang East Canal was opened up, connecting multiple rivers such as the Qiantang River, Cao'e River, Yaojiang River, and Yongjiang River. The rivers intersected, with a large natural drop, and were connected by the construction of dams, gates, and gates [10].

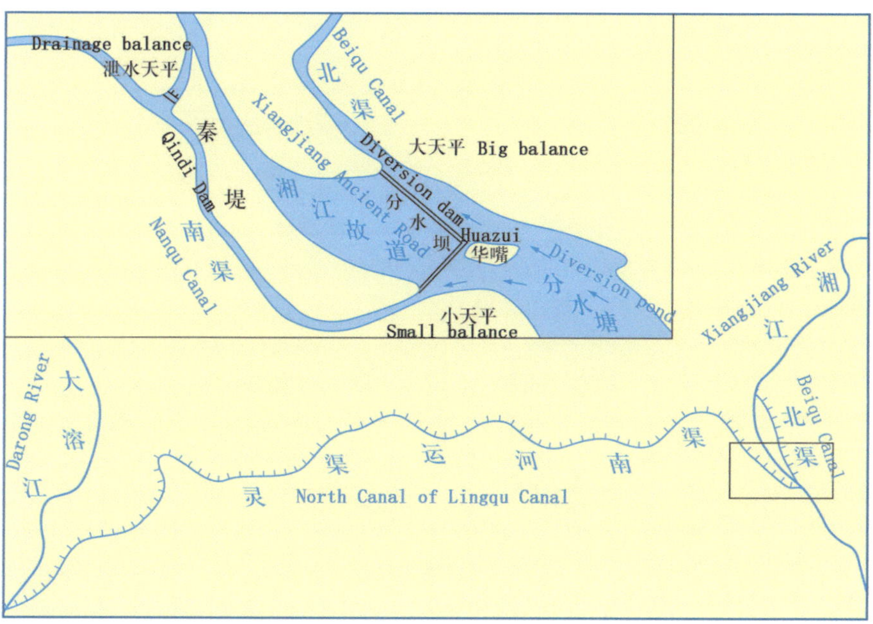

Fig. 6.3 Plan sketch of Lingqu canal

Yan Dai is a low dam built at a certain distance in the river channel to adapt to a certain water level difference and maintain the depth of the waterway. Although the Yan Dai ensured the water volume and depth of the channel, ships were unable to pass through. So people made the upstream and downstream of the weir into a relatively gentle slope, and the dam crest into a smooth circular arc shape, which was dragged by human or animal power. Later, it developed into using winches to pull ships up and down through the weir. It is actually the most primitive navigational building—the inclined ship lift [5]. The schematic diagram of ancient ships passing through weirs is shown in Fig. 6.4.

After the establishment of the Qin Dynasty, there was still a position of Si Kong in the ruling institutions, responsible for inspecting and supervising the maintenance of embankments and dredging of rivers and canals throughout the country. In addition, positions of Director and Assistant Director of Waterways Administration were established in local organizations of Jun County, responsible for specific implementation. The transportation organization was the responsibility of the local government. The Western Han Dynasty basically inherited the Qin system and made some improvements, while the Eastern Han Dynasty retreated. Overall, various political and economic systems and measures are still in the initial and exploratory stage. The water transport of grain system is only in its embryonic form, without customization,

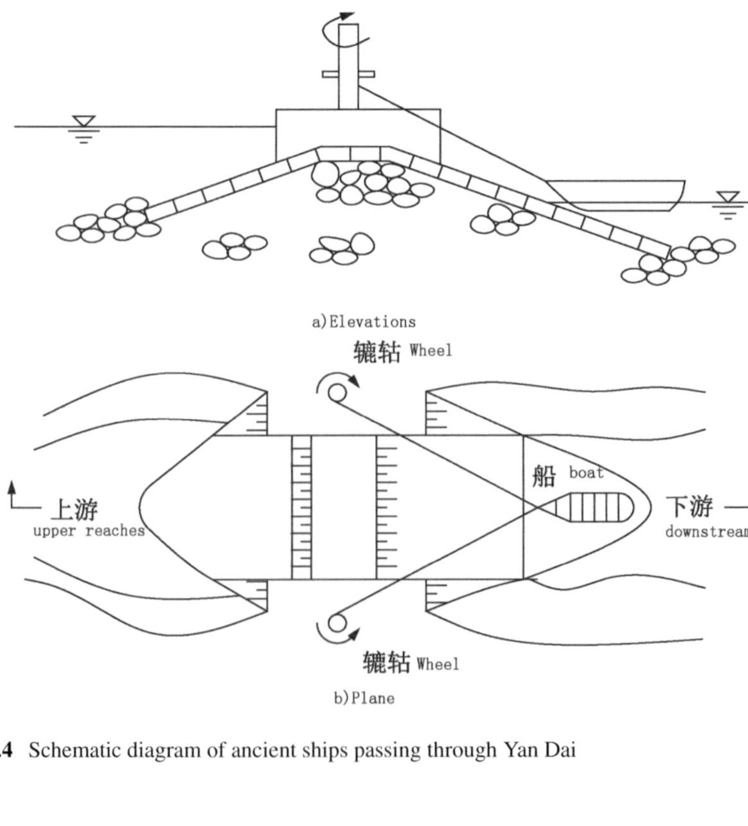

Fig. 6.4 Schematic diagram of ancient ships passing through Yan Dai

lacking unified organization and planning, without dedicated departments, and has not formed an independent economic system [5].

Role of Canals in the Formation of a Unified Empire [11, 13]

The canal transport system, as a product of centralized governance, is closely tied to it. During periods of strong, unified dynasties, canal transport flourished; however, during times of division and fragmentation, it fell into decline and disrepair.

Emperor Qin Shi Huang inaugurated the era of centralized rule, shifting the pillars of governance from hereditary nobles scattered across regions to a professional bureaucratic team loyal to imperial authority. To satisfy the material needs of this vast bureaucracy, the state levied taxes and established a salary system. It pursued policies of external fight against the barbarians and internal development, creating a strong material foundation. By drawing resources from localities, the state weakened regional power bases and ensured that the supreme ruler enjoyed all worldly riches. The most direct reason for the construction and maintenance of canals during this period was to support military campaigns aimed at unifying the country and consolidating power, by transporting soldiers, provisions, and other military supplies. The Han dynasty continued the Qin dynasty's canal transport system, which flourished despite various challenges. The canals of the Qin and Western Han periods connected economic and political centers, facilitating Qin's unification of Lingnan and subsequently reinforcing the north–south unity. They also played a crucial role in strengthening political, economic, and cultural ties between the north, south, east, and west. However, the canal transport system did not achieve significant development during the Wei, Jin, and Southern and Northern Dynasties. Among these, the Cao Wei regime placed more emphasis on dredging and constructing canals, which was strategically important for consolidating northern unity and establishing Ye City (now Linzhang, Hebei) as the capital [4].

The water transport system established during the Qin and Han periods provided essential conditions for the creation and development of centralized canal transport, offering great convenience and laying the foundation for future canal systems. During Emperor Wu of Han's reign, the amount of grain transported via canals increased from hundreds of thousands of dan to over a million dan, and eventually to 6 million dan. The advantages of large water transport capacity and low costs became evident, gradually making it the dominant mode of transport. According to the "Discourses on Salt and Iron: Basic Principles," land transport was "troublesome and costly, with goods often damaged or not worth the expense." [14]. The "Records of the Grand Historian: Biography of Zhang Yi" noted that water transport allowed "large ships loaded with grain to travel from Wenshan, floating down the Yangtze… covering over 300 li in a day, without the exertion of oxen and horses," [1] highlighting the superiority of water transport [5]. The construction of canals also promoted the development of handicrafts to some extent. First, it directly stimulated the shipbuilding industry. Second, the shipbuilding industry had high demands for iron, textiles, wood, and paint, thereby indirectly promoting these industries. During the Warring States and Qin-Han periods, the handicraft industry in the northern canal regions was highly developed, whereas the southern canal regions were relatively backward. By the Wei, Jin, and Southern and Northern Dynasties, the gap in handicraft development levels between the northern and southern canal regions had narrowed, with canals facilitating technological exchanges and communication.

Sui, Tang, and Song Dynasties: The Formation of a Canal Network

Mr. Bai Shouyi, a scholar who systematically studied the history of Chinese transportation, pointed out that ancient Chinese transportation entered a new era during the Sui and Tang dynasties, characterized by the large-scale excavation and use of canals [4]. During the Sui and Tang periods, canals were referred to as "caoqu" or "caohe" (grain transport canals) and "yunqu" (transport canals). It was only during the Song dynasty that they began to be called "canals" (yunhe). The Grand Canal of the Sui and Tang periods was composed of the Tongji Canal, the Shanyang Canal, the Meng Canal (Jiangnan Canal), and the Yongji Canal. These four canals had a total length of over 4,000 li and a width of more than 30 m (Fig. 6.5) [15, 16].

Tongji Canal. Constructed in the first year of Emperor Yang of Sui's Daye era (605 AD), it was over a thousand li long and 40 paces wide (approximately 55 m). During the Tang dynasty, it was called the Guangji Canal. According to the "Tang Geographic Records Map," it drew water from the Yellow River at Banzhu (near Niukouyu, northeast of Panshui Town in present-day Xingyang, Henan) and flowed southeast along the old course of the Bian River to Junyi (present-day Kaifeng, Henan). From there, it diverged from the old course, flowing southeast through Ningling (southeast of present-day Ningling, Henan), Songcheng (south of present-day Shangqiu, Henan), Yongcheng (north of present-day Yongcheng, Henan), Qixian (southeast of present-day Suzhou, Anhui), and Xiaqiu (present-day Sixian, Anhui), entering the Huai River across from present-day Xuyi. The Tongji Canal likely utilized the ancient Honggou and the old course of the Bian River.

Shanyang Canal: Constructed during the seventh year of Emperor Wen of Sui's Kaihuang era (587 AD), and later further dredged by Emperor Yang of Sui, it was over 300 li long and 40 paces wide (approximately 55 m). The starting point was moved from Mokou in Chuzhou (present-day Huai'an, Jiangsu) to Shanyang. From there, it flowed south through Baoying, Gaoyou, and Jiangdu, entering the Yangtze River via the Yangzi. The Shanyang Canal likely utilized the old course of the ancient Hangu Canal.

Meng Canal (Jiangnan Canal): Building upon the foundation of the ancient Jiangnan Canal, Emperor Yang of Sui straightened and deepened it in the sixth year of the Daye era (610 AD). Further renovations were undertaken during the eighth year of the Xianyun era (813 AD) in the Tang dynasty. It spanned over 800 li long and was

Fig. 6.5 Plan view of the grand canal during the sui and tang dynasties

over 10 zhang wide (approximately 40 m). Originating from Runzhou and Dantu, it diverted water from the Yangtze River southeastward, passing through Wujin, Wuji, Suzhou, Jiaxing, and Yanguan, before reaching Hangzhou in the southwest [4].

Yongji Canal: According to the "Annals of Emperor Yang" in Volume Three of the "Book of Sui," Emperor Yang of Sui ordered over a million men and women from various counties in northern China to excavate the Yongji Canal in the fourth year of the Daye era (608 AD) [17]. The canal diverted water from the Qin River, reaching south to the Yellow River and north to Zhuo County. It stretched over 2,000 li in length and was 170 feet wide (approximately 57 m). To the south, it connected to the Zhuozhou Canal through the Yellow River, while to the northeast, it passed through Weixian, Neihuang, Guantao, Linqing, and Dezhou before reaching Tianjin. It then turned northwest, branching into the Yongding River and extending directly to Zhuo County. The excavation of the Yongji Canal facilitated connectivity among the four major water systems of the Hai River, Yellow River, Huai River, and Yangtze River.

Apart from the Grand Canal, the Sui, Tang, and Song dynasties undertook the dredging of many rivers and canals, among which the most significant project was the restoration of the Han and Wei Canals. Emperor Wen of Sui dredged the Han and Wei Canals in the fourth year of the Kaihuang era (584 AD) and renamed it Guangtong Canal. It was later renamed Fumin Canal and Yongtong Canal. The Guangtong Canal diverted water from the Wei River northwest of Daxingcheng, followed the ancient canal route of the Han dynasty eastward, and reached the Yellow River at Tongguan, spanning over 300 li in length. Additionally, many granaries were established along the Tongji Canal, Yongji Canal, and Guangtong Canal for transportation and storage purposes. Among the notable ones were Liyang Granary in Liyang, Xingluo Granary (later renamed Luokou Granary) in Gongyi, Huiluo Granary, Hanjia Granary, and Heyang Granary in Luoyang, Yongfeng Granary and Guangtong Granary in Huayin, Taiyuan Granary and Changping Granary in Sanmenxia [15].

The series of canals excavated during the Sui and Tang dynasties stretched from the capital city of Chang'an in the west to Zhuo County in the north and to Yuhang in the south, spanning over 4000 li in total. These canals interconnected the five major water systems of the Hai River, Yellow River, Huai River, Yangtze River, and Qiantang River, linking the major cities and towns of Chang'an, Dongdu, Zhuo County (Youzhou), Xunyi (Bianzhou), Liangzhou (Songzhou), Shanyang (Chuzhou), Jiangdu (Yangzhou), Wu County (Suzhou), and Yuhang (Hangzhou). This network strengthened the connections between various regions.

During the Song dynasty, the Tongji Canal was referred to as the Bian River, which held a more significant position; the Shanyang Canal was known as the Huainan Canal, equivalent in importance to the Bian River; the Meng Canal was designated as the Zhejiang-West Canal, particularly crucial during the Southern Song period; and the Yongji Canal was designated as the Imperial Canal. These canals underwent continuous management during the Northern and Southern Song periods, forming the main north–south water transport routes. Besides these main water transport routes, other canals also played vital roles: the Huimin River referred to various canals southwest of Kaifeng in the Northern Song, including Min River and Cai River; the Guangji River (also known as the Wuzhang River) connected to the Bian River, starting northeast of Bianjing, passing through Dingtao, Jizhou, and Hecai Town (southwest of present-day Yuncheng, Shandong), and reaching Yunzhou along the Liangshan Reservoir, and traversing Qilu along the Ji River (North Qing River) to reach the sea; the Jinshui River was an aqueduct west of Kaifeng City during the Northern Song; and the Zhejiang-East Canal referred to several sections of canals between the Qiantang River and the Yao River [10]. The Bian River, Huimin River, Guangji River, and Jinshui River (some say the Yellow River) were collectively known as the Four Rivers of Grain Transport, forming a canal network centered around the capital Kaifeng, as depicted in Fig. 6.6. Despite the Southern Song being confined to a corner, due to the convenient access of boats and ships to both the Left River and Right Lake in Lin'an, canals continued to play an essential role. While maintaining and utilizing the Huainan and Zhejiang-West Canals, the Southern Song vigorously improved the Zhejiang-East Canal, which became the lifeline supporting the Southern Song regime.

The management system of canal transport during the Sui dynasty roughly followed that of the Qin and Han dynasties, but its level was relatively low. Canal transport management was only a duty of local officials or centrally appointed officials and had not yet formed an independent management institution. Starting from the Tang dynasty, the responsibilities of canal transport management gradually separated from local institutions and became an important part of the national economic sector, gradually forming a complete system of canal transport. The Tang dynasty can be regarded as a watershed in the ancient Chinese canal transport management.

During the Tang dynasty, the system inherited from the Sui dynasty was maintained, with the Ministry of Works and its affiliated Water Department(s) serving as the central institution responsible for implementing policies regarding waterways throughout the empire. The Tang dynasty transformed the Sui-established Capital Water Authority into the Capital Water Supervision Bureau, appointing two Water Commissioners. Underneath this bureau were two offices, the Boat and Oar Office and the River and Canal Office, which were responsible for the maintenance of canals, weirs, and reservoirs, as well as the management of water usage in the capital region

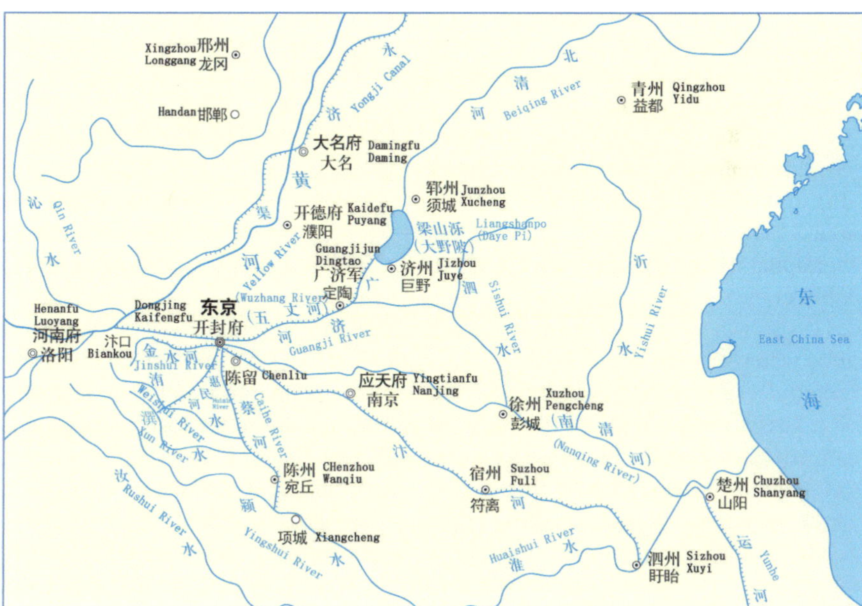

Fig. 6.6 Plan view of the four rivers of canal transport in Kaifeng during the northern song dynasty

and the regulation of boat traffic. The Capital Water Supervision Bureau operated independently of the six other government ministries. In the early Tang period, canal affairs were generally overseen by officials from the Ministry of Revenue and the Water Department under the Ministry of Works. Only during times of grain shortages in the capital or significant military campaigns would other officials be dispatched to assist or coordinate. As the scale and importance of canal transport expanded, Water and Land Transport Commissioners were established, often held concurrently by local commissioners or high-ranking officials [5].

During the Song dynasty, the canal transport management system inherited from the Sui and Tang dynasties saw significant development, particularly during the Northern Song period. It was during the Northern Song that an independent and comprehensive canal transport management institution was formally established. At the central government level, the Ministry of Works was established, with the Minister of Works overseeing various aspects of construction and infrastructure, including water management. The Ministry of Works had several subordinate departments, including the Water Department, which was responsible for matters related to rivers, canals, ferry crossings, boat traffic, and canal transport. The Three Offices of the Song dynasty were responsible for overseeing canal transport. Additionally, the central government established a specialized canal transport management institution called the Transport Department, which was primarily responsible for planning canal transport routes, allocating grain quotas for canal transport, selecting canal transport officials, supplying and distributing transport ships, and organizing bulk cargo transportation. In each region, there were Transport Offices and District Transport Offices, also known as Transport Departments, which were responsible for canal transport requisition and distribution. These offices were dispatched by the central government. Along the canal transport routes, there were also institutions such as the Transport Coordination Office and the Distribution Office, whose main responsibility was to supervise the progress of canal boats to the capital. Local officials also assisted in managing canal transport duties [5].

In terms of canal transportation technology, for canals with large-scale transportation and busy traffic, simple weirs and dams were no longer sufficient to meet the navigational needs. In order to facilitate the passage of heavily loaded fleets and to flexibly control water usage, the Tang dynasty developed the technologies of sluice gates (as shown in Fig. 6.7) and single-lock gates. In areas with steep gradients and limited water sources, the technologies of compound locks and multistage ship locks emerged. Despite these advancements, significant amounts of water were lost each time a vessel passed through the lock. To further reduce water loss, water-saving engineering, such as the pound lock, emerged during the Song dynasty. Pound lock, also known as the pound gate, involved the construction of a reservoir adjacent to the lock, where water discharged from the lock chamber was stored. When the water level in the lock chamber was insufficient, water from the reservoir was pumped back into the lock chamber using water wheels. The emergence of such ship locks reflected the advanced level of canal navigation technology during the Song dynasty.

Fig. 6.7 Sluice gate on a postage stamp

Strengthening the Unified Empire Through Canals [4, 6, 13, 18]

The Sui and Tang dynasties established the North–South Grand Canal centered around Luoyang, which for the first time connected the central capital of the Central Plains with the economically burgeoning regions of the Three Wu in Jiangnan and the northern regions, greatly strengthening political and military control over the northern and southern regions and exerting far-reaching influences on subsequent historical developments [4].

Despite the lack of explicit records regarding the volume of canal transport during the Sui dynasty, its immense scale and abundant granaries were renowned in history. According to the Tang scholar Du You, the granaries and reserves during the Sui dynasty stored "as much as tens of millions of dans, and at least not less than several million dans. " [19] The recently excavated transit granary sites from the Sui dynasty were also of considerable scale, indirectly reflecting the prosperity of canal transport during the Sui dynasty. During the Sui dynasty's expeditions to Liaodong, it is recorded that one of the military campaigns involved the transportation of up to 3.4 million people along with provisions and other military supplies, demonstrating the enormous transport capacity, with the Yongji Canal playing a crucial role. However, during the peasant uprisings towards the end of the Sui dynasty, canal transport was also interrupted.

Building upon the foundation laid by the Sui dynasty, the Tang dynasty established an unprecedented unified empire, which had a tremendous demand for canal transport, naturally leading to its high importance. In the early Tang period, when centralization was just established, expenditures were modest, and the demand from the East was still limited, with canal transport volume reaching only 200,000 dans per year. As the centralized political power expanded, and bureaucratic institutions continued to grow, during the reign of Emperor Xuanzong, the canal transport volume reached its peak, with Jianghuai rice shipments alone reaching an astonishing 7 million dans within three years. After the reigns of Emperor Suzong and Emperor Daizong, the demand for supplies from the southeast to the Guanzhong region became particularly strong. During the heyday of the Tang central government's rule, canal transport entered a splendid era, becoming the backbone of the central government. However, during the period of the Five Dynasties and Ten Kingdoms, when the country was in a state of turmoil and division, canal transport was on the verge of paralysis and even came to a halt.

During the Song dynasty, central authoritarian rule was unprecedentedly strengthened, with concerted efforts made to weaken local powers, ultimately creating a political situation where "military forces, financial resources, and territorial defense were all under the direct control of the ruler." [20] Economically, the policy of "abolishing the regional military governors and leaving only the provincial administrations, while consolidating grain, textiles, and currency in the capital" [32] was implemented as an iron law and consistently enforced for a long period. This greatly increased the demand for canal transport, making the Bian River the foundation of the state. Building upon the foundations of previous dynasties, the canals were reorganized and renovated, forming a radial canal network centered around the capital Kaifeng. This facilitated the rapid development of canal transport and led to an unprecedented period of prosperity and flourishing.

By the end of the Northern Song dynasty, with the rampant corruption in governance, the normal order of canal transport was severely disrupted, leading to a significant decline in the volume of canal transport. After the fall of the Northern Song dynasty, canal transport across the country fragmented and disintegrated. During the Southern Song period, canal transport in the capital Lin'an (present-day Hangzhou) was limited to the regions of Zhejiang and Jiangsu, with only over 1 million dans of grain transported annually, far below the level during the Northern Song dynasty. This once again demonstrates the direct correlation between the level of canal transport development and the degree of national unity, stability, and strength.

Canal Economy Reaches Its Zenith [3, 21]

During the Sui, Tang, and Song dynasties, China's economy and culture were at the forefront of the world, with advanced foreign exchanges. Particularly during the Tang dynasty, the open foreign policy led to unprecedented interactions with regions such as Korea, Japan, Southeast Asia, the Indian Peninsula, Central Asia, West Asia, Europe, and Africa. The Tang dynasty not only served as the center for economic and cultural exchange among neighboring countries but also made significant contributions to the development of world civilization.

During this period, the role of canals in maintaining national unity and social stability, and in creating favorable conditions for social and economic development, somewhat diminished. Instead, their direct functions in promoting the rise and development of a commodity economy and facilitating social and cultural exchanges became more prominent, with their influence further enhanced. The canals also played a crucial role in regional development and balanced regional growth. Additionally, they continued to promote the development of handicrafts such as shipbuilding.

During the Sui and Tang dynasties, the opening of the Grand Canal between the north and south primarily promoted trade between these regions. The Grand Canal became a transportation hub connecting the east, west, north, and south, gradually attracting a large number of merchants and goods to key water and land junctions, leading to increasing commercial prosperity. However, considering the situation during the Tang dynasty, there were relatively few direct commercial activities within the canal transport industry itself during this period.

The first documented instances of commercial activities related to canal transport appear during the Song dynasty. According to Volume 182 of the "Complete Collection of Illustrations and Writings from the Earliest to Current Times: Chapter on Food and Goods," which cites "Xishan Yuhua": "(During the Song dynasty) the canal transport from Jinghu North and South delivered rice to Zhenjiang and Yangzhou, and the boatmen would buy private salt to sell upon their return, making substantial profits." However, even in the Song dynasty, such occurrences were rare. As Emperor Taizong of Song remarked to his officials: "It is rare for those working on the boats, such as helmsmen and sailors, to engage in trade, but as long as it does not interfere with official duties, there is no need to investigate further" [22].

During the Sui, Tang, and Song dynasties, the canals directly promoted the development of the shipbuilding industry. After the opening of the Grand Canal between the north and south, the rise of canal transport and the development of waterborne transportation led to a new era in shipbuilding for both the government and the private sector. This era saw advancements in the scale, quantity, variety, and technology of shipbuilding, with the Song dynasty reaching a world-leading level in shipbuilding technology. Additionally, the shipbuilding industry stimulated the development of related industries such as iron, textiles, wood, and paint.

The canals promoted commerce and handicrafts, driving the economic development of regions along their banks. This was notably evident in the emergence of new commercially vibrant cities like Kaifeng, Lin'an, Yangzhou, Hangzhou, Suzhou, Huai'an, and Bianzhou [21]. Additionally, the canals significantly facilitated the southward shift of the economic center, a process that began during the Southern and Northern Dynasties.

During the Sui, Tang, and Song dynasties, due to the shifting political landscape in inland Asia, the Silk Road became difficult to traverse. Consequently, maritime transport increasingly became the main channel for political, economic, and cultural exchanges. The Grand Canal, serving as a vital link between the capital and its coastal regions, played an even more prominent role in the history of Sino-foreign exchanges.

The Flourishing of Canal Culture

During the Sui and Tang dynasties, the nation was powerful and economically prosperous, with not only clear and effective governance but also inclusive national policies promoting unity and equality among all ethnic groups. In terms of ideological and cultural policies, the state honored Confucianism, Buddhism, and Taoism equally, advocated for civil service examinations, encouraged literary and artistic endeavors, and facilitated cultural exchanges with foreign countries, creating fertile ground for the coexistence and flourishing of diverse cultures. During this period, there was a liberation of thought and a sense of confidence among the people. Literature and the arts flourished, reaching the pinnacle of traditional Chinese culture and leading the world: the literary scene, particularly poetry, was exceptionally brilliant; the arts, including calligraphy and painting, achieved remarkable success; and the era also saw the creation of the magnificent Buddhist art treasure trove, the Dunhuang Mogao Caves [23].

During the Tang Dynasty, poetry and poets were as numerous as stars, reaching historical heights. Along the Grand Canal, especially in canal cities, there was economic prosperity and cultural flourishing, coupled with numerous scenic spots and historical sites, attracting a large number of literati and scholars. Many Tang poets traveled along the Grand Canal to admire the splendid landscapes of their homeland, and some even served as officials and lived in canal cities. The positive interaction between the Grand Canal and poets gave rise to a plethora of poems with distinct canal characteristics, which have been passed down through the ages.

During the Song Dynasty, outstanding painters further developed the art of painting [24]. Zhang Zeduan's "Along the River During the Qingming Festival" (Fig. 6.8) represents the pinnacle of figure painting in the Song Dynasty. This painting vividly depicts the urban landscape of the capital city of Dongjing (present day Kaifeng, Henan Province) in the twelfth century Northern Song Dynasty, as well as the living conditions of people from all walks of life at that time. It serves as a testament to the prosperity of the capital and reflects the splendid epic of urban society along the Grand Canal. It holds significant value for studying the commerce, handicrafts, folk customs, architecture, and transportation along the Grand Canal.

Fig. 6.8 "Along the river during the Qingming festival" (partial)

Of particular note is the depiction of the scenes along the Bian River in the painting. The Rainbow Bridge over the Bian River serves as the focal point of this section and the entire painting. Here, there are over a hundred figures, densely packed, lively, and dynamic. Boats come and go on the river, docks are bustling with loading and unloading activities, and both banks are lined with hotels, teahouses, shops, and stalls. It is evident that the Bian River is not only the lifeline of the state and the capital but also an important foundation for people's work and life.

Born in the prosperous Tang Dynasty city of Yangzhou, Master Jianzhen (688–763 AD) was not only a typical representative of Tang culture but also embodied the characteristics of canal culture [25, 26]. At that time, Yangzhou stood as a leading political, economic, cultural, transportation, and trade center south of the Yellow River, flourishing economically and culturally with deep international roots, nurturing Jianzhen's open and inclusive personality. Jianzhen's journey spanned 12 years, with six attempts and five failures before successfully sailing east to propagate Buddhism, administer precepts, and spread Buddhist teachings. Additionally, Jianzhen introduced Tang Dynasty architecture, sculpture, Chinese language, literature, calligraphy, and medicine to Japan, profoundly influencing Japanese culture.

Yuan, Ming, and Qing Dynasties: From Prosperity to Decline

After the Yuan Dynasty conquered the Southern Song Dynasty, they continued to rely on the old Grand Canal for water and land transportation. However, this ancient canal, which had existed since the Sui and Tang Dynasties, had become increasingly impassable over time due to various changes. With the establishment of the Yuan capital in Dadu (present-day Beijing), the need for a water transportation route connecting the northern political center with the southern economic center became particularly urgent. Therefore, the authorities began to gradually repair the Grand Canal, focusing mainly on the Tonghui River, Huitong River, and Jizhou River north of Jinan. These repairs were carried out during the reign of Emperor Shizu of Yuan from 1271 to 1294. As a result, the Grand Canal was shortened by over 1800 miles. With these renovations, the overall layout of the Grand Canal was finalized (see Fig. 6.9) [5, 22, 27].

The Jizhou River was constructed in the 19th year of the Zhiyuan era (1282 AD) and completed the following year. People excavated a canal of over 150 li from Jizhou (present-day Jining, Shandong) to Xucheng Anshan (present-day southwestern Dongping, Shandong), and built dams to divert the Wen River and Si River into the Grand Canal. Subsequently, several locks and dams were constructed to further regulate the water flow of the canal. After the completion of the Huitong River, the Jizhou River was also commonly referred to as the Huitong River.

The Huitong River was initiated in the 26th year of the Zhiyuan era (1289 AD) and completed in the same year. It originated from the south of Xucheng Anshan (present-day Dongping, Shandong), where it connected with the Jizhou River in the west, and diverted water from the Liangshan Reservoir to flow northward. It ran from Shouzhang in the northwest to Dongchang (present-day Liaocheng, Shandong), then northwest to Linqing, where it joined the Wei River, spanning a total length of over 250 li. The Huitong River had a higher terrain along its riverbed, with multiple locks installed along the route to control the water flow. Additionally, pivotal points were established to divert water from the Wen River and Si River for transportation purposes.

The Tonghui River was initiated in the 29th year of the Zhiyuan era (1292 AD) and completed the following year. It originated from Baifu Village, drawing water from the Shenshan Spring (now the site of Baifuquan Park in Changping District, Beijing). It then turned westward and southward, roughly following the route of the present-day Jingmi Diversion Canal. Along its course, it collected water from various sources such as Shuangta, Yuhe, Yimu, and Yuquan, before flowing into the Wengshan Reservoir (now Kunming Lake). From there, it followed the Chang River (now Gaoliang River) to Dadu and Yimen (now Xizhimen) before entering the city. To the south, it converged at Jishuitan (now Shichahai), and to the southeast, it exited from Wenming Gate (now Chongwenmen). Continuing eastward, it reached Goryeo Village in Tongzhou before joining the Bai River, with a total length of over 164 li.

After the Yongle Emperor of the Ming Dynasty relocated the capital to Beijing, he also heavily relied on this north–south transportation artery. Therefore, a policy of

Fig. 6.9 Overall layout of the grand canal during the Yuan dynasty

"actively dredging and ensuring canal transport" was adopted for canal construction. In order to prevent siltation caused by Yellow River floods and ensure smooth canal transport, some changes were made to the old waterways.

The dredging of the Huitong River began in the ninth year of the Yongle reign (1411), extending from the north of Jinan to the Linqing section where it meets the Wei River, with a total length of 385 li (about 192.5 km). Along the way, 15 ship locks were added, with the most significant highlight being the construction of the Nanwang water diversion project. From the sixth year of the Jiajing reign (1527) to the thirty-second year of the Wanli reign (1604), for nearly 80 years, the Ming Dynasty successively opened the Nanyang River and the Jiahe River, comprehensively diverting the canal between Jinan and Xuzhou to the south.

During the Qing Dynasty, there were essentially no significant initiatives in the construction of the Grand Canal, with efforts mainly focused on ensuring its smooth operation. Apart from ongoing efforts to control the Yellow River, the primary actions involved repairing and rebuilding locks and dams, dredging waterways, reinforcing embankments, and manually regulating water levels.

The "Preface to the Canal and Waterways" in the Yuan Dynasty's "History of the Yuan Dynasty" records: "During the Yuan Dynasty, an internal institution, the Capital Water Management Bureau, was established, while various external river and canal offices were set up to promote water conservancy and prioritize the maintenance of river embankments." [28]. The Capital Water Management Bureau of the Yuan Dynasty, "with a third-rank official rank, was responsible for managing rivers and canals, as well as embankments, bridges, and water conservancy facilities." [29]. It had subordinate offices for river patrols and river defense supervision. In contrast, the Ming Dynasty's water management agency was the Capital Water Bureau, subordinate to the Ministry of Public Works, with broader responsibilities compared to the Yuan Dynasty's Capital Water Management Bureau. According to the "Treatise on Officials" in the "History of the Ming Dynasty": "The Capital Water Bureau was responsible for managing rivers, marshes, reservoirs, bridges, boats, textiles, deeds, and weights and measures." [28]. This indicates that, in addition to rivers, canals, embankments, and bridges, the Ming Capital Water Bureau's responsibilities expanded to include boat and cart manufacturing, as well as textiles, deeds, and weights and measures. In the Qing Dynasty, water officials included the Capital Water Bureau (similar to the Ming Dynasty, under the Ministry of Public Works), as well as the Grand Minister of Direct River Channels and Ditches and the Imperial Censor responsible for river channels. Additionally, there was an overseer for river channels responsible for the management of waterways. The Capital Water Bureau managed both public and private water affairs related to rivers, canals, boats, and navigation routes, with broader responsibilities than the Ming Dynasty's Capital Water Bureau. The Grand Minister of Direct River Channels and Ditches supervised the management of the five city rivers and ditches in the capital, while the Imperial Censor oversaw road ditches, and the overseer of river channels was in charge of river and canal management [4].

Canal Strengthening the Unified Empire [16, 18, 21]

Compared to the Sui, Tang, and Song dynasties, the Yuan, Ming, and Qing dynasties fell far behind in terms of the excavation and maintenance capabilities of the canal, as well as its functions and shipping capacity. The significant differences may be attributed to changes in hydrology and meteorology in the northern river sections [4].

The Yuan Dynasty inherited a centralized political system and placed importance on canal transportation. However, the geographical conditions in North China, with its higher terrain and insufficient water sources, coupled with the siltation of the

Yellow River leading to multiple blockages, resulted in limited navigable waterways during the early Yuan period. Only portions of the Grand Canal in Jiangnan and Huainan, along with sections of the Imperial Canal, were navigable. To address these challenges, the Yuan Dynasty employed various measures such as excavating new canal sections, straightening existing routes, and dredging waterways. These efforts aimed to connect the north and south via the shortest possible route, linking the political center in the north with the economic hub in the south. With the unification of the country under the Yuan Dynasty, China did not face prolonged periods of fragmentation, and the Grand Canal played a crucial role in this achievement.

The term "cao yun" refers to the inland water transportation system that conveyed official grain, military supplies, and other essential bulk goods from the south to the north through the Beijing-Hangzhou Grand Canal to the central ruling capital. It encompasses the development of canals, shipbuilding, and the collection of official and military grains. In the early Ming Dynasty, a system similar to the Han Dynasty's large fiefdoms was adopted, and the centralized system of governance was not fully implemented. Moreover, with the capital located in Nanjing (formerly known as Yingtianfu), the demand for canal transportation was limited, and there was also reliance on maritime transportation. These factors significantly impacted the traditional canal transportation centered around the capital.

After Emperor Chengzu of the Ming Dynasty ascended the throne, he vigorously implemented the policy of reducing the power of feudal lords to strengthen centralized rule. Simultaneously, the relocation of the capital to Beijing brought about fundamental changes in the location of grain requisition and its transportation routes. This shift weakened the role of maritime transportation, previously dominant alongside river transport. In the 15th year of the Yongle reign (1417), during the tenure of the Chief Grain Transport Officer Chen Xuan, the construction of the Qingjiangpu River channel project was initiated to ensure the adequate water supply of the Huai River, thereby completing the entire Grand Canal. During the Yongle era (1403–1424), grain transport to the capital increased rapidly from 2 to 4 million dan (unit of grain measurement), establishing this quantity as the standard for canal transport. In the second year of the Jingtai reign (1451), the central government began appointing Chief Grain Transport Officers stationed in Huai'an, responsible for managing nationwide grain transport affairs and ensuring the smooth operation of the Grand Canal. These officers had considerable authority and their own armed forces and naval units. They oversaw the collection of grain along the coast, the transport of grain-laden boats northward, inspection and dispatching, and ensuring the security of the transport process. To ensure the accuracy of grain transport, the central government also deployed inspectors along the canal route in Huai'an, Jining, Tianjin, and Tongzhou. However, after the Wanli period (1573–1620), as the Ming Dynasty began to decline, coupled with frequent flooding of the Yellow River and Huai River, the Grand Canal was severely affected, leading to a significant reduction in grain transport. As peasant uprisings erupted across the country towards the end of the Ming Dynasty, canal transport gradually declined.

At the beginning of its rule, the Qing Dynasty placed significant emphasis on grain transport (water transport of grain) and swiftly restored the grain transport system to

Beijing. In the second year of the Shunzhi reign (1645), the Qing government adopted a system modeled after the Ming Dynasty and set a fixed quota of 4 million dan, maintaining this standard until the mid-nineteenth century. During the Qing Dynasty, especially in its early period, grain transport fully met the needs of Qing officials and the imperial court, ensured military grain supplies, and played a significant role in balancing regional harvests and providing disaster relief. This contributed significantly to national unity and social stability, leading to the prosperity seen during the Kangxi and Qianlong eras, often referred to as the "Kang-Qian Golden Age." The Qing government, following the Ming system, appointed a River Transport Governor and a Grain Transport Governor, each based in Huai'an, as the highest administrative officials responsible for grain transport and river management. They oversaw grain transport administration in eight provinces: Shandong, Henan, Jiangsu, Anhui, Jiangxi, Zhejiang, Hubei, and Hunan. Their duties included managing grain transport, inspecting transport officers, supervising shipbuilding, inspecting empty returns, and urging grain debt collection (see Fig. 6.10). Both the River Transport Governor and the Grain Transport Governor held the official rank of zheng erpin (second rank), and those concurrently holding titles such as Minister of War or Vice Censor-in-Chief of the Censorate were considered congyipin (first rank), equating their status with provincial governors. Grain transport revenue accounted for over two-thirds of the national fiscal revenue, supporting half of the Qing Dynasty's national income. However, in the late Qing period, the central government's authority weakened due to the Opium Wars, the Taiping Rebellion, and the increasing power of local governors, leading to the decline and eventual collapse of the grain transport system.

Fig. 6.10 Former site of the office of the grain transport commissioner

Rise and Fall of the Canal Economy [5, 22, 27]

During this period, the canal system continued to play a crucial role in maintaining national unity and social stability primarily through its transportation functions, creating favorable conditions for socioeconomic development. However, its influence diminished somewhat in this regard. Its direct functions shifted towards further promoting economic development and social and cultural exchange along the canal regions, enhancing its role in these aspects. Additionally, the canal system continued to stimulate the development of industries such as shipbuilding.

The Grand Canal during the Yuan, Ming, and Qing dynasties, along with the regions along the Yangtze River, were among the wealthiest, most abundant in resources, and concentrated in goods. The canal primarily provided convenient transportation conditions, resolving issues of external commodity circulation for these regions and towns. Merchants, artisans, farmers, and those involved in canal transport all benefited, greatly promoting the prosperity and development of the commercial economy.

In the Ming Dynasty, economic development led to a widespread shift in values towards pursuing profit. Influenced by this notion, canal transport personnel became extensively involved in commercial activities. In the fourth year of the Jingtai reign (1453), an official responsible for grain transportation in Ruizhou Prefecture, Jiangxi Province, reported: "Many military ships carry private goods and engage in buying and selling at markets and docks. They stay for prolonged periods, causing civilian ships to be stranded." In fact, the Ming government allowed canal transport personnel to carry a certain amount of tax-exempt goods. During the Jiajing reign, the government had to increase the permitted quantity of local products carried by canal transport personnel to 40 dan, and in the sixth year of the Wanli reign (1578), it was further increased to 60 dan.

The Qing government formulated several policies in the canal transport sector favorable to the circulation of goods, adapting to changes in societal conditions. These policies objectively facilitated the development of the commodity economy, leading to a growing demand for transportation services. Consequently, restrictions on the weight and variety of goods carried by canal boats, apart from grain shipments, were gradually relaxed.

During the Yuan, Ming, and Qing dynasties, the Grand Canal significantly stimulated the shipbuilding industry, especially the development of vessels suitable for both river and sea navigation. The number and variety of ships increased substantially. In the Yuan Dynasty, Jinan served as a crucial center for ship management and construction. By the Ming Dynasty, with the smooth operation of the Beijing-Hangzhou Grand Canal, water transport thrived, with an estimated 8,000–10,000 vessels navigating the canal. To meet the demands of canal transport, shipbuilding industries emerged, with numerous private workshops and large-scale government shipyards such as Longjiang (located in present-day Nanjing), Qingjiang (in present-day Huai'an), and Weihe (in present-day Linqing, Shandong). In the Qing Dynasty, while river transport declined and the number of canal ships decreased, the rapid increase in civilian vessels ensured that the total ship count on the canal remained much higher than during the Ming Dynasty. During this period, the canal also facilitated the development of commerce and handicraft industries, further driving economic growth in the canal-adjacent regions. Particularly after Beijing became the capital, the functionality of the Grand Canal in northern Jiangsu became prominent, leading to the prosperity of various industrial and commercial cities such as Xuzhou, Jinan, Liaocheng, Linqing, Dezhou, Tianjin, and Beijing, thereby fostering the development of numerous trading towns.

Maturation of Canal Culture

The Grand Canal has nurtured many aspects of traditional Chinese culture, among which the quintessential Chinese opera, Peking Opera, is the most famous. It can be said that without the Grand Canal, there would be no Peking Opera [30].

In the 45th year of the Qianlong reign (1780), in January, Emperor Qianlong embarked on a southern tour along the Grand Canal, summoning various local opera troupes to gather in Yangzhou for a grand performance during his visit. Troupes specializing in different opera genres, such as Nan Kun, Bei Gou, Dong Liu, and Xi Bang, converged to entertain the Emperor with Kunqu, Gaoqiang, Bangzi, Huangmei, and Liuqiang operas. After Qianlong left Yangzhou, over a dozen opera troupes from across the country did not immediately depart but continued to perform for the local residents. Eventually, they decided to travel north to Beijing along the Grand Canal, performing as they journeyed. In the spring of the 47th year of Qianlong's reign (1782), a diverse ensemble representing various regional opera styles arrived in the capital city. They performed in Beijing's theaters, engaging in mutual exchange and fusion, laying the groundwork for the birth of Peking Opera. By the 55th year of Qianlong's reign (1790), the "Four Great Huizhou Troupes" from Yangzhou—Three Qing, Four Xi, Chun Tai, and San He—returned to Beijing via the Grand Canal to celebrate Qianlong's 80th birthday. Over several decades of development, the Huizhou Troupes gradually transitioned from a diverse array of regional opera styles to a harmonious unity, primarily centered on Pihuang opera, incorporating elements of Kunqu, Chuiguan, and Bozi opera, thus giving birth to Peking Opera (Fig. 6.11).

The canal cities also gave rise to the "Eight Eccentrics of Yangzhou" [24]. The "Eight Eccentrics of Yangzhou" [24], also known as the Yangzhou School of Painting, were a group of artists associated with the city of Yangzhou during the Qing dynasty. While historical records mention eight individuals, some argue that they functioned more as a collective rather than a precise number of members. The Eight Eccentrics were known for their mastery of flower-and-bird painting, although some excelled in figure or landscape painting. Renowned across the country, they made significant contributions to the history of Chinese painting. What set them apart was their unconventional approach to art, drawing inspiration from ordinary life and emphasizing individual expression. Their remarkable talent and unconventional methods

Fig. 6.11 Performance of Peking opera

led people to dub them as "eccentrics." While not all members of the group were natives of Yangzhou, they were primarily active in the region, and their emergence was closely tied to the unique historical and cultural landscape of Yangzhou, as well as the vibrant cultural atmosphere of the canal city.

The Grand Canal greatly facilitated the formation and development of canal tourism culture. On one hand, the canal provided convenient water transportation, reducing travel difficulties and increasing comfort. On the other hand, after thousands of years of development, the canal left behind rich cultural relics, legends, and stories, which became valuable tourism resources. These two aspects complemented each other, resulting in mutual enhancement. In ancient times, those who toured the Grand Canal were primarily emperors, generals, and literati. Among them, the southern tours of emperors left a number of traces along the Grand Canal in China, such as the Tianning Temple in Yangzhou, the Dragon King Temple Palace in Suqian, as well as numerous imperial docks and roads. These are all important resources for the development of the modern tourism industry along the Grand Canal in China [30].

During the Ming and Qing dynasties, urban commodity economy flourished, and tourism along the Grand Canal became popular. The Jiangnan Canal region became a paradise on earth, with many tourist destinations emerging in cities such as Hangzhou, Yangzhou, Suzhou, and Wuxi. With the development of urban commercial economy, early representatives of "mass tourism" began to appear among tourists, and famous attractions such as Shouxihu Lake, Tiger Hill, Huishan Mountain, Jinshan Mountain,

Fig. 6.12 Drawing of prosperity of Suzhou

Pingshantang, and West Lake attracted visitors all year round. The painting "Prosperous Suzhou" by the Qing dynasty court painter Xu Yang (shown in Fig. 6.12) vividly depicts the prosperity of that time [30].

The canal also served as an important conduit for cultural exchange between China and foreign countries. In the eighth year of the Yuan Dynasty's Zhiyuan era (1271), the Italian traveler Marco Polo visited China and traveled along the canal, exploring cities such as Dadu (present-day Beijing), Linqing, Jinan, Dongping, Huai'an, Nanjing, Zhenjiang, Yangzhou, Suzhou, and Hangzhou. His book "The Travels of Marco Polo" extensively documented the transportation along the canal, the towns along its banks, local customs and traditions, and the economic conditions. In the first year of the Yuan Dynasty's Yanhu era (1314), another Italian, Odoric of Pordenone, traveled from Hangzhou to Jinling (present-day Nanjing) and Yangzhou, and then proceeded northward along the canal through places like Jinan and Linqing before reaching Dadu. In his work "The Travels of Friar Odoric," he described the customs, geography, and economy of these cities. These two travelogues left behind rich and valuable historical materials, playing a prominent role in cultural exchanges between the East and the West [21].

In addition, the canal transportation industry of the Qing Dynasty also developed a unique culture of industry deity worship: the industry deity of canal transportation was the Four Great Kings of Gold Dragons. Since the Ming and Qing Dynasties, waterborne workers from the Jianghuai region to northern China, especially boatmen and canal transportation practitioners on the Grand Canal and Yellow River, widely worshiped them, earning them the title of "Northern Waterway Deities." Boatmen's work chants were also an important expression of Qing Dynasty canal transportation culture. They were folk songs naturally created by boatmen on the canal to synchronize their movements, increase labor interest, and improve labor efficiency, embodying the history of the Grand Canal and the hard work of boatmen [5].

Typical Canal Engineering Techniques

The hydraulic engineering achievements of the Yuan Dynasty were at the forefront globally during that period. In terms of the layout of ship locks, due to significant changes in terrain, important ports and harbors were interconnected by 2–3 ship locks, forming an organic combination. For instance, both Linqing and Jining consisted of three locks, while five locations including Qiji, Acheng, Jingmen, Jingou, and Gutou had two locks in combination. The distance between the combined locks was generally 2–3 li. This effectively turned the river section between the two locks into a large lock chamber. By utilizing the alternating opening and closing of the gates at both ends, the water level of the canal was maintained, allowing for the smooth passage of more boats while conserving water. This achieved a cascading navigation system, solving the navigation difficulties caused by significant terrain changes. Additionally, the Yue River was an artificially divided waterway constructed during river management efforts. This technique was used to divert the main stream's water flow to protect embankments while allowing boats to pass through. Due to its crescent shape, it was named the Yue River.

In the Ming Dynasty, there was a saying, "The Grand Canal stretches three thousand miles, with Nanwang as its highest point." The Grand Canal was connected north and south by the Nanwang Water Diversion Project. Located in Nanwang Town, Wenshang County, Jining City, Shandong Province, this project has a history of 600 years. The Nanwang Water Diversion Project is a significant achievement in the transportation management of the Jining Canal area. It is the most technologically advanced project along the entire Beijing-Hangzhou Grand Canal, an important component of the World Cultural Heritage of the Grand Canal, and a great achievement in the history of the Beijing-Hangzhou Grand Canal in China.

The Nanwang Water Diversion Hub was a project initiated by Song Li, the Minister of Works, in the ninth year of the Yongle reign (1411). It was based on the recommendation of Bai Ying, an elderly resident of Wenshang County. The project began by constructing the Daicun Dam in Wenshang to divert the Wenshui River. Next, a channel was excavated to divert water from the Wenshui River to the Nanwang Water Diversion Point. Then, spring water was collected and directed to supplement the source by gathering and channeling various mountain springs from the northeast of Wenshang County into the Quan River. Finally, a stone-paved slope was built at the "T"-shaped water mouth where the Xiaowen River enters the canal, with a diversion mouth (fish mouth) constructed to separate the water flow north and south (Fig. 6.13). The Nanwang Water Diversion Hub dredged the Three Lakes to serve as a water hub, constructed sluices and dams to regulate water flow, ensuring smooth navigation, comparable to the Dujiangyan Irrigation System. The location of the dam site was chosen wisely, as Daicun was an ideal high point, conforming to the natural law of water flowing downhill. The diversion and storage were well-planned, making it convenient to use while effectively managing water resources.

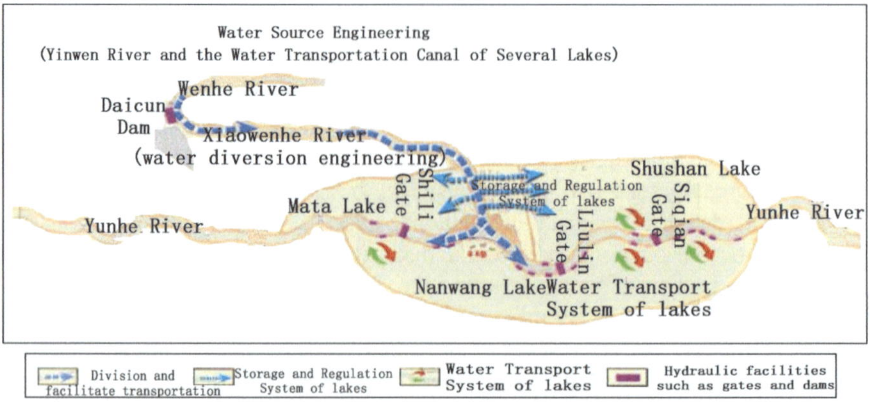

Fig. 6.13 Schematic diagram of the Nanwang water diversion hub

Grand Canal's Glorious Revival in Modern China

Since the founding of the People's Republic of China, especially with the reform and opening up, along with the rapid development of the economy and society, China's inland river transportation construction and development have achieved tremendous success, with the Yangtze River, Pearl River, Huai River, and other water systems as the mainstay. By the end of 2020, the total navigable mileage of inland river waterways reached 127,000 km, with 16,100 km of high-grade waterways. The volume of inland waterway freight transportation and cargo turnover reached 3.815 billion tons and 1.6 trillion ton-kilometers respectively. The infrastructure and transportation scale of river and port transportation rank among the top in the world, and have become an important part of the comprehensive transportation system, playing a crucial role in promoting land development, guiding optimized industrial layout, serving external opening-up, and promoting the comprehensive utilization of water resources. Within the national inland river transportation system, the Grand Canal still serves as the only waterway that spans north and south transportation, playing an important role in shipping. The volume of cargo transportation on the Grand Canal south of Jinin exceeded 9 billion tons, with the freight volume in the Jiangsu section alone reaching 530 million tons, accounting for about 20% of the total comprehensive transportation volume in the province, equivalent to the transportation capacity of eight Beijing-Shanghai expressways. It can be seen that the Grand Canal still plays a major role in modern transportation. Figure 6.14 shows the present appearance of the Beijing-Hangzhou Grand Canal.

In recent years, the Central Committee of the Communist Party of China and the State Council have attached great importance to the protection, inheritance, and utilization of the Grand Canal culture. In June 2014, the Grand Canal was successfully inscribed as a World Cultural Heritage Site, marking the beginning of a new chapter in its protection. The World Heritage Site of the Grand Canal in China includes 27

Fig. 6.14 The present appearance of the Beijing-Hangzhou grand canal

sections of the canal, 56 heritage sites, and 4 related remains. For specific details, please refer to Table 6.1.

In August 2021, the Leading Group for the Construction of National Cultural Parks issued the "Construction and Protection Plan for the Grand Canal National Cultural Park," which integrated cultural and cultural resources along the Grand Canal in eight provinces (direct-administered municipalities). Following the concept of "the canal as the line, the city as the pearl, stringing pearls along the line, and lines forming surfaces," it optimized the overall functional layout, deeply interpreted the cultural value of the Grand Canal, vigorously promoted the spirit of the Grand Canal era, increased control and protection efforts, strengthened thematic exhibition functions, promoted the integration of culture and tourism, enhanced traditional utilization levels, advanced key project implementation, and aimed to develop the Grand Canal National Cultural Park into a bright name card that showcases China's image, demonstrates Chinese civilization, and highlights cultural confidence in the new era.

In order to protect and utilize the Grand Canal effectively, in April 2022, the Ministry of Water Resources launched the "2022 Full-line Water Replenishment Action Plan" for the Beijing-Hangzhou Grand Canal. On April 28 of the same year, with the opening of the regulating gates in Shandong and Tianjin, water flow was restored along the entire length of the Beijing-Hangzhou Grand Canal for the first time in a century. However, at the same time, China's inland waterway development faces some challenges: the layout of waterways is still incomplete, with a relatively low proportion of waterways suitable for thousand-tonnage vessels, and the scope of accessibility needs to be expanded. The quality of inland waterway transportation services needs to be further improved. The maritime-river transportation in

Table 6.1 World heritage projects of the grand canal in China

Location	Canal	Heritage sites	Related relics
Beijing	2: Tonghui River, Beijing Old City Section; Tonghui River, Tongzhou Section	4: Yuhe Ancient Canal, Chengqingshang Lock (Wanning Bridge), Chengqingxia Lock (Dongbuya Bridge), Shichahai	–
Tianjin	1: Tianjin Sanchakou (confluence of the North and South Canals and the Hai River)	–	–
Heibei	1: South Canal, from Cangzhou to Dezhou	2: Xiejia Dam in Dongguang Lianzhen, Cangzhou, and Huajiakou Rammed Earth Weir in Jingxian, Hengshui	–
Shandong	6: South Yunhe River, Dezhou Section; Huitong River, Linqing Section; Huitong River, Yanggu Section; Nanwang Water Conservancy Hub; Huitong River, Weishan Section; Middle Yunhe River, Taierzhuang Section	14: Linqing Canal Chaoguan; Yanggu Ancient Lock Group (Jingmen Upper Lock, Jingmen Lower Lock, Acheng Upper Lock, Acheng Lower Lock); Daicun Dam in Dongping; Xingtong Doumen Site in Wenshang; Xujiankou Doumen Site in Wenshang; Shili Lock in Wenshang; Liulin Lock in Wenshang; Siqianpu Lock in Wenshang; Dragon King Temple Site in Nanwang; Brick-lined Canal Embankment in Wenshang; Lijian Lock in Weishan County	–
Henan	4: Tongji Canal, Zhengzhou Section; Tongji Canal, Nanguan Section in Shangqiu; Tongji Canal, Xiayi Section in Shangqiu; Yongji Canal (Wei Canal), Jinxian and Xunxian Sections	–	3: Huanjia Warehouse Site in Luoyang; Huiluo Warehouse Site in Luoyang; Liyang Warehouse Site in Junxian County
Anhui	1: Tongji Canal, Sixian Section	1: Liuzi Bridge Site	1: Liuzi Canal Site

(continued)

Table 6.1 (continued)

Location	Canal	Heritage sites	Related relics
Jiangsu	6: Middle Yunhe River, Suqian Section; Huaiyang Canal, Huai'an Section; Huaiyang Canal, Yangzhou Section; Jiangnan Canal, Changzhou Urban Area Section; Jiangnan Canal, Wuxi Urban Area Section; Jiangnan Canal, Suzhou Urban Area Section	22: Longwang Temple Palace in Suqian; Governor's Office of Transportation in Qingkou; Shuangjin Lock in Huai'an; Qingjiang Dam in Huai'an; Hongze Lake Dike in Huai'an; Liu Village Drainage Lock in Baoying; Yucheng Post in Gaoyou; Shaobo Ancient Dike in Jiangdu; Shaobo Wharf in Jiangdu; Shouxihu Lake in Yangzhou; Tianning Temple Palace and Chongning Temple in Yangzhou; Geyuan in Yangzhou; Wanglumen Residence in Yangzhou; Yanzong Temple in Yangzhou; Lushaoxu Residence in Yangzhou; Qingming Bridge Historic Cultural District in Wuxi; Panmen in Suzhou; Baodai Bridge in Suzhou; Shantang Historic Cultural District in Suzhou (including Huqiu Tower); Pingjiang Historic Cultural District in Suzhou; Wujian Ancient Towing Path in Wujiang	–
Zhejiang	6: Jiangnan Canal, Nanxun Section; Jiangnan Canal, Jiaxing to Hangzhou Section; Zhejiang Eastern Canal, Xiaoshan to Shaoxing Section; Zhejiang Eastern Canal, Shangyu to Yuyao Section; Zhejiang Eastern Canal, Ningbo Section; Sanjiangkou in Ningbo	13: Nanxun Ancient Town; Changhong Bridge in Jiaxing; Chang'an Lock in Jiaxing; Fuyicang in Hangzhou; Fengshan Water City Gate Site in Hangzhou; Qiaoxi Historic Street in Hangzhou; Gongchen Bridge in Hangzhou; Guangji Bridge in Hangzhou; Guitang Wharf in Xixing; Bazhi Bridge in Shaoxing; Bazhi Bridge Historic District in Shaoxing; Gujian Road in Shaoxing; Ningbo Qing'an Guild Hall	–

the Hai River Basin is relatively limited, and it lacks adaptability to domestic and international trade. In particular, the effective integration of the Yangtze River and Pearl River high-grade waterway networks into the canal system has not yet been achieved, with few outlets to the sea, resulting in the absence of a systematic water transportation network.

The *14th Five-Year Plan for National Economic and Social Development of the People's Republic of China and the Outline of the Long-Term Goals for 2035* proposes "research on the interconnection of canals across water systems such as the Pinglu Canal," while the "Outline of the National Comprehensive Three-Dimensional Transportation Network Plan" proposes the layout of the national high-grade waterway network known as the "Four Verticals, Four Horizontals, and Two Networks," in which the Hunan-Guangxi and Jiangxi-Guangdong Canals are important components of the "Two Verticals."

The canal interconnection project includes the restoration of the Beijing-Hangzhou Grand Canal north of the Yellow River, the Jiangxi-Guangdong Canal, the Hunan-Guangxi Canal, the Pinglu Canal, and other cross-water-system interconnection. Promoting the interconnection of the north and south water systems' canals is an inevitable requirement to better leverage the comparative advantages of water transportation. It serves the major national strategies, promotes the coordinated development of regional economies, and is an objective need to accelerate the construction of a strong transportation nation and build a modern comprehensive transportation system. In the new era, the interconnection of the north and south water system canals will not only optimize the spatial layout of our country's inland waterway channels but also significantly improve the structure of the channel grades, filling in the gaps in the development of inland waterways. It will undoubtedly become a landmark project in the construction of a strong transportation nation and one of the "super projects" in the great rejuvenation process of the Chinese nation.

References

1. Sima Qian. Shi Ji. Beijing: Zhonghua Book Company, 1959.
2. Lu Yong. The Rise of Water Conservancy and the Emergence of Great Powers: The Development and Implications of Military Water Conservancy during the Spring and Autumn and Warring States Periods [J]. Journal of Jianghai Studies, 2017(6):8.
3. Bai Shouyi. General History of China (Volume III, Part 1). Shanghai: Shanghai People's Publishing House, 2000.
4. Bai Shouyi. History of Chinese Transportation. Beijing: Tuanjie Publishing House, 2001.
5. Shandong Canal Shipping History Compilation Committee. History of Shandong Canal Shipping. Jinan: Shandong People's Publishing House, 2011.
6. Chen Bixian. History of the Grand Canal of China. Beijing: Zhonghua Book Company, 2001.
7. Ministry of Transport of the People's Republic of China. History of Chinese Water Transport (Ancient Times—1840). Beijing: China Communications Press, 2021.
8. Shang Shu (Book of Documents). Annotated by Wang Shishun and Wang Cuiye. Beijing: Zhonghua Book Company, 2012.
9. Zhou Li (Rites of Zhou). Chen Shuguo, editor. Changsha: Yuelu Bookstore, 1989.

10. An Zuozhang. History of Chinese Canal Culture (Middle and Lower Parts). Jinan: Shandong Education Press, 2001.
11. Bai Shouyi. General History of China (Volume V, Part 1). Shanghai: Shanghai People's Publishing House, 2000.
12. Zhu Xie. Anthology of Historical Materials on Chinese Canals. Nanjing: Jiangsu People's Publishing House, 2017.
13. Chen Feng. Grain Transport and Ancient Society. Xi'an: Shanxi People's Publishing House, 2000.
14. Huan Kuan. Salt and Iron Discussions. Shanghai: Shanghai People's Publishing House, 1974.
15. Bai Shouyi. General History of China (Volume VI, Part 1). Shanghai: Shanghai People's Publishing House, 2000.
16. Qian Mu. Economic History of China. Beijing: Beijing United Publishing Company, 2020.
17. Wei Zheng. Sui Shu (History of the Sui Dynasty). Beijing: Zhonghua Book Company, 2019.
18. Bai Shouyi. General History of China (Volume VII, Part 1). Shanghai: Shanghai People's Publishing House, 2000.
19. Du You. Tong Dian (Comprehensive Mirror for Governance). Hangzhou: Zhejiang Ancient Books Publishing House, 2007.
20. Ye Shi. Shuixin Ji (Collection of Shui Xin's Writings). Shanghai: Shanghai Ancient Books Publishing House, 1987.
21. An Zuozhang. History of Chinese Canal Culture. Jinan: Shandong Education Press, 2001.
22. Wu Qi. Grain Transport and Chinese Society. Wuhan: Central China Normal University Press, 1999.
23. Ma Haiyan, et al. 100 Major Events That Shaped Chinese History. Beijing: China Book Press, 2004.
24. Wang Bomin. A General History of Chinese Painting. Beijing: Life, Reading, and New Knowledge Publishing House, 2008.
25. Bai Shouyi. General History of China (Volume VI, Part 2). Shanghai: Shanghai People's Publishing House, 2000.
26. Yu Qiuyu, et al. Knowing. Shanghai: Shanghai Jinxiu Wenzhang Publishing House, 2009.
27. Bai Shouyi. General History of Transportation in China (Volumes VIII, IX, X). Shanghai: Shanghai People's Publishing House, 2000.
28. Zhang Tingyu, et al. Ming Shi (History of the Ming Dynasty). Beijing: Zhonghua Book Company, 1974.
29. Song Lian, et al. Yuan Shi (History of the Yuan Dynasty). Beijing: Zhonghua Book Company, 1976.
30. Jiang Shili. Cultural History of the Grand Canal of China. Beijing: China Building Materials Industry Press, 2019.
31. Zheng Minde. Research on the Grain Transport and Storage System along the Beijing-Hangzhou Grand Canal during the Ming and Qing Dynasties. Beijing: China Social Sciences Press, 2015.
32. Tuo Tuo, et al. Song Shi (History of the Song Dynasty). Beijing: Zhonghua Book Company, 1985.

Chapter 7
Crossing Business and Travel: Ancient Tea Horse Road and Horse Caravan Transportation

The Tea Horse Road is the highest-altitude ancient road network carved out in the towering mountains and rugged terrain, serving as a millennium-old passage for the ancient Chinese imperial court's tea-horse trade with neighboring regions. "Sichuan tea flows into various foreign markets, and horses from the north often travel thousands of miles," as described by Wu Zeng during the Southern Song Dynasty in his "Nenggai Zhai Manlu" record of the Xining and Yuanfeng periods (1068–1085) of the Northern Song Dynasty [1]. China boasts many famous ancient roads, such as the Silk Road, the Tang-Tibet Ancient Road, the Qinling Ancient Road, the Tea Horse Ancient Road, the Grassland Road, the Eight Passes of the Taihang Mountains, the Ancient Roads of Western Beijing, the Ancient Roads of Huizhou and Hangzhou, the Jiasang Ancient Road, the Musk Road, and the Meiguan Ancient Road. Each ancient road has been endowed with different characteristics and functions by history, with the Tang-Tibet Ancient Road serving regional political and diplomatic purposes, and the Qinling Ancient Road mainly utilized for political and military endeavors. The Tea Horse Ancient Road, on the other hand, primarily served commercial purposes, becoming a vital route for business and travel intersections.

Background of the Tea Horse Road

The term "Tea Horse Road" was coined by scholars such as Mu Jihong in the 1990s to describe a concept encompassing ancient trade routes primarily centered around Yunnan, Sichuan, and Tibet, where commodity exchange, particularly involving tea and horses, occurred. Subsequently, they refined the concept to specifically denote all routes within China's southwestern region and between the southwestern region and neighboring countries, facilitating tea and horse trade circulation [2]. This concept of the "Tea Horse Road" excludes ancient trade routes in the northwest region. The transportation along this trade route primarily involved horses, mules, and people,

© China Communications Press Co., Ltd 2025

H. Lu (ed.), *A Brief History of Transportation in China*,
https://doi.org/10.1007/978-981-97-9851-3_7

with caravans serving as the economic organization for transporting goods. Therefore, the "horse" in the "Tea Horse Road" refers both to the horses used for trading tea and those utilized in caravan transportation. Mu Jihong and others conducted field surveys of seven routes, including the Snowy Range Road, Tribute Tea Road, Horse Acquisition Road, Lijiang-Burma-India Road, Yunnan-Vietnam Road, Old Yunnan-Southeast Asia Road, and Tea-Picking Road, confirming their historical authenticity [2].

The geographical features of China determine the inevitability of tea-horse trade.

The regions of Tibet, Xinjiang, and Inner Mongolia in China have abundant grasslands suitable for horse breeding, making them major horse-producing areas. People in this region mainly consume tsampa, dairy products, butter, and beef and mutton, and they drink tea to aid digestion. Pu'er tea, black tea, and red tea contain significant amounts of caffeine, theophylline, theobromine, and probiotics, which help digest meat. Tea also stimulates gastric acid secretion, promotes intestinal peristalsis and relaxation of the intestinal wall, and shortens the transit time of meat in the intestines. As mentioned in the "Compendium of Materia Medica Addendum", "Pu'er tea is bitter and acrid, capable of resolving greasiness and the toxicity of beef and mutton… Its bitterness eliminates phlegm and dispels flatulence, and it clears the intestines and promotes bowel movements… Pu'er tea paste, as dark as lacquer, is the best remedy for sobering up; the greener, the better; it aids digestion, resolves phlegm, and invigorates the stomach. Puyu Tea (namely, Pu'er tea), steamed and formed into cakes, is marketed in the Western Regions and is most effective in transforming substances." [3]. Dietary habits have led people in these regions to rely on tea.

The southwestern region is a major tea-producing area. It receives abundant rainfall, has suitable soil thickness, and its high mountains and dense forests are conducive to the growth of tea trees. However, the rugged terrain makes transportation difficult, hindering the large-scale export of tea leaves. It wasn't until the time of Emperor Qin Shi Huang that the central government began large-scale operations in the southwestern region. Following the Qin dynasty's conquest of the southwest, its financial resources greatly increased, laying the foundation for its hegemony. With the establishment of accessible roads, tea from the southwestern region could play an important role, and Qin merchants seized the opportunity to gain a dominant position in the tea trade.

The majority of the North China Plain is cultivated land with high grain yields, dense population, and developed economy. Consequently, capitals of Central Plain dynasties were often established here. Through prolonged conflicts with northern nomadic tribes, the Han Chinese realized the importance of cavalry in warfare. The "Book of Later Han" mentions, "Horses are the foundation of armored soldiers, and they are crucial for the state's defense. In times of peace, they serve to maintain social order, and in times of crisis, they help in overcoming distant threats." [21] Therefore, the Central Plain regimes had a significant demand for horses. However, the cultivated areas lacked extensive pastures necessary for horse breeding, so the demand for warhorses had to be met by regions with suitable grazing lands. During times of confrontation between northern nomadic regimes and Central Plain regimes, horses were considered strategic assets and were not readily sold or traded. Thus, the Central Plain regimes relied on Qinghai-Tibet Plateau and the Western Regions to obtain horses, establishing stable economic and trade relations between the southwestern region and the Qinghai-Tibet Plateau. However, when Central Plain dynasties originated from the northeast or the Mongolian Plateau, they could directly obtain horses from the source, weakening the economic and trade relations between the southwestern region and the Qinghai-Tibet Plateau.

Horses, tea, and other goods linked the Qinghai-Tibet Plateau, the Western Regions, Mongolia, the southwestern, central-southern, North China, and Northwest regions into a cohesive commercial network. Part of this commercial network revolved around the circulation of horses and tea. Geographic features determined the characteristics of local products, which in turn influenced people's demand. As a result, relatively stable trade interactions emerged among various regions.

The most challenging area for tea-horse trading lies in the Tibetan-Yi Corridor. The "Tibetan-Yi Corridor" is the border area where the Han-Tibetan and Yi-Tibetan meet, located at the junction of Sichuan, Yunnan, and Tibet today. The mountain ranges and rivers in this area run north to south, and the valleys between high mountains, due to the erosion and deposition of water, have relatively gentle terrain. The passages connecting the southwestern region with Tibet are formed in these relatively flat areas. During the Sui and Tang Dynasties, tribes such as Tubo, Tangut, Dongnu, and Bailanqiang were active in areas such as Qinghai-Tibet Plateau and Bashu. After the Tubo tribe conquered tribes like Supi and Yangtong, they dominated the Qinghai-Tibet Plateau and engaged in a hundred-year-long struggle with the Tang Dynasty. The Tibetan-Yi Corridor was situated in the area of contention between the two sides, serving as the stage for activities of the Sino-Tibetan languages and Tibeto-Burman language groups. For example, the Naxi people in Yunnan, originally from the northwest, settled in the border areas of Yunnan, Sichuan, and Tibet through the Tibetan-Yi Corridor and interacted with surrounding ethnic groups such as the Han, Tibetan, Bai, Yi, and Pumi. The passages formed in the Tibetan-Yi Corridor include routes such as the Xuanniu Trail, Wuchidao, Jinsangmileng Road, Yongchang Road, and Longshu Road, which are collectively referred to as the "Tea Horse Ancient Road" based on their historical characteristics [4].

During times of intense conflict, strict laws prohibited private tea smuggling. In periods of peace, governmental control over the tea-horse trade would weaken, as reflected in the saying, "Sichuan tea is not used to trade for horses; only pearls and jade are traded, hence the government's horse management becomes ineffective." [5]. Private trade would consequently flourish. The channels and roles of trade between private tea vendors and government-managed tea-horse trade differed significantly. Official roads were generally controlled by the government and had checkpoints. The government engaged in tea-horse trade through these official roads, while private operations mainly occurred outside of these routes. During the Qing Dynasty, a "three vertical and one horizontal" backbone transportation network was constructed in the southwest region of China. The "three vertical" referred to the Sichuan official road, Yunnan official road, and Guilin official road, while the "one horizontal" referred to the Yangtze River official road, which was a water route. Private tea vendors primarily functioned to collect tea leaves at the starting point and sell them

Fig. 7.1 Scene of the ancient tea-horse road

at the endpoint. To gather tea leaves, a capillary-like transportation network formed between mountains and wilderness, ultimately converging onto official roads, where government-organized caravans transported them over long distances (see Fig. 7.1).

Spatial Distribution of the Ancient Tea-Horse Road

The broad concept of the Ancient Tea-Horse Road refers to the majority of trade routes used for tea-horse transactions in the southwestern and northwestern regions, including the Sichuan-Tibet line, Yunnan-Tibet line, Tangu ancient road, Qinghai-Tibet line (Tangfan ancient road), and Sichuan-India (Shu-Yuandu) road. These routes have served tea-horse trade at different times. This concept has a broad extension, encompassing roads in the northwest region as well. The Tangu ancient road is not in the southwestern region, but its historical function was tea-horse trade. Many dynasties established tea-horse markets in the northwest, so including this northwestern road used for tea-horse exchanges in the Ancient Tea-Horse Road is reasonable. Additionally, routes like the "Dian-Bi ancient road," "Dian-Yue ancient road," "Honghe ancient road," categorized by transportation lines, and "salt transport ancient road," "copper transport ancient road," "horse purchasing ancient road," "tea picking ancient road," categorized by the types of goods transported on the road, are all within the broad scope of the "Ancient Tea-Horse Road." The spatial relationship between the Ancient Tea-Horse Road and other ancient roads is illustrated in Fig. 7.2.

In a strict sense, the Ancient Tea-Horse Road mainly consists of three main routes: the Yunnan-Tibet Tea-Horse Road, the Sichuan-Tibet Tea-Horse Road, and

Fig. 7.2 Spatial relationship between the ancient tea-horse 3. Road and other ancient roads

the Shaanxi-Kang-Tibet Tea-Horse Road. These three main routes form a vast transportation network, connecting Sichuan, Yunnan, Tibet, and Shaanxi into an interconnected whole, extending outward to Bhutan, Nepal, and even as far as West Asia and Europe. The primary routes for the Tea-Horse Road entering Tibet are the Sichuan-Tibet and Yunnan-Tibet Tea-Horse Roads [6].

The Sichuan-Tibet Tea-Horse Road starts from Ya'an in Sichuan and extends to Kangding. Kangding serves as a pivotal transportation hub where multiple major traffic arteries converge. During the Ming and Qing dynasties, two main routes emerged from Ya'an to Kangding: the "Main Tea Route" and the "Minor Tea Route." The "Main Tea Route" traverses from Ya'an through Yingjing, crosses the Daxiang Mountains, passes through Feiyueling, and reaches Kangding via Luding. The "Minor Tea Route" leads from Ya'an through Tianquan, crosses the Ma'anshan Mountains, and reaches Kangding via Luding. After reaching Kangding, the road further branches into two routes, north and south. The northern route extends northward from Kangding through Daofu, Luhuo, Garze, Dege, and Jiangda, reaching Chamdo, where it merges with the Yunnan-Tibet line. This northern route corresponds to the current northern line of the Sichuan-Tibet Highway. The southern route extends southward from Kangding through Yajiang, Litang, Batang, Mangkang, and Zogang, reaching Chamdo. This southern route corresponds to the current southern line of the Sichuan-Tibet Highway [7]. The notable cities along this route are Ya'an, Kangding, Chamdo, and Lhasa. Ya'an is situated at an altitude of 1500 m, Kangding at 2560 m, Chamdo at 3240 m, and Lhasa at 3650 m, resulting in a total altitude difference of 2150 m. The domestic route of the Sichuan-Tibet Tea-Horse Road spans over 3100 km.

Starting from the Ming Dynasty, the Sichuan-Tibet Tea Road began to take shape officially. By the second year of the Chenghua reign (1466 AD), the government explicitly stipulated that envoys from the four kings of Wusitang Zanshan, Chanjiao, Chanhua, and Fujiao, as well as nearby Wusitang regions, should all enter through Sichuan when presenting tributes. During the Ming Dynasty, the Tea-Horse Administration Offices were established in Yazhou and Diaomen, where millions of kilograms of tea were transported annually to Kangba Tibetan Region and then to Dbus-Gtsang, thus extending the Tea Road from Kangba Tibetan Region to Dbus-Gtsang. The exchange of tribute envoys between Dbus-Gtsang and the central government further facilitated the smooth operation of the Tea Road. The Sichuan-Tibet Tea Road, pioneered by tea trade, simultaneously became an official route, replacing the position of the Qinghai-Tibet Road.

In fact, the Yunnan-Tibet Tea-Horse Road has three main routes: one starting from Neijiang's Helizhen, passing through military garrison area Tacheng and various other points including Pengzilan, Adeqiu, Tianzhuzhai and Maofagong before reaching Tibet; another beginning from Jianchuan's joint military garrison area Weixi and merging with the first route before entering Tibet; and a third route traversing Xishuangbanna, Pu'er, Dali, Lijiang, Zhongdian, Deqin, Mangkang, and Chaya before reaching Chamdo. From Chamdo, the tea is distributed to other parts of Tibet, and further transportation routes extend to Lhasa and beyond, even crossing the Himalayas into India's Garhwal region and reaching Kolkata. Thus, the Yunnan-Tibet Tea-Horse Road served as a crucial trade artery connecting ancient China with South Asia, with Xishuangbanna as its starting point at 553 m altitude and Lhasa as its endpoint at 3650 m altitude, spanning over 3800 km domestically.

The Yunnan-Tibet Tea-Horse Road emerged during the Tang Dynasty, closely associated with the expansion of the Tibetan Empire and its trade activities with the Kingdom of Nanzhao. In the third year of Emperor Gaozong's Yifeng reign (678 AD), Tibetan influence reached the northern region of Erhai Lake in Yunnan. In the first year of Emperor Gaozong's Yonglong reign (680 AD), the Tang Dynasty established the Shenchuan Governorate, while the Tibetan Empire appointed officials in Nanzhao, imposing taxes and levying labor services on the Bai and Wu tribes. Trade between the two sides flourished significantly, with tea-horse trade being a crucial component. The trade route between Nanzhao and the Tibetan Empire roughly corresponds to the present-day Yunnan-Tibet Highway, starting from Dali in modern Yunnan, heading north to Jianchuan and Lijiang, continuing along the northern bank of the Jinsha River via Tieqiao City, passing through Benzilan to Yulai City, advancing to Yanjing, and then proceeding north along the Lancang River to Ma'ergan (modern-day Mangkang in Tibet) and Zogang. From there, two routes lead to Tibet: one passing through Basu and Chaya to Chamdo, and the other going directly from Basu to Pomi, passing through Nyingchi before reaching Lhasa.

The Shaanxi-Kang-Tibet Tea-Horse Road, known locally as the "Tang Ancient Road," originated during the Qin and Han Dynasties. Serving as the trading hub was Kangding, with the dominant force being the Qin merchants. Initially, the route from

Kangding utilized the ancient road between Shaanxi and Sichuan. It was through this route that the Qin Dynasty conquered the regions of Hanzhong and Shu, connecting Shaanxi and Sichuan as one entity. Leveraging their advantageous political position, Qin merchants flourished and "trod" this commercial thoroughfare. The tea traded by Qin merchants included tea from southern Shaanxi, Yunnan, Sichuan, and Xiang (produced in Anhua), with Sichuan tea being predominant. The prosperity of this road was attributed to the superiority of northwest horses as warhorses due to their tall stature, swift pace, and enduring stamina, making them the preferred choice for military use throughout various dynasties. Northwest horses were in high demand, while Sichuan tea was abundant, leading to bustling trade and numerous merchant caravans along the road. During the Song Dynasty, the rise of Western Xia intensified conflicts with the Song Dynasty, making military affairs in the northwest a top priority. Emperor Shenzong established horse-buying offices in Xiguan, Minzhou, Tongyuanjun, and Yongningzhai regions, acquiring 15,000–20,000 horses annually. Subsequently, tea markets were established in Sichuan and Shaanxi, with horse-buying fields set up in Shaanxi. Tibetan horses arrived in Shaanxi via the Tangfan Ancient Road. Later, the court decreed, "Famous mountain tea from Yazhou, designated exclusively for horse trading," [8] "resulting in an increase in Tibetan horses arriving" [5]. The Shaanxi-Kang-Tibet Tea-Horse Road served as a bridge and link between ancient Tibet and the inland regions, stretching nearly 4000 km in total.

Among the three main routes, the Sichuan-Tibet Tea-Horse Road was the earliest to open and had the largest transport volume, playing a more significant role in history. After reaching Kangding, the endpoint of the Shaanxi-Kang-Tibet Tea-Horse Road, the roads leading towards Yunnan and Tibet utilized the Sichuan-Tibet Tea-Horse Road. The success of the Shaanxi-Kang-Tibet Tea-Horse Road in facilitating cross-regional trade was primarily due to the substantial capital of the Qin merchants, allowing them to purchase tea on a large scale and benefit from economies of scale. Typically, the caravans from Sichuan and Yunnan did not possess such large capital and engaged in low-cost commercial exchanges. Therefore, in Kangding, it was said that "there is no commerce without the Qin merchants."

In addition to the three main trunk lines mentioned above, the Tea-Horse Road also had numerous branches and subsidiary lines, forming a vast transportation network. These branches include the Ya'an-Songpan-Gannan branch, the northern branch of the Sichuan-Tibet Road passing through the former Dengke County (now Dege County, Garze Prefecture, Sichuan Province)—Yushu in Qinghai—Xining—Taozhou (now Lintan County, Gannan, Gansu Province), and others.

The historical "Tangfan Ancient Road" (now known as the Qinghai-Tibet Line) was a route that started from Chang'an, passing through Gansu and Qinghai to reach Lhasa in Tibet. From Lhasa, it extended further to India and Nepal, serving as the path to Tibet over 1300 years ago, spanning over 3000 miles. The origin of this road dates back over 6000 years. During the Sui Dynasty, an official road leading to Tibet was established, while the Ya'an main road served as the passage for Sichuan tea to enter Tibet. From Ya'an westward, it passed through Duogongba, Luyongguan, Luguan River, through Hutouxia, Luo Jia, Luo Li, Cangziba, went upward along Shunyingjing River, crossed mountains and hills, and reached Lizhou (now Yidong

Town, Hanyuan County, Ya'an, Sichuan). In the early Tang Dynasty, the main route for transporting tea from the mainland to Tibet was the Qinghai-Tibet Road, which served as the primary channel for communication between the Tibetan region and the Central Plains. Since the Tang Dynasty, inland travelers have mainly entered Tibet via the Qinghai-Tibet Road, with historical records stating, "In the past, this road was the main postal route, as it was the earliest to be opened, and since the Tang Dynasty, all routes have been through this road." [9]. Tea-horse trade was conducted along this route, thus encompassing the broad sense of the Tea-Horse Road. While the Tangfan Ancient Road was a "political highway," it was commonly used by commercial trading teams, primarily engaged in tea-horse trade [2, 7].

The development of passages between Sichuan and Yunnan was also a challenging process, with routes including the Wuchi Road, the Maoniu Road, the Yongchang Road, and the Nanyi Road, all integral parts of the "Shu Yuandu Road." The "Shu Yuandu Road" is divided into two main branches, the Southern and Western branches. The Southern branch includes the Minjiang Road and the Wuchi Road, running from Chengdu to Yibin, Zhaotong, Qujing, Kunming, Chuxiong, Dali, Baoshan (now Yongchang County, Jinchang, Gansu), Tengchong, Guyong, Myanmar (Kingdom of Shan), and India (Yuandu). The Western branch, also known as the Maoniu Road, was constructed along the ancient route taken by Sima Xiangru as he traveled south through the Maoniu Qiang tribe's territory. It runs from Chengdu to Qionglai, Lushan, Lugu, Xichang, Yanyuan, Dayao, Xiangyun, Dali, and joins with the Southern route. While the primary function of the Shu-Yuandu Road was not tea-horse trade, it transported tea leaves used in such trade along its route.

The Wuchi Road and the Maoniu Road (Lingguan Road) of the "Shu-Yuandu Road", deserve special attention.

The historical evolution of the Wuchi Road is as follows: The Wuchi Road originated from Bodao (now Cuiping District, Yibin, Sichuan) and ended in present-day Qujing, Yunnan. During the reign of King Zhao of Qin, Zhao Ze ordered General Sima Cuo to lead an army southward from the "Jinniu Road" to conquer the Shu Kingdom. After the conquest, it was renamed the Qin Shu Commandery, and Li Bing was appointed as its governor. To construct a road leading to the Yungui Plateau, Li Bing employed methods such as burning wood to break rocks and using water to explode them, carving out a small path from Bodao County all the way to Zhuti County (now Zhaoyang District, Zhaotong, Yunnan), approximately 600 li in length [10]. Most sections of the road were only 5 Qin Chi wide (approximately 1.16 m), hence the name " Wuchi Road ". After the unification of the seven states by Emperor Qin Shi Huang, he ordered Chang'e to "roughly connect the Wuchi Road, appointing officials in various states (referring to the southwestern tribes)… Over a decade, Qin conquered… and when Han was established, they all abandoned these states and opened up Shu for colonization… The Ba and Shu people would sometimes secretly engage in trade, acquiring horses, Bodao people, and cattle, enriching the Ba and Shu regions." [11]. This means that Chang'e extended the route all the way to Wei County (now Qujing, Yunnan), adding approximately 600 km to the distance. After the road was opened, administrative institutions were established, breaking through tribal limitations, and the trade of horses, Bodao people, and cattle flourished.

In the sixth year of the Jianyuan era of the Western Han Dynasty (135 BC), it was recorded that "troops were dispatched to build roads in Ba and Shu, from Bodao to Zangke River (now the Beipan River in northern Guizhou)." [11]. After the Kingdom of Nanyue surrendered, the Wufeng Road was extended to the Beipan River, connecting with the Yelang Road, linking Bie County (now Zunyi, Guizhou) with Zhuti County and Weixian. The Wuchi Road was later extended to Jinning District in the southern part of present-day Kunming, Yunnan, covering a distance of 1800 li (approximately 900 km), and the road width was expanded from "Five Chi" to "more than ten Chi." [10].

The Maoniu Road, also known as the "Lingguan Road," "Lingguan Road," and "Xi Yi Road," starts from Chengdu and passes through Linchong (now Qionglai, Sichuan), Lingguan (now Lushan, Sichuan), Yandao (now Ya'an, Sichuan), Zedu (now Hanyuan, Sichuan), Lugu, Xiangying Ancient Fortress (now Xide, Sichuan), Qiongdu (now Xichang, Sichuan), and Yanyuan. After crossing the Jinsha River, it enters Yunnan through Qingling (now Dayao, Yunnan), Dabonong (now Xiangyun, Yunnan), and Yeyu (now Dali, Yunnan). When Zhang Qian was in Daxia (now Afghanistan), he learned that a route from southwestern Sichuan through Yuandu (now India) could reach Daxia. Emperor Wu of Han believed that this route could form alliances, resist the Xiongnu, expand trade routes, and accumulate wealth. He was determined to spare no effort to open an official road from southwestern China to Daxia at all costs. However, due to local resistance, they could only reach Erhai Lake [10]. In the twelfth year of Emperor Ming of the Eastern Han Dynasty (69 AD), the Ailao people surrendered, and the final section of the road from Yongchang to Myanmar, known as the "Yongchang Road," was opened. The "Road to the Kingdom of Yuandu," planned by Emperor Wu of Han, was finally opened along its entire length [9].

The route of the "Road to the Kingdom of Yuandu" is as follows: Merchants from the Lingguan Road and Wuchi Road converge in Dali, then proceed along the Bonan Road, crossing the Yongping Bonan Mountain, and crossing the Lancang River. From the ancient ferry of Lanjin, they enter Yongchang (now Baoshan, Yunnan) and follow the Yongchang Road. Afterward, they cross the Nu River and reach Dianyue (now Tengchong, Yunnan). After exiting China, they arrive in Myanmar and India, and can further travel to South Asia, West Asia, and even Europe and Africa.

Historical Evolution of the Tea-Horse Trade

During the Han, Jin, Sui, Tang, Song, Yuan, Ming, and Qing dynasties, there were significant differences in the prosperity and government involvement in the tea-horse trade. Due to variations in political geography among these dynasties, as well as differences in the targets of struggle and the areas under governance, the nature of the tea-horse trade also varied.

During the reign of Emperor Suzong of Tang, from the first year of Zhengde (756 AD) to the first year of Qianyuan (758 AD), the Tang Dynasty established tea-horse trade with the Uyghurs. After the Tang Dynasty opened up the Western Regions, the primary commodities exchanged in the tea-horse trade were Tang tea and horses from the Western Regions. The "Feng's Sketch Book" records: "In previous years, the Uyghurs came to the court with large herds of famous horses, traded them for tea, and returned, which was quite surprising." [11]. At that time, as the Tibetan Empire was expanding eastward, it was becoming a potential threat to the Tang Dynasty. The horse trade between the Tang Dynasty and Tibet was thus a sensitive political issue. As Wei Zheng pointed out: "Now, as the Khagan position is not yet stabilized, if we go to various countries to trade horses, they will surely perceive our intention to be focused on horses rather than on the establishment of the Khagan. Once the Khagan is established, he will not hold any gratitude, and the various tribes will hear of it, see us value money over loyalty. We may not necessarily obtain the horses, but we will first lose our integrity." [12]. Therefore, in the early Tang Dynasty, the center of tea-horse trade was in the northwest region, where tea production was low, and a large quantity of tea came from Sichuan tea.

During the Northern Song Dynasty, the main adversaries were the Western Xia, Liao, Jin, and Yuan dynasties, making it nearly impossible to obtain horses from these northern adversaries. Tea-horse trade mainly occurred in the Shaanxi-Gansu region during the Northern Song Dynasty, where the northwest horse market was the core pillar of the state's horse policy during the reign of Emperor Shenzong. The tea for trading horses came from Sichuan. The monopoly of tea trade began in the seventh year of Xining (1074 AD) and continued into the early years of the Southern Song Dynasty. The court appointed Li Qi and Pu Zongmin to "personally procure tea from Sichuan, to supply the Qin-Feng and Xixia regions, and to extensively purchase horses along the route." [13]. They established tea and horse bureaus in Chengdu and Qinzhou (present-day Tianshui, Gansu), transporting most of Sichuan's annual production of 30 million catties of tea to regions such as Hedong, Shaanxi, Hexi, Jingdong, and Chuanxia, setting up hundreds of tea-selling stations and dozens of horse-buying stations.

Later, "the passes of Shaanxi were all lost, and trade became impossible" [14]. The new pathway to obtain horses was through the Tibetan region. Early on, the government engaged in tea-horse trade with Tibetan tribes, primarily in Diaomen (present-day Tianquan County, Ya'an, Sichuan), but the quantity was limited. At this time, the Tibetan plateau was divided, and the conflict with the Central Plains region ceased. Obtaining a large number of war horses from the Tibetan region became an important choice for the Central Plains government, which accelerated the rapid development of tea-horse trade between the Central Plains, Tibet, and the southwestern region. In the seventh year of Xining (1074 AD), after Wang Shao recaptured Hezhou, he reported to Emperor Shenzong: "Western people mostly bring fine horses to the border, and their only desire is tea." [15]. Emperor Shenzong then sent Li Si to Sichuan to organize tea-horse trading affairs, exchanging tea from the mainland for good horses. "In the early Song Dynasty, managing Sichuan tea, mutual markets were established in Yuande, Weide, and Shun counties to trade horses with barbarians… Most of them were good horses, and the majority of them were traded for profit. The Song Dynasty indirectly showed its distant kindness, and it also served as a means of curbing them." [15]. During the Tiansheng reign of Emperor Renzong (1023–1032), "more than 34,900 horses were still obtained from the Tibetan." [8]. Between the Xiangfu reign of Dazhong (1008–1016) and the Tiansheng reign, "the Qin, Wei, Huan, Jie, Lin, Wenzhou, Huoshan, Baode, and Kekui army markets sold 22,000 horses annually to supplement the deficiencies in the capital and the frontier." [15]. The Song Dynasty established systems such as the tea monopoly, tea trade permit, and tea-horse bureaus to cut off private tea trading channels to increase imperial revenue and meet the demand for high-quality warhorses. These trade management systems served the political purpose of stabilizing the border, known as "the frontier barrier, which is no match for various tribes. They rely on tea for a living, so strict laws are used to prohibit it, and horses are exchanged to reward them, to control their lives and strengthen the frontier barrier of China." [16].

In the early Yuan Dynasty, tea-horse trade was monopolized by the government, and official regulations explicitly stated that "those who privately purchase tea are guilty of the same crime as smuggling salt" [17], cracking down on private operations. With Mongolia, Tibet, Qinghai, Gansu, and the Western Regions all incorporated into the Yuan Dynasty, horses came from various sources, and tea-horse trade in the southwestern and Tibetan regions was not prioritized by the government. The Mongols' demand for tea was given priority, with the government often raising the price of tea in the Tibetan region. Tibetan people had to pay higher fees for tea, which often led to dissatisfaction. Later, Zhang Tingrui, in charge of tea affairs, changed the tea system and introduced private tea merchants to conduct free trade, with the government only collecting tea taxes. This policy met the needs of the Tibetan region and promoted the development of the tea industry in the southwestern region.

During the Ming Dynasty, horses were crucial military resources for defending against Mongol armies. However, since there was no trade route for tea and horses with the north, Zhu Yuanzhang ordered: "Use Sichuan tea to exchange for foreign horses to supply military needs." [16]. Horses from the Tibetan region became an important source of military horses, leading to the restoration of the tea-horse policy. "The Mongols are fond of cheese. Without tea, they suffer from illness. Therefore, since the Tang and Song Dynasties, the law of exchanging tea for horses has been in place to control the Qiang and Rong peoples, with the Ming system being particularly strict. There are official teas and commercial teas, all stored for border trade in horses." [16]. The tea-horse law became a national system, establishing a comprehensive government-operated tea-horse trading system. The tea monopoly institutions became more systematic, and the establishment of the military garrison system and the Buddhist clergy system promoted the prosperity of tea-horse trade. The tea-horse trading system prohibited private tea from entering Tibet, ensuring that the government could exchange for horses while achieving the goal of border control. "His Majesty carefully managed the border defense, using tea to exchange for horses, stabilizing the hearts of the tribal people, and strengthening China. Yu Xin, Minister of Revenue, once said Using 3 million catties of Han Zhong tea from Shaanxi, we can obtain 30,000 horses, the same as Sichuan Song and Mao Tea. The prohibition of trafficking must be strictly enforced." [16].

In the fourth year of the Hongwu reign of the Ming Dynasty (1371), the emperor accepted the suggestion from the Ministry of Revenue: "In Hanzhong, Jinzhou, Shiquan, Hanyin, Pingli, and Xixiang counties of Shaanxi, there are 45 hectares of tea gardens with over 860,000 tea plants. In Sichuan, there are 315 households cultivating tea with over 2.38 million tea plants. It is appropriate to decree that every ten plants, one is to be taken by the government. For ownerless tea gardens, soldiers shall be sent to pick, with eight out of ten taken to exchange for foreign horses." [17]. Tea offices were established in tea-producing areas to set taxation rates, with Shaanxi producing over 26,000 catties and Sichuan reaching 1 million catties. Tea-horse bureaus were established in Qin, Zhao, He, and Ya provinces, "from Diaomen, Li, Ya to Duogan, Wusizang, covering over 5,000 li of tea-growing areas. In the regions beyond the mountains, including various tribes in the western regions, all sell horses." [16].

The Golden Letter Patent system, also known as the Forty-One Golden Letters, was a significant measure in the establishment of the Ming Dynasty's tea-horse system. Tea-horse bureaus were established in places such as Tianshui, Lintan, Linxia, Zhangye in Gansu, Xining in Qinghai, Tianquan in Sichuan, and others. Tea-producing areas were designated as unified purchase zones, while Tibetan horse-producing areas were assigned purchasing tasks and issued golden letters as tokens of trust. "Golden Letter Patents were made, and Duke Cao Li Jinglong was sent to the frontier to negotiate with various tribes, with the seal reading 'Imperial Decree' above, 'Official Dispatch of Accordance' on the left, and 'Anyone who doubts shall be executed' on the right… Forty-one in total… Lower numbered golden letters were given to the various tribes, while higher numbered ones were kept by the Tibetan internal department as contracts. Officials were dispatched every three years to reconcile the documents… There was one route from Hezhou and another from Diaomen." [16]. Using this method, Tibet received over 500,000 catties of tea, while the court obtained 13,800 horses. Here, political, economic, and military objectives merged into one. During the Hongwu period of the Ming Dynasty, one high-quality horse could be exchanged for up to 120 catties of tea. Later, as tea prices fell slightly, during the Wanli period of the Ming Dynasty, one high-quality horse could be exchanged for 30 baskets of tea, a medium-quality horse for 20 baskets, and a lower-quality horse for 15 baskets.

The Qing Dynasty largely continued the tea-horse policy implemented in the Qinghai-Tibetan region from the Ming Dynasty. During the Qing Dynasty, there was a policy of "attracting merchants to transport tea and attracting local tribes to provide horses." Tea-horse bureaus were established in Qinghai and Gannan regions, including Xining, Taozhou (now Minxian County, Dingxi, Gansu), Hezhou (now Linxia, Gansu), Zhuanglang (now Yongdeng County, Lanzhou, Gansu), and Ganzhou (now Lanzhou, Gansu). At this time, with the social economy not yet fully recovered, tea and horses were mainly organized by the government for tea-horse trade. The Qing Dynasty tea regulations classified tea into three categories: "official tea" stored for border trade in horses, "commercial tea" used for levying taxes, and "tribute tea" for imperial use [18]. Official and commercial teas were exchanged for border horses in Shaanxi and Gansu, with tea warrants issued by the central government within the five tea-horse bureaus in the northwest to organize trade, supplying the main source of tea to various ethnic groups in Qinghai, Gansu, Tibet, Mongolia, and other regions. The Southwest Tea Horse Ancient Road was not the main direction of tea-horse trade at that time. Later, with the increase in private tea merchants, the government's control over the tea-horse border management policy began to loosen. With the participation of private trade, a situation emerged where tea was cheap and horses were expensive in tea-horse trade. In the 13th year of Emperor Yongzheng's reign in the Qing Dynasty (1735), the official tea-horse trade system was terminated.

The prosperity and decline of the Tea Horse Ancient Road reflect the fortunes of the central empire established with agricultural regions as its core. During periods of turmoil, such as increased demand for good horses in the central plains, government control over the Tea Horse Ancient Road would strengthen. In times of peace, when the military's demand for warhorses significantly decreases, it is inevitable that horses become cheap and tea becomes expensive. When the central government was established by the Han ethnicity, as in the periods of the Han, Song, and Ming Dynasties, the Tea Horse Ancient Road thrived. Conversely, when the central government was established by nomadic peoples from the north, as in the Yuan and Qing Dynasties, the Tea Horse Ancient Road was tend to be relatively quiet.

Caravan on the Tea Horse Ancient Road Network

Tents, firearms, and steeds evoke a sense of wilderness, romance, and rugged beauty, which is the essence of caravan life in the minds of many. Yet, herding horses and mules, setting up gongs and cauldrons, and pitching tents constitute their daily reality—hard, exhausting, and fraught with danger. Which aspect represents the true face of their lives? Over the years, caravans have thrived along the Tea Horse Ancient Road. Nestled deep in the mountains, they remain cut off from information, unaware of the changes in the Han and Tang dynasties, indifferent to the movements of the celestial bodies. Together with their horses and mules, they tread paths step by step, carving out routes and shaping generations of lives. In reality, their fate is closely linked to the outside world. In times of chaos, their business booms; in times of peace, life carries on quietly. They are the "current" of that era in space and the "current" traversing through history. From this "current," one can feel the pulsations of history and the prosperity of nations.

Caravans were a commercial operation group. In ancient China, there was a highly developed official transportation system known as the postal relay system, and there were no national roads. The primary mode of transportation was horses, with a history dating back over 2,000 years since the Qin Dynasty. The roads in the southwestern mountainous areas were rugged, and transporting goods relied solely on horses. Moreover, the mountainous regions were dangerous, with bandits and wild beasts lurking about, necessitating travelers to form groups for survival. Thus, caravans emerged as a mutually supportive organization, with large caravans often consisting of hundreds or even thousands of horses and mules. Additionally, caravans had the traditional guild characteristic of self-defense with weapons.

The caravans responsible for transportation and the trading firms responsible for buying and selling are interdependent in a mutually beneficial relationship. Large trading firms may organize their own caravans to transport their goods, which can lower costs. Similarly, large caravans may carry their own goods to seize opportunities for profit. However, caravans and trading firms generally have a division of labor and cooperation relationship, especially for smaller trading firms, their demand for caravans is even more significant. This division of labor and cooperation relationship benefits both parties in expanding their business operations.

Caravans are divided into professional caravans and joint venture caravans: professional caravans are larger in scale, with organized personnel and clear division of labor, resulting in higher operational efficiency; joint venture caravans are formed by several smaller caravans pooling together, providing mutual assistance but with slightly less coordination. Caravans adopt a purely commercial operating model, with clear responsibilities for each cargo transport. The leader of the caravan, known as the "head of the pot," is the operator of the caravan responsible for major matters such as the delivery of goods and funds. The caravan drivers are responsible for caring for the mules, goods, and cargo.

Caravans place great emphasis on member selection, often based on family, ethnicity, religion, neighbors, and companionship, forming bonds such as kinship family caravans, neighborhood joint caravans, and temporary joint caravans. The relationship between caravan drivers and the leader of the caravan is particularly close within kinship family caravans and neighborhood joint caravans, facilitating command and cooperation, with mutual respect making long-term coexistence easier.

The "Guo Tou" is the core figure of the caravan, responsible for various purchases and expenses, as well as handling external affairs. During outdoor meals, the "Guo Tou" takes charge of serving and allocation. The name "Guo Tou" originated from the fact that both the "Guo Tou" and the caravan's followers, known as "Ma Jiao Zi", eat from the same pot, hence the name "Guo Tou". Most "Ma Jiao Zi" come from humble backgrounds and have an employer-employee relationship with the "Guo Tou", obeying their commands. Through their own efforts, "Ma Jiao Zi" may eventually own one or two mules and earn a portion of the transportation fee. With vision and capital, they can also invest in some goods and potentially earn profits. However, the business of "Ma Jiao Zi" entails significant risks, and there are not many who dare to engage in this profession. To encourage "Ma Jiao Zi", major merchants and caravans along the Yunnan-Tibet route promise that if a "Ma Jiao Zi" drives a caravan for a period of three years, they will be rewarded with a mule. The expenses are covered by the merchant, and all profits earned belong to the "Ma Jiao Zi", thus allowing them to share both the benefits and risks with the caravan.

Caravans are a resilient group accustomed to enduring hardships in the wilderness. The Ancient Tea Horse Road is the highest-altitude trade route in the world, characterized by extreme cold, thin air, and unpredictable weather changes. A popular folk saying goes: "In the first three months, the mountains are sealed with snow; in the fourth, fifth, and sixth months, it's so rainy you'll cry; in the seventh, eighth, and ninth months, travel is slightly easier; in the tenth, eleventh, and twelfth months, you crawl like a dog." It's easy to imagine the tremendous hardships of making a living along this trade route. Most of the time, caravans spend their nights outdoors, carrying tents and provisions, while also arming themselves for self-defense against bandits and wild animals, covering a distance of 20–30 km per day.

Humans are the organizers and controllers of caravans, while horses serve as the transportation means. Suitable horses for transportation include the Dali horse, Dianchi horse, Tibetan horse, Dongshan horse, and Tengyue horse. Horses traveling mountain roads need to have good endurance and should not be too tall to ensure stability. When marching, the mules line up in a queue, with the lead mule and second mule being the best at recognizing the way. The lead mule wears a decorated headstall with protective goggles, tassels, red ornaments, and a noseband, and its saddle is equipped with bells. The lead mule carries a large copper bell, while the second mule carries "second hairpins". They often have the same coat color. "The lead mule runs, the second mule follows," creating a picturesque scene along the entire caravan route.

Ancient Towns and Cities on the Tea Horse Road

The Ancient Tea Horse Road thrived due to the tea-horse trade, and many significant towns along the way prospered because of it. Places where horse caravans rested often evolved into horse inns or market towns, while locations for tea-horse trade mostly developed into larger towns. In these horse inns, travelers could feed their own horses, safeguard their belongings, and find affordable accommodations, making them ideal places for horse caravans to rest for the night.

The towns along the Ancient Tea Horse Road in the Sichuan-Tibet route include Ya'an, Luding, Kangding, Daocheng, Luhuo, Garze, Dege, Jiangda, Yajiang, Litang, Batang, Mangkang, Zuogong, and Chamdo. On the Yunnan-Tibet route, the towns include Heilizhen in Neijiang, Xunde Tacheng, Bengzilan, Adeqiu, Tianzhuzhai, Maofagong, Xishuangbanna, Pu'er, Dali, Lijiang, Zhongdian (now Shangri-La in Yunnan), Deqin, Mangkang, and Chaya. These towns can be mainly classified into the following four types:

Market-oriented towns are usually located at the intersections of ethnic regions and major roads, serving as dedicated trading centers for horses and tea. Service-oriented towns, on the other hand, are often situated along the routes, typically at the intersections of multiple roads. These towns are larger and serve as resting places for travelers, where merchants and drivers can rest, eat, and exchange information. They typically have more inns, restaurants, and taverns, and their distribution is related to the distance covered by travelers in a day. Transfer-oriented towns are primarily located in tea-producing areas, where scattered tea leaves need to be collected. Merchants from various places gather here, forming small towns. The size of these towns depends on the scale of the tea-producing area. Hybrid towns combine elements of the above types, and their development often hinges on their strategic geographic location and favorable natural resources, leading them to evolve into larger cities.

Tea-horse trade is a significant factor contributing to the formation of the aforementioned towns, but their emergence is the result of multiple factors working together. Below is a brief introduction to the main towns along the Tea Horse Ancient Road.

Luding. Luding is a typical place that prospered due to Sichuan tea trade, falling under the category of the service-oriented city, but its rise occurred relatively late. The Dadu River, as a natural barrier, was difficult to traverse, so Luding was merely a "western frontier village" in the late Ming and early Qing dynasties, with few traces of merchants. It wasn't until the completion of the Iron Chain Bridge in the forty-fifth year of Kangxi's reign (1706) that opened up the transportation artery of the Dadu River, that merchants from other regions began to gather in Luding for trade, leading to its prosperity. By 1930, there were over 30 merchants in the area, making it the gateway to the western part of Dadu River and an important checkpoint for tea from the southern route to enter Dajianlu (modern-day Kangding, Garze, Sichuan), becoming a transfer point for goods between the inland and Kangding.

Kangding, also known as "Dazhezhu," "Dazhe Duo," and "Dajianlu," is a comprehensive city. With the southward shift of Han-Tibetan trade, merchants from various regions outside the pass and Tibet transported local products to Kangding in exchange for tea and cloth, making Kangding a center for border tea trade. During the Yuan Dynasty, this area was a desolate wilderness, but it gradually formed a village in the Ming Dynasty. In earlier times, there was a mutual distrust between the Han Chinese and Tibetans, with trade occurring at intersections. However, with the development of Kangding, this barrier was broken. In the early days of Kangding's rise during the Ming Dynasty, it was a gathering place for mule drivers to cook and stay overnight. During the Qing Dynasty, Kangding became the hub of southern border tea trade, thriving with tea markets and evolving into a large market-oriented city. It attracted both Han and tribal traders, and as it was the convergence point of multiple routes, it became a major city on the western frontier.

Chamdo, positioned at the hub of trade and commerce, is known as the "Pearl of Eastern Tibet" and is a comprehensive city. Situated at the confluence of the Jinsha River, Lancang River, and Nu River, it forms the Three Rivers Gorge, where "two mountains enclose one river, and two rivers enclose one mountain." This area serves as the gateway to Tibet. The ancient Tea Horse Roads from Sichuan and Tibet and from Yunnan and Tibet intersect here, while the north–south passage serves as a corridor for ethnic groups. Chamdo lies precisely at the crossroads of east–west and north–south major transportation routes. During the Qing Dynasty, the Minister stationed in Tibet, Song Yun, described Chamdo as the most critical area in the junction of Sichuan, Yunnan, and Tibet, a strategically vital region facilitating connections to Sichuan in the east, Yunnan in the south, Tibet in the west, and Qinghai in the north. Surrounding cities include Nyingchi, Nagqu, and Lhasa, making Chamdo a pivotal point for the distribution of tea throughout Tibet.

Dali is a comprehensive city inhabited by 13 ethnic groups for generations, and it is situated at the confluence of four major river systems: the Jinsha River, Lancang River,

Nu River, and Yuan River (Red River). During the Tang Dynasty, Dali was connected to Chengdu in the east, extended to Myanmar and India in the west, reached Jiaozhi (Northern Vietnam) in the south, and went northward to Tubo (Tibet). With Xiaguan of Dali as its center, it had extensive connections. The Wuchi Road, the Maoniu Road, and the Yongchang Road intersect in Dali. These three roads belong to the "Shu Yuandu Road", indicating that Dali is the core city of the southwestern international thoroughfare. The beginning of the Tea Horse Road from Yunnan to Tibet lies in the tea-producing areas of Xishuangbanna and Simao. Merchants transported tea westward to Dali and then northward to Lijiang, Zhongdian, Weixi, and Deqin before entering the Tibetan region via Sichuan. Therefore, Dali serves as a hub on the Tea Horse Road. Although not the origin of Yunnan tea, Dali's strategic location as a transportation hub and its frequent tea-horse trading activities have contributed to its prosperity, making it a pivotal trading center for transporting Yunnan tea northward to Tibet.

Pu'er is a service-oriented city and serves as the largest oasis on the Tropic of Cancer. It is hailed by the United Nations Environment Programme as "the world's paradise, and the paradise's world". Currently, it has a tea plantation area of 3.18 million mu (approximately 2120 km^2), resembling an ocean of tea. Pu'er is one of the important production areas for the renowned Pu'er tea and is also one of the larger tea-producing regions in China. The Tea Horse Road in the Pu'er region includes the Banjupao Tea Horse Road and the Nakeli Tea Horse Road, which are important sections of the "Official Horse Road" leading to Beijing, with a history of over 1000 years.

The caravans traversing the ancient Tea Horse Road take eight to nine months to deliver tea from its place of origin to consumers. Exposed to wind, sun, and rain along the journey, Pu'er tea undergoes natural fermentation, transforming into the mellow-flavored aged Pu'er tea known as "ripe Pu'er." Local Yunnan production consists only of "raw Pu'er," while consumers outside the region perceive Pu'er tea as "ripe Pu'er," unaware that what they are drinking is actually "aged" Pu'er tea. The horsemen transporting the tea do not disclose this truth, and tea drinkers find the tea enjoyable, believing it aids in digesting oily foods. Currently, both "raw Pu'er" and "ripe Pu'er" have their own merits, each possessing a distinct aroma and flavor profile. Whether from the place of origin or elsewhere, tea drinkers make their choices based on personal preference [19].

In addition, there are numerous small towns along the Tea Horse Road, such as Yiwu Ancient Town in Xishuangbanna, Lushi Ancient Town in Lincang, Shaxi Ancient Town in Dali, Shangli Ancient Town in Ya'an, Shuhe Ancient Town in Lijiang, and Songgui Ancient Town in Dali. Countless ancient towns contribute to the splendid civilization along the Tea Horse Road.

Supply Routes During the War of Resistance

During the War of Resistance against Japan, the Camel Route was well-known. General Claire Lee Chennault led the Flying Tigers to fly over the Himalayas, transporting strategic supplies from India to China, which effectively supported China's war against Japan. In 1970, Russian scholar Peter Goullart mentioned in "Forgotten Kingdom" the caravan transportation, which refers to the Tea Horse Road in the southwest region. During the war, when all routes into China were blocked, caravan transportation used 8,000 mules and 20,000 yaks, with long-distance caravans arriving in Lijiang almost every week. Few people realized the magnitude and importance of these caravans shuttling between India and China. This convincingly demonstrated to the world that even if all transportation methods failed, horses were always ready to form new bonds among dispersed peoples and nations [20].

During the War of Resistance against Japan, the Nationalist Government relocated to Chongqing, around which the main forces of the resistance were concentrated. Chongqing, located in the southwest region, faced severed land transportation routes, making the road from India through Burma to Chongqing crucial. The people of Yunnan Province took nine months to open up the Burma Road. In 1941, after the outbreak of the Pacific War, Japanese forces, aiming to sever Chongqing's supply lines, deployed four divisions totaling 100,000 troops and 250 combat aircraft, launching a three-pronged attack on Burma. This marked the beginning of the Western Yunnan resistance by China. The Japanese attempted to block military aid from the Allies to China, capture Kunming, and encircle Chongqing from the rear. In late October 1943, the counteroffensive against the Japanese in northern Burma achieved initial success. In September 1944, towns such as Songshan, Tengchong, Longling, and Mangshi were successively liberated. In January 1945, victory was achieved in the Western Yunnan Campaign.

The Burma Road became a focal point of contention, sometimes open, sometimes closed, yet the Tea Horse Road through the jungle never ceased. During the later stages of the War of Resistance against Japan, the Tea Horse Road became the primary international route in the rear of the Greater Southwest, with approximately 30,000 caravans traveling annually between Yunnan, Tibet, and India. The Tea Horse Road symbolized the indomitable spirit of the Chinese people, supporting the great struggle of the Chinese people against aggression during the difficult period of the war.

Historical and Cultural Value of the Tea Horse Road

After the founding of the People's Republic of China, the Communist Party of China led the Chinese people in constructing National Highway 318 (Sichuan-Tibet Line), National Highway 109 (Qinghai-Tibet Line), National Highway 214 (Yunnan-Tibet

Line, including the Tangfan Ancient Road), and National Highway 219 (Xinjiang-Tibet Line). In terms of railways, the Qinghai-Tibet Railway, Sichuan-Tibet Railway, Yunnan-Tibet Railway, Gansu-Tibet Railway, and Xinjiang-Tibet Railway were constructed. In terms of airports, Lhasa Gonggar International Airport, Nyingchi Mainling Airport, Chamdo Bangda Airport, Ali Kunsha Airport, and Shigatse Peace Airport were built. Today, people can travel to Tibet through various means of transportation, while the Tea Horse Road is submerged in the dust of history. Today, as we pay tribute to this history of over a thousand years, we can only sigh with emotion at the hardships of our ancestors, the vastness of history, and the great achievements of humanity in the new era. In the past, the Tea Horse Road was not only a trade route and a means of livelihood but also a path for religious dissemination, ethnic migration, ethnic integration, and border stabilization.

Path of Religious Propagation

The Tibetan Buddhism once spread along the Tea Horse Ancient Road, covering the Qinghai-Tibet Plateau and surrounding areas. It branched into two dissemination routes: one passing through Tibet, Qinghai, Gansu, Inner Mongolia, Liaoning, Jilin, Heilongjiang, Beijing, Hebei, Henan, and Jiangsu; the other route traversing Tibet, Yunnan, Sichuan, Guangxi, and Hubei. The dissemination of Bon religion and Tibetan Buddhism facilitated exchanges among the Naxi, Bai, Tibetan, and Han ethnic groups in northwest Yunnan, contributing to their mutual development.

Path of Ethnic Migration

Throughout history, ancient tribes such as Shu, Tu, Qiong, Guan, Sou, Kunming, Ailao, Dian, Yelang, and Dianyue thrived in the southwestern region. Leveraging the routes of the Tea Horse Ancient Road, for thousands of years, various ethnic groups including the Han, Tibetan, Yi, Naxi, Lisu, Hani, Jinuo, Qiang, Pumi, Bai, Nu, Jingpo, and Achang have continuously interacted and migrated.

Path of Ethnic Integration

The Tea Horse Ancient Road is a vast network spanning across southwestern China, extending all the way to the Central Plains, northwest, northeast, Mongolia, the Western Regions, and even expanding into Myanmar, India, Europe, and Africa, forming an even broader network. With the commercial exchanges among various ethnic groups along the route, intermarriage between the Han people and minority ethnic groups in the border areas gradually developed into a trend. More than 20 ethnic groups including the Tibetan, Yi, Bai, Hani, and Blang have achieved communication and integration through this Tea Horse Ancient Road.

Path of Securing and Stabilizing the Borders

The immense demand generated by the tea-horse trade spurred rulers to expand into the southwest, helping the region emerge from a state of wilderness. The flourishing tea-horse trade facilitated significant advancements in transportation, consequently fostering substantial growth in urban areas and the economy. This laid the foundation for stability and consolidation in the southwestern borderlands of China.

From a Macroscopic Geographic Perspective: Tea-Horse Ancient Road

From a macroscopic geographic perspective, the "Circum-Himalayan" region is a hub of civilizations. The Himalayas serve as the source of over ten major rivers, which in turn form numerous floodplains downstream. On these plains or plateaus, there are developed agricultural and pastoral industries, giving rise to various civilizations throughout history, including the Central Plains civilization, Mongolian civilization, Central Asian civilization, Persian civilization, Indian civilization, and Southeast Asian civilization. These civilizations, all stemming from the water sources of the Himalayas, are characterized by river-based cultures and share certain similarities. Among them, the plain where the Chinese civilization is situated is the largest, with the most abundant water sources and the richest and most fertile soil. This uninterrupted civilization undoubtedly shines as the most brilliant pearl within this circle. The establishment of transportation routes within the Himalayan Civilization Circle was a remarkable feat of humanity. The merging of the "Shu Shendu Road" and the northern Silk Road gave birth to a commercial route encircling the Himalayas. Significant contributions were made by Emperor Qin Shi Huang and Emperor Han Wu Di, who opened up routes to Myanmar, India, and Daxia with the famous phrase "to penetrate the southern mountains and cross the Lancang River." Along these routes, the main traded goods included tea, Shu cloth, Qiong staff, silk, jade, and lacquerware. The Tea Horse Road of Yunnan and Tibet, the Sichuan-Tibet Tea Horse Road, and the Tang-Tibet Ancient Road were radial routes pointing towards the summits of high mountains. The combination of circular and radial transportation routes is a characteristic feature of this transportation network. The rivers on this network nurture the Eastern civilization around the Himalayas, which rivals the Western civilization around the Mediterranean Sea.

References

1. Wu Zeng. Nenggaizhai Manlu (Random Notes from the Ability-to-Change Studio). Jinan: Shandong People's Publishing House, 2020.
2. Wang Fengyun, Wei Mei. The Formation and Dissemination of the Concept of 'Ancient Tea-Horse Road' [J]. Guangxi Ethnological Studies, 2020(6):8.
3. Zhao Xuemin. Supplements to the Compendium of Materia Medica. Beijing: China Press of Traditional Chinese Medicine, 2007.
4. Xianba, Ma Luyue. Tea as a Medium: A Study on the Exchanges between the Qinghai-Tibet Region and the Central Plains Dynasties [J]. Journal of Social Sciences of Jiamusi University, 2021, 39(6):4.
5. Ma Duanlin. Wenxian Tongkao (Comprehensive Examination of Documents). Beijing: Zhonghua Book Company, 2011.
6. Luo Yong. Ethnic Relations in Western Yunnan During the Han Dynasties and the Opening of the Bonan Ancient Road [C]// 2016. Journal of Chinese Middle Ages History, Volume 6. Beijing: The Commercial Press, 2016.
7. Liu Chaofan. A Review of Research on the Ancient Tea-Horse Road in the Past Three Decades [J]. Agricultural Archaeology, 2021(2).
8. Li Tao. Continuation of the Comprehensive Mirror for Aid in Government. Beijing: Zhonghua Book Company, 1979.
9. Chen Guanxun. Records of Tibet. Chengdu: Bashu Publishing House, 1986.
10. Sima Qian. Shi Ji. Beijing: Zhonghua Book Company, 1959.
11. Feng Yan. Fengshi Wenjianji Jiaozhu (Feng's Records of What I've Heard and Seen, Annotated). Annotated by Zhao Zhenxin. Beijing: Zhonghua Book Company, 2005.
12. Song Qi, Ouyang Xiu, Fan Zhen, et al. The New Book of Tang. Beijing: Zhonghua Book Company, 1975.
13. Qiu Jin. Supplement to the Great Learning's Derivation of Meaning. Changchun: Jilin Publishing Group, 2005.
14. Tan Fangzhi. The Sale of Yunnan Tea in Tibet [C]// Frontier Political Discussions, Volume 3. 1944.
15. Tuo Tuo, et al. History of the Song Dynasty. Beijing: Zhonghua Book Company, 1985.
16. Zhang Tingyu, et al. History of the Ming Dynasty. Beijing: Zhonghua Book Company, 1974.
17. Ji Huang. Qinding Xu Wenxian Tongkao (Imperially Ordered Continuation of Comprehensive Examination of Documents). Taipei: The Commercial Press, Ltd. (Taiwan), 2008.
18. Zhao Erxun, et al. Draft History of the Qing Dynasty. Beijing: Zhonghua Book Company, 1998.
19. Qin Guoqiang. A Historical Account of Chinese Transportation. Shanghai: Fudan University Press, 2012.
20. Isabel F. Crook. The Forgotten Kingdom. Kunming: Yunnan People's Publishing House, 2007.
21. Fan Ye. Hou Hanshu (History of the Later Han Dynasty). Annotated by Li Xian, et al. Beijing: Zhonghua Book Company, 1997.

Chapter 8
Urban Roads: Inheritance and Development of Traditional Chinese Urban Planning Wisdom

The concept of urban planning in ancient China has been an integral part of Chinese civilization for a long time. In dealing with the relationship between humans and nature, it fully embodies the ancient philosophical concept of "harmony between man and nature", with ideas such as "imitating heaven and following the laws of the earth", "establishing order through rituals", and "using local materials according to local conditions" being the most typical. Under the guidance of these excellent ideas, Chinese urban construction has produced numerous masterpieces in the history of world urban planning. The city of Chang'an in the Tang Dynasty had a profound influence on the design of Japan's capital city; Beijing is an outstanding example of world urban planning; Hangzhou, with its beautiful scenery and long history, is an excellent representative of ancient cities. The planning of the Qiantang New City, transitioning from the West Lake era to the Qiantang River era, has been highly praised by the international planning community. In modern times, the introduction of Western urban planning ideas has had a significant impact on the traditional Chinese urban planning system. However, after a series of absorptions and integrations, fruitful results have still been achieved. In modern times, especially since the 21st century, with urbanization entering a new stage, under the guidance of new development concepts, China's urban planning and construction field has witnessed many new explorations and bold innovations. These not only inherit and carry forward China's traditional construction concepts but also point out the direction for the future development of Chinese cities.

This chapter starts with the close relationship between cities and transportation, and generally narrates the concepts and methods of urban planning and urban road construction in ancient, modern, and contemporary China in chronological order. Firstly, using Chang'an (now Xi'an), Beijing, and Lin'an (now Hangzhou) as examples, it elaborates on the ancient Chinese urban planning concepts and their manifestations. Secondly, using cases such as Dalian, Nanjing, and Bagua City in Tekes County, it introduces the struggles and integration of traditional Chinese construction thoughts in the face of Western planning concepts in modern times, as well as

© China Communications Press Co., Ltd 2025

H. Lu (ed.), *A Brief History of Transportation in China*,
https://doi.org/10.1007/978-981-97-9851-3_8

the main characteristics displayed. Lastly, combining planning cases of Xiong'an New Area and Beichuan New County, it discusses the new concepts and methods of current urban and road planning.

Traditional Chinese Urban Planning Thought System

Erlitou Site: The Earliest Urban Road Network

The world-renowned Erlitou Site is the earliest discovered capital site in China to date. Its grid-shaped road network can be considered the earliest urban road network in China [1].

The Erlitou site, located in Yanshi City, Luoyang, Henan Province, dates back approximately 3800–3500 years, corresponding to the Xia and Shang dynasties mentioned in ancient texts. Since its discovery in 1959, archaeologists have uncovered a series of significant findings through excavations. These include large-scale palace complexes, urban layouts, and workshop remains, as well as a wealth of artifacts such as pottery, stone tools, bone tools, and valuable bronze, jade, turquoise, ivory, and lacquerware items. Key discoveries at the site encompass: palace and palace system represented by the rectangular palace city and large rammed-earth palace buildings; funeral ritual system revealed by the burial practices of the aristocracy; national sacrificial system represented by altars; system of utensils demonstrated by national treasures including the bronze and jade ritual vessels and turquoise dragon. Therefore, the Erlitou site is regarded as the earliest confirmed capital site of a large-scale state with centralized power in China [1].

Through the archaeological excavations at the Erlitou site, it can be observed that the central area of the site is laid out in a "nine-square grid" pattern, with the city's roads distributed in a grid-like arrangement. The nine-square grid and the grid-like roads divide the site into multiple grid areas (Fig. 8.1) [1]. With ongoing archaeological excavations in recent years, the extent of these grid-like roads has become even more extensive, and the nine-square grid layout and grid-like main road network of the central area of the Erlitou site have become clearer. According to archaeological experts, the Erlitou site is a meticulously planned, large-scale, and unprecedented capital site, which serves as the origin of many ancient Chinese urban and political systems.

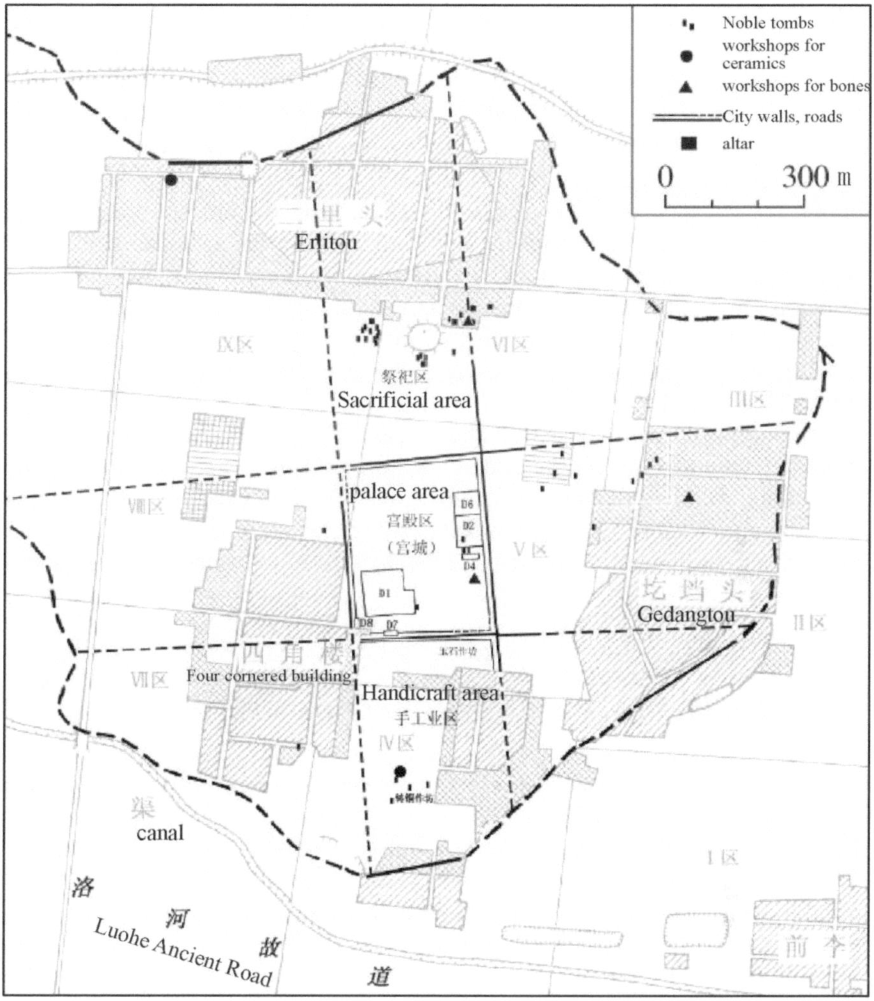

Fig. 8.1 Plan layout of the Erlitou site

Origin of the Concept "Harmony Between Heaven and Humanity

Although the Erlitou site has revealed some early forms of ancient Chinese city planning, systematic and theoretical documentation appeared later. It was not until the Zhou dynasty that coherent written records began to emerge.

The core concept of ancient Chinese city planning, "unity of heaven and humanity," originated from ancient Chinese philosophy and addressed the relationship between humans and nature, representing an important aspect of Eastern holistic thinking [2]. This idea primarily stemmed from the "Book of Changes" and "Tao Te Ching", expressing a principle of harmonious coexistence between humans and nature. The Zhou Yi (Book of Changes), in the Qian Hexagram under the section "Wen Yan", mentions: "The great man aligns his virtue with that of Heaven and Earth, aligns his brightness with that of the sun and moon, aligns his order with that of the four seasons, aligns his fortune and misfortune with that of spirits. He precedes Heaven, and Heaven does not contradict him; he follows Heaven, and complies with the timing of Heaven." Here, "Heaven and Earth" refers to the external natural world of humans, while "human" refers to humans themselves and their creative activities. This phrase highlights the concept and realm of "the unity of Heaven and Human," signifying that humans should harmonize with the external natural laws, not violate the timing of Heaven, and act according to the laws, thereby achieving "harmony between Heaven and Human."

The concept of "unity of heaven and humanity" had a profound impact on the construction of ancient cities, particularly in the regulatory systems and methods for city planning recorded in three ancient texts: the "Book of Changes", the "Rites of Zhou", and the "Guanzi". These texts laid the foundation for traditional Chinese city planning wisdom and provided concentrated interpretations of the "unity of heaven and humanity" concept.

Influence of the "Book of Changes" on Urban Planning

The Western Zhou period laid the foundation for the concept of ritualistic urban planning, with the "Book of Changes" making significant contributions in this regard. Also known as the "Yijing," it is said to have been compiled by King Wen of Zhou and served as a guide to interpreting divination results. The "Book of Changes" mainly consists of sixty-four hexagrams and three hundred eighty-four lines, each with its own explanations (line statements), used for divination purposes. It summarizes the laws of phenomena such as heaven, earth, mountains, rivers, wind, rain, thunder, and hail, creating two basic images of yang and yin lines. Through permutation and combination, it forms eight single hexagrams and sixty-four compound hexagrams, reflecting the evolutionary mechanism from simplicity to complexity. Although it inherited the superstitious traditions of the Shang Dynasty, it underwent significant improvements, reflecting the simple thoughts of the ancient people's understanding of nature and society.

The influence of the "Book of Changes" on ancient urban planning ideas mainly encompasses two aspects [3]:

Firstly, it applied the concept of "observing objects to derive analogies" to the handling of material space, establishing the planning method of "imitating heaven and following the earth," which involves applying abstract thinking to the real material environment. According to the "Records of Wu County," during the Spring and Autumn Period, the King of Wu established his capital in Helu (present-day Suzhou, Jiangsu Province) and commissioned Wu Zixu to build the city. Wu Zixu "observed the earth and tested the waters, imitated heaven and followed the earth, and built a large city… with eight land gates to imitate the eight winds of heaven; and eight water gates to conform to the eight trigrams of the earth" [4]. The ancient Chinese belief in the "round sky and square earth" theory was also reflected in architectural planning. For example, the Altar of Heaven, representing the will of heaven, was built in a round shape, while the Altar of Earth, representing the earth, was built in a square shape, and the Altar of the Sun and the Altar of the Moon were positioned in the east and west respectively. From the concept of "imitating heaven and following the earth," the notion of "favorable form" further evolved, moving away from simple imitation. These ideas had a significant impact on the development of Feng Shui in later periods.

Secondly, it manifested as a pursuit of "numbers," revealing the relationship between "numbers" and "symbols." The pursuit of "numbers" and the arrangement and combination of the eight trigrams led directly to the standardization and symmetry of spatial compositions. This turned buildings and spaces into different symbolic combinations, allowing for systematic extension. In urban architecture, ancient people were keen on seeking auspicious "numbers" and "symbols," prioritizing numbers such as three, six, eight, and nine. Additionally, because numbers represent a kind of order, the exploration of this order further evolved into the concept of "Dao," which refers to the truth behind the appearance of things, thereby achieving a transcendence of understanding.

Influence of the Ritual of Zhou on Urban Planning

"The Rites of Zhou" is an ancient text that records the official system of the Zhou dynasty. Through continuous enrichment and expansion, it has become an important document for studying the social and political systems of ancient China, as well as an important work for researching ancient urban planning ideas. The core urban planning idea reflected in "The Rites of Zhou" is to follow the ritual system, which is consistent with Confucianism, and to maintain traditional social hierarchies and religious rituals. As recorded in the "Kaogongji": "Craftsmen built the country, it is nine li square, with three gates on each side. Within the capital, there are nine north–south and nine east–west streets, each wide enough for nine carriages to travel side by side. The ancestral temples are on the left and shrines are on the right, the court faces south, and equal the size of a farmer's land." [5]. This has become an important principle adhered to in ancient urban planning and construction, as shown in Fig. 8.2 [3].

"The Rites of Zhou" exerted significant influence on ancient urban planning and architecture in two main aspects:

Firstly, it infused hierarchical thinking throughout. In the planning and construction of cities, it fully implemented the strict hierarchical concept of the national system to reflect the feudal system's hierarchy between rulers and subjects, fathers

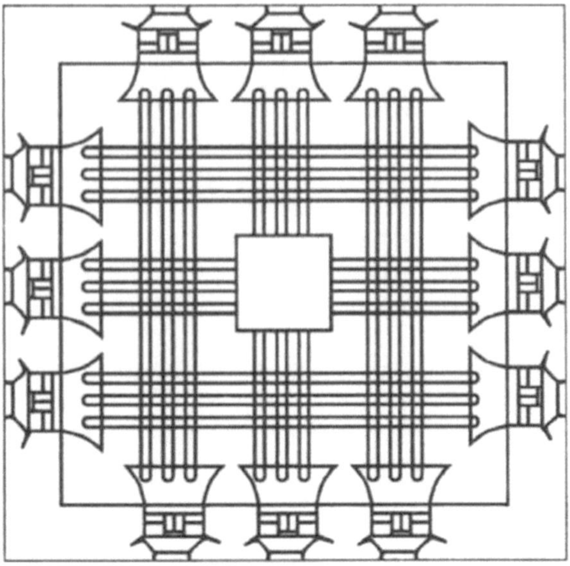

Fig. 8.2 Schematic diagram of the nine vertical and nine horizontal lines in the national classic

and sons, and superiors and subordinates. This hierarchical concept was continuously standardized, faithfully serving the ritual system, resulting in ancient cities appearing more orderly and rhythmic. For example, in the construction of capital cities, the palace and the palace for imperial ceremonies should be prominently located in the central and elevated position of the city, higher than other buildings. The capital city itself was larger in scale compared to other cities. Many ancient Chinese capital cities were built according to this principle, with the most typical examples being Chang'an City of the Tang Dynasty and Beijing City of the Yuan, Ming, and Qing Dynasties.

Secondly, it established the system of rites and music. In Confucianism, equal emphasis was placed on rites (li) and music (yue), advocating for "benevolent governance" and the "Doctrine of the Mean." "Rites" represented hierarchy and order, while "music" represented education and rhythm. In reality, the integration of rites and music reflected the idea that the material environment of the city should serve the authority's rule and the spiritual world. This was manifested in various aspects of ancient city architecture and layout, such as the arrangement of ancestral temples on the left and shrines on the right, and the ringing of bells and drums at dawn and dusk. Therefore, the influence of "The Rites of Zhou" on ancient urban planning concepts not only pertained to the external forms of city construction described in the "Kaogongji" but also involved various aspects of politics, culture, ethics, aesthetics, and more within the state and society [2].

City Planning Ideas in Guanzi

Compared to the orderliness of "The Rites of Zhou—Record of the Architect", there is also a rational and functionalist aspect in traditional Chinese urban planning theory, which is prominently reflected in the book "Guanzi".

"The Book of Master Guan" was written during the Warring States period, and is considered a political masterpiece attributed to Guan Zhong by later generations, reflecting many Daoist ideas. In terms of urban planning and architecture, "The Book of Master Guan" provides a highly systematic exposition, covering various aspects such as site selection for cities, the relationship between urban development and economy, determination of urban scale, division of urban land use functions, layout of road networks, water sources, and drainage.

The planning theories described in "The Book of Master Guan," characterized by a naturalistic perspective and functionality, stand in stark contrast, even opposition, to the Confucian "Rites of Zhou." "The Book of Master Guan" advocates for a practical approach, prioritizing substance over form, embracing diversity, and adhering to the principle of 'utilizing local materials and adapting to local conditions', free from the constraints of ancestral rituals and social norms. Therefore, it states, "City layouts need not adhere strictly to standardized shapes, nor roads to rigid guidelines." [6]. This stands in sharp contrast to the prescription in the "Kaogongji" which outlines, "Craftsmen plan a capital, it is nine li square, with three gates on each side. Within the capital, there are nine north–south and nine east–west streets, each wide enough for nine carriages to travel side by side." [5].

The contribution of "The Book of Master Guan" to the ancient Chinese urban planning concept lies in:

Firstly, it proposes the idea of conforming to natural laws. "The Book of Master Guan" suggests that urban planning should "adapt to local conditions and make the best use of natural resources." Regarding the relationship between cities and the natural environment, it states: "Whenever establishing urban cities and large towns, they should not be built at the foot of tall mountains, but must be located near large rivers. The terrain should not be too high, to avoid proximity to arid areas and ensure an adequate water supply; nor should it be too low, to avoid proximity to waterlogged areas, thus saving efforts in building embankments and channels." [6]. Choosing a city location with convenient access to water and efficient drainage is emphasized. "The Book of Master Guan" also points out in the chapter on water and land: "Water is the essence of the land, akin to the circulation of veins and arteries. Hence, it is said that water is essential material." This addresses the relationship between water and the city. These ideas played an important role in the formation and development of Feng Shui theories in later generations. An example of a city significantly influenced by "The Book of Master Guan" is Nanjing. Its terrain is complex, with the Yangtze River flowing from southwest to northeast, surrounded by mountains on all sides. Only the central area is relatively flat, making it a typical irregular capital city in ancient times.

Secondly, it emphasizes the impact of the relationship between people and land on urban development. “The Book of Master Guan” proposes: “A country where all the land has been cultivated but the people still have no savings is because the country’s territory is small and there is little arable land; whereas a country where only half of the land is cultivated but the people have surplus food and abundant grains is because the country’s territory is large and there is ample arable land.” [6]. It suggests arranging the population according to the situation of land cultivation and surplus, ensuring a balance between people and land and leaving room for expansion. This viewpoint is crucial for the development of cities in the agricultural era.

Thirdly, it introduces the concept of functional zoning. The zoning concept proposed by “The Book of Master Guan” is based on the practical needs of economic and urban development, closely related to the economic development of the Qi state at that time. As the first hegemon during the Spring and Autumn Period, the Qi state, under the leadership of Guan Zhong, fully developed commerce and industry, significantly enhancing its national strength. The capital city, Linzi, thrived due to the flourishing commerce and industry, with a large population. In its layout, the positioning of handicraft industries and commercial markets was considered to facilitate the development of various industries. Therefore, this approach represents a practical arrangement of urban functions from the perspective of socio-economic activities, reflecting the pragmatism of urban planning concepts [2].

Ancient Typical Capital Cities

In the long history of ancient society, the Chinese ancestors created a series of masterpieces in urban planning and construction. From site selection and planning to layout design, most of them underwent meticulous planning and precise design by skilled craftsmen. To a certain extent, the origin of Chinese cities is synchronous with urban planning and construction.

A survey of ancient urban literature reveals that the planning and construction of capital cities are the most concentrated embodiment of urban planning ideology. This is because capital cities are often meticulously designed and built by rulers to meet the political needs of royal power on both material and spiritual levels. However, capital cities of different eras each have their own unique characteristics. The following sections on the Han and Tang Dynasty’s Chang’an City, the Yuan and Ming Dynasty’s Beijing City, and the Southern Song Dynasty’s Lin’an City, will detail how they inherit the same construction ideology while also reflecting the shifting trends and continuous development of planning principles over time.

Capital City of chang'an in the Han and Tang Dynasties

Chang'an is a renowned city in Chinese history, witnessing the most glorious periods of two ancient dynasties, namely the Han and Tang dynasties. However, geographically, the Han and Tang capitals of Chang'an were located in two different positions (see Fig. 8.3). The Han capital of Chang'an was situated northwest of the present-day Xi'an, at the northwest foot of Longshouyuan, extending northward to the banks of the Wei River, while the Tang capital of Chang'an was located at the southern foot of Longshouyuan, extending southward to the shores of the Qujiang Pool.

At the beginning of the Han Dynasty, when Liu Bang established the Han Dynasty in 202 BC, Han Chang'an City only had one imperial retreat palace left from the Qin Dynasty, which was the Changle Palace. After Liu Bang, the founding emperor of the Han Dynasty, moved the capital to Chang'an, he officially resided in the Changle Palace in 200 BC and appointed the Prime Minister Xiao He to be in charge of the construction of Chang'an City. According to "Records of the Grand Historian," it is recorded that he "constructed the Weiyang Palace, established the Eastern and Northern Gates, the Front Hall, the Armory, and the Granary." [7]. To the west of the Changle Palace, Xiao He built the Weiyang Palace, marking the beginning of the construction of Han Chang'an City. Subsequently, during the reign of Emperor Hui of Han, a city wall was built around the outskirts of Chang'an City within five years (194–190 BC), forming the basic outline of the city. By the time of Emperor Wu of Han, the scale and layout of Chang'an City were further improved. According

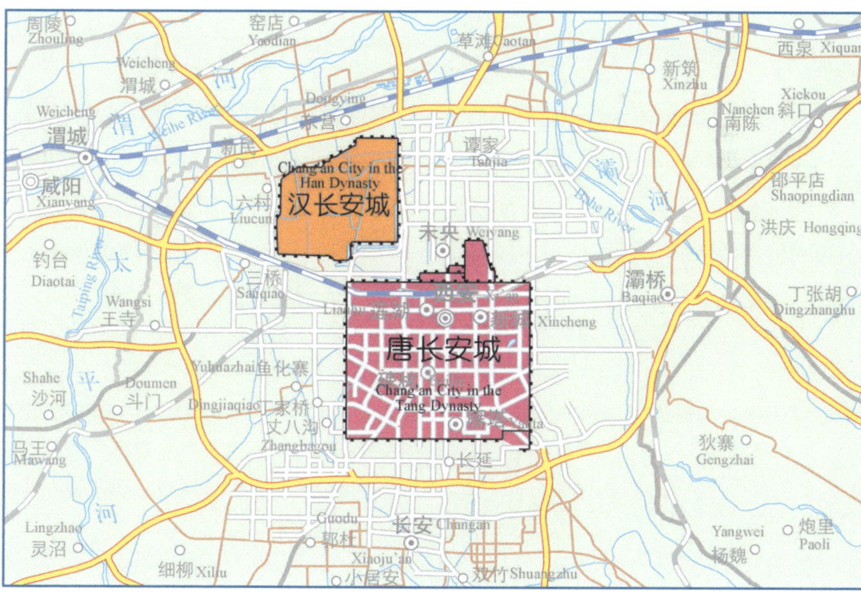

Fig. 8.3 Locations of Han Chang'an and Tang Chang'an

to the records in the “Maps/Documents of the Sanfu Regio,” it states, “Chang’an City is thirty-two li and eighteen steps in length and width, covering a land area of 973 hectares, with eight main streets and nine side streets, three palaces and nine governmental offices, three temples and twelve gates, nine markets and sixteen bridges.” [8]. Additionally, Emperor Wu of Han also built the Jianzhang Palace outside the city. By the end of the Western Han Dynasty, Wang Mang built the Nine Temples in the southern suburbs of Chang’an for the purpose of worshipping his ancestors. Thus, the construction of Han Chang’an City was completed, as shown in Fig. 8.4.

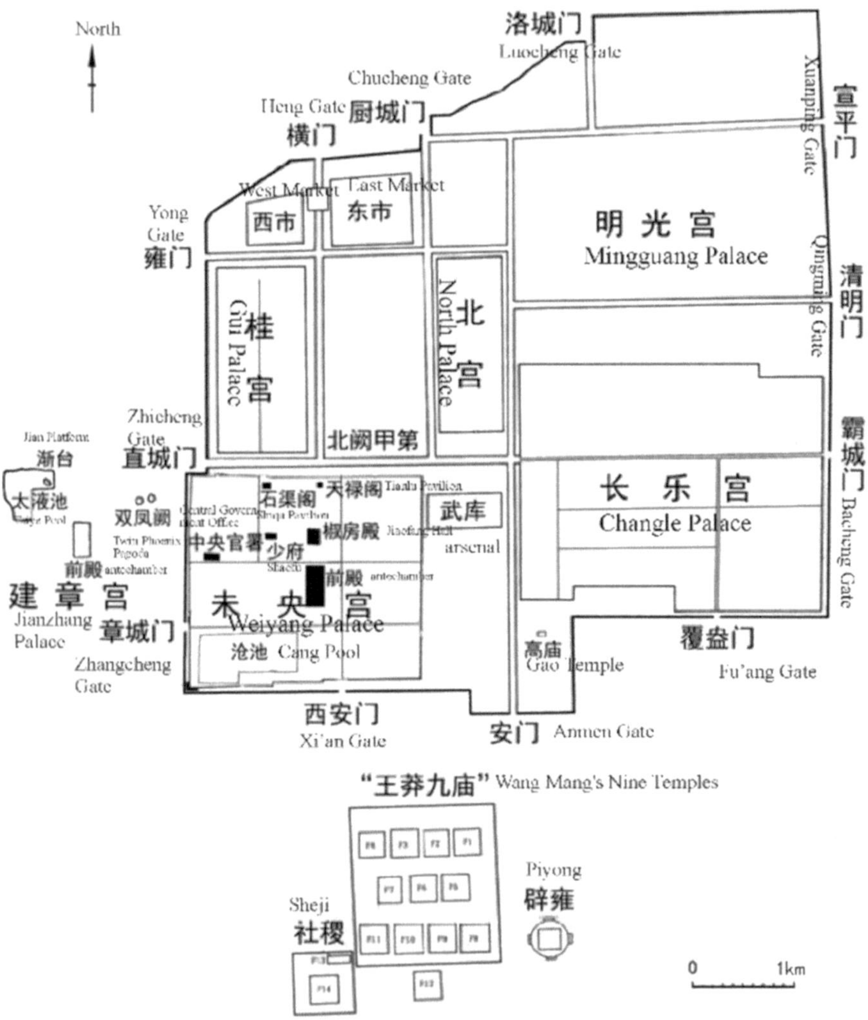

Fig. 8.4 Plan of Han Chang’an City

The road layout within the Han Chang'an City did not form a regular grid pattern typical of the well-organized field system, mainly due to the phased construction. Over the centuries of its formation, Han Chang'an City suffered from a lack of comprehensive planning, with unclear urban axes and road networks, appearing somewhat lacking in overall coherence. Nonetheless, it still reflects the ancient considerations in urban construction. Firstly, Han Chang'an City was built on the southern bank of the Wei River, following the course of the river, with the northern city wall showing a "tortuous and winding" pattern [9], reflecting the harmonious relationship between the city and nature. Secondly, Han Chang'an City was known in ancient times as the "Dipper City", with special significance. Ancient people worshipped the Gig Dipper, believing it was associated with emperors, symbolizing stability, order, and the well-being of the nation and its people. However, this hopeful vision turned to ashes at the end of the Western Han Dynasty. Due to natural disasters and human calamities, continuous warfare, and socio-economic crises, in the third year of the Chimei uprising (25 CE), the Chimei Army invaded Chang'an, burning palaces, looting extensively, and ultimately leading the once glorious city to decline.

The construction of the Tang Chang'an City was built on the foundation of the Sui Daxing City. In the first year of the Kaihuang era (581 AD), during the establishment of the Sui Dynasty, Emperor Wen of Sui established his capital in Guanzhong, avoiding the former site of Han Chang'an City, and chose to build a new capital in the southeast direction, known as "Daxing City" (refer to Fig. 8.5). Daxing City was designed by the architect Yuwen Kai and was an ideal royal city that closely followed the principles outlined in "Kaogongji". The layout of Daxing City was square, with the entire city centered around a main axis aligned with the southern gate of the Imperial Palace, Imperial City, and Outer City. The Imperial Palace, Imperial City, and residential areas were distinctly separated and relatively isolated, ensuring both security and practicality. However, before Daxing City could be fully completed, the Sui Dynasty collapsed.

When the Tang Dynasty replaced the Sui Dynasty, Li Yuan did not destroy the capital city but instead expanded it, making it more in line with the description in the "Konggongji". The design of the blocks in Tang Chang'an City reflects the concept of "unity of heaven and humanity". The city's 25 main streets running east–west and north–south divide the entire city into two markets and 108 blocks. Among them, Zhuque Street serves as the boundary, dividing the city into eastern and western parts: the eastern part belongs to Wannian County, which should have had 55 blocks, but due to the Qujiang Scenic Area occupying two blocks in the southeast corner of the city, it actually has 53 blocks; the western part belongs to Chang'an County, with one market and 55 blocks. All the blocks are neatly symmetrical and evenly distributed along the central axis, embodying the changes of seasons and the rituals of the Zhou Dynasty. This layout is unique to Tang Chang'an City (see Fig. 8.6).

The layout of the Tang Chang'an City was tightly regulated, inheriting the traditional planning system of the previous dynasties while pioneering a new urban planning system, thus reaching the pinnacle of ancient Chinese urban construction.

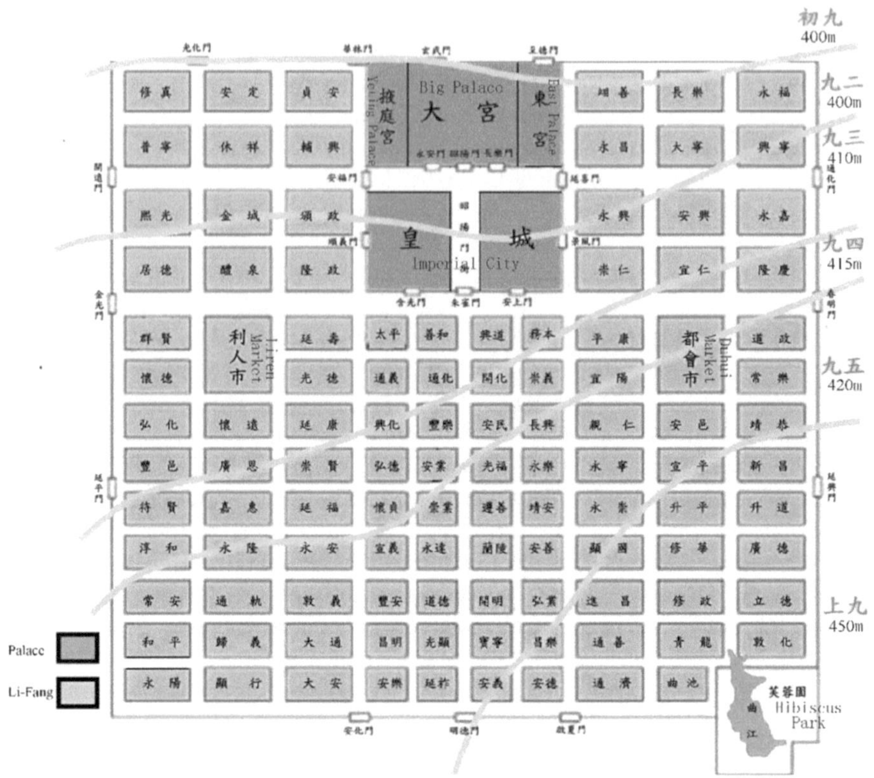

Fig. 8.5 Plan of Sui Daxing City

Yuan and Ming Dynasties' Beijing City

The Ming Dynasty Beijing City was built on the foundation of the Yuan Dynasty's Dadu City. In the eighth year of the Yuan Dynasty (1271), Kublai Khan established the Yuan Dynasty and moved the capital to Beijing the following year, marking the beginning of the urban construction of Dadu City. The site of Dadu City is located in present-day Beijing, extending from the remains of the Yuan Dadu Earthen Wall in the north to Chang'an Street in the south, and from the Second Ring Road in the east to the west. The street layout of the capital city established at that time laid the basic urban pattern for present-day Beijing.

The construction of Yuan Dadu was overseen by the official Liu Bingzhong from the Imperial Secretariat, with the water management expert Guo Shoujing serving as the supervisor of the Imperial Waterways. They constructed the Tonghui Canal from Yuan Dadu to Tongzhou, connecting the Grand Canal from Yuan Dadu to Hangzhou. The layout of Yuan Dadu was rectangular, with a shorter width from east to west and a longer length from north to south. The city wall had a total length of 60 li and

Fig. 8.6 Tang Chang'an City plan

240 paces (approximately 2.86 km), with 11 gates: three each on the south, east, and west sides, and two on the north side. The Imperial City was located slightly south of the city center, surrounded by a red wall known as the "Xiao Wall," with the main gate named Lingxing Gate and thousand-step corridors on both sides. Outside the east wall of the Xiao Wall was the canal for transporting goods. The layout of the Imperial City did not follow a symmetrical east–west axis based on the main palace, but rather centered around the Taiye Pool, with three palaces arranged around it—the Grand Internal Palace, the Longfu Palace, and the Xingsheng Palace—reflecting the Mongolian tradition of settling near water. The main gates of the Grand Internal Palace were the Chengtian Gate to the south, the Houzai Gate to the north, the Donghua Gate to the east, and the Xihua Gate to the west. In front of the Chongtian

Gate was the Jinshui River, with the Zhou Bridge spanning it. The main hall of the Grand Internal Palace was the Daming Hall, where the emperor conducted affairs of state and resided. It had a width of 11 rooms, with a corridor connecting to the rear hall. Behind the Daming Hall was the Yanchun Pavilion, the residence of the empress. Additionally, there were other buildings such as the Yude Hall, the Imperial Treasury, the Falconry House, and the Sheepfold (see Fig. 8.7).

To the south and southeast of the imperial city lies the administrative district, while to the north, there is a bustling market area adjacent to a water pond. The east is occupied by the imperial ancestral temple, and the west is dedicated to the state shrines. Additionally, each side of the city wall has three gates (with the exception of the northern wall), conforming to the records in the 'Kaogongji' regarding 'nine

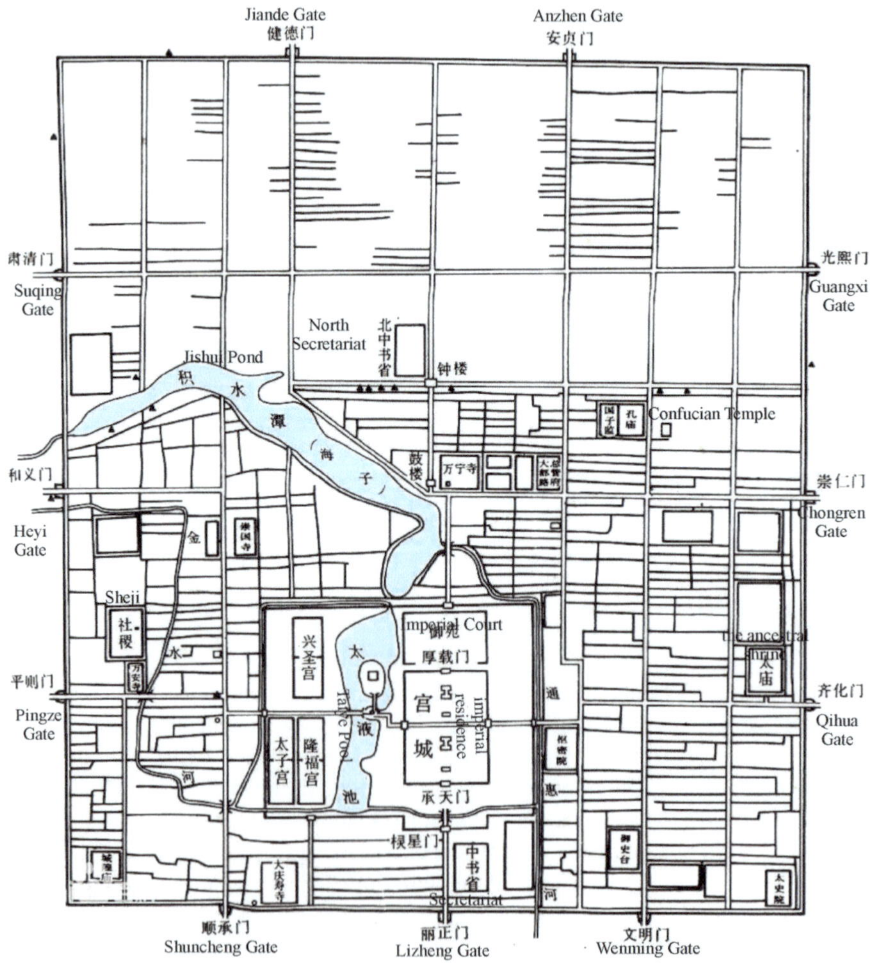

Fig. 8.7 Plan of Yuan Dadu

north–south streets, nine east–west streets,' 'the market in the rear,' and 'ancestral temples on the left and altars on the right' [5]. To the west of the central pavilion lies the Drum Tower and Bell Tower, which serve as the city's timekeeping facilities. To the west of the Drum and Bell Towers is the Water Accumulation Pool, the terminus of Yuan Dynasty canal transportation and the busiest commercial area in the entire city. Here, there are markets for rice, noodles, satin, fur hats, jewelry, ironware, geese, and ducks, as well as pavilions like the Prospect Lake Pavilion and the Ten Thousand Springs Garden.

The streets of Great Capital of the Yuan Dynasty were meticulously planned, with main roads generally connecting the opposite city gates. Marco Polo's travelogue records: "The entire city is designed with straight-line planning. Overall, all streets are straight, leading directly to the city walls. If a person stands on the city gate and looks straight ahead, they can see the opposite city gate on the opposite wall. On both sides of the public streets in the city, there are various shops and stalls… The entire city is laid out in a square shape, like a chessboard [23]." Although Great Capital of the Yuan Dynasty had three gates on the south side and two gates on the north side, there was a wide Imperial Road, with remnants discovered through surveys measuring up to 28 m wide, running north from the Li Zhengmen gate through the central axis of the Imperial City, passing through the Chongtianmen, Damingmen, Daming Palace, Yanchunmen, Yanchun Pavilion, Qingning Palace, and Huozai Gate, directly to the central pavilion.

The basic layout of the streets in the Great Capital of the Yuan Dynasty consisted of numerous east–west alleys arranged at equal distances on both sides of the main north–south thoroughfares. The main streets were approximately 25 m wide, while the alleys were 6–7 m wide. Heading north from the Guangxi Gate, there were a total of 22 east–west alleys, and the same number existed between the Chongren Gate and the Qihua Gate. It can be seen that the uniform format of arranging 22 alleys between adjacent pairs of city gates was a consistent feature of the urban planning in the Great Capital of the Yuan Dynasty.

In the first year of Ming Hongwu (1368), Zhu Yuanzhang established the Ming Dynasty in Nanjing. In the same year, they seized the Great Capital of the Yuan Dynasty, demolishing all the existing palaces inside, aiming to eradicate the so-called "royal aura." Subsequently, the Military General Xu Da made some adjustments to the capital, and particularly after the decision by the Ming Emperor Zhu Di to move the capital, extensive construction work took place. This construction lasted for 14 years, starting from the fourth year of the Yongle era (1406) until the eighteenth year (1420), when it was mostly completed, and the following year, the capital was officially relocated to Beijing. The most significant adjustment made to the Great Capital of the Yuan Dynasty during this phase was the "reduction of its northern city by five li," [24] moving the northern city wall south to the position of the current northern wall of Beijing (utilizing the narrowest part of the Jishuitan (water accumulation pond) in the northwest to inwardly taper), thereby reducing the area of the city by one-third compared to the Great Capital of the Yuan Dynasty. The north city wall still had two gates, the East Gate called "Anding" and the West Gate called "Desheng." The Guangxi Gate at the north end of the east wall and the Suqing Gate at the north end of the west wall were abolished, reducing the number of gates from 11 to 9. Additionally, the south city wall was expanded southward by one kilometer, positioning the Imperial City in the center of the city, emphasizing the supreme authority of the emperor. A west–east street was formed on the basis of the original south wall (nowadays known as Chang'an Street), creating a T-shaped square with the main avenue in front of the Imperial Palace, reminiscent of the layout of the historical Cao Wei Ye City.

In the thirty-second year of the Jiajing reign (1553), due to pressing circumstances, plans were made to construct an outer city outside the inner city walls to defend against external invasions. However, due to the exorbitant costs involved, only the southern city wall was erected. According to the "Ming Huidian" (Collected Statutes of the Ming Dynasty): "In the thirty-second year of Jiajing's reign, a heavy wall was built, enclosing the southern side of the capital, with towers at the eastern and western corners only. It was twenty-eight li long, with seven gates; namely, Yongding Gate, Zuo'an Gate, You'an Gate in the south, Guangqu Gate, Dongbian Gate in the east, Guangning Gate, and Xibian Gate in the west." [10]. This outer city gradually developed into a commercial and residential area, with streets and lanes mostly irregularly arranged, winding or slanting rather than neatly aligned. Only the north–south thoroughfare from Zhengyang Gate to Yongding Gate was straight. With this, the basic layout of Beijing city became relatively stable. After the fall of the Ming Dynasty, the Qing Dynasty continued to use the layout of Ming Beijing City with little alteration (Fig. 8.8).

The renovation and construction of Ming Beijing City exhibit the following characteristics [11]:

Adherence to Traditional Feudal System: It adhered to traditional feudal systems like "Selecting the Center for the Imperial Palace," "Ancestral Temples on the Left and Shrines on the Right," "Front Court at the Front and Market at the Back," and "Court at the Front and Residential Area at the Back." The Imperial Palace was situated at the center, forming the main layout, with a rigorous arrangement reflecting clear political intentions. In front of the Imperial Palace, the Temple of Heaven was on the left, the Altar of Land and Grain on the right, and outside the city in the four directions were the Altars of Heaven (south), Earth (north), Sun (east), and Moon (west). North of the Imperial Palace, outside Xuanwu Gate, there was a market every fourth day of the month, known as the Inner Market.

Separation of Political and Economic Zones: A layout pattern emerged where the political function zone was placed in the north of the city and the economic activity zone in the south, forming the "Market in the South, Palace in the North" layout. The political function zone held dominance, with the layout of city governance, palaces, government offices, and official religious and cultural facilities following traditional feudal ceremonial thoughts, inheriting and developing the traditional planning of capitals from past dynasties, becoming a typical representative of traditional urban planning and construction in China. The economic activity zone mainly consisted of residential areas, commercial markets, guildhalls, gardens, and folk religious buildings, emphasizing adaptability to local conditions and exhibiting more flexibility in its spontaneous formation.

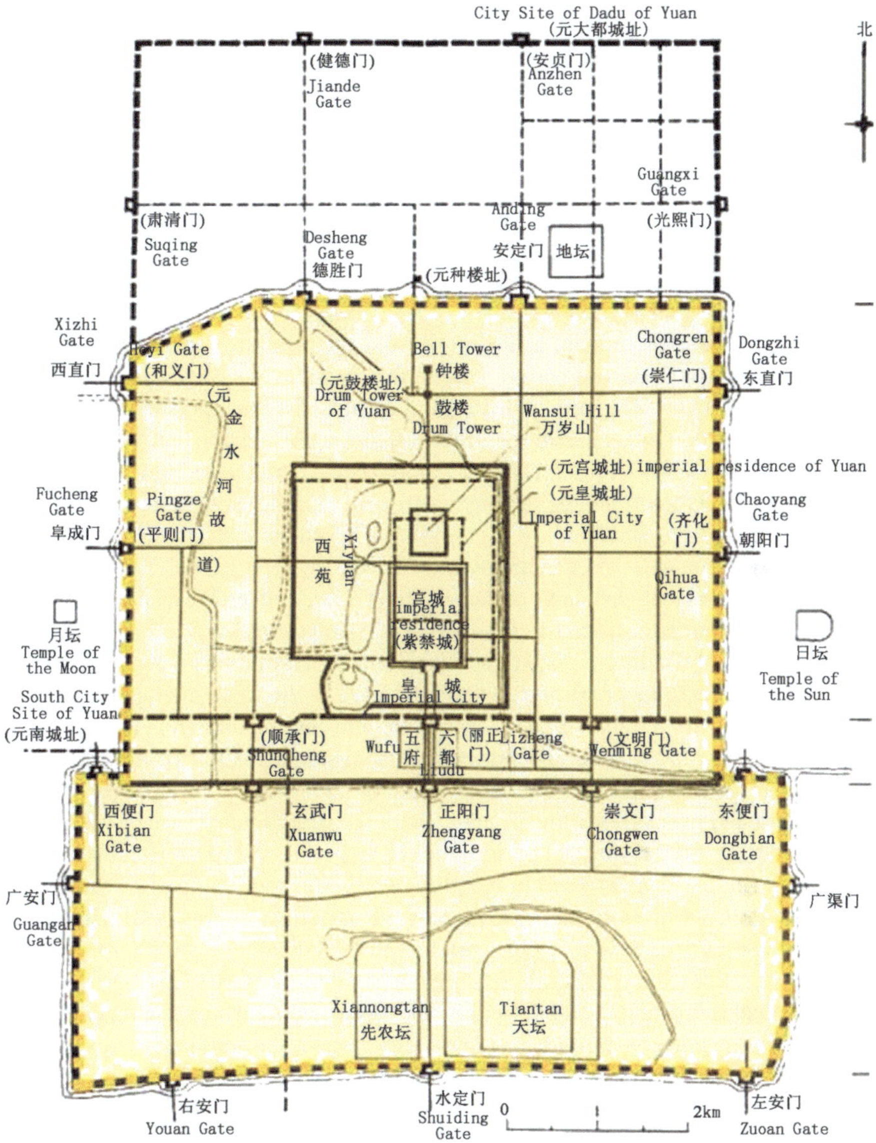

Fig. 8.8 Plan of Beijing City in the Ming Dynasty

Emphasis on the Central Axis: The central axis was highlighted, starting from Yongding Gate in the south, passing through Zhengyang Gate, Daming Gate, Chengtian Gate, Duan Gate, Wumen Gate, the Imperial Palace, Jingshan Park, and extending to the Bell and Drum Towers, in the north, reaching the central point of the two northern gates. This 8-km-long axis served as the main axis of the entire city, guiding the layout of various functional zones within the city and surrounding suburb altars. Along this axis, architectural elements such as gates, towers, and palaces were arranged using artistic techniques like elevation variations and spatial contrasts, creating a rhythmic and orderly composition with varied spatial changes, emphasizing the dominant position of the central area in urban spatial organization.

Lin'an City of Southern Song Dynasty

Hangzhou, located in Jiangnan, suffered little damage from wars and was prosperous as early as the Five Dynasties period. During the Northern Song Dynasty, Hangzhou was the capital of the Zhejiang and Zhezhou circuits. Su Dongpo served as a local official in Hangzhou twice during his lifetime and was involved in dredging the West Lake. When the Southern Song Dynasty established its capital in Hangzhou, it was renamed Lin'an Prefecture, and the city underwent significant development.

Influenced by the surrounding topographical features such as the West Lake, Phoenix Mountain, and the Qiantang River, Lin'an City (Hangzhou) takes on a long and narrow shape along the banks of the Qiantang River and the West Lake shoreline, stretching from north to south. Its urban layout is characterized by "facing south " and "adapting to local conditions," a unique arrangement determined by its geographical environment. Due to its proximity to the West Lake to the west, the Qiantang River to the southeast, the Grand Canal to the north, and the mountainous terrain to the south, Hangzhou has a natural landscape with higher elevation in the south and lower elevation in the north. Following the traditional layout principle of "occupying high ground and overlooking low ground," both the subordinate cities of the Wuyue Kingdom in the Five Dynasties period and the imperial city of the Southern Song Dynasty in Lin'an were situated in the hilly area in the southern part of the city, at the foot of Phoenix Mountain. Consequently, the imperial city was positioned at the southernmost point of the entire city, resulting in the layout of "facing north", while government offices and neighborhoods were located to the north, leading visitors to enter from behind, a practice colloquially known as "riding the dragon backward."

The planning and construction of Lin'an City respected traditional rites, planning the imperial palace area symmetrically and orderly. Meanwhile, there was a strong emphasis on environmental planning and design. The road network of Lin'an City was closely integrated with the waterway network, presenting a natural and organic form. Street layout adapted to the terrain, harmonizing well with the surrounding mountains and rivers. Simultaneously, leveraging the characteristics and advantages of a water town, the city centered around the government office, with rivers serving as the main arteries. Combining various small and large rivers in the outskirts, it formed a circular waterway transportation network, becoming a crucial element in organizing transportation. The restoration map of Southern Song Lin'an City is depicted in Fig. 8.9.

Fig. 8.9 The restoration map of Lin'an (Capital of the Southern Song Dynasty) [25]

Urban Planning and Road Systems in Modern and Contemporary Times

This section summarizes the characteristics of urban planning and road construction in China from the Opium War in 1840 to the establishment of the People's Republic of China in 1949.

With the invasion of modern Western imperialism and the development of domestic capitalism, there were significant changes in the material elements and functional structure of Chinese cities. The traditional closed spatial structure of cities and the system of blocks and markets gradually disintegrated, shifting towards an open structure. In terms of planning ideology, Western planning concepts and techniques became popular and were introduced into traditional Chinese cities. Initially, colonialists forcibly imposed urban planning, as seen in cities like Dalian and Qingdao. Subsequently, traditional Chinese urban planning began to imitate and integrate Western planning concepts, with Nanjing being the most typical example, and the Bagua City of Tekes County in Xinjiang also noteworthy.

Urban Planning in Modern Dalian

Before the 1880s, Dalian was a small fishing village called Qingniwa. During the Self-Strengthening Movement, the Qing government selected Lüshunkou at the tip of the Liaodong Peninsula as the base for the Beiyang Fleet's construction. In the twenty-fourth year of the Guangxu reign (1898), Tsarist Russia, under the pretext of protecting China from aggression, sent warships into Dalian Bay and Lüshunkou, forcibly signing a lease agreement with the Qing government for the lease of the Liaodong Peninsula. They named the area "Dalini," meaning "faraway" in Russian. In the thirty-first year of Guangxu's reign (1905), Japan defeated Russia and renamed Dalini to Dalian.

After forcibly leasing Dalian for 25 years, the Russian Empire formulated a development plan in the twenty-fifth year of the Guangxu reign (1899). Russia's aim was to establish an eastern seaport for expansion. They proposed to develop Dalian into an international free trade port and the seaport for the Middle East Railway, making port and railway construction the focus of urban planning. They planned Peter Street (now Coastal Street) to divide the port area from the city area, with the south side of the street designated for the city and the north side for the port. The city area was further divided into three districts: the European district, the Chinese district, and the administrative district. Separations were established between the city districts, connected by railway bridges and roads. Due to Russia's focus on port construction and the principles of proximity for residence and management, there was a lack of comprehensive consideration for the long-term rational layout of the city: First, the port area was narrow, with the administrative district located on an isolated highland, separated from the city by a railway, hindering long-term development. Second, the urban area was blocked by the port docks and railway lines, preventing access to the seaside, with the entire shoreline dedicated to production and transportation. Third, the planned residential area for Chinese residents was rudimentary and blatantly discriminatory.

The urban road system adopted a Baroque-style design, consisting of circular squares and radial roads, which formed the overall framework of the plan. It was evidently a direct emulation of the planning styles of St. Petersburg and Paris at the time, with a strong emphasis on formalism. The roads were classified into four levels: boulevards, main roads, side roads, and alleys. The boulevard, mainly represented by Peter Street, was 85 m wide, while the main roads were 25 m wide. The main sections of the roads were paved with cobblestones, while side roads were mostly laid with crushed limestone and compacted, and alleys were often unpaved dirt roads.

The city center was a large square (nowadays known as Zhongshan Square), with a diameter of 213 m, surrounded by 10 radial roads, exhibiting strong classical formalism. Major public buildings for administration, commerce, finance, postal services, culture, and entertainment were mostly located around the square. By the 31st year of Guangxu's reign (1905), when the city was occupied by Japan, the main roads and buildings planned at that time were essentially completed.

After Japan occupied Dalian, they decided to maintain its status as a "free port" and inherited the previous Russian planning. In the first year of the Xuantong era (1909), Japan based the "Dalian Urban Planning" on the urban planning during the Russian period. They did not make any changes to the parts of the Russian plan that had already been implemented but continued to refine or adjust the parts that had not yet been formed. For example, the area around Sanba Square was designated as a high-end residential area. The original planned neighborhoods were relatively large, so the new plan included the subdivision of block roads, dividing them into smaller blocks. The squares and straight roads in the area south of Renmin Road and Zhongshan Road were preserved, and curved roads were straightened. Subsequently, the "Dalian City Building Regulations" were formulated, detailing requirements for building setbacks from road boundaries, building heights along streets, building density, and building functions. As the population grew and land became increasingly scarce, Japan drafted the "Urban Expansion Plan" in 1919, which directed the city's development westward, expanding the land area from 16.7 square kilometers to 35.6 square kilometers. The implementation of these plans during this period formed the basic framework of Dalian, profoundly influencing the city's future planning and construction [12].

Due to the need for economic exploitation, Japan continued to advance and complete the construction of Dalian, and industrial and urban development correspondingly progressed. After Japan occupied the Northeast in 1931, they revised the urban planning of Dalian, aiming for a population of 1.22 million and an area of 416 square kilometers. The city mainly expanded to the west, with grid-like roads used in flat areas. Beyond People's Square, neatly planned grid-like main roads formed large squares, divided by block roads into several orderly arranged small squares. The implementation of these planning schemes laid the basic structure of the road network in the current old urban area of Dalian [12].

Capital Planning of Nanjing

In early 1928, with the Northern Expedition on the verge of victory and the political situation gradually stabilizing, the construction of the capital was put on the agenda. In December of the same year, the "National Capital Design and Technical Committee," led by Sun Ke, hired American architect Murphy and engineer Gulicz as consultants to begin drafting the overall urban plan for Nanjing. By the end of 1929, the "Capital Plan" was officially announced by the Nationalist Government. This was the first relatively systematic urban planning in Nanjing's history and the most complete urban planning of the Republican era. The planning concepts and methods presented therein had a significant influence on the development of modern urban planning in China.

The "Capital Plan" emphasized "scientific rationality" and "nationalism", guiding its principles by "adopting the principles of European and American science while incorporating the advantages of our national art." It adopted European and American planning concepts at the macro level while employing traditional Chinese architectural forms in micro-level architectural design [5]. For instance, in road design, the form and criteria directly borrowed from the small grid pattern of Europe and America, but did not adopt the Washington "diagonal" road layout, as it would inconveniently divide related land parcels (see Fig. 8.10). Murphy explicitly stated that in recreating the new capital on the basis of ancient Nanjing, it should have "Chinese flavor", strongly advocating the Forbidden City as "the most beautiful architectural complex in the world." To design truly Chinese-style architecture, it was necessary to start with preserving the appearance of traditional Chinese architecture from the outset, rather than initially constructing Western-style architecture and then incorporating Chinese elements. As for interior facilities, they could primarily be arranged according to modern functional needs. These ideas reflected the integration of Chinese and Western cultures in the field of urban planning and architecture at that time.

The transportation plan occupies a significant portion of the "Capital Plan," covering various aspects such as highways, railways, water transport, and air transport, as well as the form and standard requirements for road construction. For detailed content, please refer to the cited reference [14].

In the "Capital Plan," the internal roads of the city are classified into four types: main roads, secondary roads, ring roads, and boulevards (refer to Fig. 8.11). In the old city area south of the Drum Tower, roads crisscross forming a network, but their standards are too low. Apart from improving existing routes as much as possible, new roads are opened to replace overly winding and short ones, forming a unified road network system. Roads in the new development area are uniformly planned, adopting a grid-like pattern. Boulevards are constructed between parks, serving as extensions of the parks and connecting them with each other. Particularly notable is the ring boulevard constructed along the Ming city wall, prioritizing pedestrian movement.

Additionally, the plan proposes the conversion of the Ming city wall into a ring road, following the example of elevated highways abroad. Since the city wall has lost its defensive value, it could be refurbished for vehicle traffic, offering panoramic views of both the city and its surroundings. However, this proposal clearly overlooks the need for preserving the integrity of the ancient capital and the potential harm to the wall structure. Fortunately, after negotiations, this plan was ultimately not

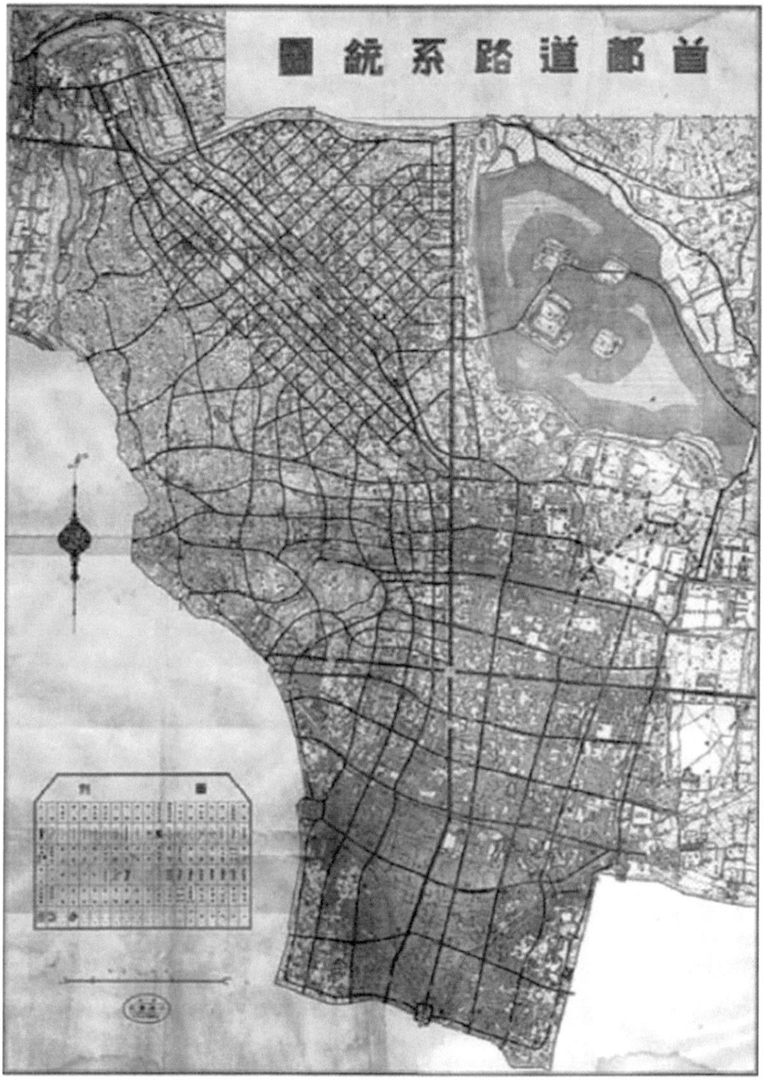

Fig. 8.10 Road system map from the "capital plan" [13]

implemented, allowing the Nanjing city wall to be preserved to this day. In contrast, cities such as Guangzhou, Shanghai, Wuhan, Changsha, and later Beijing, have all demolished their original city walls at different times.

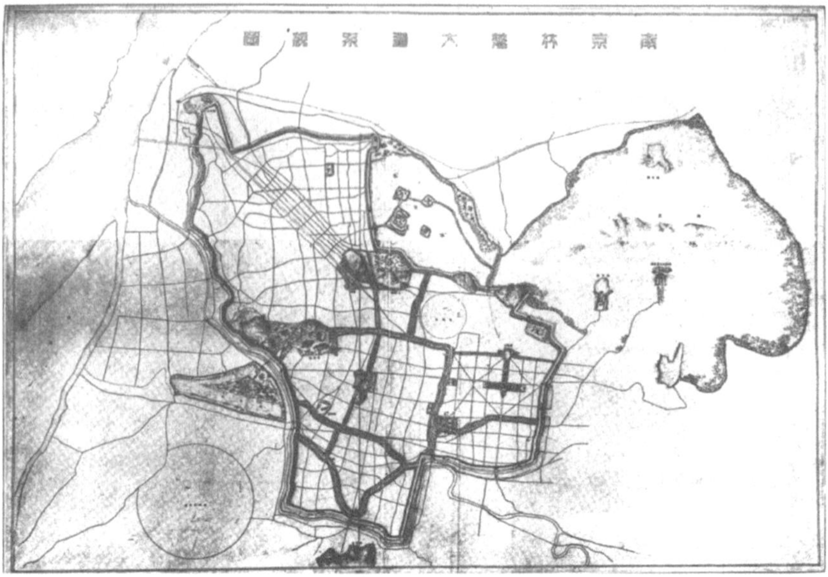

Fig. 8.11 "Capital plan": Nanjing boulevard system map [14]

Bagua City of Tekes County

In the history of urban construction in China, the Bagua City in Tekes County, Xinjiang, stands out as a uniquely shaped example. Tekes County was established in 1937, with the site personally selected by Qiu Zongjun, the then Ili Reclamation Commissioner. The city's layout was designed based on the Bagua diagram from the "I Ching" (Book of Changes), earning it the name "Bagua City." It is also referred to as a "solidified I Ching," showcasing the tangible embodiment of I Ching culture. The overall layout of the Bagua City in Tekes is nearly circular, with a central plaza park. Eight main streets radiate outwards from the center in all directions. Four ring roads encircle the center, forming three concentric circles at intervals of 350 m. This creates a pattern of one ring with eight streets, two rings with sixteen streets, three rings with thirty-two streets, and four rings with sixty-four streets. All roads are interconnected, providing seamless access throughout the city (see Figs. 8.12). In 1996, the relevant authorities removed all traffic lights within the city, making Bagua City a traffic light-free city. In 2007, Tekes County was officially recognized as a National Historic and Cultural City, and its unique urban layout has since garnered increasing attention.

The most notable feature of the Bagua City in Tekes is its adherence to the ancient I Ching (Book of Changes) and feng shui theories. However, instead of the traditional rectangular layout commonly seen in ancient Chinese urban planning, it adopts a rare circular radial spatial form. There are two differing perspectives on this phenomenon: one view holds that this design is a masterstroke inspired by I Ching and traditional

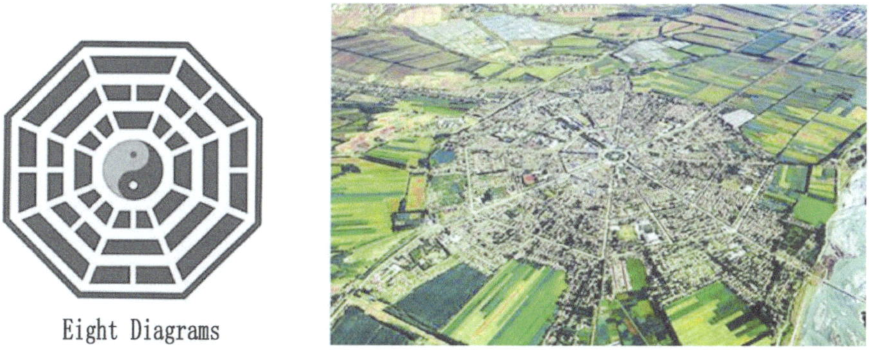

Fig. 8.12 Bagua diagram and the Bagua City of Tekes

planning theories, making it not only a miracle in the history of ancient Chinese planning but also a marvel in the history of world architecture [15]; the other view suggests that the layout of Bagua City resembles the ideal cities of the Western Renaissance, and this similarity is no coincidence. Its formation was guided by traditional Chinese I Ching culture while also being influenced by Western planning ideas, making it a product of modern Chinese urban planning design under the impact of Western thought [16].

A New Phase in Urban Road System Planning

In December 2015, after a 37-year interval, the Central Committee of the Communist Party of China convened the Urban Work Conference once again, explicitly emphasizing the need to "respect the laws of urban development." Urban and economic development complement and promote each other. Urban development involves the aggregation of rural populations into cities and the conversion of agricultural land into urban construction land on an appropriate scale. Population and land use must be aligned, and the scale of cities should match the carrying capacity of resources and the environment. It is essential to understand, respect, and follow the laws of urban development, correct the guiding principles of urban development, and effectively carry out urban work.

The "Several Opinions of the Central Committee of the Communist Party of China and the State Council on Further Strengthening Urban Planning, Construction, and Management," released in February 2016, further clarified the need to "understand, respect, and follow the laws of urban development." It emphasized implementing the principles of "functional, economical, green, and aesthetically pleasing" architecture, promoting the block system, and establishing the urban road layout concept of "narrow roads and dense road networks."

The above speeches and documents reflect the latest requirements of China's "city-building concepts" and road network planning and construction in the new era. Planning and design schemes, such as the Xiong'an New Area plan and the reconstruction plan for the new Beichuan County, are the best interpretations of these concepts.

Master Plan for Xiong'an New Area

In April 2018, the "Hebei Xiong'an New Area Master Plan" (referred to as the "Xiong'an Plan") was officially released. The Xiong'an Plan adheres to a global perspective, international standards, Chinese characteristics, and high positioning. It focuses on creating a concentrated area for the relocation of Beijing's non-capital functions, establishing "Xiong'an Quality," forming a national model for promoting high-quality development in the new era, cultivating a new engine for a modern economic system, and building a high-level socialist modern city [17].

The "Xiong'an Plan," through a meticulous comparative study of urban planning methods both domestic and international, examines the inheritance and future development of China's traditional urban planning concepts and spatial genes from a global perspective and international standards. It analyzes urban planning thoughts embodying Chinese wisdom, such as "harmony between heaven and humans," "manifesting ritual through structures," "leveraging time and geography," and "flexibility within norms." The plan extracts outstanding Chinese spatial genes with enduring value, including "mountain and river orientation," "rectangular layout," "central axis symmetry," and "alley and street systems." These concepts are integrated with the unique environment of the area near Baiyangdian (Lindian area) in the North China Plain, further uncovering the city-water relationship gene of "elevated platforms surrounded by lower land." This provides scientific support for establishing the framework of the initial development area of the Xiong'an New Area, reflecting Chinese characteristics [18].

During the formulation of the master plan, the city's mountain and river orientation was first determined based on the "grand historical perspective" and the "grand landscape perspective." The city is centrally located within a square-shaped layout, with north–south and east–west axes extending in four directions, forming a grand mountain and river pattern. The city is backed by the Yanshan Mountain Range to the north, bordered by the vast Baiyangdian to the south, facing the peaks of the Taihang Mountains to the west, and connected to the Xiongwan Bay of Bohai Sea to the east.

In the initial development area of the new district, guided by local conditions and taking advantage of the natural features such as mountains, rivers, forests, fields, and lakes, the overall spatial layout of "north city, central garden, south lake" was clearly defined. The spatial genes of "square-shaped layout," "central axis symmetry," "street alleys and lanes," and "high platforms surrounding lowlands" were integrated into the urban spatial development blueprint. This resulted in the formation of a spatial intention characterized by "one central city, two axes, five clusters, ten scenic gardens, a hundred flower fields, a thousand-year forest, and ten thousand hectares of lakes (Figs. 8.13 and 8.14)." The square-shaped layout reflects the traditional Chinese urban planning concept, while the extending axes incorporate contemporary urban spirit. Each cluster is strategically guided by the terrain, combining standardization with flexibility. These spatial organizational methods together create a cityscape characterized by "Chinese elegance, scenic lakes, and innovative trends." (Figs. 8.13 and 8.14).

The central square of Xiong'an New Area and its orthogonal axis system collectively embody the inheritance of traditional Chinese urban wisdom at the macro scale. Through the orthogonal axes, the five major clusters adopt a unified grid

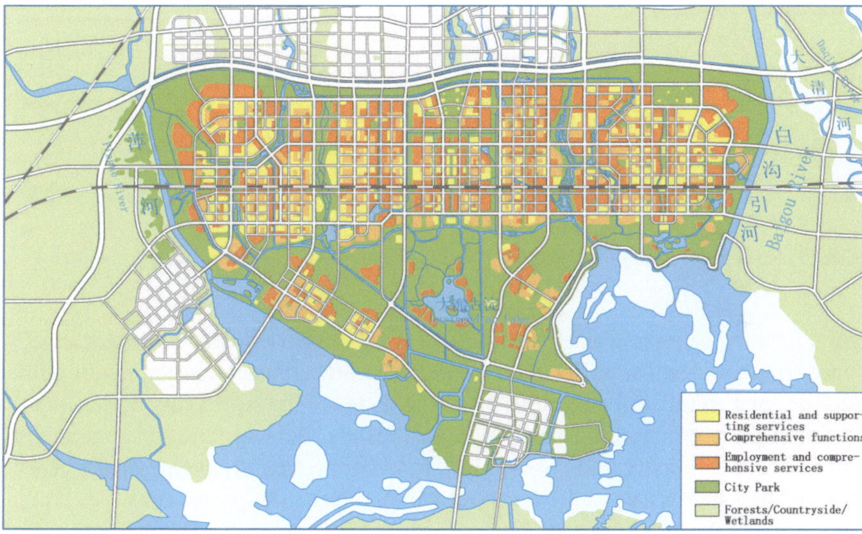

Fig. 8.13 Planning scheme for the starting area of Xiong'an new area [17]

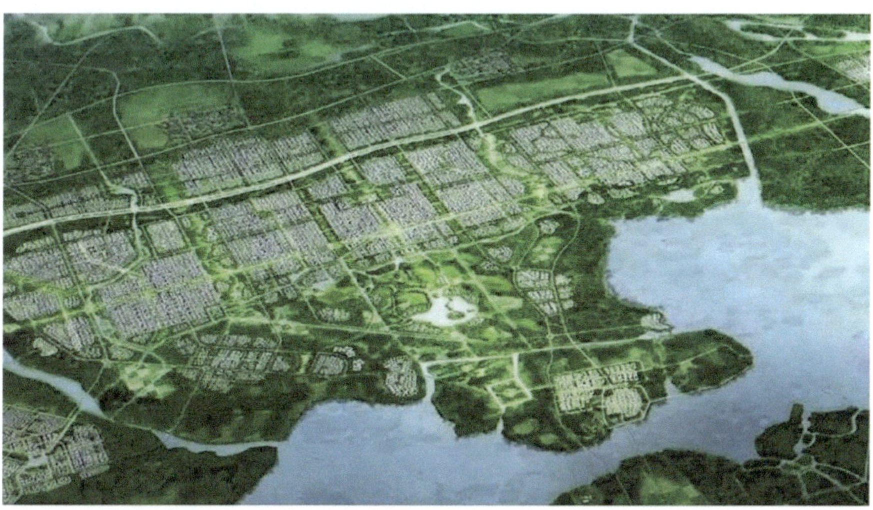

Fig. 8.14 Planning intentions for the starting area of Xiong'an new area [17]

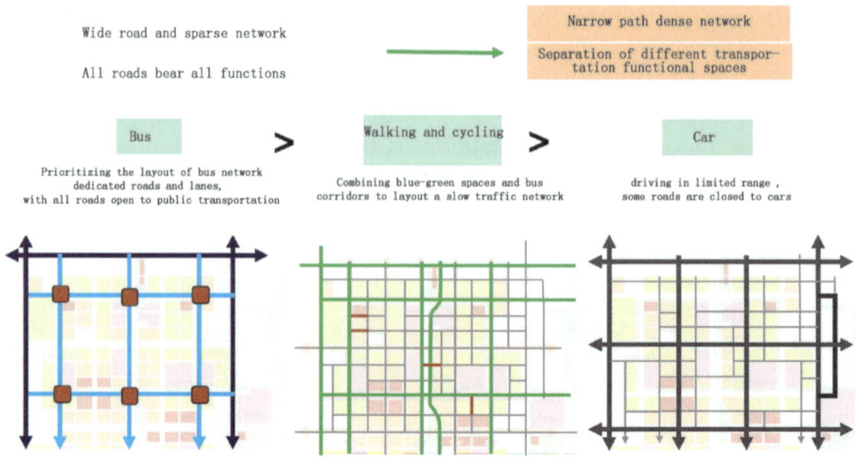

Fig. 8.15 Road traffic planning concept for Xiong'an new area

layout, forming a dynamic spatial form with standardized elements. Simultaneously, connections are established between the public centers of each cluster, major public buildings, and significant natural and cultural nodes of the North City, Central Park, and South Dian, integrating the five major clusters, the city, and the surrounding nature both functionally and spiritually. These connections converge at the central square, enhancing its role as the symbolic center of urban spirit, the core area crucial to the image of the new area, and the key carrier of conveying the development values of Xiong'an [18].

In terms of transportation planning, the concept prioritizes and encourages pedestrian and bicycle traffic, while ensuring comprehensive public transportation and effectively guiding and controlling private car usage. The road network density is scientifically planned, aiming for a density of 10–15 km per square kilometer in the starting area, with road widths designed reasonably to meet the construction requirements of "small blocks and narrow streets with dense networks" (see Fig. 8.15). In terms of traffic organization, it adheres to the "public transport + slow traffic" orientation, innovates management concepts, and explores the potential of dense networks through the spatial separation of different traffic functions. Smart technology is utilized steadily to maintain a proactive and open attitude, with caution exercised in the use of complex engineering measures, particularly those tailored for private cars.

Planning and Design of New Beichuan County City

Beichuan County is the only autonomous Qiang ethnic county in China. The epicenter of the 2008 Wenchuan earthquake completely destroyed the county seat located in Qushan Town. The Central Committee swiftly organized rescue and reconstruction efforts and put forward the requirement to "rebuild a new Beichuan." By 2011, the post-disaster reconstruction project of Beichuan's new county city was completed comprehensively. The new county city covers an area of 3.2 square kilometers and accommodates the resettlement of 30,000 disaster-affected and relocated residents. The planning and construction process of the new Beichuan County city embodies the overall requirements of "safety, livability, prosperity, distinctiveness, civilization, and harmony."

The site selection for the new Beichuan County city learned from the lessons of the previous Qushan Town site. The original Qushan Town was surrounded by high mountains, with many houses built close to steep cliffs and the terrain being low-lying, earning it the nickname "the bottom of a pot" among the locals, which was not conducive to disaster avoidance and rescue efforts. The geographical constraints forced Beichuan to expand within the "pot bottom" by cutting through mountains and ridges, exacerbating geological instability and leaving hidden risks that could worsen the consequences of earthquake disasters [19]. The site selection for the new county city prioritized safety as the primary consideration. After careful study and argumentation, the southeast part of Anchang Town was identified as the optimal location, leading to the county city's relocation of 23 km to the south.

The overall planning of the new county city has established a "surrounded by mountains and water" landscape pattern, providing an excellent living environment. The spatial structure forms an "ecological corridor, leisure belt, facility ring, and landscape axis," providing a clear, orderly, and efficient urban spatial pattern for urban development in the context of rapid reconstruction. In terms of functional layout, the new county city consists of the central comprehensive area of Pingba, the southern industrial park, the northern hilly leisure tourism development area, and the western urban expansion area, forming a relatively compact urban form. This layout controls the travel distance of the majority of residents within walking distance, maximizing the efficiency of public facilities and services (Fig. 8.16) [20].

In terms of road transportation, It adhered to not blindly copying the development model of large cities and propose to construct a road transportation network that is pleasant in scale and efficient to access. It has established a small-town street pattern with a dense network of roads and small blocks, thereby improving accessibility by walking [21]. The main road red lines have a width of primarily 20 m, and the spacing between core area roads does not exceed 200 m. The relatively narrow street spaces provide residents with appropriately scaled social spaces. Road cross-section designs are based on road functions and fully consider the differences in traffic demands between different roads and sections. More than 30 road cross-sections have been determined, mainly featuring small scales, which are more suitable for the traffic characteristics of small county towns (Fig. 8.17). In terms of traffic organization, we prioritize guiding motor vehicle travel reasonably to ensure priority for slow-moving traffic. Strict separation is maintained between the slow-moving traffic space and motor vehicle traffic space on all main roads. The road transportation planning in Beichuan fully embodies the people-oriented and fine design concepts and proposes integrated traffic engineering design methods for planning, design, construction, and management [22], which can serve as a reference for subsequent urban road transportation planning.

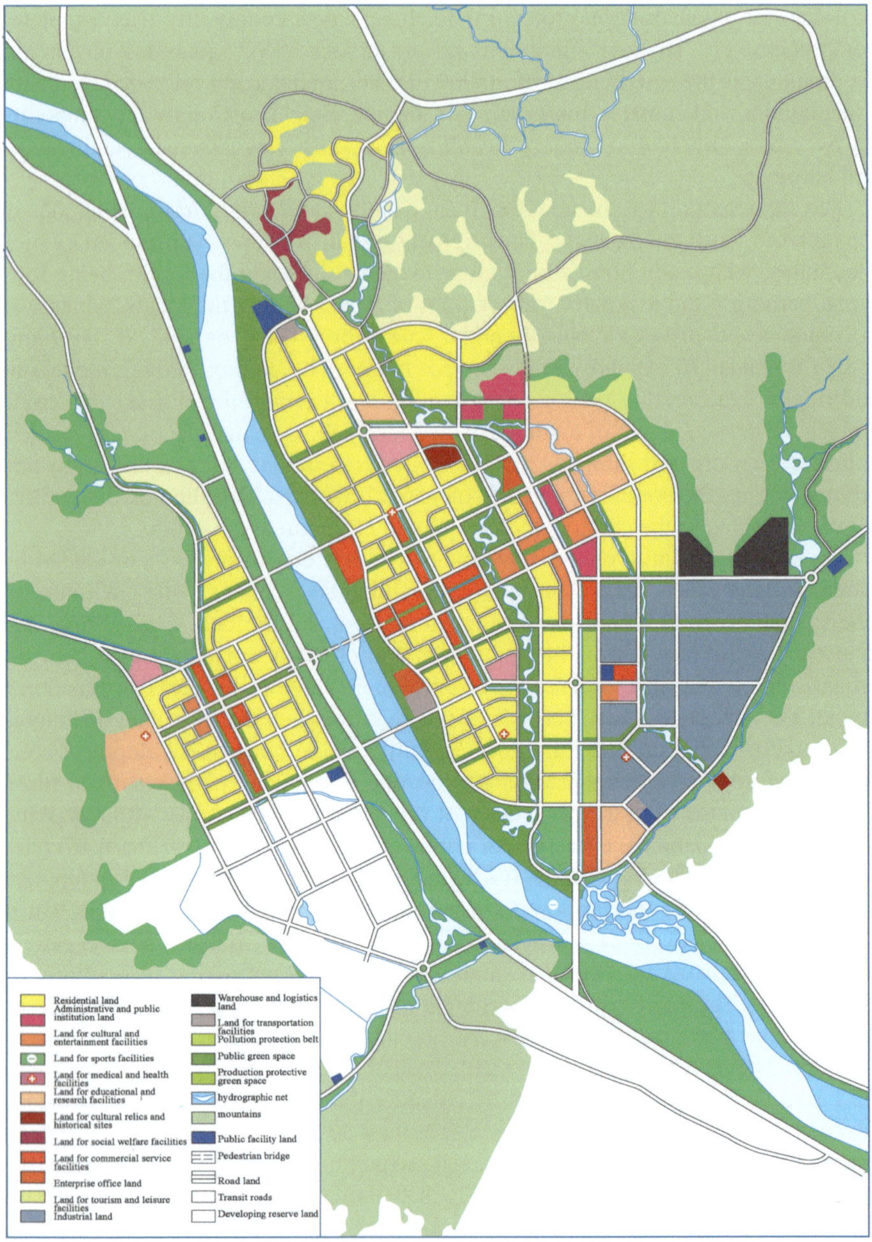

Fig. 8.16 Land use layout plan of the New Beichuan county town

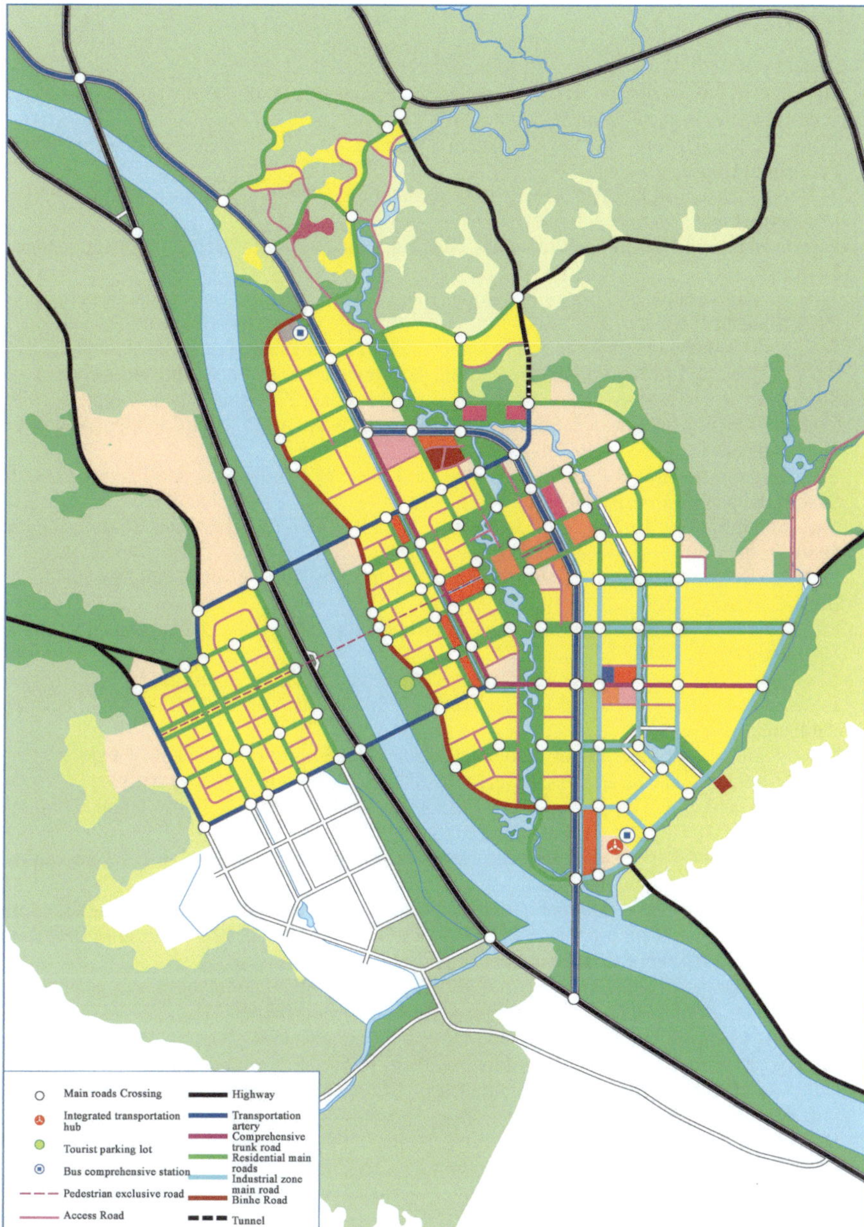

Fig. 8.17 Road functional layout of the New Beichuan county town [17]

References

1. Du Jinpeng. Discovery and Research of Sacrificial Remains at the Erlitou Site in Yanshi [J]. Cultural Relics of the Central Plains, 2019(04): 56–70.
2. Wang Dehua. History of Chinese Urban Planning. Shanghai: Shanghai Ancient Books Publishing House, 1987.
3. Li Hequn. History of Urban Planning and Construction in China and Abroad. Beijing: Peking University Press, 2018.
4. Fan Dacheng. Records of Wu Prefecture. Nanjing: Jiangsu Ancient Books Publishing House, 1986.
5. Rites of Zhou (Zhou Li). Edited and Annotated by Chen Shuguo. Beijing/Changsha: Yuelu Book Society, 1989.
6. Guanzi (Master Guan's Works). Annotated by Fang Xuanling, Supplementary Annotations by Liu Ji, Edited by Liu Xiaoyi. Shanghai: Shanghai Ancient Books Publishing House, 2015.
7. Sima Qian. Records of the Grand Historian (Shi Ji). Beijing: Zhonghua Book Company, 1959.
8. Collation and Explanation of Sanfu Huangtu (Illustrated Maps of the Three Capitals). Annotated by He Qinggu. Xi'an: Sanqin Publishing House, 2006.
9. Song Minqiu. Illustrated Records of Chang'an. Edited by Xin Deyong and Lang Jie. Xi'an: Sanqin Publishing House, 2013.
10. Shen Shixing. Ming Hui Dian (Code of Ming Dynasty). Beijing: Zhonghua Book Company, 1989.
11. I Ching (Book of Changes). Translated and Annotated by Yang Tiancai. Beijing: Zhonghua Book Company, 2022.
12. Liu Changde. A Hundred Years of Dalian Urban Planning: 1899–1999. Dalian: Dalian Maritime University Press, 1999.
13. Xue Bing. History of Nanjing City. Nanjing: Southeast University Press, 2015.
14. Edited by the Office of Capital Design Technical Specialists. Capital Plan. Nanjing: Nanjing Publishing House, 2018.
15. Geng Shunkai and Tian Zijian. Xinjiang Bagua City: Exploring the Path of Famous City Conservation Amid Cultural Fusion [N]. China Culture Daily, February 2, 2012 (7).
16. Liu Quan and Shi Yiting. Exploring the Origin of Tekes Bagua City's Urban Planning Morphology [J]. City Planning Review, 2016, 40(05): 82–90+112.
17. Hebei Provincial Party Committee and Hebei Provincial People's Government. The Outline Plan for Hebei Xiong'an New Area [Z]. April, 2018.
18. Xiong'an China. Interpretation of Xiong'an New Area's Fangcheng: Inheritance and Development of Traditional Chinese Urban Planning Wisdom [EB/OL]. [2022-04-23]. http://www.xiongan.gov.cn/2019-06/06/c_1210153121.htm.
19. China Academy of Urban Planning and Design. The Laws of Urban Development: Knowledge and Practice. Beijing: China Architecture & Building Press, 2016.
20. China Academy of Urban Planning and Design. Comprehensive Planning of Beichuan New County (2008–2020) [R]. Beijing: China Academy of Urban Planning and Design, 2009.
21. China Academy of Urban Planning and Design. Special Planning and Design of Road Traffic in Beichuan New County [R]. Beijing: China Academy of Urban Planning and Design, 2010.
22. Dai Jifeng, Yin Guangtao, Zhao Jie, et al. Design Methodology and Practice of Integrated Traffic Engineering in Beichuan New County [J]. City Planning Review, 2011, 35(S2): 56–60.
23. Marco Polo. The Travels of Marco Polo. Translated by Liang Shengzhi. Beijing: China Literature and History Publishing House, 2008.
24. Chen Xun and Peng Shi. Comprehensive Records of the Universe (Huan Yu Tong Zhi). Beijing: Bibliographical Documentation Publishing House, 2014.
25. Fu Shulan. The Formation History of Hangzhou as a Scenic City: A Study on the Evolution of the Morphological Relationship between West Lake and the City. Nanjing: Southeast University Press, 2015.

Chapter 9
Silk Road: Common Wealth and Great Unity

The Silk Road refers to the ancient trade and cultural exchange routes that began in ancient China and connected Asia, Europe, and Africa. It began in ancient China. The Silk Road in a narrow sense refers only to the land-based Silk Road, whereas the Silk Road in a broader sense is divided into two parts: the land-based Silk Road and the Maritime Silk Road. The Silk Road is a link between the East and West civilizations of trade and cultural exchanges, as well as an important link of human civilization convergence and fusion, from ancient times to the present day. It also emphasizes peace and cooperation, openness and tolerance, mutual learning and mutual learning, mutual benefit, and the win–win road. Since the Han, Tang, and Song dynasties, China has spread commodities and technologies such as silk, porcelain, tea, iron smelting and farming abroad through land and sea routes, while bringing back from abroad species and goods not available in China, forming economic and trade ties and cultural exchanges, upgrading the social productivity of the countries along the routes, improving their living standards, and making significant contributions to the advancement of civilization. The Silk Road connected the four major civilizations of Asia and Africa—Chinese, Indian, Ancient Babylonian, and Ancient Egyptian—as well as the central regions of Greek and Roman civilizations, which are regarded as the cradle of civilizations worldwide.

The German geographer Ferdinand von Richthofen (Ferdinand von Richthofen) used the term "Silk Road" [1] for the first time in his book "China" in 1877, which is still recognized by Chinese and foreign historians today. The Silk Road is not only a great channel of cultural exchange and integration opened in physical space, but also symbolizes the courage to explore the unknown outside world and the indomitable spirit. Since the beginning of the Western Han Dynasty, Emperor Wu of the Han Dynasty sent Zhang Qian to the Western Regions to open the Silk Road under extremely difficult conditions. Since then, successive emperors of the Han Dynasty have vigorously developed the vast Western Regions, opening postal routes in the north and south of the Hexi Corridor and the Taklamakan Desert, connecting the Central Plains with Central Asia, West Asia, South Asia, and Europe. With

© China Communications Press Co., Ltd 2025

H. Lu (ed.), *A Brief History of Transportation in China*,
https://doi.org/10.1007/978-981-97-9851-3_9

the passage of time, this cultural bridge and trade artery running through Asia and Europe continued to develop and evolve, profoundly changing the pattern of world development and the course of human civilization until now.

The Arduous Opening of the Silk Road and the Han and Tang Strategies in the Western Regions

As early as before the Western Han Dynasty, there was a trade between the Western Regions and the Central Plains, and the main commodity was jade. A large number of jade artifacts were unearthed in the tomb of Fuhao in the Yin Ruins of Anyang, most of which were seed jade produced in Hetian, Xinjiang, indicating that at least in the middle of the Shang Dynasty, the Central Plains had already traded large quantities of jade with the faraway Hotan, Xinjiang, and that this was the predecessor of the Silk Road, the "Jade Road". The Jade Road, with Hetian jade as its bridge, not only laid the foundation for the subsequent opening of the Silk Road, but also played an important role in the dissemination of oriental culture and art, and in the communication between the East and the West in terms of economic, scientific, and technological, and cultural exchanges.

The bamboo book Tale of King Mu, Son of Heaven (Fig. 9.1), discovered during the Western Jin Dynasty, outlined the general route of the Jade Road. In 2002, when the Chinese Academy of Sciences organized a field survey of the Hetian Jade Mines, Baihe Jade Mines, and Jade Transportation Route in the Kunlun Mountain Range, a large number of jade specimens and remnants of the activities of the ancient prehistoric people were discovered along the way, confirming the records of Tale of King Mu, Son of Heaven, which provided strong evidence for the Jade Road.

In fact, the Silk Road is the "advanced version" of the Jade Road, in which the merchants gradually evolved goods from single jade into silk, jade, spices, horses and other commodities. From the Shang Dynasty to the Spring and Autumn and Warring States Periods, east to west, active trading, laying a solid foundation for the real formation of the Silk Road. During the Spring and Autumn and Warring States Periods, many commodities from North Africa and West Asia also arrived in China through various transportation routes, but during the Qin and Han Dynasties, the Xiongnu rose suddenly, and the trade route through the Hexi Corridor was blocked, so the important task of opening this trade route between the West and the East fell on the shoulders of Emperor Wu of the Han Dynasty and Zhang Qian.

The story of Zhang Qian's mission to the Western Regions is well known, and he himself has been celebrated as one of the pioneers of the Silk Road and an outstanding diplomat. In the third year of the reign of Emperor Wu of the Han Dynasty (138 BC), Zhang Qian was recruited as an envoy to the western region of the Da Rouzhi, with the intention of joining forces with the other side to fight against the Xiongnu. Zhang Qian was recruited as an emissary and set off from Chang'an with more than a hundred men, but was unexpectedly captured by the Xiongnu in the

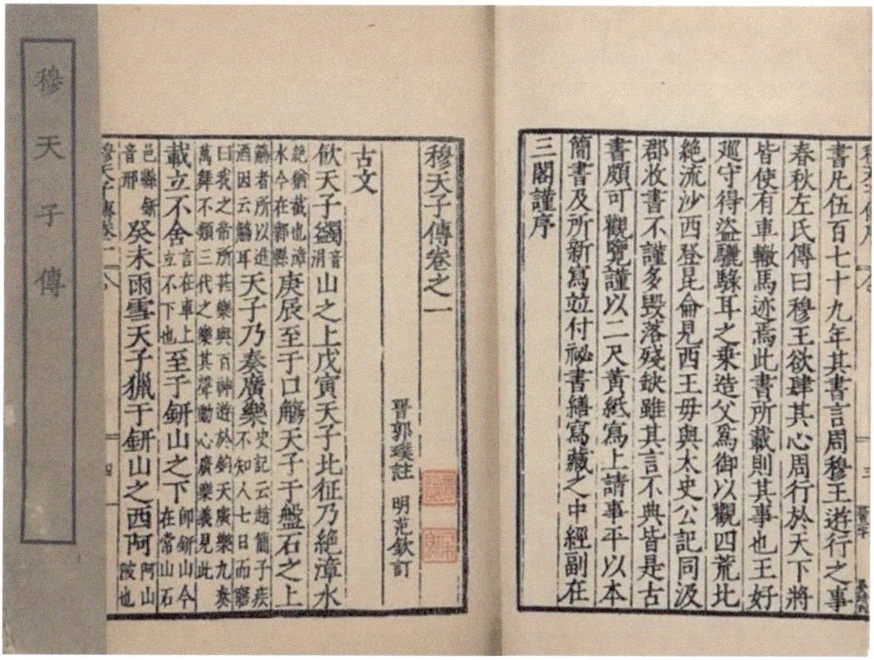

書凡伍百七十九年其書言周穆王遊行之事春秋左氏傳曰穆王欲肆其心周行於天下將皆使有車轍馬迹焉此書所載則其事也王好巡守得盜驪騄耳之乘造父為御以觀四荒北絕流沙西登昆侖見西王母與太史公記同汲郡收書不謹多毀落殘缺雖其言不典皆是古書頗可觀覽謹以二尺黃紙寫上請事平以本簡書及所新寫並付祕書繕寫藏之中經副在三閣謹序

穆天子傳卷之一

晉郭璞註 明范欽訂

古文

飲天子蠲山之上戊寅天子北征乃絕漳水庚辰至于□觴天子于盤石之上天子乃奏廣樂載立不舍至于鈃山之下癸未雨雪天子獵于鈃山之西阿

Fig. 9.1 Excerpts from the tale of King Mu, son of heaven

Hexi Corridor and detained under house arrest for more than ten years. During this period, Zhang Qian did not disappoint the emperor's order and kept his honor, he sought an opportunity to escape and traveled westward through Dawan (present-day border area of Uzbekistan, Tajikistan, and Kyrgyzstan), Kangju (present-day territory of Uzbekistan and Tajikistan), and arrived at the Da Rouzhi (present-day territory of Uzbekistan), and then went to Daxia (present-day territory of Afghanistan and Uzbekistan). However, at this time, the Da Rouzhi had already moved westward for a long time and did not want to fight with the Xiongnu, so Zhang Qian failed to achieve his goal and stayed here for more than a year before returning, and was unfortunately detained by the Xiongnu for more than a year on the way back. In the third year of Emperor Wu's Yuanshuo reign (126 BC), Zhang Qian took advantage of the Xiongnu's civil unrest to escape back to Chang'an. After arriving at Chang'an, Zhang Qian reported the situation in the Western Regions to Emperor Wu of the Han Dynasty, who was so satisfied that he appointed Zhang Qian as Taizhong Daifu, and later as the Marquis of Bowang. Zhang Qian's mission to the Western Regions was intended to unite with the Da Rouzhi in the fight against the Xiongnu, and although it failed to achieve its original goal, the mission broke the isolation of the Central Plains to the west for many years, promoted cultural exchanges, and enabled the civilization of the Central Plains to spread outward rapidly through the Silk Road.

In the fourth year of Emperor Wu's Yuanshou reign (119 BC), Zhang Qian was sent on his second mission to the Western Regions. At this time, the Han Dynasty had taken control of the Hexi Corridor, and Zhang Qian, considering the actual situation of the conflict between the countries in the Western Regions and the Xiongnu, suggested that the Wusuns should be brought together to resist the Xiongnu, which was the famous strategy of "breaking the right arm of the Xiongnu". At the same time, Zhang Qian also proposed to strengthen friendly relations with the countries and tribes in the Western Regions. Emperor Wu of the Han Dynasty adopted these suggestions. Later, Zhang Qian led a mission to persuade the king of Wusun to return to the east, but it was unsuccessful. He sent his men to Dawan, Kangju, Da Rouzhi, Daxia, Anxii (Parthian Empire, today's Iran), and Sindhu (today's India), and so on, and from then on, the line of communication between the Han Dynasty and the Western Regions was formally set up. In the second year of Yuanding of Emperor Wu of Han Dynasty (115 BC), Zhang Qian returned to Chang'an from his second mission to the Western Regions and was honored with the title of Daxingling. He died in Chang'an in the following year and was buried in his hometown of Hanzhong, Shaanxi Province.

Although Zhang Qian's mission to the Western Regions did not achieve the intended purpose of combating Xiongnu, the actual impact it had was extremely great and of great historical significance. Before the Han Dynasty, the westernmost border of the Central Plains was only Lintao in Gansu Province, and the vast Western Regions beyond the Yumen Pass were hard to be reached by the political and cultural power of the Central Plains. Zhang Qian's first mission to the West, for the first time to expand the scope of influence of the Central Plains to the Cong Ridge (now the Pamir Plateau) around the Hexi Corridor up to the Western Region of the vast area of the increasingly strengthened links with the Central Plains, but also with Central Asia, West Asia and even Europe's direct contacts have gradually become closer. In particular, the Han Dynasty and the contemporaneous Roman Empire began to have a preliminary contact. Sima Qian's recorded Zhang Qian's achievements, describing his mission to the Western Regions as "Zao Kong" [2] (Pei Gui, a scholar of the Tang Dynasty, quoted the explanation of Su Lin, a scholar of the end of the Eastern Han Dynasty: "Zao is to open up; Kong is to pass through. Qian opened the way to the West."). This evaluation, undoubtedly, is extremely pertinent.

Zhang Qian's mission to the Western Regions was not only an arduous diplomatic activity, but also a fruitful scientific investigation. Zhang Qian carried out field investigation and research on the western region, not only personally visited the countries in the western region and Central Asia, such as Dawan, Kangju, Da Rouzhi, Daxia and so on, but also from these places to extend the understanding of the customs and development of the Wusun, Yancai, Anshi, Jiaozhi, Sindhu and other countries. After returning to Chang'an, Zhang Qian made a detailed report to Emperor Wu of the Han Dynasty, explaining the location, population, politics and military power of Central Asia, West Asia, and South Asia, which was recorded by Sima Qian and has been handed down to this day. Zhang Qian's mission to the Western Regions opened the vision and horizons of the Han Dynasty, and formed the most informative and reliable record of these regions for the first time in China, leaving extremely valuable information for the study of the history of the above regions and countries.

After Zhang Qian successfully opened the Silk Road, Emperor Wu of the Han Dynasty launched the Battle of Henan, the Battle of Monan, the Battle of Hexi and the Battle of Mobei, which completely dismantled the Xiongnu's attempts to block the communication between the East and the West. In the fourth year of Emperor Xuan Di's Shenjue reign (58 BC), the once invincible Xiongnu split into five parts, with the major monarchs fighting among themselves, and the Xiongnu changed from a unified empire to two parts, the South Xiongnu surrendered to the Han Dynasty, while the North Xiongnu faded away in the long river of history. The Silk Road, however, became the main channel of foreign exchange for the successive dynasties of the Central Plains, and there was an endless stream of emissaries and merchants coming to China from the Western Regions.

With the flourishing of the Silk Road and the deepening of trade, Chang'an, the capital of the Western Han Dynasty, gradually developed into the world's leading metropolis at that time. Here, merchants gathered, stores stood in abundance, and trade flourished with a wide array of goods. According to the record of "Sanfu Huangtu", "There are nine markets in Chang'an, with two hundred and sixty-six paces on each side. Six markets were on the west side of the road, three markets were on the east side of the road, and every four miles was a market." [3]. The East and West market is the most prevalent among Chang'an nine markets (Fig. 9.2).

The end of the Eastern Han Dynasty to the Wei Jin and North and South Dynasties period was full of war and chaos, many big families of the Central Plains and scholars moved to the Hexi Corridor, so that the cultural level of the Hexi Corridor has been unprecedentedly improved, also laid the foundation for the prosperity of the Silk Road. After the unification of the Emperor Wendi of Sui Dynasty, Sui and Tang dynasties contacted the chiefs of various countries through the Hu merchants in the western region to manage the trade, which opened the curtain of further prosperity of the Silk Road. The great integration of peoples and the continuous development of the frontier made the communication between the East and the West more unimpeded.

The Tujue are an ancient nomadic people in the north of China, rising in the warring period of the North and South Dynasties, and their sphere of influence used to be five to six thousand miles from the west of the Liaohai in the east to the West Sea (i.e., the Caspian Sea) in the west, from the north of the desert in the south, and to the North Sea (Lake Baikal) in the north. There were several big battles between the Tujue and Sui Dynasties, but they were never defeated. Later, the Tujue were divided into two parts, the East and the West, with the East Tujue being the stronger one. During the Tang Dynasty, the Tujue fought dozens of major battles with the Tang Dynasty for more than a hundred years to compete for the interests of the border. During the wars, the Tang Dynasty turned from appeasement and defense to offense, coupled with the internal division and civil unrest of the Tujue, the Tang Dynasty completely defeated the Eastern and Western Tujue in the fourth year of Tang's Zhenguan reign (A.D 630) and the second year of the Tang's Xianqing reign (A.D 657), capturing the Eastern Tujue Jili Khan and the Western Tujue Shabaluo Khan and the Tujue Khanate perished.

In the 14th year of the Tang's Zhenguan reign (640 AD), the Western Protectorate was set up in Gaochang (southeast of Turpan City, Xinjiang Uygur Autonomous Region), and in September of the same year, the Anxi Protectorate was set up in Jiaohe (northwest of Turpan City, Xinjiang Uygur Autonomous Region), to guard

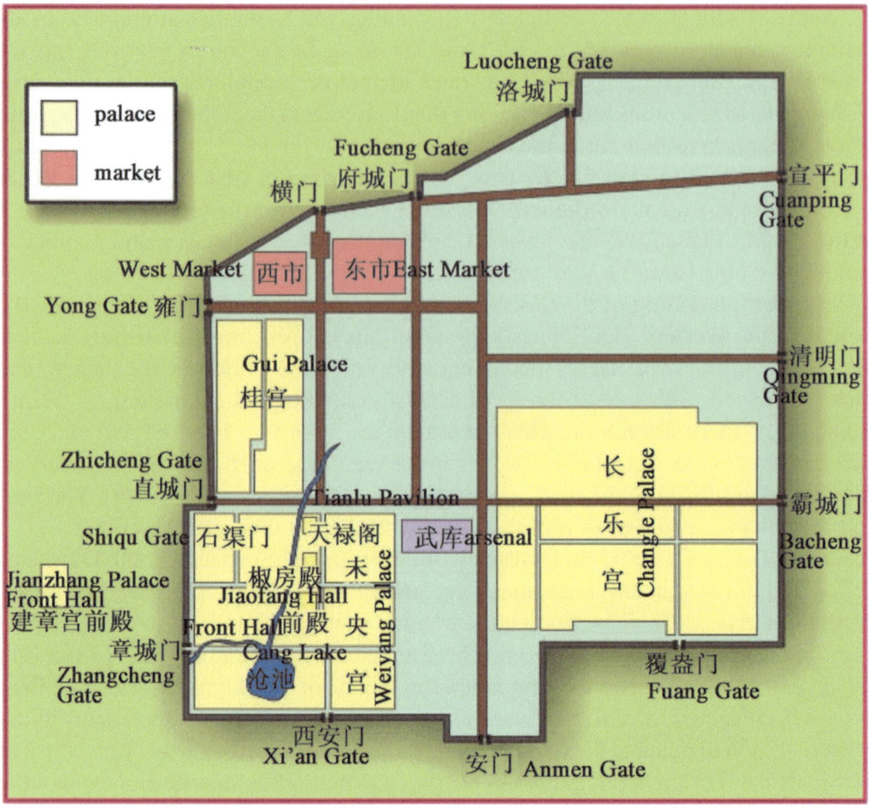

Fig. 9.2 Location map of the east city and west market of Chang'an City, Han dynasty

against the Western Tujue. In the 22th year of the Tang's Zhenguan reign (648 AD), the Anxi Protectorate was further moved westward to Qiuci (present-day Kuche City, Xinjiang Uygur Autonomous Region). The Anxi Protectorate was not only the highest administrative organ of the Tang Dynasty in the Western Region, but also the highest military commanding organ, and its structure was well-developed, with a fixed number of officials and specialized positions, and all military and civil affairs had specialized institutions responsible for them. The capital guard is the main official of the Protectorate, ranked from the second grade. The Tang Dynasty army not only defends against the Tubo army and the Western Tujue remaining attack on the Western Region, maintains the long-term stability of the Western Region, but also guarantees the smooth flow of the Silk Road. Until 14th year of the Tang's Tianbao reign (755 AD), after the Anshi Rebellion, the Tang government decline from the prosperity, the control ability of the Anxi Protectorate is not as strong as it once was, the Tubo seized the opportunity to launch a full-scale attack, the 6th year of Tang's Zhenyuan reign (790 AD), the Anxi Protectorate lost contact with Tang Dynasty, the Silk Road is gradually lost the protection of the official.

Starting with the Song Dynasty, a series of wars were triggered by nomadic peoples from the north moving southward, combined with the gradual shift of the Song Dynasty's economic and cultural center to the south, the Maritime Silk Road was gradually opened, eroding the importance of the Land Silk Road and disrupting it for a time. Following the establishment of the Yuan Empire, the overland Silk Road provided a new opportunity for development. In the Ming and Qing Dynasties and the Republic of China, due to the closed-door policy, coupled with the political turmoil in modern China and the serious desertification along the way, the Silk Road was no longer as prosperous as before, but the cultural and economic exchanges were not cut off. In the new era, with the "Belt and Road" initiative, the ancient Silk Road has ushered in a more glorious spring.

East–West Trade Culture Mingling and Sublimation and Religious Propagation

The Silk Road is the axis of convergence and integration of maritime, nomadic, and agricultural civilizations. The Silk Road creatively opened the East–West corridor in history, and for the first time constructed a world transportation network across Asia and Europe, with its crisscrossed and well-connected routes, building a major artery connecting the civilizations of the East and the West in ancient times, which is called a miracle in the history of transportation in the world. Land Silk Road will be the Central Plains, the Western Regions (in the narrow sense of the Western Regions is the Yumen Pass, Yangguan, west of the Cong ridge, east of the Cong ridge, north of the Kunlun Mountains, south of Lake Balkhash area, that is, the Han Dynasty Western Region Protectorate; the broader sense Western Regions, including the Onion Mountain west of Central Asia, the Roman Empire and so on, covering the present-day Afghanistan, Iran, Uzbekistan, to the Mediterranean coastline), and the Arabian, the Persian Gulf is closely linked together. After centuries of continuous efforts, the Silk Road stretched westward to the Mediterranean region, becoming a road of friendship for economic and cultural exchanges between Asia, Europe and Africa.

The greatest contribution of the Silk Road is to greatly promote the great circulation of goods in ancient times, taking the lead in the realization of Asia and Europe, the East and the West, trade and economic communication, and the realization of a large number of goods, science and technology, and cultural and artistic exchanges between the East and the West. In this region, China's silk, porcelain, pottery, tea, and a lot of precious goods through the land route to Central Asia, South Asia, West Asia, Europe, and Africa. At the same time, other countries of astronomy, arithmetic, medicine, and other scientific knowledge, as well as music, dance, sculpture, painting, architecture and other cultural and artistic also into the vision of Chinese civilization. According to incomplete statistics, in terms of goods, China exported a large number of crops such as grain, sorghum, rice, ginger, peaches, pears, tea, and cinnamon, turmeric, camphor, rhubarb, musk and other traditional Chinese medicines; in terms of handicrafts, in addition to familiar silk and porcelain, there are quite a lot of lacquer, iron and copper; in terms of technology, sericulture and silk weaving technology, papermaking, ironmaking and so on.

Silk is one of the main luxury goods traded in the countries along the route. As early as in the Hemudu site, archaeologists have found textile tools, which can be inferred that the use of silk at least not later than the period of Liangzhu [4]. In ancient Greece, wearing silk has become a symbol of wealth and status, "Homer's epic" has a specific description of this, and the source of this silk things called " Serica country" [5]. In Ancient Greece.

In the Parthenon, the statue of Athena was clothed in a transparent silk robe (Fig. 9.3).

Fig. 9.3 The goddess Athena of the Parthenon wearing silk decorations [6]

According to historical records, the introduction of Chinese silk into ancient Greece can be roughly divided into three stages: the first stage is a sporadic introduction, the time is the classical era of Greece (about the 5th century to the 4th century); the second stage is a small-scale introduction, the time is Alexander's eastern conquests and the subsequent Hellenistic era (about the late 4th century to the 1st century); the third stage is a large-scale introduction, the time is the period of the Byzantine Empire (ca. 4th century and later) [7]. Silk trade had existed for a long time before Zhang Qian's mission to the West, it was a cultural exchange channel between the East and the West dominated by the nomads living in the vast Eurasian steppe, but the real large-scale transmission and trade was still after the Silk Road was really established.

China's tea, tea tree, tea drinking customs and tea-making technology, also with the Silk Road cultural exchanges and commercial trade to the world (Fig. 9.4). About the 5th century (Northern and Southern Dynasties period), China's tea began to export to Korea, Japan, and then by the southern sea route to South Asia, Southeast Asia. 10th century, due to trade exchanges, China's brick tea along the Silk Road was spread to Central Asia. In the Song and Yuan Dynasties, China's foreign trade ports increased rapidly, ceramics and tea has become the main export commodities. 16th century, the Portuguese merchant ships came to China, tea began to sell to the West. 1610, the Dutch from Macao tea and transit into Europe. 1662, the Portuguese Princess Katharine, who is addicted to tea, married to the King of England, King Charles II, and advocated the royal family to drink tea, led to the nation's tea drinking. In 1669, the British East India Company began to directly engage in the import of Chinese tea business, and further strengthened the European court aristocrat's tea drinking culture. Tea trade made the trade income of the countries along the route increased dramatically, led to the prosperity of the economy along the route, and the entrepot trade also made great development, which changed the way of life in Europe from then on.

In addition, the Silk Road also promoted the interactive dissemination of science and technology, which extensively and profoundly contributed to the production progress and social change of the countries along the route. Before the industrial revolution in Europe, the Four Great Inventions of ancient China and iron smelting technologies were successively introduced to the West through the Silk Road, which became an important factor in promoting the change of the capitalist mode of production. Ancient Chinese papermaking was introduced to Korea in the 4th century, to Japan in the 7th century, and to Central Asia in the 8th century. According to the 11th-century Arabic Book of "Book of Curious and Entertaining Information", in 751, when the Tang Dynasty fought with the Arabian Abbasid dynasty in Tantalus (present-day Kazakhstan), the Abbasid dynasty utilized captured Tang soldiers skilled in papermaking to open a paper factory in Samarkand (present-day Uzbekistan). Since then, "Samarkand paper" was famous for its exquisite and suitable quality, and at the end of the 8th century, the Abbasid dynasty successively opened paper mills in Baghdad and Damascus, which once became the main production area of paper for Europe. Afterwards, paper-making technology was spread to North Africa and Europe, and became one of the most important technologies in the world.

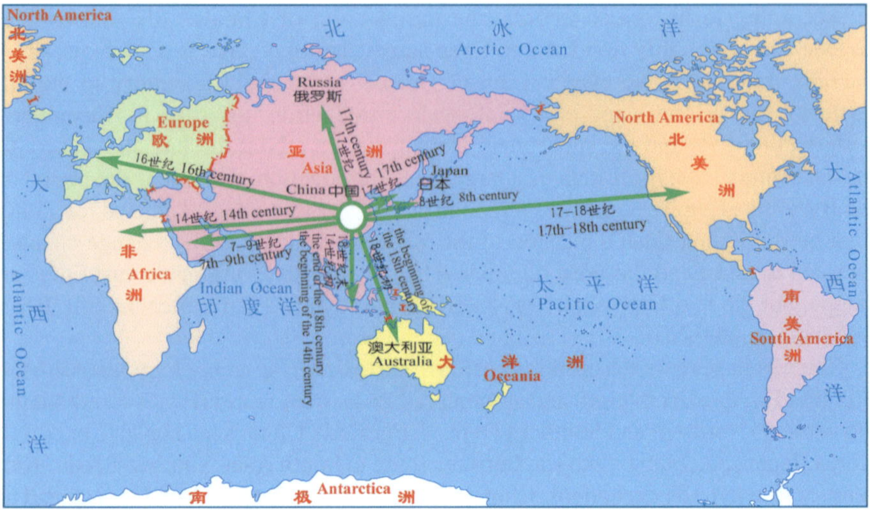

Fig. 9.4 The spread of Chinese tea to the world

Ancient Chinese printing was another important technology that spread westward along the Silk Road. As early as the 7th century, engraved printing was invented in China, and a large number of woodblocks and some paper products used for engraved printing have been found in Turpan and Dunhuang. In the Northern Song Dynasty, Bisheng invented movable-type printing, and around the 13th century, printing was brought back to Europe by Europeans who came to China along the Silk Road. 15th century, German inventor Johannes Gutenberg printed a Bible with movable-type printing, triggering a media revolution in the West, and rapidly advancing the development of science and society in the West. 1466 In 1466, the first printing press in Europe was established in Italy, and printing technology spread rapidly throughout Europe. The British scholar Francis Bacon praised the Chinese gunpowder, compass, and printing press, and believed that these three inventions changed the face of the whole world. Marx said: "Gunpowder, the compass and the printing press—these are the three great inventions that heralded the advent of bourgeois society. Gunpowder blew the chivalric class to pieces, the compass opened the world market and established colonies, and the printing press became the instrument of Protestantism, the means of scientific revival in general, and the most powerful lever for the creation of the necessary preconditions for spiritual development." [8].

In fact, the spread of Chinese inventions to the West provided important conditions for the Renaissance and the emergence of capitalism in Europe. In addition to commodities and technology, the exchanges between the East and the West also included a variety of cultures, such as music, song and dance, astronomy and calendars, literature and language, clothing, and customs. The folk music of the countries along the ancient Silk Road spread, influenced and learned from each other, and the organic fusion of local music forms and playing skills not only became the representatives and symbols of the national and regional cultures of the countries along the route, but also deeply engraved in the literature and opera, songs and dances of the countries and nationalities along the route, and in all aspects of folk life. Cultural exchanges, blending and interaction have always accompanied the development of the ancient Silk Road. The Silk Road has formed a unique Silk Road culture and civilization while closely linking various civilizations, making a monumental contribution to the development of world civilization and the progress of mankind.

The pipa is an important instrument of Chinese folk music, and has an important place in the history of music and literature in ancient China. The pipa was introduced to China through the Silk Road from Persia via the Western Regions during the Northern and Southern Dynasties. Not only did it become the main instrument in the nine-part or ten-part music in the Sui and Tang dynasties, but it also created several pipa genres, especially the "Song on the Fortress", "Sunset Pipers and Drummers", "Ambush on Ten Sides" and other famous songs that have been handed down to the world. Bai Juyi, Yuan Zhen, Su Shi, and others have all left their classical pieces lamenting the pipa for ages.

Since Zhang Qian's mission to the West officially opened the Silk Road, political, economic, and cultural exchanges between countries along the Silk Road have been very frequent, of which the spread of religious ideas is an important element. During the Han Dynasty, Buddhism from ancient India was introduced into China along the Silk Road, and then Islam, Christianity and other religions also entered the Central Plains along this path. The spread of these foreign religions played an important role in the development of Chinese thought and civilization.

Buddhism was the first foreign religion to be welcomed into China. Since its entry into China, Buddhism has been integrating with the local Chinese culture, eventually forming a religious culture with Chinese characteristics. In the 10th year of Yongping (67 AD) of the Eastern Han Dynasty, Emperor Ming of the Han Dynasty sent an emissary to seek Buddhist teachings, and welcomed the Indian monks, Kasyapa Matanga and Dharmaratna, and established the Baima Temple in Luoyang, which was the earliest Buddhist temple ever recorded in China. The route of Buddhism's introduction to the Central Plains is shown in Fig. 9.5.

During this period, the countries in Western Regions were the transit point for Buddhist exchanges between India and China, and many monks came to translate scriptures and preach the law through the Silk Road, leaving a large number of Buddhist rock carvings and caves along the Silk Road, distributed in the north and south of the Tianshan Mountains and the Hexi Corridor area, which are still remaining today.

Due to the language and living customs of the gap, most of the early translation of the Buddha's scriptures are distorted, it is difficult to truly convey the original meaning of the teachings, so some people with insight hope to go to India to seek the true scriptures of Buddhism. History's most famous monk Xuanzang for the real scriptures, had along the Silk Road to visit India, in the Indian Buddhist center of the Nalanda Temple under the master of the preceptor to learn Buddhism. It took Xuanzang 17 years and 50,000 miles to travel to and from India, passing through 138

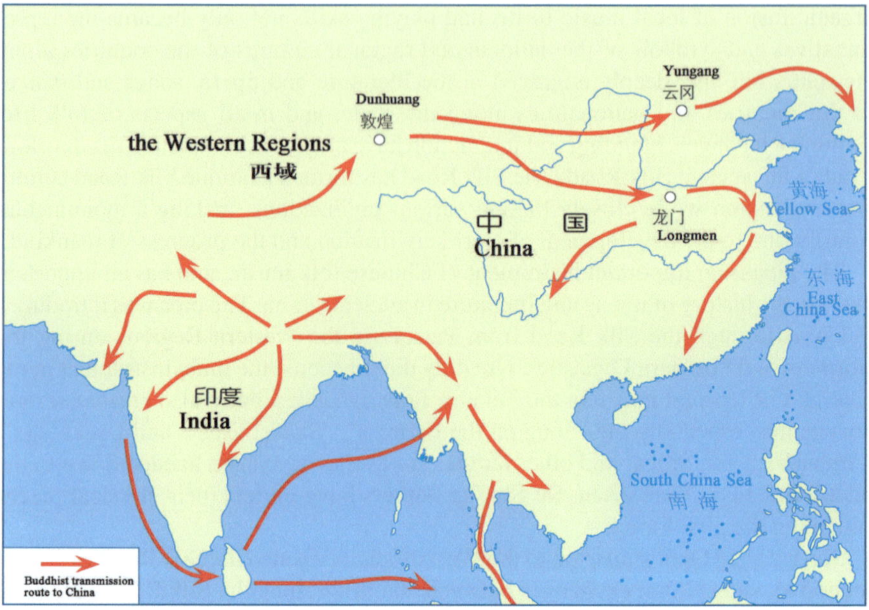

Fig. 9.5 Routes of Buddhism into the central plains

countries of all sizes and bringing back a total of 520 Buddhist classics of all sizes and 657 Buddhist classics, making an outstanding contribution to the translation of the Chinese Buddhist scriptures and the development of Buddhism.

Islam was founded by the Arab Muhammad in the 7th century and spread with the establishment and expansion of the Arab Empire. After the Arab Empire conquered Persia, its territory bordered directly with that of the Tang Dynasty, which facilitated the introduction of Islam into China. At that time, the land transportation route between the Arab Empire and the Tang Dynasty was still the Silk Road. During the reign of Emperor Gaozong of the Tang Dynasty (650–683 AD), the caliphs of the Arab Empire sent envoys to the Tang Dynasty to make goodwill, and thereafter Arabs began to enter the territories under the jurisdiction of the Tang Dynasty to do business and settle down, thus gradually spreading Islam into China. During the Northern Song Dynasty, the Karakhan Dynasty, established by the migratory migrants from the west and covering parts of present-day Central Asia and the southern part of China's Xinjiang Province, made Islam the state religion. From then on, Islam had a stronghold in Xinjiang, China. During the Yuan Dynasty, Xinjiang came under the rule of the Chagatai Khanate, and a high proportion of Arabs, Persians and Central Asians who believed in Islam were stationed in the northwestern part of the country, while some of them were stationed in the central plains of China and the southern part of the country, thus spreading Islam to a wider area. The Yuan Dynasty re-opened the Eurasian continent transportation line, the Silk Road is unobstructed, the people of all ethnic groups who believe in Islam also came one after another, which also accelerated the spread of Islam, so there is a "Yuan's Hui people are all over the world" said. The development of Islam and the route of its introduction into China are shown in Figs. 9.6.

The entry of Nestorianism into China can be regarded as the first spread of Christianity in China. Nestorianism was founded by the Syrian cleric Nestorius in 428–431 AD, and was one of the Christian sects, but was expelled by the Holy See as a heresy. Nestorianism was officially introduced to China in the ninth year of the Tang Dynasty (635 AD), when the Syrian missionary Aroben came from Persia along the Silk Road to Chang'an, where he was greeted by the prime minister Fang Xuanling and met with Emperor Taizong of the Tang Dynasty in his palace. Tang Taizong approved of Nestorianism, and subsequent Tang emperors supported its development by building many Nestorian temples, and in the second year of the Tang's Jianzhong era (781 AD), a monument to the popularity of Nestorian Christianity in China was erected in Daqin (Fig. 9.7). As the earliest record of the popularity of Christianity in China, it describes the development of Nestorianism.

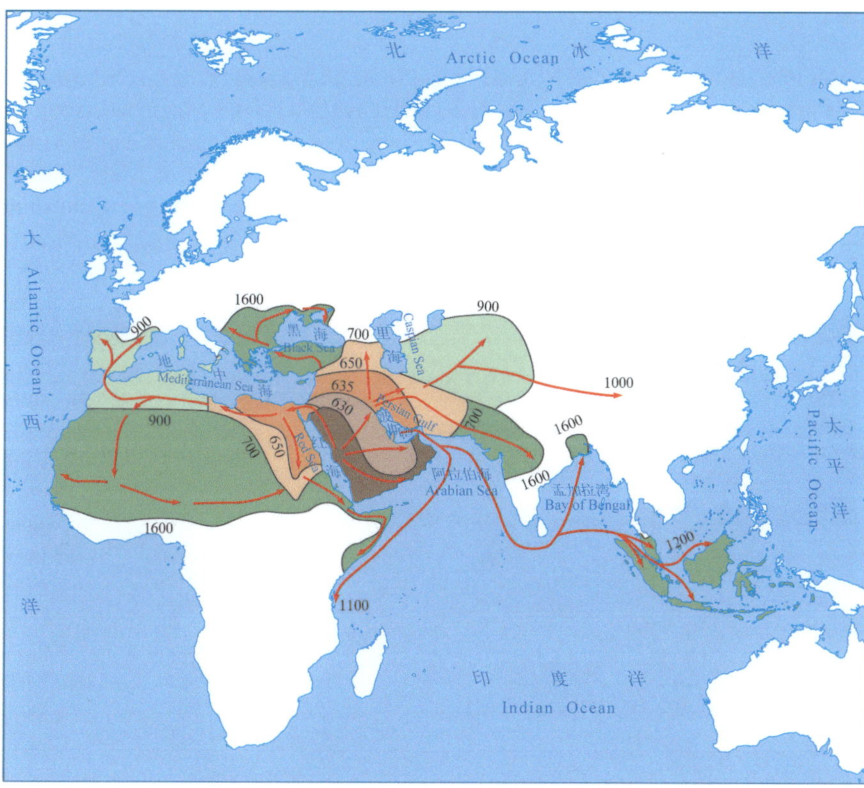

Fig. 9.6 Routes of Islamic development and introduction into China. *Note* The numbers in the figure are approximate years of propagation

Fig. 9.7 Da Qin Jingjiao popular China Stele now in the Xi'an Beilin Museum

Manichaeism was founded by the Persian Mani in the middle of the 3rd century. In the first year of Yanzhai of the Wu Zhou Dynasty (694 AD), a Manichaean named Miusius traveled along the Silk Road into Tang to spread Manichaeism and was summoned by Empress Wu Zetian, which is the earliest record of Manichaeism in China. At present, there are the only remaining Manichean sites in Quanzhou (Fig. 9.8).

Fig. 9.8 Manichaean ruins at Caoan, Jinjiang, Quanzhou, China

World Connections on the Maritime Silk Road

After the Anshi Rebellion of the Tang Dynasty (755–763 AD), the eastern section of the Silk Road came to a standstill. The Tang Dynasty lost the Western Regions, the Song Dynasty lost the Hexi, and Tubo crossed the Kunlun Mountains to the north and invaded most of the Western Regions. China's northern region has been at war for many years, silk, porcelain production is declining, merchants also only seek to protect themselves and do not want to travel. After the Tang Dynasty, China's economic center gradually shifted to the south, so the relative stability of the southern foreign trade increased significantly, led to the prosperity of the Southern Silk Road and the Maritime Silk Road, Chengdu and Quanzhou also gradually become the southern economic power house. During the Yuan and Ming dynasties, the route was mainly a religious and cultural road, and the amount of goods passing through was small; meanwhile, the manufacturing techniques of silk and ceramics had been mastered by India and the West, and the demand for Chinese goods from foreign countries declined. While the western part of the Silk Road was opened and controlled by the Persian Empire, the corridor to the west was also cut off by the Ottoman Empire around the 15th century, followed by the Qing Dynasty, which practiced the policy of locking up the country. As a result, the Silk Road remained in a state of decline for more than 1,200 years.

However, in addition to the land Silk Road, there was also the famous Maritime Silk Road, which was a sea route for transportation, trade and cultural exchanges between ancient China and foreign countries. The term "Maritime Silk Road" was first proposed by French sinologist Emmanuel-Édouard Chavannes in 1913 [9]. The Maritime Silk Road was formed in the Qin and Han dynasties, flourished in the Tang, Song and Yuan dynasties, and gradually declined during the Ming and Qing dynasties, making it one of the oldest known maritime routes in the world. China's Maritime Silk Road is divided into two routes: the East China Sea route and the South China Sea route, with the South China Sea navigation as the center (Fig. 9.9).

The East China Sea route mainly goes to the Korean Peninsula and the Japanese islands through Liaodong Peninsula after leaving the sea from East China. In Tang Dynasty, the sea trade between China, Korea and Japan through Shandong Peninsula and Jiangsu and Zhejiang coasts gradually emerged; in Song Dynasty, Mingzhou (now Ningbo, Zhejiang Province) became the main port of sea trade between China, Korea, and Japan. The starting point of the South China Sea route was mainly in Guangzhou and Quanzhou. This route from China through the Central and Southern Peninsula and the South China Sea countries, through the Strait of Malacca across the Indian Ocean, into the Red Sea, connecting East Africa and Europe, China and foreign trade and cultural exchanges on the sea channel, promoting the common development of countries along the route.

The main ports of the Chinese Maritime Silk Road were Guangzhou, Quanzhou, and Ningbo, with several feeder ports in Nanjing, Wenzhou, Fuzhou, Zhangzhou, Putian, Jiangmen, Yangjiang and Beihai. Before the Sui and Tang dynasties, the

Fig. 9.9 The south China sea route of the maritime silk road

Maritime Silk Road had just been opened, and it was only used as a supplementary route to the land-based Silk Road. But in the Sui and Tang dynasties, due to the continuous wars in the Western Region, the land silk road was blocked, and the maritime silk road rose rapidly. Tang Dynasty shipbuilding and navigation technology developed rapidly, China to Southeast Asia, the Straits of Malacca, the Indian Ocean, the Red Sea and even the African continent of the shipping routes have been opened and extended, the Maritime Silk Road ultimately replaced the land-based Silk Road, became the main channel of China's foreign exchanges. According to the "New Book of Tang—Geography" records, in the southeast coast during the Tang Dynasty, there is a route connecting Southeast Asia, the northern Indian Ocean countries, the Red Sea coast, Northeast Africa and the Persian Gulf countries, known as the "Guangzhou Sea Channel" [10], this is also the official history for the earliest title of the Maritime Silk Road. At that time through this channel to the export of goods are mainly silk, porcelain, tea and copper and iron, imported mainly spices, flowers and other exotic treasures for the court to enjoy. This situation has continued until the Song and Yuan dynasties.

During the Song Dynasty, shipbuilding and navigation techniques were further improved, and the compass was widely used in navigation, greatly enhancing the voyaging ability of Chinese merchant ships. The Song dynasty maintained long-term friendly relations with the southeastern coastal countries, and Guangzhou remained the largest port for overseas trade. The continuous development of the maritime silk road in the Song Dynasty greatly increased the financial income of the imperial court, promoted the development of economy and cities to a certain extent, and provided convenient conditions for cultural exchanges between China and foreign countries. The Song Dynasty adopted the policy of emphasizing commerce and encouraging overseas trade, and the countries and regions that traded with China covered Asia,

Africa, Europe and the United States, and the development of the Maritime Silk Road entered its peak.

Existing studies have pointed out that the late of Song Shenzong reign was a more developed period of commerce and trade in ancient China, with 70% of the total tax revenue coming from industry and commerce and the remaining 30% from agriculture [11]. This not only promoted the growth of national revenue and the improvement of people's material and cultural living standards, but also promoted the economic development, scientific and technological progress and cultural exchanges of the countries and regions along the route. The prosperity of trade and commerce cannot be separated from the golden trade route of the Maritime Silk Road. Quanzhou's historical position in overseas transportation was also established during this period. Among the three major ports, merchants from Japan and the Korean Peninsula hope that the main port of the Song Dynasty as far north as possible, while the Arab world and the South China Sea countries with a greater volume of trade hope that the port as far south as possible, the result of the compromise between the two forces is in the middle of the Quanzhou. In the second year of the Northern Song Dynasty (1087), Quanzhou officially established the Maritime Trade Office (Fig. 9.10), which managed the overseas trade and related affairs of the ports of Quanzhou. Since then, with the geographical advantage of radiating from the north and the south, Quanzhou's trading position quickly surpassed that of Ningbo and Guangzhou, becoming the largest port in China.

In the Yuan Dynasty, the main bulk commodities supporting the Maritime Silk Road were changed from silk to porcelain, and the countries along the route began to refer to China by the name of ceramics (China). Yuan Shizu to Yuan 14th years (1277), Quanzhou City's Maritime Trade Office was reconstructed, Quanzhou's overseas transportation trade moved into the second golden age. Quanzhou's maritime trade

Fig. 9.10 Ruins of Quanzhou City maritime trade office

extended from Japan in the east to Persia, Arabia, Africa in the west and Southeast Asia in the south, exporting many ceramics, silk, tea, gold, and silver. In those days, the merchant ships gathered here, and the merchants praised the largest port in the world, which is unparalleled in the world, namely the Spurgeon's Nest (i.e., Quanzhou).

The fifteenth to eighteenth centuries were a time of great change in human history. The geographical discovery by Europeans opened a new era of navigation and world maritime trade. The maritime expansion of European merchants changed the character of the traditional Maritime Silk Road, which was characterized by peaceful trade, and commercial activities began to have the background of military. During this period, Zheng He's voyages to the West achieved the pinnacle of the Maritime Silk Road (for details, see Chap. 10, "Zheng He's voyages to the West during the Ming Dynasty").

The Return of the Western Region and the Great Idea of "One Belt, One Road" and Its Future Development

In the Silk Road's thousand-year history of silence, China's political center is also gradually moving east. Since the Tang Dynasty, Chang'an is no longer as a political center, the capital gradually to Kaifeng, Hangzhou and other central and eastern areas of the transfer, the needs to operate business activities in the Western Region reduced significantly. Until the 24th year of the Qing dynasty of Qianlong (1759), the Qing dynasty suppress Dzungar Khanate and the revolt of the Altishahr Khojas, unified the Western Region. The Qing government clearly announced that "Dzungar has been pacified, and all the old nomads are in my territory." [12], including the north and south of the Tianshan Mountains, Altay Mountains, all the way to the Pamir Plateau and Lake Balkhash to the east and south of the vast western region under the effective jurisdiction of the Qing dynasty. The Qing people to the new return to the Western Region, also known as "Xinjiang" or "Western Region Xinjiang".

The Opium War of 1840 broke the traditional Chinese political system, and China under the Qing Dynasty became an object of invasion and plunder by Western powers; a series of unequal treaties forced China to become a semi-colonial and semi-feudal society, and it was humiliatingly incorporated into the system of modern international relations. The invasion of Aguba, the occupation of Ili by Russia, and the rivalry between Britain and Russia in Xinjiang worsened the situation in the region. With the loss of China's territory and sovereignty, the Xinjiang region was plunged into an increasingly crisis.

During the 13th year of Tongzhi (1874), Japan invaded the Taiwan area, and the southeastern sea defense also appeared to be in crisis, so the Qing government disagreed on sending troops to recover Xinjiang, resulting in a "sea defense" and "fortress defense" controversy. The focus of the dispute is to recover Xinjiang or give up Xinjiang. To Shanxi-Gansu Governor Zuo Zongtang as the representative of the party advocating the recovery of Xinjiang, put forward "the east is the sea defense, the west is the fortress defense, the two equal importance" strategic thinking, but that the primary task is to focus on the "fortress defense". In the 1st year of Guangxu (1875), the Qing government appointed Zuo Zongtang as the Imperial Envoy, supervising the military affairs of the Northwest, with fully responsible for the restoration of Xinjiang. Zuo Zongtang accepted the appointment, integrated troops, raised money, formulated battle plans, after a year and a half, the Qing army eliminated the Agubai regime, recovered the whole territory of Xinjiang except for Ili, ensuring the territorial integrity of China and maintaining the unity of the country. In the 8th year of Guangxu (1882), the Qing government formally took over Yili and finally accomplished the task of recovering Xinjiang.

"He who does not plan for the whole world is not enough to plan for a moment; he who does not plan for the whole situation is not enough to plan for a region." The western part of China is one of the main sources of Chinese civilization, and strategizing the western part of the country has been a major national agenda for generations. In September and October 2013, during his visits to Central Asia and Southeast Asia, President Xi Jinping put forward the major initiatives of building the "Silk Road Economic Belt" ([1]) and the "21st Century Maritime Silk Road" (2), which have attracted great attention from the international community. The major initiative of "Silk Road Economic Belt"[2] has gained great attention from the international community. The "Silk Road Economic Belt" will promote the economic integration of Southeast Asia and Northeast Asia, which will eventually merge and lead to Europe, forming a general trend of economic integration of the Eurasian continent. The "21st Century Maritime Silk Road" will connect the three continents of Europe, Asia and Africa and the Silk Road Economic Belt by sea, forming a closed loop encompassing both sea and land.

"The Belt and Road Initiative responds to the requirements of the times and the aspirations of all countries to accelerate their development, and is a huge development platform that is rich in content, broad in scope and highly inclusive. "The Belt and Road Initiative is an important way for China to deepen reform and opening up in the new era and promote common prosperity in the world, creating a solid win–win environment for building a community of human destiny. "Countries along the Belt and Road Initiative, as an important region accounting for more than 60% of the world's total population and about 30% of the global economy, account for about 1/4 of the world's trade volume and are important investment destinations in the world. As the most important participant in the "Belt and Road" initiative, China has set up more than 110 economic and trade cooperation zones overseas, nearly half of which are in countries along the "Belt and Road" route, and 20 national-level overseas industrial parks are also located here, driving many Chinese enterprises to "go global". The

[1] See "Dream, coming from the depths of history—Remembering President Xi Jinping's visit to four Central Asian countries and the building of the Silk Road Economic Belt", People's Daily, September 13, 2013, 02 edition.

[2] See "Raising the Sail of the Asia–Pacific Ship—Remembering President Xi Jinping's Attendance at the 21st APEC Leaders' Informal Meeting", People's Daily, October 09, 2013, 02 edition.

"Belt and Road" program has driven a large number of Chinese enterprises to go out. Over the past 10 years since the "Belt and Road" initiative was put forward, China's cumulative direct investment in the countries along the route has exceeded 150 billion U.S. dollars, and through China's complete modern industrial system and production and construction capacity, it has helped the countries along the route to develop the manufacturing industry and build infrastructure. Through the construction of the Belt and Road, China has raised the level of opening of various regions in the country, expanded the field of opening to the outside world, promoted systematic opening up, constructed a wide circle of friends, explored new paths to promote common development, and realized mutual benefits and win–win results with the countries along the routes. It has realized mutual benefits and win–win situation with the co-constructing countries.

Despite the accelerating evolution of unprecedented global changes, the fierce competition caused by the new round of scientific and technological revolutions, as well as industrial transformation. The unprecedented impact of global issues such as climate change, pandemic prevention, and others is increasing the complexity of the international environment for the joint construction of the Belt and Road Initiative. However, the contemporary theme of peace and development remains unchanged, as does the general direction of economic globalisation, and the strategic development of the international landscape is favourable for China. The construction of the "Belt and Road" is mainly centered on policy communication, facility connectivity, smooth trade and financial integration, and people-to-people communication, and the construction of transportation infrastructure and facility connectivity is the basic guarantee for the joint construction of the "Belt and Road". On the basis of respecting the sovereignty and security concerns of the countries concerned, and with the joint efforts of all countries, an all-round, multi-layered and composite infrastructure network with railroads, highways, shipping, aviation, pipelines and comprehensive spatial information networks at its core is accelerating its formation, and the cost of inter-regional transactions of commodities, capital, information and technology has been greatly reduced, effectively promoting the orderly flow and optimal allocation of cross-regional resource factors and realizing mutually beneficial cooperation and win–win development. Mutually beneficial cooperation and win–win development have been realized. The six major international economic cooperation corridors, including the New Asia-Europe Continental Bridge, China-Mongolia-Russia, China-Central Asia-West Asia, China-Central South Peninsula, China-Pakistan and Bangladesh-China-India-Myanmar, have linked the Asian economic sphere with the European economic sphere, and have played an important role in the establishment and enhancement of the partnership of countries' interconnection and the construction of a large, efficient and smooth Asian-European market. These corridors have played an important role in establishing and strengthening the connectivity partnership among countries and building a large, efficient and smooth Asia-Europe market. With the signing of bilateral international road transportation agreements between China and Kazakhstan, Uzbekistan, Turkey and other countries, as well as multilateral international road transportation agreements such as China-Pakistan-Haji, China-Hazard-Russia and China-Japan-Ukraine, the infrastructure construction in

Fig. 9.11 Vientiane railway station on the China-Laos railway

Central Asia and West Asia has been continuously improved. The Kunman Highway and the China-Laos Railway (Figs. 9.11) have all been completed, and projects such as the China-Thailand Railway are advancing steadily.

With the passage of time, under the joint efforts of all parties, the joint construction of the Belt and Road will surely make steady progress, become the Road of Peace, Prosperity, Openness, Greenness, Innovation, Civilization, and Integrity, and promote economic globalization in the direction of more openness, inclusiveness, universally beneficial, balanced and win–win situation, so that the thousand-year-old Silk Road can be written with even greater splendor in the new era. The Silk Road will have a broader splendor in the new era.

References

1. Ferdinand von Richthofen. Diary of Richthofen's Travels in China [M]. E.Thyssen, Selected and edited. Li Yan, Wang Yanhui, Translation. Beijing: Commercial Press, 2016.
2. Sima Qian. The Records of the Grand Historian [M]. Beijing: Zhonghua shubu, 1959.
3. Anonymous. Yuan ben Tokyo menghua lu yuan ben sanfu huangtu [M]. Beijing: National Library Press, 2018.
4. Weiss. Chinese silk weaving technology beginning of the age of the initial exploration – and the beginning of the age of Chinese sericulture problem [J]. Journal of Zhejiang Silk Institute of Technology. 1993(3):26–32.
5. Homer. The Iliad [M]. Luo Niansheng, Wang Huansheng, Translation. Beijing: People's Literature Publishing House, 2015.
6. Topfu Art Museum. Ancient Greek Mythological Characters Goddess Athena Athena Parthenon in the Parthenon of Ancient Greece [EB/OL]. (2017-09-04) [2022–04–25]. http://www.fjam.org.cn/html/yishuzhishi/211.html.
7. Zhang Dainian, Fang Keli. Introduction to Chinese Culture [M]. Revised edition. Beijing: Beijing Normal University Press, 2004.

8. Marx. Machines. Natural forces and the application of science [M]. Institute for the History of Natural Science, Translation. Beijing: People's Publishing House, 1978.
9. Chavannes. Western Turkestan [M]. Feng Chengjun, translation. Shanghai: Shanghai Academy of Social Sciences Press, 2016.
10. Ouyang Xiu, Song Qi. New Tangshu [M]. Beijing: Zhonghua shuju, 1975.
11. Jia Daquan. A preliminary study of the tax structure of the Song dynasty [J]. Social science research, 1981(3):9.
12. Wang Kaixi. Qing history factual records [M]. Beijing: Oriental Publishing House, 2018.

Chapter 10
Maritime Area: The Rise of Maritime Transport and Economic Development

China has a long history of maritime transportation. Maritime transportation is not only the driving force of China's economic, social, and cultural development, but also an important part of China's long history and splendid culture. From the Qin Dynasty, Xu Fu's east voyage opened the prelude to navigation; the Han and Tang Dynasties, coastal navigation is getting better and better; the Song and Yuan Dynasties, overseas trade greatly promote the development of navigation and economic prosperity; the Ming Dynasty, Zheng He commanded seven expeditionary treasure voyages to the West to promote China's friendship with Southeast Asian countries and the development of trade and economic cooperation; the Qing Dynasty's sea ban and the hard sail of modern sea trade, to the booming development of the contemporary maritime transportation, a long history of maritime transportation has left a lot of unforgettable stories and infinite reverie to people. Today, China has become a big maritime country and is striding towards a strong maritime country. Among the world's top ten ports with cargo and container throughput, China occupies eight and seven seats respectively; China's shipping volume accounts for 1/3 of the world's shipping volume, container throughput accounts for 35% of the world's container throughput, and the scale of shipping fleet capacity ranks the second in the world; the global network of shipping services continues to be perfected, the waterway international transportation routes to and from more than 100 countries and regions, the liner shipping connectivity index ranks the first in the world, and the efficiency of terminal handling operations is one of the world's leading. The efficiency of terminal loading and unloading operations is the first in the world. Overseas port and shipping service network is constantly improving, and has made an important contribution to the stable development of the global industrial chain and supply chain. Especially under the impact of the current COVID pandemic, the ubiquitous Chinese shipping routes and efficiently operated shipping networks are contributing Chinese wisdom and strength to the construction of a global maritime supply chain system that covers the whole world and is highly intelligent, safe and reliable.

© China Communications Press Co., Ltd 2025

H. Lu (ed.), *A Brief History of Transportation in China*,
https://doi.org/10.1007/978-981-97-9851-3_10

Mazu Culture

Mazu is the "god of navigation" and "the god of peace" with characteristics of oriental culture, and is the most important folk belief in China's maritime culture (Fig. 10.1). The belief in Mazu has a history of more than 1,000 years, and has profoundly influenced the development of China's maritime navigation, maritime transportation, maritime trade, and maritime diplomacy. Mazu culture has become a core component of China's maritime culture and an important symbol of maritime civilization.

Ma Zu was born as a human being and died as a god. The historical Ma Zu, surnamed Lin Mo, was born in the first year of the Northern Song Dynasty (960) [1] in Putian, Fujian Province, to a family of officials. Since childhood, she received a good education, and was fond of reading books on medical science, sea, meteorology, and navigation, and was capable of rescuing people from shipwrecks and helping those in distress. According to Nanning Liao Pengfei compiled "Sheng Dun Ancestral Temple Reconstruction of Shun Ji Temple Records", "the legend of Tongtian Goddess also, surnamed Lin, Meizhouyu people." [2]. Huang Gongdu, a poet of the Song Dynasty, wrote a poem "Title of Shunji Temple": "The dead wood is the source of the spirit of the east of the sea, and the palace is a rocky peak in the clear sky. I am not tired of

Fig. 10.1 Mazu, the god of navigation (Photo by the author)

mixing with old women, and I can still be effective in my country's endeavors after death. There are no droughts or floods in ten thousand households, and children sing and dance all the time. It is rumored that the favorable scenery still exists today, and that the masts of thousands of miles have a wind of faith." [3].

The culture of Mazu has a long history and has survived for thousands of years. During the Song and Yuan dynasties, maritime shipping and trade flourished. It was believed that the safety of the sea could not be separated from the blessing of the gods. In the fifth year of the Xuanhe reign of the Northern Song Dynasty (1123), the Shengdun Temple, a branch of the ancestral temple of the Mazu Temple in Meizhou, was granted the title of "Shunji" for the first time by the imperial court. Since then, the imperial courts of the Song, Yuan, Ming, and Qing dynasties have honored and bestowed titles on Mazu 36 times, including 14 times in the Song Dynasty, 5 times in the Yuan Dynasty, 2 times in the Ming Dynasty and 15 times in the Qing Dynasty. Emperor Gaozong of Song Dynasty and Emperor Xiaozong of Song Dynasty named Mazu as "Lady Linghui", Emperor Guangzong of Song Dynasty named Mazu as "Concubine", Emperor Shizu of Yuan Dynasty named Mazu as "Heavenly Concubine", and Emperor Kangxi of Qing Dynasty named Mazu as "Holy Mother of Heaven". Heavenly Mother" "Queen of Heaven" [4]. Since the Yuan Dynasty, the word "Huguo" has been added to the title of Mazu, meaning that Mazu has evolved from a local sea goddess to a national sea goddess, and has been included in the national ceremonies.

Since modern times, A-Ma culture has been gradually sublimated into a humanistic spirit of benevolence, amity, relief of suffering, compassion, health and peace, freedom, and peace. In 2009, Mazu culture was inscribed on the UNESCO Intangible Cultural Heritage of Humanity List, making it a spiritual treasure shared by all humankind.

With Meizhou as its starting point, the Mazu faith has continued to expand and penetrate in all directions. From the coast of China to inland ports and inland cities, from China to the Korean Peninsula and the Japanese islands, and from Southeast Asia to Europe and the United States, Mazu beliefs have spread to more than 40 countries and regions around the globe, with more than 10,000 Mazu temples and more than 300 million followers. Figure 10.2 shows a panoramic view of the Mazu Temple on Meizhou Island.

Mazu culture has spread rapidly around the world with the migration of seafarers, merchants, and crews from overseas and the migration of Chinese. Since the 1970s, Mazu temples have been built in Washington, New York, California, and Hawaii in the United States, Sydney and Melbourne in Australia, São Paulo in Brazil, and Cape Town in South Africa, among other cities. In France, Mazu is honored as the Goddess of World Peace at the "True One Church" in Paris [6].

Fig. 10.2 Panoramic view of Mazu Ancestral Temple on Meizhou Island [5]

Navigation Activities of the Qin Dynasty

During the Spring and Autumn and Warring States Periods, ancient China already had foundation of shipbuilding and navigation technology. In particular, the coastal states of Qi, Wu and Yue had developed into maritime powers that "could not abandon their oars for a day" [7]. In 221 BC, "the six kings were ended, and the four seas were united" (Du Mu's "A-Fang Palace Fu"), the king of Qin merged the six kingdoms and established the first great unification dynasty in Chinese history. By then, the Bohai Sea and the southeast coast of China had been incorporated into the Qin Dynasty.

To consolidate the centralization of power, Qin Shihuang used to pass through the sea waterway and went on five inspections of Qin's territory, four of which were to the eastern coastal areas, and made sea navigation.

Qin Shihuang twenty-eight years (219 BC), during Qin Shihuang's first tour of the East, he climbed Zou Yishan (now Yishan), Taishan, then from the south coast of the Bohai Sea sailed through Huangxian (today's Yantai Longkou City, Shandong Province, east of the county), the county (today's Yantai Fushan District, Shandong Province), and so on up to Chengshantou. According to the Records of the Grand Historian, "it through Bohai to the east, over the yellow, the Pang, Qiongcheng mountains, climbed the Fu, set up a stone extolling the virtues of Qin and go", and then continued along the Shandong Peninsula sailing south to the Luangxie (present-day Huangdao District, Qingdao, Shandong, Luangya Town), "Luangxie, Dale of the stay for three months" [8]. Qin Shihuang stayed in Langya for three months, building the Langya Terrace and carving a stone to praise Qin's virtue. After a year, Qin Shihuang's start the second tour of the East to Zhifu (present-day Yantai Zhifu District, Shandong Province), and once again along the Shandong Peninsula by boat tour of the South China Sea down to Langya (Fig. 10.3).

Qin Shihuang thirty-two years (215 BC), the first emperor's third east tour to Jieshi (today's Qinhuangdao Beidaihe District, Hebei Province), and "engraved Jieshi Gate". Qin Shihuang's eastern tour, on the one hand, is to visit the northern naval

Fig. 10.3 Qin Shi Huang's eastern tour [9]

fortress, strengthen the centralization of power, "can follow the Luan River deep into the hinterland of Jin and Tang area, and along the coast of the eastern voyage to the Liaodong Peninsula, it is essential an important sailing base of the northern throat" [10]; on the other hand, Qin Shihuang search for immortality, specifically send people into the sea to seek immortal and immortal medicine.

Qin Shihuang 37 years (210 BC), the first emperor's last tour of the East. He first pass the Yunmeng (today's Hubei Xiaogan Yunmeng County) to Huiji (today's Zhejiang Shaoxing), and then to the north through the Jiangshi (today's Zhenjiang, Jiangsu Province) across the Yangtze River after the coastal voyage to reach the north of Luangya, achieve the river-sea link from the Yangtze River to the Yellow Sea and the Bohai Sea.

Qin Shi Huang's several tours to the sea directly contributed to the development of Qin's navigation technology and created Xu Fu's voyage to Japan (Fig. 10.4).

Fig. 10.4 Xu Fu's eastern voyage [1]

Xu Fu (recorded as "Xu Shi"), known as Junfang, was a famous alchemist of Qi at the end of the Warring States period, who knew astronomy, navigation, and medicine. During Qin Shihuang's first eastern tour to Langya and stayed there, Xu Fu visited the city of Langya and wrote a letter to Qin Shihuang.

"Xu Shi of Qi wrote a letter saying that there were three sacred mountains in the sea, called Penglai, Fangzhang and Yingzhou, where the immortals lived. Xu Shi asked for boys and girls to seek for them. The Emperor sent Xu Shi thousands of boys and girls to the Eastern sea to seek the immortals" [8]. However, several years' journey to the sea was in vain. When Emperor Qin Shihuang's last eastward tour came to Langya, Xufu, because of "excuses for high costs and fear of condemnation", twice made a false statement to cheat Emperor Qin Shihuang to fund the personnel and materials to go to the sea to seek for the immortals and ask for medicines. Qin Shi Huang again provided the personnel and materials, "sent Xu Fu men and women 3,000, funded with all kinds of grains and workers", and finally "Xu Fu got the plains and wide lake, stopping the king not to come" [8]. According to the evidence, the "plain and wide lake" is the area of the Japanese islands, and Xu Fu never returned to Qin after he arrived at the Japanese islands.

From the point of view of the development of navigation technology, Xu Fu led a huge team of thousands of people to Japan, more than 1,500 years earlier than Columbus's fleet of ships to discover the American continent, which is a feat in the history of China's and even the world's navigation. The success of Xu Fu's eastward voyage fully reflected the advanced level of shipbuilding technology and navigation of the Qin Dynasty, and laid a solid foundation for the development of shipbuilding and shipping in the successive dynasties after the Qin Dynasty. At the same time, Xu Fu brought advanced technologies such as shipbuilding and navigation, iron and copper smelting, farming and textile to the Korean Peninsula and the Japanese islands, where productivity was low at that time, and successfully opened up the Northern Maritime Silk Road of China in ancient times while promoting the development and progress of the local society and culture.

Han and Tang Dynasties Maritime Voyage Activities

After the Chu-Han conflict, Liu Bang, the first Han emperor, founded the Western Han Dynasty in 202 B.C. and made Chang'an his capital. After Liu Bang ascended the throne, he adopted a policy of non-interference with the people, tranquil and uninvolved, and recuperation and rest. After decades of development, the Han Dynasty reached its peak during the reign of Emperor Wu of the Han Dynasty, with a stable political environment, a strong state, and a prosperous economy. The Silk Road on land and the Silk Road on sea were opened during the period of Emperor Wu of Han Dynasty.

Since the 2nd year of Emperor Wu of the Han Dynasty (139 BC), Zhang Qian was recruited to visit the Western Regions, opening the major transportation routes from the Han Dynasty to Central Asia, strengthening the links and exchanges between the Han Dynasty and the Western Regions, and enabling merchants to conduct trade activities across the Eurasian continent. However, between China, as the supplier of silk products, and the Roman Empire, as the demand side, the two sides could not carry out direct trade activities because of the kingdoms such as the Da Rouzhi and the Sabbath. Chinese silk merchants sold their products to Daxia (present-day northwestern Afghanistan) and Sut merchants (the ancient commercial people of Central Asia), who then passed them on to Sabian merchants for eventual resale to the Romans. The price of silk changed hands several times, and it was the Sabaeans who profited the most.

According to the Records of the Grand Historian: Southwest Barbarians, Zhang Qian learned from the Daxia merchants that in the southwest of China there is a folk trade route from Chengdu through the South China Sea to the Sindhu in South Asia (India). To further expand foreign exchanges and international trade channels, reduce transit links, reduce trade costs, Yuanding 6th years (111 BC) and Yuanfeng 1st year (110 BC), Emperor Wu of the Han Dynasty consecutively pacified the South Vietnam and the East Vietnam, basically cleared the southeastern coastal shipping routes. The oceanic shipping route, later called the "Southwest Silk Road", started from Sichuan through Yelang (present-day northwestern Guizhou, northeastern Yunnan, and southern Sichuan), and traveled downstream from the Xi River to Guangzhou, and then took the sea route from Xuwen and Hepu to today's India and Sri Lanka, and then on to West Asia and even Central Asia.

At the beginning of the Tang Dynasty, the country was politically stable and economically prosperous, with the reign of Zhenguan and the reign of Kaiyuan both occurring at this stage of history. In addition to military and economic needs, sea transportation during the Sui and Tang dynasties mainly communicated with the Bohai Kingdom and Japan, as well as the South and Indian Oceans.

The capital of the Tang Dynasty was in Chang'an, while the economic center had gradually moved southward to the Jianghuai area. The supply of military food and materials to the central and northern regions could not be fully covered by the Grand Canal, and a large part of it had to rely on the sea transportation route. According to Du Fu's poem "Xi You": "The army is in full use in Yulan, and the supply is also laborious. The Wu Gate transferred the corn and silk, the route go through Penglai on the sea." [11]. Wu Gate refers to the present Suzhou, the canal ships through the mouth of the Yangtze River northward along the coast of Jiangsu and Shandong, through Cheng Shantou, Penglai, Bohai Bay into the Haihe River, arriving at the city of military grain (Fig. 10.5, located in the lower reaches of the Haihe River near the sea in the Tang Dynasty).

In the early Tang Dynasty, the state was strong and created a prosperous economy, a complete system of government and a brilliant civilization, attracting countries around the world to learn. Residing in the Songhua River area of the Sume Mohe (Da jorong after receive the title from Tang Xuanzang, remove Mohe in their title, named as Bohai State) to the annexation; Heishui Mohe, Goryeo, Silla, Baekje, Japan, and other countries have come to pay tribute. According to Tang Yan "Bohai State" said, Bohai State for the tribute to the Tang Dynasty, specifically built a sea channel from the western capital of Bryan Province (now Linjiang City, Jilin Bayshan) by sea

Fig. 10.5 Ruins of Jungang City [12]

to Dengzhou (now Penglai District, Yantai, Shandong Province), and then transferred to the land to Chang'an to pay tribute. Since the first year of Shenlong (705 years) to the fourth year of Kai Cheng (839 years), Bohai sent envoys to pay tribute to Tang 40 times [13]. In addition to the political purpose, the countries in Northeast Asia and the Tang Dynasty used the sea transportation to pay tribute to each other by "customs clearance", "return trade", "market trade", "private trade" and "private tribute", and cultural dissemination and exchanges in the form of sending foreign students and taking the examination for the imperial examinations to become government officials.

According to records, Japan has sent more than 10 envoys to Tang. At first, the route used was the relatively safer "north road", basically along the near land coastline bypass navigation, from Dengzhou into the Tang Dynasty, the voyage takes about 50 days; later the route changed to the "south island road", from Kyushu along the Ryukyu Islands southward, across the East China Sea into the mouth of the Yangtze River, and then through the Canal, Yellow River, Weihe River to Chang'an, the voyage period is shortened by nearly half (Fig. 10.6).

At the same time the Tang Dynasty was ushering in a golden age, the Arab Empire emerged on the other end of the Eurasian continent, referred to as "Da Shi" (meaning 'the Arab lands') in the context of the Tang and Song dynasties. The Arabs transported silks, porcelains, and Indian spices, which Tang traders had shipped to Yemen, further to the Red Sea or Mediterranean coasts. By the ninth year of the Tianbao era in the Tang Dynasty (750 AD), with the establishment of the Abbasid Dynasty, the Arab Empire entered its golden age. During this period, groups of Arab merchants sailed eastward to China, encountering the westbound sea ships of the Tang Dynasty. Historical records indicate that during the reign of Emperor Daizong of the Tang Dynasty (762–779 AD), over 4,000 Arab merchant ships arrived in Guangzhou each year, attracting a large influx of foreign merchants from West Asia and even Africa.

Jia Dan in Tang Dynasty wrote "Huang Hua Si Da Ji" recorded in detail the routes and voyages from Guangzhou, passing through Vietnam, the Malay Peninsula, and Sumatra, crossing the Indian Ocean to India and Sri Lanka, and extending to the countries along the Persian Gulf. He also detailed the geographical directions, names,

Fig. 10.6 Route of the Japanese mission to Tang Dynasty

key islands, mountain ranges, rivers, and cultural customs of numerous countries and regions encountered along the journey [14].

The Tang Dynasty established the Guangzhou Sea Route, which was divided into three segments. The first segment started from Guangzhou, proceeded along the Gulf of Tonkin, then southward down the Indochinese Peninsula, across the Gulf of Siam (now the Gulf of Thailand), and continued down the Malay Peninsula to the southeast of Sumatra, reaching Java. The second segment began in Singapore, took a northwestward course through the Strait of Malacca, used the Nicobar Islands (part of present-day India) as a stopover, continued to the southern tip of the Indian Peninsula, sailed along India's west coast to the Strait of Hormuz, and reached Baghdad, the capital of the Abbasid Caliphate. The third segment departed from Basra, located at the head of the Persian Gulf, exited the Strait of Hormuz, traveled westward along the southern coast of the Arabian Peninsula, passing Bahrain, Oman, Yemen, and others, reached the Red Sea's mouth, crossed the Bab al-Mandeb Strait, and sailed down to the East African coast. This marked the first well-documented transcontinental voyage by the Chinese to the African continent [15]. During this time, the Tang Dynasty experienced a phase of widespread economic prosperity, enhancing maritime exchanges due to trade demands and cultural interactions, elevating maritime communication to a new level. Figure 10.7 depicts a restored model of a typical Tang Dynasty ship, the "Black Stone."

Fig. 10.7 Restoration model of "Black Stone", a typical ship of the Tang Dynasty [16]

After extensive maritime experience, navigation technology during the Han and Tang dynasties had evolved into a systematic discipline. The most prevalent navigation techniques were terrestrial navigation (or pilotage) and astronomical navigation. Terrestrial navigation or pilotage is the method and technique of navigating based on the landscape's topography and waterways. In contrast, when a ship sails at night or in open seas out of sight of land, it must rely on observing the sun, moon, and stars to determine direction and ascertain the ship's position. This is the astronomical navigation technique of the Han and Tang periods. The Western Han Dynasty text "Huainanzi" notes: "Those who are lost at sea, unsure of their directions, will find clarity upon seeing the Big Dipper and the North Star." [17].

In the astronomical category of "The Book of Han—Treatise on Literature," there are as many as 136 volumes dedicated to astronomical navigation. Ancient Chinese astronomy divided the sky into 28 starry regions, known as the 28 constellations, and mapped terrestrial regions in alignment with these constellations. Although the specific contents of these books are no longer accessible, the ability of Han Dynasty navigators to complete ocean voyages to countries like India and Sri Lanka using navigational techniques sufficiently demonstrates that the astronomical navigation technology of that time had reached a high level of scientific sophistication.

The Use of Compasses in the Song Dynasty

The Song Dynasty represented a significant era of maritime exploration in ancient China. While there might not have been significant expansion in oceanic voyages, it witnessed an epoch-making innovation in navigation technology—the application of the compass on ships.

As is widely recognized, the compass is one of the four great inventions of ancient China, resulting from the ancient laborers' long-term empirical understanding of magnetism, and was historically referred to as "Sinan." However, modern scholars argue that the "Sinan" mentioned in ancient texts is not a magnetic pointer; it likely

Fig. 10.8 Ancient "Sinan" and the compass during the Song Dynasty [19]

refers to something else, given that the concept of a magnetized spoon serving as a compass is not practical. The practically usable, artificially magnetized compass was developed during the Tang and Song dynasties (Fig. 10.8) [18].

Before the Song Dynasty, various orienting devices (such as compass car and Sinan, etc.) were not actually applicable to navigation. During the Song Dynasty, through extensive experimentation and refinement, people began to rub steel needles against natural magnets to endow the needles with magnetism. Only these artificially magnetized steel needles could be regarded as the formal inception of the compass [20]. The Song Dynasty's innovations in artificial magnetization and magnetic needle mechanisms established the essential technological groundwork for employing the compass in maritime navigation.

Before the adoption of the compass for navigation, early Song Dynasty mariners primarily depended on landmarks and celestial navigation, along with assessments of water conditions, to ascertain their positions. However, these methods, based on natural phenomena and human experience, had their limitations. As maritime commerce flourished during the Song Dynasty, and as voyages extended over increasingly longer distances, the need for navigation instruments unaffected by natural conditions became more pressing. The advent of the compass addressed this critical need.

The use of the compass for navigation was a significant innovation in maritime technology. The earliest documented instance of humans using the compass for ship navigation comes from Zhu Yu, a geographer of the Northern Song Dynasty, in his work "Pingzhou Ketan": "Navigators understand geography; by night they observe the stars, by day the sun, and in overcast or foggy conditions, they rely on the compass." [21]. Research indicates that "Pingzhou Ketan" details Zhu Yu's observations between 1101 and 1103 (during the reign of Emperor Huizong of the Northern Song Dynasty) in Guangzhou. This evidence suggests that the use of the compass for navigation by the Chinese dates back to no later than 1103. This record demonstrates

Fig. 10.9 Nautical map and compass [24]

that by the Northern Song Dynasty, the compass was already in use for navigation, marking the earliest known use of the compass in the history of world navigation. Xu Jing's "Xuanhe Fengshi Gaoli Tujing" also mentions the use of the compass, stating: "At night, when the sea was impassable, and the stars were obscured, the floating compass needle was relied upon to discern north from south." [22].

Over a century later, in Zhao Rushi's "Zhu Fan Zhi" (completed in 1225) during the Southern Song Dynasty, a passage vividly illustrates the critical role of the compass in navigation: "For ships coming and going, the compass serves as the guiding principle, meticulously observed day and night. The slightest deviation could mean the difference between life and death." [23]. This passage also indirectly highlights that Southern Song maritime technology had advanced to a level where even the most minor discrepancies could not be overlooked, as even a minuscule error could lead to being miles off the intended destination during an ocean voyage.

From a scientific and technological history perspective, the use of the compass not only granted humanity the capability for all-weather navigation but, along with the rudder and sails, provided the three essential elements for open-sea voyages: navigation by compass, steering with the rudder, and efficient sail usage (Fig. 10.9).

The application of the compass in navigation played a significant role in advancing geographical explorations and maritime commerce. The German astronomer Temple stated in *The Creative Spirit of China*: "Europe would never have embarked on the voyages that led to major geographical discoveries without adopting navigational technologies like the compass from China." [25].

During the Song Dynasty, Arab traders, often referred to as the "Sea coachman" played a pivotal role as intermediaries between China and European countries. It was through these Arab merchants that Song Dynasty's compass technology was transmitted to Europe. The introduction of the compass had a profound impact, directly contributing to the Age of Discovery in the 15th to 17th centuries. This era enabled humanity to forge global connections across continents and oceans for the first time, ushering the Western world into the capitalist era and triggering significant historical and social transformations.

Quanzhou, the World Center of Maritime Commerce and Trade in the Song and Yuan Dynasties

Quanzhou is on the World Heritage List as a center of world maritime commerce during the Song and Yuan dynasties. The Song and Yuan dynasties were a prosperous period in the history of Chinese maritime trade.

Starting from the Tang Dynasty, there was a significant emphasis on both land and sea trade, encompassing not only transcontinental land routes but also maritime routes extending eastward and westward. During the Song and Yuan dynasties, due to factors such as the shift of the geopolitical and economic center southward and the development of the maritime and shipbuilding industries, the importance of oceanic trade markedly increased. Quanzhou emerged as a pivotal hub for China's maritime trade, facilitating connections and interactions with the "Northern Sea," "Eastern Sea," and "Western Sea", thus contributing to the establishment of a marine economic sphere within the Eastern world.

The Song and Yuan dynasties marked an era when China's waterborne transport was fully developed, and maritime trade reached its zenith. Throughout Chinese history, the Song dynasty placed a significant emphasis on commerce, leading to unparalleled economic growth. Notably, the Jiangnan region played a crucial supportive role in the national economy. After the Yuan Dynasty consolidated national power, the reliance on the south for financial and grain support remained unchanged. To ensure a continuous supply of grain from the south to Beijing, the Yuan Dynasty highly prioritized waterway transportation and implemented supportive policies, facilitating the rapid development of the Quanzhou port.

Since its emergence in the Tang Dynasty, the port of Quanzhou flourished during the Song Dynasty and reached its zenith in the Yuan Dynasty. In the Song era, Quanzhou forged trade connections with over 70 countries and regions, facilitating maritime links that spanned from Japan in the east, to the nations of the South China Sea in the south, and extended to Persia, Arabia, and East Africa in the west. The principal imports were spices and medicines, while exports were dominated by silk and porcelain. During the Yuan period, Quanzhou's port saw further expansion, with its international trade network encompassing nearly a hundred countries and regions. The trade routes remained primarily focused on what was historically referred to as the Western Ocean, maintaining a route stability similar to that of the Song Dynasty. At that time, Quanzhou was not only a pivotal international trading port but also a major hub for the exchange of various domestic and foreign goods. Imports through Quanzhou included 58 types of spices, 12 types of precious items and curios, 27 types of industrial raw materials, 19 types of textiles, 9 types of metals, 6 types of utility goods, and 7 types of foodstuffs. Exports comprised 54 varieties of silk textiles, 41 types of ceramics, and 63 kinds of metals, miscellaneous goods, and medicines, reaching 64 countries and regions.

During the Song Dynasty, Quanzhou's overseas transportation and trade reached unprecedented levels of prosperity. The port of Quanzhou was acclaimed as "one of the world's largest trading ports," gaining international fame alongside the port of Alexandria in Egypt. Quanzhou's silk, porcelain, and copper and iron products were renowned for their advanced craftsmanship and superior quality overseas. In the second year of Yuanfeng in the Northern Song Dynasty (1087), Quanzhou established the Maritime Trade Office, and subsequently set up guesthouses called Laiyuan Post, to accommodate tribute envoys and foreign merchants. To promote overseas trade, the Maritime Trade Office and local officials in the Song Dynasty would specially organize ceremonies to pray for favorable winds or offer sacrifices to the sea at the beginning of the sailing season, wishing for the ships' smooth and safe voyages. "The Geography Records" of the Song History notes that during the Chunyou period (1241–1252), Quanzhou had a total of 255,758 households and 1,329,940 inhabitants [26]. By the mid-Southern Song Dynasty, Quanzhou had surpassed Lin'an Prefecture to become the largest city in the world. In terms of taxation, Quanzhou was a land of plenty. The Maritime Trade Office alone collected over 2 million guan in taxes annually, representing about 5% of the Southern Song Dynasty's total annual fiscal revenue.

During the Yuan Dynasty, Quanzhou's trade with overseas had expanded to nearly a hundred countries and regions. The Liusheng Pagoda, still standing along Quanzhou's coast, serves as a relic of the ancient lighthouse that once guided ships to and from the harbor. At that time, the harbor, surrounded by the ubiquitous Paulownia trees and thus also referred to as "Paulownia City", was bustling with endless rows of ships, and bulk goods were stacked like mountains. It was commonly said: "Quanzhou is the metropolis of the seven Min regions, a deep pool of foreign merchandise, exotic treasures, and the dwelling of affluent traders and magnates from various lands, acclaimed as the foremost in the world." [27].

The advancement of maritime transport spurred enhancements in infrastructure development. The people of Quanzhou innovated a novel bridge foundation technique known as the "raft foundation." This involved placing stones along the bridge's central axis at the riverbed and expanding them outwards to form a short stone platform across the river bottom to serve as the foundation, subsequently building piers on top to enhance the site's stability. The design of the piers was also distinctive, employing long stones laid in an interlocking pattern with protruding ends to divert water flow and lessen the impact on the piers. Another pioneering method was the use of oyster cultivation for foundation reinforcement. By cultivating a large number of oysters beneath the bridge, leveraging their strong adhesion and rapid reproduction, the bridge's foundation and piers were bonded into a robust unit, effectively encasing the foundation in an inseparable "Golden Bell Jacket." This technique, merging biology with bridge engineering to stabilize structures, was a global first at the time. For the bridge deck construction, the "floating girder" technique was innovatively employed, utilizing the principle of water buoyancy. During low tide, stone beams were transported to precise locations between piers on rafts, and at high tide, the rising waters were used to adjust and position the stones onto the piers. The Luoyang Bridge was built using this "floating girder" technique, incorporating stone beams up to 11 m long (Fig. 10.10).

Fig. 10.10 Quanzhou Luoyang bridge

Zheng He's Seven Voyage to the West

Zheng He's voyages to the West represent seven consecutive large-scale maritime expeditions undertaken from 1405 to 1433 during the early Ming Dynasty, spanning regions from East Asia and the Indian subcontinent to the Arabian Peninsula and East Africa. These voyages are considered the largest of their kind at the time and a monumental achievement in maritime history. From the third year of Yongle (1405) to the eighth year of Xuande (1433), Emperors Zhu Di and Xuanzong of the Ming Dynasty commissioned Zheng He to command an impressive fleet of over 200 ships with more than 27,800 crew members, visiting over 30 countries and regions across the Pacific and Indian Oceans. Zheng He's initial expedition predated Diaz's rounding of the Cape of Good Hope and Columbus's discovery of the Americas by over 80 years. The combined distance covered in Zheng He's seven expeditions to the Western Seas totaled more than 70,000 nautical miles, equivalent to thrice the Earth's equatorial circumference.

Zheng He's fleet undertook seven voyages to the Western Seas, navigating through the East China Sea, the South China Sea, and the Gulf of Thailand, then passing the Strait of Malacca to enter the Andaman Sea (currently known as the Burma Sea) and the Bay of Bengal. The journey continued around South India to the Arabian Sea and the Persian Gulf, entering the Red Sea from the Gulf of Aden, and even reached the farther maritime regions north of Madagascar in East Africa. The fleet visited over 30 coastal nations and islands, including Java, Sumatra, Sulu, Pahang, Chenla (Cambodia), and Siam in Southeast Asia; Guri and Punggara in India; Adan, Tianfang, Zufar, and Hormuz in the Arabian Peninsula; as well as locations in East Africa like Mogadishu and the port of Beira in Mozambique. Six of Zheng He's voyages occurred during the Yongle era (1402–1424), with the seventh expedition taking place in the Xuande period (1426–1435). The initial voyages reached as far as Calicut on India's southwest coast, the fourth extended to the Strait of Hormuz in the Persian Gulf, and the final voyage ventured to the Arabian Peninsula and Kenya in East Africa. The fleet also navigated the vicinity of the Madagascar Strait, and there are accounts of a few ships circumnavigating the Cape of Good Hope into the Atlantic Ocean [28].

Qing Dynasty's Sea Ban Policy

To fend off Zheng Chenggong's fleet's attacks, the Qing government adhered to the Ming Dynasty's maritime restrictions, enacting a series of "maritime prohibition decrees" (including the "Sea Prohibition Decree" and the "Sea Relocation Decree") as the Qing forces progressed along the coastlines and islands. According to statistics compiled by Xu Xiangmin, during the early Qing period, the Shunzhi and Kangxi emperors issued at least 13 such decrees, spanning from the fourth year of Shunzhi (1647) to the seventeenth year of Kangxi (1678) [29].

In the twelfth year of Shunzhi's reign (1655), the Qing government, emulating the Ming Dynasty, issued a prohibition, followed by more detailed policies in the subsequent year (1656). As recorded in the "Qin Ding Da Qing Hui Dian Shi Li" during the Guangxu era (Fig. 10.11), "Except for those with official licenses to venture overseas, any officials or civilians found building large ships with more than two masts for trafficking prohibited goods abroad, or renting out large ships for overseas voyages and sharing profits from foreign goods, shall be referred to the Ministry of Justice for separate sentencing. Small single-masted boats are permitted for local residents to fish or collect firewood along the nearby coast, without interference from military or government officials." [30] At this time, the Qing government's maritime restrictions were not overly strict, mainly limiting the size of sea vessels and requiring the proper licensing. Sea vessels with more than two masts were also allowed to engage in overseas trade. Coastal fishermen were permitted to obtain licenses for single-masted sloops for livelihood fishing near the shore.

In the 13th year of Shunzhi (1656), shortly after the aforementioned policies were implemented, the Qing government issued decrees to coastal provinces and regions (including Fujian, Guangdong, Zhejiang, Jiangnan, Shandong, Tianjin, etc.), reiterating the importance of ship licensing. Unauthorized maritime activities were declared capital offenses. "Henceforth, any merchant or civilian vessels that secretly set sail to trade items such as grain and goods with rebels, regardless of their official or civilian status, shall be reported and subjected to the death penalty." [30].

By the 18th year of Shunzhi (1661), the Qing government intensified its scrutiny of merchant and fishing vessels at sea, with ships exceeding the regulated size facing severe penalties. As detailed in "Guangdong New Words," Volume 2, "Discourse on Relocation to the Sea," "Individuals with ships exceeding the allowed dimensions were seized and executed." [31]. Concurrently, the Qing government enacted the "Sea Relocation Order" in Fujian, moving coastal residents inland. "Observations at Sea" records: "The Emperor commanded Minister Su Na Hai of the Ministry of Household to Fujian to relocate seaside residents inland, 30 li (approximately 15

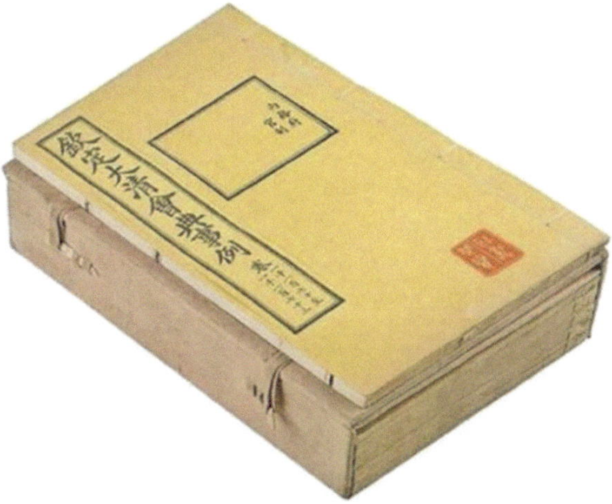

Fig. 10.11 Examples of the Qin Ding Da Qing Hui Dian [29]

km) away from the coast, with villages, fields, and homes completely abandoned and burned." [32]. In the first year of Kangxi (1662), this relocation distance was extended to 50 li (approximately 25 km). "Discourse on Relocation to the Sea" notes: "Coastal residents were ordered to move 50 li inland to cut off aid to Taiwan and eliminate potential threats." [31].

In the 7th year of Kangxi (1668), the government reaffirmed the maritime ban, prohibiting ocean-going trade by sea vessels, with violators facing capital punishment and the implementation of collective punishment. "The Authorized Code of Qing" Volume 776 records: "Officials, soldiers, and civilians found violating the sea ban and engaging in prohibited overseas trade shall, in addition to the prescribed penalties, see their local militia leaders who conspired or knowingly allowed such acts beheaded; those aware but not reporting will be hanged; and those unaware will receive one hundred lashes and be exiled 3,000 li away." [30]. Between the 11th and 17th years of Kangxi (1672–1678), the Qing government continually issued the "Sea Relocation Order." "The Authorized Code of Qing" Volume 629 records: "Residents of islands were ordered to relocate to the mainland to prevent the harboring and aiding of rebels." [30].

The Qing Dynasty's "sea ban" policy and related legal measures impeded societal progress. Jiang Liangqi noted in "Donghua Lu": "The coastal residents of Fujian were relocated inland, forbidden from maritime trade, with the relocated populace exempted from labor taxes." [33]. This policy led to the displacement and unemployment of many merchants and fishermen, disrupting their livelihoods. However, the early Qing's sea ban was not enforced indefinitely. In the 22nd year of Kangxi (1683), the Zheng family in Taiwan capitulated, ending the standoff across the Taiwan Strait. In the 23rd year of Kangxi (1684), the Qing government formally lifted the sea ban, proclaiming: "Now that the empire is unified and the world is at peace, with the Manchu and Han people united, we decree the opening of maritime trade to showcase the wealth and prosperity of our governance." [34]. From then on, the Qing

Fig. 10.12 Maritime trade in the Bohai Bay region after the lifting of the sea ban [35]

Dynasty's sea ban policy was abolished, rejuvenating maritime trade and fisheries in the coastal regions (Fig. 10.12).

Embargo

After World War II, driven by animosity and apprehension towards socialist states, Western capitalist countries led by the United States implemented what was termed a "blockade and embargo" policy. This was aimed at undermining the economic development of socialist countries to ultimately suppress the socialist system. In November 1949, the United States, the United Kingdom, France, and other nations established the "Coordinating Committee for Multilateral Export Controls" in Paris, specifically to enforce this "blockade and embargo" against the socialist bloc.

At this juncture, newly established PRC was in dire straits, with extensive reconstruction needed. The urgency of economic revival and infrastructure development was paramount, particularly the acquisition of advanced foreign technology and

Fig. 10.13 Hong Kong ports during the "blockade and embargo" period [36]

equipment. Unexpectedly, China became a principal target of the capitalist countries' "blockade and embargo" policies. From 1950, the intensity of these policies by the United States and other capitalist nations against China increased, significantly hampering the country's efforts to restore its national economy. The impact of the "blockade and embargo" is exemplified by the state of the Hong Kong port during this period, as depicted in Fig. 10.13.

From February to June 1950, the U.S. banned countries like the United Kingdom from exporting strategic materials to newly established China, enacting the "Measures for the Control of Strategic Materials" in March (covering over 660 types of controlled items including machinery, transportation vehicles, metal products, and chemical raw materials) and the "Export Control Act of 1950" in June. Following the Korean War's outbreak on June 25, 1950, the U.S. intensified export restrictions. By November, the list of strategic materials under U.S. control expanded to over 2,100 types.

From December 1950 to February 1951, the United States implemented a comprehensive "blockade and embargo" policy against China and froze its foreign exchange. Simultaneously, the US collaborated with Taiwan's authorities to blockade coastal cities and ports such as Shanghai and Tianjin, preventing merchant ships from all countries from entering or leaving Chinese ports.

A key goal of the United States' "blockade and embargo" policy against China in the 1950s was to support the military offensive in the Korean War, attempting to exert political and economic pressure. This aimed to induce severe economic difficulties, potentially leading to a collapse that would cause political instability and social unrest within China. The ultimate objective was to enable the resurgence of reactionary forces both domestically and internationally, to overthrow the people's democratic dictatorship under the leadership of the Communist Party of China, and to strangle New China in its cradle [36].

Faced with the stringent "blockade and embargo" measures by Western countries, the Party and the State assessed the situation and established appropriate foreign trade policies and guidelines. They conducted thorough research and planning on organizational structures, trade practices, directions, and the foreign trade system, guiding the Chinese people in a flexible and astute fight against the "blockade and embargo."

Against the background of the "blockade and embargo", the new China was unable to obtain ships from traditional maritime powers such as the United Kingdom and the United States. In order to break the international economic blockade and trade embargo, on June 15, 1951, New China and the Republic of Poland in Eastern Europe established a joint venture ocean shipping enterprise—Sino-Polish Shipping Co. This was the first Sino-foreign joint venture in New China and opened a new chapter in China's ocean transportation business.

In its initial stages, the Sino-Polish Shipping Co. (CPSL) faced significant challenges. To maintain secrecy, it operated under the guise of a Polish ocean shipping company. In just over a year following its inception, CPSL transported equipment for 52 factories back to New China. Under the challenging circumstances of newly established China, CPSL provided substantial transportation support, significantly contributing to the nation's economic construction, industrial advancement, and material provisioning.

In the end, thanks to the efforts of many parties, the Chinese Government broke through the "blockade and embargo" policy, and the volume of foreign trade increased year by year, successfully accomplishing the task of restoring the national economy.

Breaking Through the "Blockade" and Going Global

On the eve of New China's founding, the ships of the China Merchants Steam Navigation Company were under the control of the Kuomintang authorities. As the call for national liberation sounded, crew members on these ships, motivated by patriotic zeal, raised the banner of rebellion. Notably, vessels such as "Hai Liao" (Fig. 10.14) and "Hai Xuan," among 13 others from the China Merchants fleet, returned to the embrace of the motherland, thereby providing a solid foundation in both resources and talent for the development of New China's shipping industry. Following the establishment of New China and the national economy's revival, the country's maritime sector progressively overcame the "blockade and embargo" imposed by hostile forces at sea, achieving significant development.

In the initial years after New China was founded, the country's maritime fleet was limited in variety, with small tonnage and outdated technology, accounting for less than 0.3% of the global total. During the 1950s, China primarily relied on chartering foreign vessels for its ocean-going transportation needs. In 1959, in response to the anti-Chinese political incidents in Indonesia and the absence of Chinese ocean-going ships, China was compelled to charter over ten ships from the Soviet Union and Chinese merchants abroad, often at steep prices, to evacuate overseas Chinese. Due to the difficulties in managing chartered vessels, the Ocean Shipping Bureau proposed purchasing two second-hand passenger ships for direct management using the charter

Fig. 10.14 Hai Liao sailing to Dalian port [37]

funds. With the State Council's approval, China commissioned a Czech company to acquire two nearly decommissioned ocean-going passenger ships, "Guanghua" (Fig. 10.15) and "Xinhua." Concurrently, the Ministry of Transportation's Ocean Shipping Bureau and its Guangzhou Office initiated the establishment of China Ocean Shipping Company, tightly integrating the preparations for launching ocean voyages with the efforts to repatriate overseas Chinese.

On April 27, 1961, with approval from the State Council's Foreign Affairs Office, the China Ocean Shipping Company (COSCO), New China's inaugural state-owned international ocean shipping enterprise, was officially established. This marked the beginning of a new era in the development of China's international maritime industry

Fig. 10.15 M/V Guanghua [38]

and laid a solid foundation for its globalization [39]. At that time, COSCO's fleet consisted of four vessels: "Guanghua," "Xinhua," "Heping," and "Friendship," with a combined deadweight tonnage of 22,600 tons. The next day, the "Guanghua," having undergone comprehensive repairs, embarked on its maiden voyage to Indonesia to repatriate overseas Chinese and embassy personnel, becoming the first ocean-going vessel to fly the five-star red flag. To celebrate "Guanghua's" triumphant return, Marshal Chen Yi joyfully penned in "Man Jiang Hong—Visiting the Guanghua Ship," the lines: "For a century, the seas bore no trace of me, yet now we rise to thrive and surge forth." [40].

Starting in the 1960s, China methodically began to build its independent ocean shipping industry. In early 1964, the Ministry of Transportation and the People's Bank of China entered into a ship-purchasing loan agreement, acquiring 20 ocean-going vessels totaling 249,000 deadweight tons (DWT), which significantly accelerated the growth of China's ocean-going fleet [41]. By 1965, COSCO owned 63 ocean-going ships, amassing 600,000 DWT, and had established multiple international routes to countries and regions across Asia, Europe, and Africa. After 1966, the burgeoning development of ocean transport faced severe setbacks due to the dissolution and merger of coastal port and navigation management organizations and the disruption of transportation command, inflicting considerable damage on China's maritime industry and further widening the gap with maritime-advanced nations.

After 1969, the national economy improved, and the contradiction between the demand for foreign trade shipping and the capacity of ocean transportation became obvious again. Under the sustained attention of Premier Zhou Enlai and the efforts of the shipping industry, China's maritime transportation industry gradually eliminated interference from 1971 onwards, maritime regulations were restored and revised, maritime command and control agencies were restored and rebuilt, and the ocean transportation industry gained greater development.

Simultaneously, the relatively relaxed international climate of the early 1970s provided favorable conditions for establishing China's north–south maritime route. In 1972, the north–south route was successfully inaugurated, breaking the hostile forces' blockade of China's southern regions that had lasted for over 20 years. This development played a crucial role in facilitating and expanding material exchanges between the north and south, and significantly bolstering the growth of coastal transportation in southern China [41].

By 1975, China's ocean-going fleet comprised 330 vessels, totaling 5.38 million deadweight tons. In the subsequent year, the carrying capacity of China's ocean-going fleet constituted 70% of the volume transported in foreign trade by our ships [41], effectively ending the long-standing reliance on chartering foreign vessels. This development laid a solid foundation for the comprehensive revitalization of China's international maritime industry and foreign trade.

After more than 20 years of entrepreneurship and development, in 1978, the Ministry of Communications directly under the coastal shipping enterprises completed 49.8 million tons of freight, 11.6 times as much as in 1952; completed cargo turnover of 38.666 billion tons—miles (about 19.333 billion tons—kilometers), 18.8 times as much as in 1952.

Since the Third Plenary Session of the Eleventh Central Committee of the Communist Party of China, China's economic system has undergone major and profound changes, the national economy and foreign trade have achieved sustained and rapid development, and China's shipping industry, with the goal of building a unified, open, competitive and orderly shipping market, has continued to deepen its reforms, opened up to the outside world, and realized rapid development.

The Rise of Containerized Transport

The Rise of Containerized Transport in the World

A review of the development history of global container transportation can be summarized into four stages: the nascent and initiation phase, the rapid growth phase, and the maturation phase.

The period from 1830 to 1956 marks the nascent phase of container transport. In 1801, British Dr. James Anderson first proposed the concept of container transport. By 1830, the British railways saw the introduction of containers for coal transport [42]. In 1845, freight carriages resembling modern containers began to appear on British railways, representing the earliest form of container transport. In 1880, the United States constructed its first inland waterway container ship. By 1900, British railways had adopted relatively simple container transport systems. In 1917, container transport was trialed on American railways.

The years from 1956 to 1966 marked the initiation phase of container transportation. In October 1957, the United States' Pan-Atlantic Steamship Company converted six C-2 type general cargo ships into fully equipped container ships with compartments, signifying the advent of maritime container transport. In 1961, the International Organization for Standardization (ISO) initiated discussions on standardizing container sizes, designating the 20-foot container (TEU) as the standard unit [13].

1966—the end of the 1980s for the rapid growth of container transportation. 1966 May, the United States Sea-Land Transportation Company's container ship arrived in the port of Rotterdam, the Netherlands, which marked the emergence of the world's container transportation [42].

From the late 1980s to the present marks the maturation period of container transportation. During this era, containerized multimodal transport experienced rapid growth, with developed countries largely achieving integrated multimodal systems. In the 1990s, shipping companies increasingly formed consortia [42].

The Rise of Containerized Transport in China

Compared to global container transportation, China's container transport began later but has experienced the fastest development pace.

The 1950s marked the nascent and exploratory development phase of container transportation in China, which originated with the railways. In 1955, the Ministry of Railways established a general office for container transportation operations, selecting six stations in Beijing, Tianjin, Shenyang, Harbin, Jinan, and Shanghai to initiate container transport services. By 1956, China had commenced international container transport with Central and Eastern European countries. The maritime transport sector also conducted short-term trial runs using railway containers in 1956, 1960, and 1972 [42].

The 1970s marked the beginning of the development phase for China's container transportation. In 1973, China inaugurated its first international maritime container route. The China Ocean Shipping Company (the precursor to COSCO Shipping Group) and other entities initiated a pilot program for international container transport on general cargo liners between Tianjin, Shanghai, and the Japanese cities of Osaka, Kobe, and Yokohama. In September 1973, the Shanghai to Yokohama, Osaka, and Kobe route was launched, utilizing general cargo ships to transport small containers. The "Bohai No. 1" vessel, loaded with small containers in Kobe, Japan, arrived at the Port of Tianjin. By the late 1970s, China's container transportation had gradually embarked on a path of sustained development.

The 1980s marked a period of steady development for China's container transportation. In 1973, Tianjin Port handled its first international container. In January 1981, China's inaugural specialized container terminal company, Tianjin Port Container Company, was officially established. Later that year, in December, the container terminal at Tianjin Port's Third Harbor Basin was formally accepted by the state and put into operation (Fig. 10.16). By the end of 1986, COSCO had established 12 international container shipping routes, including those to the USA, Europe, Japan, and Australia. By this time, 16 Chinese ports were engaged in international container handling, achieving a throughput of 627,000 TEUs.

The 1990s marked a period of rapid development for China's container transportation. In 1991, the nation had 101 dedicated container vessels, an international container shipping volume of 2.29 million TEUs, 23 specialized container berths with an annual designed capacity of 1.95 million TEUs, and a national port international container throughput of 2.17 million TEUs. By 1993, a total of 58 regular container liner routes had been established, connecting China with Western Europe, Southeast Asia, North America, Japan, Australia, New Zealand, the Persian Gulf, the Mediterranean coast, and Hong Kong. In 1997, the number of dedicated container vessels had grown to 1080 with a capacity of 300,000 TEU slots. The national vessels' international container shipping volume reached 3.57 million TEUs, supported by 65 specialized container berths with an annual designed capacity of 10.03 million TEUs, and the national port container throughput surged to 10.77 million TEUs.

Fig. 10.16 In 1981, China's first specialized container terminal was built in Tianjin Port [43]

Since the 21st century, China's container transportation has reached its mature development phase. After joining the World Trade Organization in 2001, China's foreign trade volume and port container throughput have seen steady growth. By 2009, these figures had reached $2.2073 trillion and 130 million TEUs, respectively. In 2002, China's port container throughput surpassed that of the United States for the first time, ending the latter's 46-year streak as the world leader in this metric.

China's port container throughput from the start to 1 million TEUs took 16 years, from 1 million TEUs to 10 million TEUs took 8 years, from 10 million TEUs to 50 million TEUs took 7 years, from 50 million TEUs to 100 million TEUs took 3 years. In terms of development speed, it can be said that the world's fastest.

For Every Size of Ship in the World, China Has a Port to Match

China has consistently led the world in the construction of large specialized terminals. With the clear trend towards larger vessels globally, China's port berths have accelerated their specialization and deepening. As shipbuilding technology advances, companies aim to provide better services and gain new competitive advantages in the global maritime market by constructing larger vessels to lower transportation costs and increase efficiency. Whether it's specialized terminals for essential goods like coal, crude oil, iron ore, and containers, or cruise terminals crucial to the national economy and people's livelihoods, China's ports have always planned ahead, aligned with demand, and continuously advanced the construction of large specialized berths. This proactive approach has set the pace for the global trend towards larger terminal berths.

The crude oil terminal is equipped to handle today's largest crude oil vessels. In 1976, the Dalian New Port crude oil terminal commenced operations, becoming New China's inaugural modern 100,000-ton crude oil terminal (Fig. 10.17); it featured two berths and was the largest and deepest berth among the country's crude oil export ports at the time [44]. On March 30, 2005, the 440,000-ton Belgian "Tai'ou" tanker (Fig. 10.18) docked at the Shihua Terminal in the Daxie Port Area of Ningbo Port. Measuring 460 m in length, 82 m in width, and with a water depth of 25 m, the Shihua Terminal became the first in China capable of servicing 400,000-ton tankers.

The world's largest ore terminal was established in China, setting multiple world records. In the early 1980s, China's first 100,000-ton ore transshipment terminal was inaugurated in Beilun, Ningbo Port, marking a significant milestone in the country's port development (Fig. 10.19).

On March 9, 2013, the Dongjiakou Ore Terminal of Qingdao Port commenced operations. This terminal, China's first built for 300,000-ton vessels with considerations for 400,000-ton ships, was also the world's largest ore receiving and discharging terminal at that time. Measuring 510 m in length and with a water depth of 25 m, it boasts a design annual throughput capacity of 16 million tons. The project features the deepest dock water depth, the most advanced environmental protection system, the most sophisticated information technology operating system, and the largest quay crane [41].

Fig. 10.17 New China's first modern 100,000-ton crude oil terminal commissioning conference [44]

Fig. 10.18 Tanker Tai'ou [45]

On July 4, 2015, China's first 400,000-ton ore ship "Yuan Zhuo Hai" made its inaugural docking at the 400,000-ton ore terminal in the Dongjiakou Port Area of Qingdao Port (Fig. 10.20).

From the first domestic to the world's largest, China's journey in container terminal development is remarkable. The country's inaugural container terminal was constructed in Tianjin Xingang starting in June 1974, with the first berth officially coming into operation in December 1981 (Fig. 10.21). By the end of 1985, three additional container berths in the newly constructed fourth port pool of Tianjin Xingang were operational. With a total of four berths, this terminal, boasting an annual design throughput capacity of 400,000 TEUs, became the largest dedicated container terminal in China at the time. On August 3, 2021, the world's largest container ship, "Chang Fan" (Fig. 10.22), successfully docked at berth No. 6 in the Meishan Port Area of Ningbo-Zhoushan Port. Measuring 400 m in length and 62 m in width, with a maximum carrying capacity of 23,992 TEUs, "Chang Fan" set a new docking record at Ningbo-Zhoushan Port.

Currently, China boasts the world's largest port terminal facilities in terms of capacity, with its large specialized berths leading globally. This achievement truly fulfills the solemn pledge that "for every size of ship in the world, China has a port to match."

As of the end of 2020, China's ports lead globally with the highest number of operational berths for vessels over 10,000 tons. The share of large, deep-water, specialized terminal berths has significantly increased, with the operational efficiency of specialized terminals and the container throughput per 100 m of coastline at major coastal ports setting global benchmarks. By year-end, national ports boasted 22,142 operational berths, with coastal ports accounting for 5,461 and inland river ports for 16,681. Of these, 2,592 berths could accommodate vessels over 10,000 tons, with

Fig. 10.19 Ningbo Beilun 100,000-ton class ore terminal under construction [46]

coastal ports having 2,138 such berths and river ports 454. The country had 440 berths for vessels over 100,000 tons and 1,371 specialized berths, including 354 for containers, 265 for coal, 85 for metal ores, and 87 for crude oil, alongside 592 general bulk cargo berths and 415 for general cargo.

Tianjin Port, a modern comprehensive port and one of the world's artificial deep-water ports, is home to China's first international container liner route and the first dedicated container terminal. The port can accommodate vessels up to 300,000 tons, with a navigational channel depth of 22 m. It features 192 berths in total, including 128 berths for vessels over 10,000 tons. Among these, there are 23 large-scale, deep-water, specialized container handling berths, capable of servicing the world's largest container ships (Fig. 10.23).

Qingdao Port boasts the world's most advanced and largest specialized deep-water terminals, including the largest 400,000-ton ore terminal, a 450,000-ton crude oil

Fig. 10.20 China's first 400,000-ton ore ship "Yuan Zhuo Hai" first berth on 400,000-ton ore terminal at Dongjiakou port area of Qingdao Port [47] on July 4, 2015

Fig. 10.21 China's first specialized container terminal completed in Tianjin Xingang at the End of 1981 [48]

Fig. 10.22 "Chang Fan" smoothly berthing at Meishan Port of Ningbo-Zhoushan Port [49]

Fig. 10.23 Tianjin port container terminal berths [50]

terminal, a container terminal capable of accommodating the world's largest 24,000 TEU vessels, and a dedicated terminal for the world's largest 227,000-ton cruise ships. Notably, the Qianwan Container Terminal features 24 specialized berths for deep-water container vessels (Fig. 10.24), with a quay length of 8,651 m and a berth depth of 20 m, enabling around-the-clock operations for ultra-large container ships exceeding 24,000 TEUs.

Fig. 10.24 Berths of Qingdao port container terminal [51]

Ningbo-Zhoushan Port stands as the world's first mega-port to surpass an annual cargo throughput of over 1 billion tons, serving as a crucial container hub, the largest iron ore transit, and crude oil transshipment base in China. As of January 2020, the port boasts over 620 operational berths, including nearly 170 large berths for vessels over 10,000 tons, and more than 100 large and super-large deep-water berths for ships over 50,000 tons. The core port area's main navigation channel has a depth exceeding 22.5 m, allowing 300,000-ton vessels to freely enter and exit, while super vessels over 400,000 tons can navigate with the tide. Ningbo-Zhoushan Port is China's busiest for mega-vessels' arrivals and departures (Fig. 10.25).

Fig. 10.25 Ningbo-Zhoushan port container terminal [52]

The construction of large-scale and intelligent ports in China has ensured global leadership in service efficiency. The continuous improvement of the throughput capacity of large-scale and specialized berths, coupled with the accelerated construction of automated terminals, has made China's port facilities the world's leading in terms of comprehensive efficiency, and has made a positive contribution to the smooth flow of the international logistics chain and the reduction of comprehensive logistics costs.

Shanghai Shipping Exchange released in 2021 the world's major ports of ocean-going mainline international container ships average time in port, berthing time data (Table 10.1) shows that in 2021 China's ports overall efficiency performance is good, loading and unloading operational efficiency is the world's leading. Comprehensive efficiency performance is good for the port of Kaohsiung (1.03 days), Hong Kong (1.33 days), Xiamen (1.35 days), Guangzhou (1.49 days); loading and unloading efficiency is higher for the port of Kaohsiung (0.83 days), Guangzhou (0.85 days), the port of Hong Kong (0.90 days), Xiamen (1.05 days), Shanghai (1.06 days) and other ports in China.

See Table 10.1.

International Containers for Ocean Trunk Lines at Major Global Ports, 2021

Protecting China to the World

China's maritime transportation connectivity has ranked first in the world for many years. With the in-depth promotion of China's all-round opening up to the outside world, especially in recent years, under the background of high-quality promotion of the construction of "Belt and Road", China's shipping configuration is constantly optimized, the route network is constantly improved, and the index of maritime liner connectivity has remained the world's first for many years in a row, and maritime transportation has played an irreplaceable role in China's opening up to the outside world and the stable development of global trade (Figs. 10.26 and 10.27). Maritime transportation has played an irreplaceable role in China's opening to the outside world and the stable development of global trade (Figs. 10.26 and 10.27. Especially under the influence of the current epidemic of Crown Pneumonia, the ubiquitous Chinese shipping routes have become the backbone of stabilizing the global industrial chain supply chain.

By the end of 2020, China's international shipping volume had accounted for one third of the global shipping volume, with the capacity of the shipping fleet reaching 314 million deadweight tons, ranking second in the world; the global shipping service network has been continuously improved, with waterway international

Table 10.1 Average time in port and at berth for ships [53]

Port	Throughput in 2021 (10,000 TEUs)	Average time in Hong Kong (days)	Average time at berth (days)
Shanghai port	4703	2.57	1.06
Singapore Harbor	3747	1.78	1.27
Ningbo-Zhoushan Port	3108	2.16	1.07
Shenzhen port	2877	2.04	1.24
Guangzhou port	2418	1.49	0.85
Qingdao port, subprovincial city in Shandong	2371	1.77	1.17
Busan port	2269	1.57	1.31
TPD	2027	1.88	1.50
Hong Kong port	1779	1.33	0.90
Port of Rotterdam	1510	2.24	1.61
Dubai Harbor	1377	1.68	1.07
Port Klang, capital of Malaysia	1374	1.84	1.22
Xiamen Port	1205	1.35	1.05
Port of Antwerp	1202	1.97	1.43
Tanjong Parapas Harbor	1120	1.89	1.13
Port Angeles, capital of Los Angeles	1067	9.49	5.94
Gaoxiong or Kaohsiung harbor, in south Taiwan	986	1.03	0.83
Long Beach Harbor	938	10.36	5.03
New York Harbor	899	2.42	1.81
Hamburg port	870	2.35	1.65

transport routes to and from more than 100 countries and regions, and the shipping liner connectivity index has steadily ranked first in the world.

China's major container ports are globally interconnected, serving as crucial strategic supports for the country's comprehensive opening up. As per official data from various port companies, Shanghai Port has led the world in container throughput for 12 consecutive years as of July 2022, with its international liner routes spanning the globe's principal shipping zones. Tianjin Port operates 130 container routes with over 550 monthly services, maintaining trade links with more than 800 ports across over 200 countries and regions. Qingdao Port offers seamless foreign trade routes worldwide and domestic routes spanning north to south, with a continuously expanding feeder network. It features over 200 container routes reaching more than

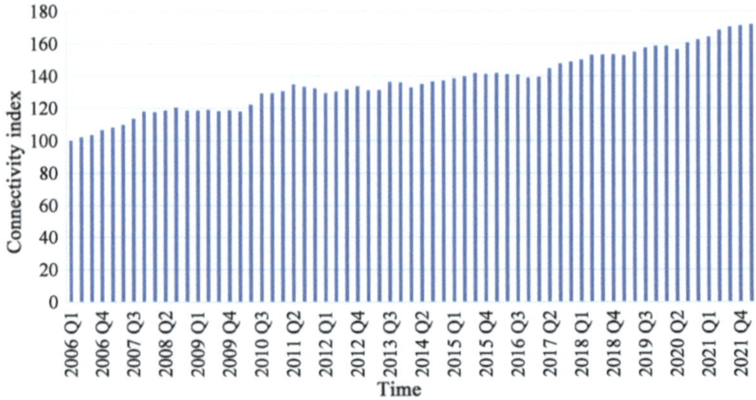

Fig. 10.26 Changes in China's maritime liner connectivity index since 2006. *Note* Data from the United Nations conference on trade and development

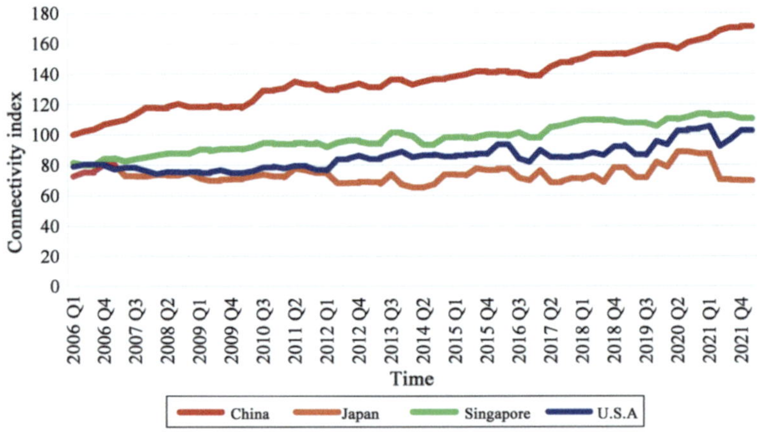

Fig. 10.27 World typical country maritime liner connectivity index since 2006. *Note* Data from the United Nations conference on trade and development

700 ports in over 180 countries, ranking first in route density among northern Chinese ports. All container routes of the top 20 global shipping companies dock at Qingdao Port. Ningbo-Zhoushan Port stands out for its "hardcore" capabilities and strategic location, backed by China's vibrant Yangtze River Delta Economic Circle. Its 260 container routes connect over 600 ports in more than 190 countries, creating a vast maritime trade network. Guangzhou Port boasts over 200 container routes, including more than 140 international trade liner routes, covering major domestic and international ports. The shipping liner connectivity index for some global ports in the first quarter of 2022 and typical ports since 2006 are depicted in Figs. 10.28 and 10.29.

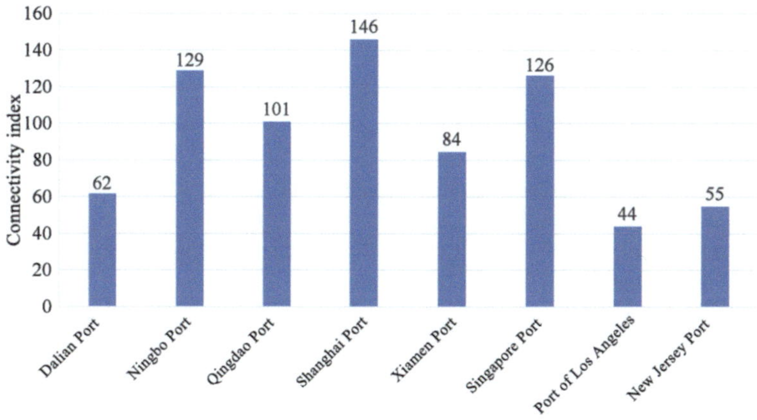

Fig. 10.28 Maritime liner connectivity index for selected world ports, Q1 2022. *Note* Data from the United Nations conference on trade and development

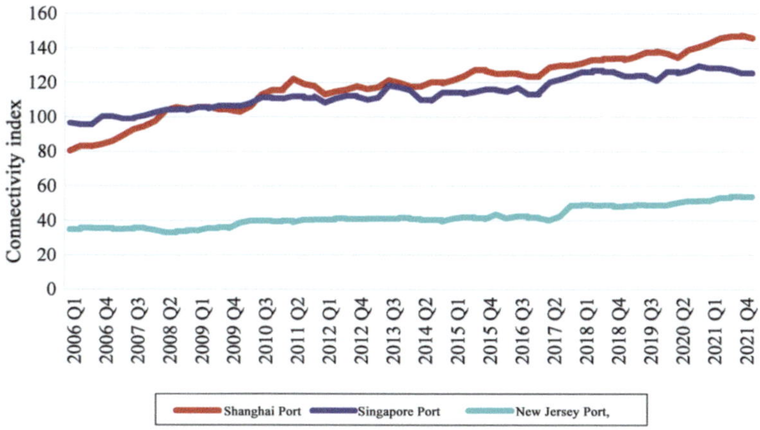

Fig. 10.29 Maritime liner connectivity index for typical world ports since 2006. *Note* Data from the United Nations conference on trade and development

The shipping service capacity of large port and shipping enterprises has been increasing. According to the data on the official website of China COSCO Shipping Group, as of the end of August 2021, China COSCO Shipping Group operates a fleet with a combined capacity of 110 million deadweight tons, ranking first in the world. Among them, the size of the container fleet of 3.08 million TEU, ranking third in the world; dry bulk fleet capacity of 44.6 million DWT, oil and gas fleet capacity of 29.36 million DWT, general cargo special fleet capacity of 4.75 million DWT, all ranked the first in the world; the Group's global sales of bunker fuel more than 28.19 million tons, ranked the first in the world; container leasing business retained 3.7 million TEUs, ranked second in the world.

The global liner capacity rankings as of February 7, 2022 are shown in Table 10.2 [53].

The accelerated pace of going out by port and shipping enterprises has strongly supported the improvement of China's international logistics supply chain network. Represented by COSCO Shipping Ports, China Merchants Ports and other enterprises, they have accelerated the pace of "going out", accelerated the deployment of overseas ports, and continuously improved the international shipping and logistics network, which has strongly supported the development of China's foreign trade and industrial "going out".

According to the data on the company's official website, COSCO Shipping's port terminal portfolio covers five major port groups along China's coast as well as ports in Southeast Asia, the Middle East, Europe, South America and the Mediterranean Sea. By the end of 2020, COSCO Shipping Port operates and manages 357 berths in 36 ports around the world, among which, there are 210 container berths, with a handling capacity of about 120 million TEUs. 2020, COSCO Shipping Port's container throughput in overseas regions will be 28.44 million TEUs, accounting for 28% of COSCO Shipping Port's total container throughput. China Merchants Port overseas business, through Terminal Link Company to complete the acquisition of eight terminals equity, port business expansion to Southeast Asia, the Middle East, Europe, the Caribbean and other places, and continue to improve the global port layout; 2020, China Merchants Port overseas areas to complete the container

Table 10.2 Global liner capacity ranking (February 7, 2022)

Rankings	Company name	Chinese abbreviation	Number of containers (10,000 TEUs)	Operation of container ships (ships)
1	Mediterranean Shg Co.	Mediterranean shipping	431.4	655
2	Maersk	Maersk line	428.0	733
3	CMA CGM group	Duffy line	319.7	569
4	COSCO group	COSCO shipping group	292.6	475
5	Hapag-Lloyd	Hapag-Lloyd (name)	175.0	250
6	ONE (Ocean Network Express)	Ocean NetLink shipping (ONE)	152.5	207
7	Evergreen line	Evergreen marine corp. (Taiwan shipping line)	147.4	202
8	HMM Co Ltd.	Hyundai merchant marine	82.2	76
9	Yang Ming marine transport corp	Yang Ming marine transportation	66.4	91
10	Zim	Star shipping	42.1	111

throughput of 28.88 million TEUs, accounting for 24% of China Merchants Port total container throughput.

References

1. Zhao Kui, Bai Mei. Tianhougong and the Fujian Association [M]. Nanjing: Southeast University Press, 2019.
2. Liao Pengfei. Reconstruction of Shunji Temple at Shengdun Ancestral Temple[M]||Jiang Wei-Pong, Zheng Li-Hang, compilation. A Compendium of Mazu Literature and Historical Materials. Beijing: China Archives Press, 2007.
3. Huang Gongdu. Zhi jia weng ji [M]. Beijing: China Bookstore, 2018.
4. Lin, Ming-Tai. The Inheritance and Development of Mazu Culture in the Countries and Regions Along the Maritime Silk Road[M]||The Editorial Committee of Southeast Weekend Forum. Fuzhou: Straits Literature and Art Publishing House, 2009:101–115.
5. Sohu. A brick and a tile, is pious pile up the source and long [EB/OL]. (2019-02-25) [2022-05-04]. https://www.sohu.com/a/297717344_417690.
6. Pan Zhenjin. The light of Mazu:Mazu in the eyes of a newsman [M]. Fuzhou: Straits Literature and Art Publishing House, 2018.
7. Gu Donggao. Spring and Autumn Annals [M]. Beijing: Zhonghua shuju, 1993.
8. Sima Qian. The Records of the Grand Historian [M]. Beijing: Zhonghua shubu, 1959.
9. History as a mystery. Japan is a descendant of Xu Fu? Where did Xufu really go? [EB/OL]. (2019-05-03) [2022-05-04]. https://www.163.com/dy/article/EE9PGVV50537071 8.html.
10. Sun Guangqi, Zhang Houquan, Sun Xiajun, et al. History of shipping in ancient China [M]. Dalian: Dalian Maritime University Press, 2015.
11. Du Fu. Du Gongbu ji [M]. Shanghai: Shanghai Ancient Books Publishing House, 2003.
12. Folk Talk Hot Spot. Tianjin's largest archaeological excavation, "three will sea port" military grain city, is the earliest seaport [EB/OL]. (2022-03-11) [2022-05-04]. https://www.sohu.com/a/529126231_120827543.
13. Tang Yan. Bohai Guo Zhi [M]. Shanghai: Shanghai Ancient Books, 1962.
14. Wu Bin. History of the Silk Road [M]. Shenyang: Shenyang Publishing House, 2018.
15. Qian Lisheng, Hu Yimin. The "Maritime Silk Road" of the Han and Tang Dynasties [J]. Navigation, 2019(01):18–20.
16. Shanghai Museum. Shipbuilding and Navigation Technology in the Tang Dynasty [EB/OL]. (2020-12-25) [2022-05-04]. https://sghexport.shobserver.com/html/baijiahao/2020/12/25/321518.html.
17. Liu An. Huai Nan Zi [M]. Xu Shen, note. Chen, Guangzhong, point of view. Beijing: Zhonghua shubu, 2016.
18. Huang Xing. Literature review of research on the history of the Chinese compass [J]. Natural Dialectics Newsletter, 2017, 39(01):85–94.
19. Xiaochi Town Guanhu Center Primary School. Seventh grade next book of history knowledge comb [EB/OL]. (2020-05-31) [2022-05-04]. https://www.sohu.com/a/398828794_99911939.
20. Guo Yemin. From Sinan to Compass: The Road to the Invention of the Compass [J]. National Humanities History, 2020(19):120–127.
21. Zhu Yu. Pingzhou [M]. Li Weiguo, Gao Keqin, editors. Shanghai: Shanghai Ancient Books Publishing House, 2012.
22. Xu Jing. Seonhwa Bongyeo Goryeo Tujing [M]. Park, K.H., labeling. Changchun: Jilin Literature and History Press, 1986.
23. Zhao Rushi. The Journal of the Tibetans [M]. Shanghai: Shanghai Ancient Books Publishing House, 1993.
24. 51 Postings. Compass wallpaper pictures [EB/OL]. [2022-05-04]. https://www.51tietu.net/p/2862908.html.

25. Temple. China's Creative Spirit - China's 100 World Firsts [M]. Chen Yuzheng, et al, Translation. Beijing: People's Education Press, 2002.
26. Tutu. History of the Song dynasty [M]. Beijing: Zhonghua shubu, 1985.
27. Wu Cheng. Wu Wenzhenggong ji [M]||Photocopied Wenyuange siku quanshu. Taipei: Commercial Press, Taiwan, 2008.
28. Liu, W.-Y. Geographic inquiry: History of geography and geoscience and technology [M]. Beijing: Modern Publishing House, 2015.
29. Xu Xiangmin. "The Legislative Purpose of the Decree on the Prohibition of the Sea–An Argument Against the Cultural Closure of the Qing Dynasty [J]. Law, 2020(01):168–182.
30. Kungang. Qing dynasty canon examples [M]||SiKuQuanQuan. Shanghai: Shanghai Ancient Books Publishing House, 2002.
31. Qu Dajun. Guangdong Xinyu [M]. Beijing: Zhonghua shubu, 1985.
32. Nguyen Min-seok. The Records of the Sea [M]. Fuzhou: Fujian People's Publishing House, 1982.
33. Jiang Liangqi. Donghua lu [M]. Bao Sitao and Xiyuan, editors. Jinan: Qilu Shusha, 2005.
34. JI Juan. Qin Ding Renwen Tongkao [M]||Photocopied Wen Yuan Ge Siku Quanshu. Taipei: Taiwan Commercial Press, Ltd. 2008.
35. Alabaster China. What do you find in this Western oil painting? [EB/OL]. (2018-11-30) [2022-05-04]. https://www.sohu.com/a/278897443_656524.
36. Dong Zhikai. China's Struggle Against "Blockade and Embargo" in the 1950s [J]. Research on Chinese Economic History,1991(01):1-19.
37. China Transportation News. "New Hailiao, Sail Away for the Country! [EB/OL]. (2020-09-28) [2022-05-04]. https://news.ycwb.com/2020-09/28/content_1193200.htm.
38. 桂工网. Guanghua ship ventures to complete new China's first evacuation [EB/OL]. (2018-05-10) [2022-05-04]. http://www.gxworker.com/a/180510/280975.shtml.
39. China Nautical Society. Chinese maritime history (modern maritime history) [M]. Beijing: People's Transportation Press, 1989.
40. Chen Yi. Selected Poems of Chen Yi [M]. Beijing: People's Literature Publishing House, 1977.
41. Li Xianqiang. Introduction to navigation [M]. Dalian: Dalian Maritime University Press, 2018.
42. Shipping Winds CNSS.A brief history of the development of world container transportation [EB/OL]. (2019-10-03) [2022-05-04]. https://www.sohu.com/a/344845995_100020595.
43. Sina Finance. Tianjin port stormy development road [EB/OL]. (2018-08-20) [2022-05-04]. http://finance.sina.com.cn/roll/20090921/20226776214.shtml.
44. Ministry of Transportation and Communications. Dalian Port | "100-year oil port" safety with [EB/OL]. (2018-09-25) [2022-05-04]. https://baijiahao.baidu.com/s?id=1612577528593279599&wfr=spider&for=pc.
45. Sohu. World's largest super-giant tanker docked in Ningbo, two national records broken at the same time [EB/OL]. (2017-08-15) [2022-05-04]. https://www.sohu.com/a/164814556_809485.
46. Sanhang Bureau. Rooted in Ningbo! Building the world's No. 1 port [EB/OL]. (2021-06-25) [2022-05-04]. https://m.thepaper.cn/baijiahao_13319421.
47. Sohu. [Diagram 70 years] Iron ore operations across three major port areas Let the mouse revolutionize the life of the shovel [EB/OL]. (2019-09-16) [2022-05-04]. https://www.sohu.com/a/341465724_99917666.
48. Visible History. Old photos New achievements of construction in the early 80s Feel the economic growth of the Republic [EB/OL]. (2019-11-15) [2022-05-04]. https://www.163.com/dy/article/EU1CDPSB0516A873.html.
49. Longde Shipman. Ningbo Zhoushan port picks up the world's largest container ship "Changfan" ship [EB/OL]. (2021-08-03) [2022-05-04]. https://www.imarine.cn/news/744564.html.
50. Tianjin. How "divine" is the port of Tianjin? [EB/OL]. (2022-03-10) [2022-05-04]. https://view.inews.qq.com/a/20220310A010C700.
51. Qingdao Port Dongjiakou Ore Terminal Co. [Industry News] Here is China's Qingdao Port|Video Interview with Li Fengli, Chairman of Qingdao Port Group [EB/OL]. (2019-06-10) [2022-05-04]. https://www.sohu.com/a/319611789_99917666.

52. Zhoushan Daily News. Good news! World's No. 1 port joins hands with world's No. 1 mining company [EB/OL]. (2020-11-13) [2022-05-04]. https://www.sohu.com/a/431725575_100193006.
53. Wisdom Research Consulting. Top 100 Global Liner Line Capacity Rankings, February 7, 2022 [EB/OL]. (2022-02-17) [2022-05-01]. http://guba.eastmoney.com/news,gssz,1142840878.html?jumph5=1.

Chapter 11
Driving Progress Through Transportation: The Role of China's Highway Development in Economic Prosperity

The road system is the most basic infrastructure within the transportation network, characterized by its vast coverage, fundamental role in carrying various modes of transport, and its long-standing history of development. As mentioned earlier, its origins can be traced back to the dawn of humanity and has evolved alongside human civilization. The historical development of roads spans over 4,000 years, dating back to when Yu the Great constructed the nine pathways. Since the establishment of New China, the nation's road sector has seen unprecedented rapid growth, particularly after the economic reforms and opening up, with highway development progressing rapidly and the total length surpassing 5 million km, making China the world leader in road mileage. The proverb "To get rich, first build roads" aptly illustrates how the swift expansion of highways has supported and spurred the consistent and rapid development of China's economy.

"A road, as defined by the *Modern Chinese Dictionary* (7th edition), has two meanings: firstly, it is 'a part of the ground designated for the passage of people, carts, and horses'; and secondly, 'a conduit between two places, encompassing both terrestrial and aquatic pathways'. In a broader sense, roads began to unfold the moment humanity's forebears first set foot from the ocean onto land. Lu Xun once expressed a profound yet simple truth: 'Originally, there were no paths on the earth, but as more people walked upon it, paths emerged.' Roads have perpetually been indispensable to life, foundational to development, and instrumental in the proliferation of civilization."

"The concept of the highway emerged concurrently with the invention of the automobile, a development that dates back just over one hundred years. The genesis of China's highways can be traced to the early 20th century, originating from the 'Official Horse Roads' of the Qing Dynasty. From the late Qing era through to 1949, the evolution of China's highways—mirroring the trials and tribulations of the Chinese nation itself—has navigated a tumultuous path."

© China Communications Press Co., Ltd 2025

H. Lu (ed.), *A Brief History of Transportation in China*,
https://doi.org/10.1007/978-981-97-9851-3_11

It was not until the establishment of New China that, under the guidance of the Party and government, the Chinese highway system experienced a rebirth. Post-reform and opening-up, the highway sector was among the first to be liberalized, heralding a developmental renaissance. By the late 1990s, a significant surge in the development of highway infrastructure effectively alleviated long-standing economic constraints, positioning the sector as a true trailblazer. Since the 18th National Congress of the Communist Party of China, the Chinese highway network has seen continuous enhancements amid rapid and stable growth, positioning it among the global frontrunners. This progress has solidified the groundwork necessary for achieving the initial Centenary Goal and advancing towards becoming a leading transport nation.

Exploration and Development of Highways in New China (1949 to the End of 1978, Before Reform and Opening Up)

On October 1, 1949, the PRC was founded, and China's highways were given a new lease of life.

At the end of 1949, the total highway network was only 80,800 km, of which only 32,000 km had road surfaces. Among these, only 300 km had hard surfaces such as asphalt or cement. Compared with the country's land area of 9.6 million m^2 and its population of 542 million, this level of highway infrastructure was virtually nonexistent. The development of New China's highways began from such impoverished conditions.

In the early stages of New China's founding, under the leadership of the State Council, the Ministry of Transport established a national system for the construction, maintenance, and management of highways. It introduced regulations for the collection and use of road maintenance fees and took the lead in constructing national defense highways, economic arterial roads, and county and township roads. As a result, China's modern highways were gradually restored and developed.

In 1950, alongside the People's Liberation Army (PLA) marching into Xizang, the construction of the Kham-Xizang Highway (Sichuan-Xizang Highway) commenced. The southern segment of the Qinghai-Xizang Highway, initiated in May 1954, was completed by the end of December in the same year. Henceforth, Xizang was equipped with a modern highway system. The successful completion and operation of the Qinghai-Xizang and Sichuan-Xizang Highways epitomized the "spirit" of the People's Republic of China, and gave rise to a distinctive ethos within the highway sector, enduring to this day—the "Two Roads" spirit of braving hardships and death, enduring perseverance, commitment to the foundational role in the nation's development, and fostering unity between the military and civilians, alongside ethnic harmony.

In 1953, the first five-year plan was implemented, marking the beginning of a new development stage for China's highways, shifting focus from wartime needs to economic construction. By the end of 1957, the national highway mileage had increased to 254,600 km, double the 126,700 km five years earlier, with paved roads constituting 47.56% of the total.

The "Great Leap Forward" movement and the "Cultural Revolution" led China to gradually move away from focusing on national economic construction. Despite not maintaining the early rapid development momentum of New China, road construction, as an indispensable infrastructure for national economy and defense, still achieved relatively fast growth overall. Especially notable during this period were the projects associated with the third front construction, including significant highways such as the Beijing to Shanxi Yuanping (Jingyuan Highway), Yunnan Chuxiong to Menghuang (Chu Meng Highway), Henan Luoyang to Lingbao (Luoling Highway), Karakorum Highway, Qingxin Highway, Tian Shan Highway (i.e., Dukou Highway), Liaoning Shenyang to Fushun (Shenfu First Class Highway), Yunnan to Xizang (Yunnan-Xizang Highway), and the expansion and upgrade of the Sichuan-Xizang and Qinghai-Xizang Highways [5].

In the 1970s, with the gradual recovery of the economy, China faced significant challenges with its limited highways, low standards, poor quality, and accessibility issues. In July 1970, the Ministry of Railways, the Ministry of Transportation, and the postal operations under the Ministry of Posts and Telecommunications merged to form the Ministry of Transportation, which established a highway division to oversee highway affairs. To address issues of mixed traffic and severe congestion, some leaders and experts within the Ministry planned the construction of fully enclosed, fully elevated highways. In 1972, the Second Highway Survey and Design Institute, commissioned by the Ministry of Transportation's Highway Division, conducted field investigations and researched solutions for the Beijing-Tianjin-Tanggu Highway. In April 1974, based on the fully enclosed, fully elevated first-class highway standards, the Second Highway Survey and Design Institute submitted their investigation report to the Ministry of Transportation's Highway Bureau (reorganized from the Highway Group in early 1973) [6].

In 1975, Deng Xiaoping, serving as Vice Premier of the State Council, took charge of the Central Committee and the State Council's daily operations, initiating a comprehensive reorganization. He emphasized the importance of embracing science and technology across all sectors to rapidly boost production. In 1975, the Ministry of Transportation and the Ministry of Railways were separated, each reverting to their original structures. In November, the Ministry of Transportation's Highway Bureau dispatched a delegation to Japan for an extensive 10-day visit, where they observed several highways, both under construction and operational, gaining detailed insights into engineering standards, construction techniques, funding sources, toll methods, and operational practices. Upon returning, they submitted a report to the Ministry of Transportation and commenced the translation and organization of the acquired technical materials.

By the end of 1977, China's highway mileage reached 855,600 km, 10.59 times that of 1949. Over these 28 years, the net increase in highway mileage exceeded 770,000 km.

Springtime for Road Development (Before Accelerated Road Construction, 1978–1998)

In 1978, the Eleventh Central Committee's Third Plenary Session made the historic decision of shifting the focus to economic development and initiating reforms and opening up. China's highways shifted their focus from national defense and third-line construction to embracing the reform and opening-up era, simultaneously making significant strides towards reaching international advanced standards.

In the middle of 1977 and the end of 1978, Ye Fei, the then Minister of Transportation, stated after two visits to Western Europe that developing highways, particularly expressways, significantly impacts industrial and agricultural growth and the achievement of the 'four modernizations.' In 1978, the Ministry of Transportation initiated the preliminary surveying and design work for the Beijing-Tianjin-Tanggu Highway in line with expressway standards. Subsequently, to support the surveying and construction of the Beijing-Tianjin-Tanggu Expressway, the Ministry of Transportation's Highway Bureau mobilized technical personnel from its directly affiliated research, design, and construction units. They systematically translated and studied international expressway technical standards, surveying, and construction techniques, leading to the development of the Beijing-Tianjin-Tanggu Expressway's technical standards.

After entering the 1980s, with the gradual deepening of reform and opening up and rapid economic development, China's highway construction ushered in the spring of development.

Road and Waterways Take the Lead in Opening up the Market to Bring in Fresh Water

The National Transportation Work Conference held on March 7, 1983, the Ministry of Transportation put forward the reform concept of 'using boats on rivers and cars on roads', signifying that road and water transportation were the first two modes in the comprehensive transportation system to open their markets to the entire society. The road and waterway transportation markets began to 'revive'.

The year 1984 was a year of far-reaching impact on the development of roads in China: on 30 November, the Ministry of Communications submitted to the State Council a report on accelerating road construction, proposing sources of funding and measures for road development: mobilizing the enthusiasm of all parties and raising funds extensively. The 54th executive meeting of the State Council held in December of the same year made three major decisions of historical significance to the development of China's roads, namely raising the road maintenance fee collection standards, imposing a surcharge on vehicle purchases, and allowing the construction of superior roads, large bridges, and tunnels through loans or collective funding, with tolls collected from vehicular traffic. These three major decisions, later summarized as "loans to build roads, charges to repay loans" 8 words, gave China's road construction a stable source of funds and a policy environment for accelerated development, and laid a good financial and policy foundation for China's road business to seize the opportunity and accelerate its development thereafter.

Guangdong Province borrowed HK$150 million in November 1984 to build four bridges on the Guangzhou-Zhuhai highway, which were completed one after the other, with the principal and interest repaid through bridge tolls. This was China's first attempt at financing the construction of a toll road.

The First Fully Implemented Highway Network Plan in New China.

In October 1978, the Ministry of Transportation organized the compilation of the "1978–1985 100,000 km National Highway Network Plan (discussion draft)", incorporating feedback from various stakeholders and adhering to the principles of integrating short-term and long-term goals, civilian and military needs, as well as balancing the feasible with the necessary. In April 1979, this led to the preliminary scheme of the "1981–1990 100,000 km National Highway Network Plan." This scheme marks a significant milestone in the advancement of China's highway development into a new era.

The "National Trunk Highway Network (Trial Scheme)", commonly known as the National Highway Network Plan, proposed 70 national highways comprising 12 Beijing radial routes, 28 north–south longitudinal routes, and 30 east–west transversal routes, with a total length of 109,200 km. This plan represented China's first comprehensive and scientifically structured highway network planning. It has been demonstrated that this planning approach, tailored to national conditions, offers a fundamentally sound balance in terms of scale, density, structural integrity, and connectivity. It essentially established the groundwork for China's future highway network planning, which integrates radial and grid patterns.

By 1990, in response to the rapid development of the national economy and the practical outcomes of the National Highway Network Plan, the Ministry of Transportation proposed the "Three Mains and One Support" strategy. This strategy encompasses the main highway framework, principal water transportation channels, key port and station hubs, and a transportation support and security system. The "main highway framework" was officially designated as the "National Highway Trunk Line System." Leveraging the National Highway Network Plan, it envisioned a "Eight Vertical and Nine Horizontal" layout, comprising 17 trunk lines. To expedite the establishment of this trunk line system and alleviate the bottlenecks in highway transportation impeding economic development, the Ministry refined and improved the plan in 1992. Consequently, a "Five Vertical and Seven Horizontal" trunk line system layout was proposed and endorsed by the State Council. This revised plan spans approximately 35,000 km and is projected to be completed over 30 years. To ensure these trunk lines serve their primary function in transportation security, they will generally adhere to expressway standards, although in sparsely populated areas of the western border and hinterland, secondary roads may temporarily suffice.

While the national highway network plan has been fully implemented, the "Five Vertical and Seven Horizontal" national highway trunk routes have served as China's first highway network plan.

Zero Breakthrough of Highway in Mainland China

The earliest highway in China was not on the mainland, but the 373.4 km Kaohsiung-Keelung Highway (named the "North–South Highway"), which was completed in 1978 and runs north–south through Taiwan.

At that time in mainland China, aside from a small number of experts who understood the technical essence of expressways, most people were unfamiliar with the concept of highways. Although the development of highway in mainland China began later, their growth rate was remarkably rapid.

On June 27, 1984, the construction of the Shenyang-Dalian Highway began. In 1986, the Shenda Highway was approved to be constructed as a first-class highway throughout its entire length. After several efforts, the road was approved to be built to highway standards. On December 21, 1984, the Shanghai (Shanghai) Jiading (Ding) Highway construction commenced. In 1988, the north and south sections of the Shenda Highway were opened to traffic, along with the Shanghai-Jiading Highway, marking the mainland's initial breakthrough in highway development. It was not until August 1990, when the 375 km Shenda Highway was fully opened, that it surpassed Taiwan's North–South Highway to become the longest highway in China, earning it the title of "China's No. 1 Highway".

On October 23, 1987, the signing ceremony for the civil engineering contract of the 142.69-km-long Beijing-Tianjin-Tanggu Highway, the first in China to be built to international standards with World Bank loans, through international tendering, and incorporating FIDIC (International Federation of Consulting Engineers) terms, took place in Beijing. Completed and opened to traffic in 1990, this highway became China's inaugural inter-provincial highway and a benchmark in Chinese highway construction. The adoption of FIDIC terms and the construction practices aligned with international norms profoundly influenced highway construction across China.

During this period, aside from the Shenyang-Dalian, Shanghai-Jiading, and Beijing-Tianjin-Tanggu highway, Guangdong Province completed the Guangzhou-Foshan highway in August 1989, and Shaanxi Province finished the Xi'an-Lintong highway in December 1990.

"During the Sixth Five-Year Plan period (1981–1985), although the construction of highways in mainland China had begun, it was not a smooth ride. Until the end of the Seventh Five-Year Plan period, there was still a great deal of controversy in society as to whether highways should be built in China, and the public perception was still severe.

Highway Refreshes Public Perception

Shenda highway opened to traffic to bring great social and economic benefits, refreshing people's perception of the highway.

The Shenda Expressway traverses the Liaodong Peninsula, linking the capital of Liaoning Province, Shenyang, with the cities of Liaoyang, Anshan, Yingkou, and Dalian, along with two major ports (Dalian and Yingkou Ports) and two international airports (Shenyang Taoxian and Dalian Zhoushuizi). This has significantly reduced the travel time between these cities. Before the highway, driving from Shenyang to Dalian took 10 h, but this was cut to 4 h upon the highway 's completion. The reduced travel distance made the coastal access for inland cities like Shenyang, Liaoyang, and Anshan comparable to that of port cities. Previously, a drive from Shenyang to the closest seaport, Yingkou's Bayuquan Port, took over 6 h, but this journey now takes only 3 h via the Shenda Highway. Consequently, every summer, tens of thousands from Shenyang would travel to Bayuquan in Yingkou for a "sea bath," a novelty in the early '90s.

"Liaoning has shrunk" is a sentiment widely shared by many in Liaoning following the full opening of the Shenda Highway. The vivid descriptions and glowing recommendations shared by users based on their personal experiences have proven to be more direct and effective than any media promotion. The term "highway" has firmly entered the public lexicon, and the erstwhile skepticism surrounding expressway construction has begun to fade. The Shenda Expressway, stretching 375 km, constituted nearly 72% of mainland China's total expressway mileage for that year.

In the early 1990s, the Hefei-Nanjing highway became famous nationwide due to an unexpected natural disaster.

In the mid-1980s, the Hefei-Nanjing Highway was initially approved as a first-class dedicated motorway but was later upgraded to highway standards. During the transition from spring to summer in 1991, when the Hefei-Nanjing Highway was planned to open, the Jianghuai Basin entered its plum rain season a month early. A flood made the not-yet-formally-opened, essentially completed Hefei-Nanjing Highway play a crucial role unexpectedly, leading the people and officials in the affected areas to affectionately dub it the "life-saving road."

Speed, comfort, and a higher capacity for disaster resilience and traffic flow are the immediate advantages experienced by early highway users, significantly alleviating societal skepticism. However, for widespread acknowledgment from government agencies, experts in relevant research institutions, numerous media outlets, and the entire society, these "intuitive feelings" about highways are insufficient. The full range of benefits of highways and dedicated automobile roads—beyond speed, efficiency, and safety, such as transforming the economic landscape along their routes, stimulating industrial development, meeting diverse societal travel needs, and promoting energy conservation and environmental protection—requires extensive scientific investigation, statistical analysis, and public dissemination to achieve comprehensive recognition and widespread agreement.

If in the mid to late 1980s highways were still in their nascent stage with limited public support, by the 1990s, they had become widely embraced, entering a phase of rapid development. Slogans like "To be wealthy, first build roads," "With highways, all industries flourish," and "Small roads lead to modest wealth, major roads to greater wealth, highways to rapid wealth" had gained acceptance by government departments at all levels, research institutions, and the media. More importantly, they were widely endorsed and circulated by urban and rural officials and the public alike. The ideological obstacles to expressway construction had vanished, becoming a thing of the past.

Peak of Highway Construction

In 1989, the Ministry of Transportation convened a national conference in Shenyang and Dalian, Liaoning, to exchange experiences in high-grade highway construction, commonly referred to as the "Shenyang Meeting". This conference aimed to disseminate the successful practices from the construction of the Shenyang-Dalian Highway. State Councilor Zou Jiahua, who attended the meeting, emphasized in his address that the question was not whether highway should be developed but that their development was imperative. He highlighted the importance of planning for both the present and the future. The Shenyang Meeting spurred the first nationwide surge in highway construction, marking a pivotal moment in the history of China's highway development.

At the end of 1989, China's total highway mileage surpassed one million km for the first time, reaching 1,014,300 km. By the conclusion of the Seventh Five-Year Plan in 1990, the national highway network extended to 1,028,300 km, including 522 km of highways. Beijing, Tianjin, Hebei, Liaoning, Shanghai, Guangdong, and Shaanxi achieved their inaugural highway milestones.

From June 18 to 23, 1993, to strategize on elevating highway construction to a new level for the millennium and to discuss accelerating policies and measures, the Ministry of Transportation convened a national highway construction conference in Jinan and Qingdao, Shandong Province, known as the "Jinan Conference." This marked the second milestone meeting in China's highway development history following the 1989 Shenyang Conference. If the Shenyang Conference addressed the question of whether expressways should be built, overcoming ideological barriers across the industry and society, and initially resolving technical standards for highways, the Jinan Conference, guided by the "Five Verticals Abd Seven Horizontals" national trunk highway plan, focused on how to construct highways. It specifically identified "two verticals, two horizontals, and three critical sections" as the construction priorities, thus igniting another surge in China's highway construction.

By the end of 1997, the total mileage of highways nationwide had reached 1,226,400 km, a net increase of 370,000 km compared with 1977; among them, the total mileage of highways had reached 4,771 km, which had already covered 23 provinces of China from east to west and from south to north [1].

Highway Construction Achievements Impressive to the World (1998–2012)

Just as China's highway sector was progressively accelerating and steadily growing, a financial crisis erupted in Southeast Asia in late 1997, significantly affecting China's economic development by 1998. In response to the Asian financial crisis and to sustain the stable and swift growth of the national economy, the Central Committee of the Communist Party and the State Council decided to adopt a proactive fiscal policy, expedite infrastructure development, and boost domestic demand.

The unforeseen Asian financial crisis, but for China's highway business has brought a rare opportunity for development, to promote the highway business into the "accelerate the construction" as the characteristics of the new stage. It is also thanks to the early 1990s sufficient project reserves, the national highway sector to bear the burden, accelerate the development of China's highway construction to further promote the climax, and in just over 10 years, at the beginning of the 21st century on the realization of highway infrastructure development by leaps and bounds, quickly among the world's most advanced, and the achievements of the world to note.

Fuzhou Meeting Kicks off Accelerated Highway Construction

From June 20 to 23, 1998, to execute the directives of the CPC Central Committee and the State Council, the Ministry of Transportation convened a national conference in Fuzhou to accelerate highway construction, known as the "Fuzhou Conference." Vice Premier Wu Bangguo attended and spoke at the conference. The conference decided to increase the total investment in highway construction for 1998 from the initially planned 120 billion yuan to 160 billion yuan, a year-on-year increase of one-third. Later, as the financial crisis deepened, the investment target was further raised to 180 billion yuan on August 23, marking a 50% year-on-year increase from 1997. This conference was the third of significant importance in the history of China's highway development, propelling the sector onto a fast-track development path.

After the "use loans to build roads, repay with tolls" policy was introduced in 1984, it was challenging for domestic banks to finance highway projects due to large investments, long payback periods, and restrictions on the duration of toll collection. Following the Fuzhou Conference, amidst pressures from the international financial crisis, domestic economic downturn, and employment challenges, and propelled by sound decisions from the CPC Central Committee and the State Council, the domestic banking sector began actively funding highway construction. Bank loans quickly became the most significant funding source for China's highway projects, eventually accounting for approximately 70% of highway construction investment, thus becoming the real "driving force" behind the rapid development of highway infrastructure.

By the end of 1998, actual highway investment reached 216.8 billion yuan, exceeding the planned amount by 80.7%. That year saw the initiation of 85,615 km of highways, including 15,932 km of expressways. Of these, 45,677 km of highways were completed within the year, with expressways accounting for 1,663 km. Including the 2,299 km from previous constructions and upgrades to meet new standards, the total expressway length amounted to 8,733 km.

Only four provinces or autonomous regions, Xinjiang, Xizang, Qinghai and Ningxia, are left without highways.

By the end of 2000, the final year of the Ninth Five-Year Plan, the national highway mileage had reached 16,300 km.

Scientific Planning Leads to Rapid Highway Extension

During the 1990s, China's highway construction primarily followed the national highway network plan and the "five verticals and seven horizontals" national trunk highway strategy. Due to the early timing of these plans, the overall coverage was limited and did not fully align with the rapid economic growth and the increasing travel demands of the population. To address the needs of building a moderately prosperous society in China at the start of the 21st century, the Ministry of Transportation initiated a preliminary study on the national expressway network plan in 2001, and officially began its development in 2002.

On December 17, 2004, Premier Wen Jiabao chaired a State Council executive meeting to discuss and provisionally approve the National Highway Network Plan. Subsequently, the National Development and Reform Commission issued this plan under the document number NDRC Transportation [2004] No. 3057. This marked China's first comprehensive highway network plan. The plan adheres to the strategy of "densification in the east, networking in the central regions, and connectivity in the west," focusing on constructing 7 radial, 9 north–south, 18 east–west routes, and 5 regional ring roads along with connecting routes centered around Beijing, collectively known as the "7918 Network." The total planned length is approximately 85,000 km, with 68,000 km for main routes and 17,000 km for regional rings and other connecting roads. It aims to complete the national expressway within the next 30 years, establishing an extensive and well-developed trunk highway network nationwide.

On February 2, 2005, the State Council's 80th executive meeting approved the Rural Road Construction Plan. The plan establishes guidelines emphasizing government leadership, layered responsibility, comprehensive planning, phased implementation, local adaptation, and category-based guidance, with equal focus on construction and maintenance for harmonious development. The overarching goal for the first two decades of the 21st century is to connect eligible townships and villages with asphalt (or cement) roads, forming a high-service rural road network to facilitate safer, more comfortable, and more convenient travel for rural residents, aligning with the broader objective of building a well-off society in all respects. There are two specific targets: one for the Eleventh Five-Year Plan period, aiming for a national rural road length of 3.1 million km by 2010; and a mid-term goal by 2020, to ensure asphalt (or cement) roads reach all qualifying townships and villages, extending the rural road network to 3.7 million km, thereby significantly enhancing the density and service quality of rural roads. This will establish a comprehensive rural road network, with county and township roads as the backbone, facilitating the comprehensive construction of a moderately prosperous society.

The Initial Network of Highways is Starting to Reap the Scale Benefits."

Accelerate the speed of highway construction in the early stages of the emergence of the speed barrier, quality barrier, investment barrier was quickly overcome.

Before the Fuzhou Conference in 1998, nearly 90% of China's provinces (including autonomous regions and municipalities) had achieved their initial expressway milestones. Following the conference, expressway construction entered a crucial phase of extensive building and network formation. During this period, the only four provinces (autonomous regions) in the western region without expressways—Xinjiang, Ningxia, and Qinghai—hastened their efforts and successively achieved their expressway milestones in August 1998, June 2000, and July 2001, respectively. By 2009, Xizang had commenced construction on its first expressway, marking the advent of expressways onto the Xizang Plateau and achieving nationwide expressway coverage across mainland China.

"The 'Two Verticals, Two Horizontals, and Three Critical Sections' are interconnected. The first decade of the new millennium marked a period of leapfrog development for China's expressways, with the fruits of accelerated construction becoming rapidly evident. The trend across the country showed steady advancement in the eastern regions, faster development in the central areas, and rapid growth in the western regions."

"In 2001, the first year of the Tenth Five-Year Plan, highway construction investment was planned at 190 billion yuan but actually surpassed 260 billion yuan, setting a new historical record. That year, 3,123 km of expressways were completed, with the total mileage reaching 19,437 km, surpassing Canada and ranking second only to the United States, thereby becoming the world's second largest. In October of the same year, the Southwest Highway Sea Outlet, spanning 1,709 km and connecting the provinces (regions) of Sichuan, Guizhou, and Guangxi, was fully operational. Henan Province completed 571 km of expressways in 2001, becoming the seventh province to exceed 1,000 km of expressway mileage."

"Over the subsequent years, China's expressway mileage continued its rapid expansion. In 2002, investment in highway construction reached 321.2 billion yuan, surpassing 300 billion yuan for the first time. On November 1, 2002, the total mileage of national expressways exceeded 20,000 km."

In 2003, the Beijing-Zhuhai (North–South) expressway, spanning 2,389 km, was fully operational. In 2004, the Tongjiang-Sanya line, extending 5,700 km, the Lianyungang-Horgos line, 3,980 km in length, and the Shanghai-Chengdu line, measuring 2,970 km, achieved full connectivity with at least second-class highway standards, except for a few segments that were not yet expressways. This accomplishment signified the realization of the "Two Verticals, Two Horizontals, and Three Critical Sections" connectivity goal. In October of the same year, annual investment in highway construction soared past 400 billion yuan to reach 440 billion yuan. The addition of 4,543 km of new expressway mileage brought the total expressway network to over 30,000 km.

The "Five Verticals and Seven Horizontals" national trunk highway plan was completed ahead of its planned schedule. By the end of 2005, the total mileage of expressways in China reached 41,000 km, maintaining its position as the second-largest in the world. Both Shandong and Guangdong provinces saw their expressway mileage surpass 3,000 km, while Jiangsu, Henan, and Hebei provinces each exceeded 2,000 km. Additionally, the number of provinces with expressway mileage over 1,000 km reached 19.

During the Tenth Five-Year Plan (2001–2005), China invested 1.95 trillion yuan in highway construction, 2.17 times the amount during the Ninth Five-Year Plan (1996–2000). A total of 24,700 km of expressways were built, 1.5 times the mileage constructed in the previous decade, averaging nearly 5,000 km per year over the five years. China's mainland, achieving its first expressway milestone in 1988, ascended to the world's second-largest expressway network in just 14 years, a development journey that took developed countries four to five decades, marking a significant leap forward.

Entering the "Eleventh Five-Year Plan" (2006–2010) period, the highway continues to maintain the momentum of rapid and steady growth on the high level of high-speed development in the "Tenth Five-Year Plan".

In 2006, the annual investment in highway construction exceeded 600 billion yuan for the first time, reaching 623.1 billion yuan. An additional 4,334 km of highways were constructed, bringing the total mileage to 45,300 km. Provinces with a total highway mileage exceeding 3,000 km increased by two from the previous year, adding Henan and Jiangsu, making a total of four. Hebei and Zhejiang each surpassed 2,000 km of highway mileage.

In 2007, an additional 8,574 km of expressways were constructed, setting a new record for annual growth in mileage. The "Five Vertical and Seven Horizontal" main national expressway network was completed 13 years ahead of schedule, achieving the full benefits of the primary network. The total length of expressways reached 53,913 km. In Henan and Shandong provinces, the expressway mileage surpassed 4,000 km; in Jiangsu and Guangdong, it exceeded 3,000 km; and in Hebei, Zhejiang, Yunnan, Hubei, Anhui, Shaanxi, and Jiangxi provinces, the mileage exceeded 2,000 km. The number of provinces with expressway mileages exceeding 1,000 km reached 21.

In 2008, the total mileage of highways surpassed the 60,000 km mark, reaching 60,302 km. The number of provinces with highway mileage exceeding 3,000 km rose to six, with the addition of Hebei and Zhejiang.

In 2009, the annual investment in highway construction reached a total of 966.8 billion yuan, an increase of over 40%, making it the year with the highest rate of year-on-year investment growth during the 'Eleventh Five-Year Plan' period. By the end of the year, the total highway mileage had reached 65,100 km. The number of provinces with highway mileage exceeding 3,000 km was increased by one with the addition of Hubei, bringing the total to seven.

"At the end of the 11th Five-Year Plan period in 2010, investment in highway construction surpassed the historical milestone of 1 trillion yuan. The highway mileage reached 74,113 km. The number of provinces with highway mileage exceeding 3,000 km increased to 11, including Shaanxi, Liaoning, Jiangxi, and Shanxi; meanwhile, the number of provinces with highway mileage exceeding 1,000 km rose to 26."

"By the end of 2010, the total mileage of national highways surpassed the historical milestone of 4 million km, reaching 4.0082 million km. Over the five years of the 11th Five-Year Plan, the investment in highway construction totaled 4 trillion yuan, which is more than double that of the 10th Five-Year Plan period. Over these five years, 33,100 km of new highways were added, with an average annual increase exceeding 6,600 km. Of the 85,000 km outlined in the National Highway Network Plan, 57,700 km have been completed, resulting in a completion rate of 67.88%; moreover, the eight inter-provincial corridors for the development of western China have been largely interconnected."

"Since 2001, when China's highways became the second-largest in the world, until the end of 2010, marking the conclusion of the Eleventh Five-Year Plan period, China's highway mileage expanded by 54,700 km, effectively establishing a national highway transportation main network. Achieving this in just nine years, this feat not only surpassed the objectives of the National Highway Network Plan by over 50%, but also precipitated the early realization of the network's comprehensive socio-economic benefits. Moreover, it established a robust foundation for a new era of exponential growth in highway infrastructure, spearheaded by highways."

Entering the 21st century, along with the rapid development of the national economy, highway reconstruction and expansion also began to speed up.

"In May 2002, the reconstruction and expansion project for the Shenyang-Dalian Expressway, nicknamed 'China's No. 1 Road' and spanning a total length of 375 km, commenced. The project extended over 348 km, widening the expressway from four lanes to eight lanes, and increased the daily traffic capacity from 50,000 vehicle trips to 150,000. Completed in 2004, the project achieved its ambitious objectives of 'top-notch design, construction, management, quality, and landscape'."

The Shanghai-Hangzhou-Ningbo Expressway, Zhejiang Province's first expressway, traverses Jiaxing, Hangzhou, Shaoxing, and Ningbo, spanning a total of 248 km. It began operating in segments in 1995 and was fully operational by 1998. In February 2003, the "four-to-eight-lane" expansion project commenced in three phases: The first phase, from Hongken in Hangzhou to Guzhu, covered 44 km and was completed and opened at the end of 2003. The second phase, from Fengjing in Shanghai to Dajing in Hangzhou, spanned 95.61 km and was completed and opened at the end of 2005. The third phase, from Guzhu in Hangzhou to Ningbo, extended 80.82 km and opened on December 6, 2007. With the lane expansion, the expressway's capacity significantly increased, allowing the design traffic flow to increase from 50,000 to 100,000 vehicles per day.

In May 2003, the Shanghai-Nanjing Expressway, Jiangsu Province's first highway that opened in September 1996, began its "four-to-eight" expansion under the "widening on both sides with partial separation" plan, covering 248.20 km. The project was completed ahead of schedule by the end of December 2005. During the reconstruction and expansion of the Shanghai-Nanjing Expressway, traffic continued uninterrupted and did not result in significant congestion. The upgraded Shanghai-Nanjing Expressway now offers a smoother driving experience.

In October 2005, the Guangdong Jingzhu South Highway's "Four-to-Six-Lane" expansion project was completed and opened to traffic. The expanded section spans 186 km, which represents 93% of the expressway's total length.

Into the "Eleventh Five-Year Plan", highway reconstruction and expansion are more common.

As one of China's first highways, the Beijing-Tianjin-Tanggu Expressway had long required reconstruction and expansion due to increasing traffic volumes and road conditions. However, given the road's critical importance, it wasn't until after the opening of the Beijing-Tianjin Expressway on July 16, 2008—a project aimed at serving the 2008 Beijing Olympic Games and providing a second major expressway connection between the municipalities of Beijing and Tianjin—that the reconstruction and expansion of the Beijing-Tianjin-Tanggu Expressway were scheduled. By the end of 2008, the entire expressway began its expansion: the Beijing section was widened to eight lanes, a 30 km stretch in the Tianjin section was expanded to six lanes, while the remainder was also expanded to eight lanes. Throughout the reconstruction and expansion, traffic on the Beijing-Tianjin-Tanggu Expressway continued without interruption.

Highway service areas are continuously being enhanced. The operation of highways involves comprehensive access control, necessitating that service areas offer essential amenities such as food, accommodation, rest, and vehicle maintenance to drivers and passengers. These services aim to ensure driving safety, maintain transportation efficiency, and alleviate the physiological fatigue experienced by drivers and passengers.

The services offered by service areas primarily encompass rest areas, parking facilities, and support services. As functional components of the highway infrastructure, service areas are crucial indicators of the maturity of modern highway systems, particularly expressway networks. The quality of management, operations, and service levels in these areas significantly reflects the overall management standards of the highway system and the professional ethos of the industry.

According to incomplete statistics, by the end of 2012, there were over 1,400 pairs of service areas on China's highways, generating a total output value exceeding 100 billion yuan. China's highway service areas have begun to develop an operational management system characterized by unique Chinese features and adapted to the country's specific conditions, encompassing management, business operations, and services.

Highway Networking Yields Huge Benefits

By the end of 2012, China's total highway mileage had surged to 4.2375 million km, representing a net increase of 3.01 million km since 1997 (this includes 1.42 million km of village roads that were officially added to the statistics in 2006). Of this, the expressway mileage reached 96,200 km, matching the world leader, the United States, and marking China's official status as a major player in highway transportation.

By the end of 2012, in China's integrated transportation system, the proportions of highway passenger volume, passenger turnover, freight volume, and freight turnover were 93.5, 55.3, 77.8, and 34.3 respectively. The only exception was in freight turnover, where highways accounted for less than the 47.0% share held by water transport. However, in the other three key measures of passenger and freight transport, highways maintained a foundational and leading role.

In the integrated transportation system, road transport has emerged as the most extensively used mode, bearing the greatest volume of societal transport and having the closest connection to people's daily lives. It boasts notable advantages such as broad coverage, high adaptability, excellent directness, flexibility, low cost of acquiring vehicles, and ease of vehicle operation. Additionally, road transport plays a fundamental role in linking various transportation modes. Highways are not only crucial within the road transport sector but also have a significant impact on stimulating the national economy. They are vital for achieving balanced national development, establishing a unified market economy, enhancing modern logistics efficiency, and improving public quality of life, making their social and economic benefits even more pronounced. As noted by the renowned scholar Hu Angang, "Transportation development has powerfully supported the achievement of the primary strategic goal of modernization."

The economic and social benefits of the initial formation of the highway network are concentrated in three areas.

The highway network plays a significant role in enhancing the road transportation industry. As the most advanced and dynamic segment of road transportation, the interconnectedness of the highway network itself greatly contributes to the improvement of the road transportation industry.

Firstly, the initial development of the highway network has elevated the overall hierarchy of the road system and optimized its structure, thereby enhancing the competitiveness of highway transportation. The daily capacity of a four-lane expressway is at least 5–10 times greater than that of a two-lane secondary road. For six to eight-lane expressways, the capacity is even higher, and the efficiency of traffic flow far surpasses that of conventional roads with mixed traffic. Operating in a fully enclosed environment and adhering to higher standards, expressways significantly increase average travel speeds while also offering superior safety and comfort compared to regular roads.

As of the end of 2012, China's expressway mileage comprised only 2.3% of the total national highway mileage, yet it handled more than 20% of the traffic volume within the highway sector. Meanwhile, 68,000 km of the national expressway network were completed, achieving 79% of the planned mileage. This early formation of the expressway backbone network surpassed planned progress, ushering in significant scale benefits.

Secondly, since the advent of expressways leading to their initial network formation, several regional integrated transportation corridors have emerged. These corridors have tightly integrated various modes of transportation, ensuring swift transfers for ports, railways, civil aviation, and inter-city rapid rail systems.

Moreover, highways have facilitated the improvement of the organizational structure of road transportation and expanded the scope of the transportation sector, leading to significant enhancements in vehicle equipment standards. Specialized transportation services such as long-distance sleeper buses, refrigerated and fresh goods transport, container transport, and oversized cargo transport have rapidly evolved from conventional transportation methods. The accelerated adoption of specialized, high-speed, and large transport vehicles has significantly boosted the overall efficiency and quality of highway transportation.

The role of highways in supporting economic development is outstanding. The impact of highway construction and development on the country's economy has been significant, mainly in the following areas.

Firstly, highways significantly contribute to stimulating and advancing the development of related industries. The direct impact of highway investments on the national economy and associated sectors is substantial. It has been estimated that the correlation coefficient between highway construction investments and Gross Domestic Product (GDP) growth is 0.4. Since 1998, with a focus on expanding highway infrastructure, substantial investments in highways have spurred the production in the steel, cement, and asphalt industries, along with related processing sectors, and have also fostered growth in engineering machinery manufacturing and leasing services. Analysis of nearly a decade of operational statistics post-1998 indicates that every yuan invested in highways yields a return of four yuan in benefits and generates numerous job opportunities, with an investment of 700 million yuan potentially creating employment for about 10,000 individuals.

Secondly, highways play a crucial role in fostering regional and along-the-route economic development, providing a significant impetus. The operation of highways has dramatically reduced spatial and temporal distances, optimizing previously dispersed productive forces and connecting different regions over broader areas. This enhances the economic interdependence of various regions, transforms the distribution of productive forces within those areas, and stimulates the growth and emergence of diverse industries along the routes, including high-tech agriculture and industry, export-oriented economies, commerce, and tourism. The gradual formation and robust economic growth of the Bohai Rim, Pearl River Delta, and Yangtze

River Delta metropolitan areas are closely tied to the comprehensive benefits realized through their respective highway networks, further elevating their status in the national economic landscape. The industrial structures within these three economic zones have fundamentally changed, with even Guizhou in the west becoming a new data center due to highway development.

Thirdly, highways are instrumental in enhancing the investment climate and broadening the scope of openness along their corridors. Global experiences underscore the importance of a conducive investment environment for economic progression, necessitating essential infrastructure such as ports, communication networks, airports, and highways. Highways, in particular, play a pivotal role by facilitating connections between cities along the routes and major urban centers, transport hubs, industrial zones, and open ports. This not only improves the investment landscape and heightens the appeal to foreign investors but also transforms the regional openness pattern, extending economic openness from coastal cities to the inland areas.

Fourthly, it is conducive to the development and utilization of land resources. After the highway opens, the effect of land appreciation along the route is very significant. Simultaneously, the construction of the highway accelerates the development and utilization of land resources in the areas along it, leading to the formation of locational advantages in these areas. This can guide changes in the way land is utilized along the highway, optimizing the land use structure for agriculture and industry and commerce, and enhancing overall economic efficiency.

Fifth, the highway facilitates the urbanization process and economic prosperity. The rural economy is largely dependent on road transport, and the opening of highways can significantly reduce the storage and transportation time for agricultural products, especially perishable ones. This ensures the timely delivery of agricultural supplies, effectively enhances the marketization and organization of the rural economy, and directly boosts the rapid development of enterprises in townships and villages along the routes, as well as the diversified adjustment of the rural economic structure. The highway network promotes the transport and development of fresh agricultural products in rural areas, creating numerous regions nationally renowned for their production of such goods. These products have become the "calling cards" of their local areas and a cornerstone industry for farmers' wealth and advancement towards a more prosperous life. For instance, Shouguang in Shandong, known as the "vegetable town", and Hainan, known as the "fruit town", could not have achieved their status without the support of an extensive highway network. The completion of the Jiangxi Taihe-Jinggangshan Highway, Shaanxi Xi'an-Yan'an Highway, and Guizhou Guiyang-Zunyi Highway has brought a continuous flow of tourists to notable revolutionary sites like Jinggangshan, Yan'an, and Zunyi, aiding the economic revival of these old regions and fulfilling the local aspiration for a better-off society. The role of highways in improving social life is obvious. The changes in the mode of transportation triggered by the construction of highways have profoundly affected all aspects of society and people's lives, subtly changing people's mindsets and modes of travel. First, highways have made outstanding contributions to disaster relief and response to emergencies; second, highways have greatly improved the safety of vehicles; third, highways have improved the mobility of the country as a whole; fourth, highways have high comprehensive benefits in terms of land use, energy conservation and environmental protection; and fifth, highways have improved the quality of life of the public.

Construction of Rural Roads Moves into the Fast Track

At the 1998 Fuzhou conference, while advocating for the acceleration of main highway construction, and the overhaul and enhancement of the road network, a decision was made to allocate 20 billion yuan from government funds at all levels for the construction of county, township, and border roads. In 1999, the Ministry of Transportation's party group prioritized the strategic study of rural highway development and the advancement of rural highway construction as a key task. It conducted a survey on rural highways and completed the report "Survey on the Current State of Rural Highway Development". The survey identified the primary issues with China's rural highways as a lack of planning, insufficient total volume, poor quality, funding shortages, and weak management. From September 4 to 19 of the same year, national seminars on rural highway construction and development were convened in Lanzhou, Gansu, and Weifang, Shandong, by the Ministry of Transportation. These meetings focused on three major tasks for improving rural highways and resulted in the "Opinions on Accelerating the Development of Rural Highways", a document jointly issued in August 2000 by the Ministry of Transportation and the National Development and Reform Commission.

After 1998, investment in rural road construction remained high, with a substantial increase in road mileage and a significant improvement in rural access to roads. In 1998, investment in county and township roads amounted to 36.3 billion yuan, in 1999, it was 33.1 billion yuan, and in 2000, 30.74 billion yuan.

On July 20–21, 2000, to implement the directive of the CPC Central Committee and the State Council that "western development must prioritize infrastructure construction, with a focus on highway construction in the near term," the Ministry of Transportation convened the Western Development Transportation Infrastructure Construction Work Conference in Chengdu (referred to as the "Chengdu Conference"). The conference clarified that in the western region, the focus would be on highways, emphasizing the development of construction projects across three levels: main national highways, inter-provincial connections, and rural access projects. A series of policy measures were also introduced..

The Chengdu Conference was another meeting of significant importance for China's highway industry, following the Shenyang, Jinan, and Fuzhou conferences. It addressed the issue of "how to accelerate" highway construction in the western region, laying the foundation for rapid development in the area to align with the national pace. Moreover, it marked a significant step toward accelerating the construction of rural highways nationwide.

At the end of 2000, the total mileage of highways nationwide had soared to 1,679,800 km, an increase of 453,400 km from the end of 1997. The average annual growth in total highway mileage exceeded 100,000 km, achieving a significant leap in scale.

In June 2001, Zhu Rongji, then Premier of the State Council, conducted an inspection in Ganzi, Sichuan Province, and decided to fully implement the road construction project connecting the county capitals within the three prefectures of Ganzi, Aba, and Liangshan. This was the first large-scale rural road construction special project undertaken in China.

Between 2000 and 2003, rural roads rapidly expanded, reducing the number of townships and administrative villages without road access from 341 and 67,800 to 173 and less than 60,000, respectively. In 2004, with the implementation of special projects such as the highway projects for revolutionary sites and old revolutionary areas, and the red tourism highway project, the construction of rural roads nationwide entered a period of rapid development. By the end of 2005, including the newly added approximately 1.4 million km of village roads to the statistics, the total length of China's rural roads—which includes county, township, and village roads—reached 3.050 million km. During the Tenth Five-Year Plan, rural road construction achieved a historic breakthrough.

From the acceleration in 1998 to the conclusion of the 11th Five-Year Plan in 2010, this period truly represents the golden era of development for China's rural roads. On October 27–29, 2010, the 2nd World Rural Roads Conference, organized by the International Road Federation (IRF) and co-hosted by the China Highway Society and the Shandong Provincial Department of Transportation, was held in Jinan, Shandong Province. The IRF presented the "Contribution to Rural Road Construction and Social Equity Development Award" to the Ministry of Transport, recognizing the Ministry's significant efforts in promoting the construction and development of rural roads in China..

By the end of 2012, the mileage of rural roadways nationwide had reached 3.678 million km, accounting for 87.49% of the total mileage of roads nationwide [2]. The proportion of townships with access to roads accounted for 99.97% of the total number of townships in China, and the number of villages with roads accounted for 99.55% of the total number of villages in China. The proportion of townships with hardened roads accounts for 97.43% of the total number of townships nationwide, and the number of established villages with hardened roads accounts for 86.46% of the total number of villages nationwide, successfully achieving the goals of the Rural Road Construction Plan eight years ahead of schedule. The rural road network has seen significant improvements in both scope and quality. The rapid expansion and capacity enhancement of the rural road network have further improved the overall service capacity and structure of the road network, while providing the most basic support and guarantee for the growth of logistics and human flow in the rural economy.

The national highway and transportation sector has devoted a decade of relentless effort to turn the slogans "To prosper, first build roads" and "ensuring farmers walk on asphalt and concrete roads" into reality.

The Rural Road Network Lifts up the Farmers' Dreams of Escaping Poverty and Achieving Prosperity

Improvements in rural road conditions have delivered tangible benefits to hundreds of millions of farmers, laid the groundwork for rural economic and social development, and have been instrumental in accelerating the construction of a new socialist countryside and fostering a harmonious socialist society. The rapid expansion of the rural road network has yielded significant economic and social benefits, evident in "five changes and five promotions".

Firstly, rural roads have revolutionized transportation backwardness and boosted farmers' income. "To get rich, first build roads"; with rural roads established, the rural economy has been revitalized, unlocking developmental potential. "Roads connect, bridging urban and rural gaps; minds open, expanding opportunities; incomes rise, transforming the landscape"—this reflects the firsthand experiences of many farmers. Without rural roads to improve transportation, the stark disparity between life inside and outside the mountainous areas could not be eradicated, nor could the challenges of transporting and selling agricultural products, including frequent losses, high costs, and low prices, be effectively addressed. Following improvements in rural transportation, Wucheng District in Zhejiang Jinhua has vigorously developed eight distinct agricultural industries, including floriculture, organic rice, livestock and poultry breeding, dairy, fruits and vegetables, tea, bamboo and bamboo shoot dual-purpose forests, and aquaculture. It has been successively named as "China's Camellia Town," "China's Osmanthus Town," and "China's Town of Southern Dairy Cattle and Dairy Products." As a result, agricultural output and farmers' incomes have increased, profoundly altering the rural landscape. Such transformations are widespread across the vast countryside of China.

Secondly, rural roads have transformed the consumption patterns and fueled domestic economic expansion. Expanding domestic demand is a cornerstone of China's long-term economic strategy, serving as a sustaining force behind stable and swift economic growth. Significant improvements in rural transportation have addressed the challenges of "difficult to buy and difficult to sell" in the countryside, enabling over 900 million Chinese farmers to increase their consumption. Additionally, rural road construction has also spurred growth in the building materials industry, including cement, steel, asphalt, and sand and gravel. Annually, rural road construction consumes an estimated 45 million tons of cement, amounting to about 11% of the country's total cement output, thereby making a positive contribution to expanding domestic demand and stimulating economic growth.

Thirdly, rural roads have transformed village landscapes and fostered cultural advancement. Well-maintained rural roads have narrowed the spatial and temporal divide between urban and rural areas, facilitating the spread of urban civilization to the countryside, thereby encouraging farmers to embrace modern cultural norms and adopt civilized, scientific lifestyles. Following rural road construction, numerous areas have initiated projects to improve water, electricity, housing, sanitation, and waste management, leading to subtle shifts in villagers' living habits and rural traditions. It can be said that rural roads are not only an essential prerequisite for rural economic development but also a key catalyst for the development of tidy villages and the cultivation of a civilized rural ethos in the new socialist countryside.

Fourthly, rural roads have transformed the relationship between officials and the populace and have fostered grassroots democracy. During the construction of rural roads, the adoption of democratic management practices has strengthened the bond between officials and the community, contributing to peace, stability, and harmony in rural areas. For instance, research in Guizhou Province reveals that at the grassroots level, officials have gained greater respect, demonstrate more enthusiasm, and the farmers' voices carry more weight, leading to fewer grievances.

Fifthly, rural roads have transformed the lives of farmers and bolstered social stability. With improved rural transportation conditions, there's a decrease in children's school dropout rates, and the convenience of accessing urban hospitals has increased. Enhanced educational and healthcare standards in rural areas have set the stage for harmonized urban–rural development, transforming the urban–rural divide, reducing regional inequalities, and ensuring social stability.

The significant progress in rural road construction has been made possible through the effective leadership of the CPC Central Committee and the State Council, the high priority given by Party committees and governments at all levels, the diligent efforts of all professionals in the highway and transportation sector, and the widespread engagement from all segments of society, particularly from the vast majority of farmers. Rural road construction has evolved from an initiative led by the transportation sector to a collective endeavor involving the entire society.

Progress Towards a Strong Highway Nation (2012–2021)

Since the 18th CPC National Congress, the pursuit of socialism with Chinese characteristics has entered a new era, with highway transportation seeing significant advancements in balance, coordination, and sustainability, setting the course towards becoming a superior, efficient, equitable, sustainable, and safer transportation powerhouse. Currently, China's highway mileage ranks among the world's highest, with technological innovation making rapid strides, the gap with global leading standards rapidly closing, and some sectors already surpassing global counterparts. Moreover, the comprehensive transport system is rapidly modernizing.

The Blueprint of a Strong Transportation Nation is Initially Presented

If the 13 consecutive Five-Year Plans are the cornerstone of China's sustained and stable development, then the detailed blueprint for the highway sector, starting with the completion of the national highway network plan in the late 1970s, through to the early 21st century with the "7918 Network", the National Highway Network Plan (2013–2030), the Outline for the Construction of a Strong Transportation Nation in 2019, and the Outline of the National Comprehensive Three-dimensional Transportation Network Plan in 2021, has been integral. In concrete implementation, these plans have worked in tandem with the Five-Year Plans and provincial plans, complementing and reinforcing each other. This layered, progressively refined "top-level design" has guided the development of China's highway sector, steadily advancing towards becoming a leading transportation nation.

With the national highway network plan set to be accelerated and completed ahead of schedule, to address issues such as incomplete coverage, insufficient main corridor capacity, network efficiency needing further enhancement, and the necessity of enhancing integration with other transportation modes, the State Council formally approved the National Highway Network Plan (2013–2030) in 2013. It was officially announced at the State Council's press conference on June 20 of the same year.

Aimed at meeting the economic and social development needs over the next two decades, the National Highway Network Plan (2013–2030) updates and expands the ordinary national highway network from the 1980s and 1990s, and also enhances and refines the National Highway Network Plan released in 2004. This represents the first national-level road network plan with a comprehensive and unified layout since the establishment of New China, encompassing ordinary national highways and expressways, composed of trunk highways with national and regional significance in politics, economy, and defense, thus outlining a blueprint for a functionally comprehensive, extensively covered, adequately capacitated, seamlessly connected, and reliably operated national trunk highway network.

It has been proven that the implementation of the National Highway Network Plan (2013–2030) has played an important leading and supporting role in supporting the comprehensive achievement of a moderately prosperous society, achieving the grand goal of socialist modernization, fostering coordinated urban–rural development, advancing the construction of a comprehensive transport system, and ensuring sustainable highway transportation.

In December 2016, the State Council Information Office released a white paper titled 'China's Transportation Development', which highlighted that since its establishment, and particularly after the initiation of reform and opening-up policies, China's transportation sector has seen historic transformations, evolving from 'bottleneck constraints' through 'initial relief' to 'basic adaptation' [3].

In 2019, the CPC Central Committee and the State Council issued the "Outline for the Construction of a Transportation Power," setting clear goals to substantially build a transportation powerhouse by 2035, including establishing a nationwide "1–2–3 Travel Circle" (1-h commute within metropolitan areas, 2-h access within city clusters, and 3-h coverage to major national cities) and a global "1–2–3 Logistics Circle" (1-day domestic delivery, 2-day delivery to neighboring countries, and 3-day delivery to major global cities). By the mid-21st century, the aim is to fully develop into a world-leading transportation powerhouse that meets the people's expectations, ensures robust security, and achieves world-leading status in infrastructure quality, technological advancements, innovation capacity, and levels of intelligence and sustainability, while elevating traffic safety, governance, civility, and global competitiveness and influence to internationally advanced standards.

In February 2021, the CPC Central Committee and the State Council issued the "Outline of the National Comprehensive Three-dimensional Transportation Network Plan," setting forth the goal to, by 2035, essentially establish a modern, high-quality, comprehensive, and multi-dimensional transportation network that is efficient, eco-friendly, smart, and secure, with the quality of transportation infrastructure and the levels of intelligence and sustainability ranking among the world's best. By the middle of the 21st century, this initiative aims to fully realize a world-class transportation infrastructure system, establishing China as a leading transportation nation and laying the foundation for a fully modernized socialist powerhouse.

The Highway Network is Becoming Increasingly Refined

Since the 18th CPC National Congress, integrated transportation has entered a new stage of accelerating the construction of a modern comprehensive transportation system.

"Guided by the National Highway Network Plan (2013–2030), provinces (including autonomous regions and municipalities) have embarked on a new wave of planning that aligns with the evolving demands of highway network development. The highway infrastructure network is continuously enhanced through its steady and rapid expansion, achieving a level of development that not only meets but exceeds the 'basic adaptability' to the national and regional economic growth."

"In 2010, China's annual highway construction investment surpassed 1 trillion yuan for the first time, breaking through 2 trillion yuan by 2017. By 2021, investment in highway construction remained at elevated levels. Consequently, the mileage of China's highways has been experiencing rapid and stable growth, with continuous enhancements in quality and ongoing improvements in the network structure."

"By the end of 2015, at the conclusion of the 12th Five-Year Plan, China's highway mileage reached 123,500 km, with the national highway network—the '7918 Network' introduced in 2004—largely completed. The country's highway infrastructure had reached a stage of basic adaptation. During this period, significant strides were made in intelligent and green transportation, solidifying China's status as a leading nation in highway transportation."

"From 2012 to 2020, China's total highway mileage and expressway mileage grew from 4,237,500 and 96,200 km to 5,198,100 and 161,000 km, respectively, marking a net increase of 960,600 km for highways and 64,800 km for expressways. This includes the completion of 113,000 km of national expressways. The construction of the national highway network, as specified in the National Highway Network Plan (2013–2030), has nearly been completed, achieving this milestone 10 years ahead of schedule [4]."

"The continuous improvement of the road network has gradually reduced the disparities in highway development, revitalizing the central and western provinces that have long been constrained by their topography and plagued by transportation challenges. Guizhou Province stands out as a prime example of this revitalization."

In 2021, Guizhou's highway network spans over 7,600 km, ranking it fifth nationwide and ensuring county-wide expressway access and village-wide asphalt road coverage. This development swiftly shed its reputation for underdeveloped transportation, positioning Guizhou as a significant land transportation hub in Southwest China. Leveraging its pleasant climate, unique geography, and beautiful landscapes, along with the enhanced transportation infrastructure, Guizhou has attracted numerous big data enterprises. Today, it stands as one of the world's regions with a high concentration of mega data centers, becoming a key pillar in the national 'East Data, West Computing' strategy. According to the "2021 White Paper on China's Digital Economy City Development," Guizhou's digital economy grew by over 15% in 2020, maintaining the top spot nationally for six consecutive years. Modern transportation has opened broad avenues for Guizhou, turning its lush landscapes into valuable assets.

Rural Roads Have Established the Firmest Foundation for Achieving the First Centennial Goal

"In 2013, General Secretary Xi Jinping emphasized the importance of "Four Good Rural Roads," stating that rural road construction should be tailored to local conditions and people-centered, aligning with the optimization of village and town layouts, rural economic development, and ensuring safe and convenient travel for peasants. He advocated for innovation in systems and policy improvements to enhance the construction, management, maintenance, and operation of rural roads. He highlighted the need to eliminate transportation bottlenecks hindering rural progress, thereby providing a stronger foundation for peasants to overcome poverty and achieve prosperity. Following the strategic directive of the 18th CPC National Congress towards a moderately prosperous society, the Ministry of Transportation set forth in the "13th Five-Year Plan for Transportation Poverty Alleviation" a commitment to ensure that no region lags behind due to transportation issues on the path to prosperity. This commitment has since evolved into a broader pledge that no individual will be left behind on the journey to national well-being. This marked the beginning of a new phase for China's rural roads, focusing on accelerated construction, enhanced quality and efficiency, and scientific development, thereby paving the way for the comprehensive achievement of a moderately prosperous society." (Fig. 11.1).

By the end of 2021, the total length of rural roads across the country reached 4.466 million km, marking a net increase of 787,600 km since 2012. More than 94% of counties and districts achieved an AAA level or higher in the integration of urban and rural transportation development. Additionally, 100% of townships and established villages with the necessary conditions had access to bus services, positioning public transportation among the leaders in national public service quality satisfaction. Rural roads have thus become the most crucial infrastructure for enabling rural prosperity and the development of a new socialist countryside, significantly contributing to the goal of equalizing public road transportation services.

On May 19, 2021, the Information Office of the State Council conducted a routine policy briefing on strengthening the construction of the county-level commercial systems. The data revealed at the briefing highlighted that China's annual production and consumption of fruits and vegetables reach up to 700 million tons, constituting about 40% of the global total. Retail sales of consumer goods in rural areas have outpaced urban areas for eight consecutive years. Every village in the country now has direct postal service, and 98% of townships have express delivery service coverage. In 2020, rural areas in China handled over 30 billion parcels (out of a national total of 66 billion), facilitating the flow of over 1.5 trillion yuan worth of industrial goods to rural areas and agricultural products to urban areas. China aims to achieve full express delivery service coverage in all administrative villages within 3–5 years. These achievements would be unimaginable without the essential foundation of "100% accessibility via hardened roads and bus services," particularly the "village-to-village" connectivity provided by rural roads.

On December 17, 2019, the Ministry of Transportation held a press conference focused on 'Poverty Alleviation'. It was noted that over the past two decades of the 21st century, the Ministry has allocated over 612 billion yuan of central funds to construct and upgrade 2.3 million km of highways to townships and established villages, achieving 100% access to paved roads for all townships and villages. China has established a rural road network exceeding 4 million km in total length, supporting the dreams of countless farmers for better living standards, wealth, and poverty eradication. China's rural roads stand as the country's true mega-projects.

The Success of China Road Epitomizes the Chinese Dream

Since 1998, with the acceleration of highway construction, China's highways have catalyzed the national dream of prosperity. In 2001, the National Development and Reform Commission and the Ministry of Transportation initiated the 'Western Counties Oil Road Project.' By 2003, the Ministry encouraged 'bringing farmers onto

a) Rural roads in Zanhuang County, Chengde and Shijiazhuang City, Hebei Province in 2007

b) Changes in Rural Highways in Anhui Province in 2007

c) Rural roads in Guizhou Province in 2007

d) Night view of rural roads in Jiangsu Province in 2008

Fig. 11.1 Rural roads in China

e) Rural Road Construction Site in Qingyang City, Gansu Province in 2015

f) Village level Highway in Gannan Tibetan Autonomous Prefecture, Gansu Province in 2015

g) Rural Road, Xidong Avenue, Wuxi City, Jiangsu Province in 2013

Fig. 11.1 (continued)

h)Rural road passenger transportation in Hebei Province, currently in operation in 2007

i)Bus service in Li Village Miao Village, Sanya City, Hainan Province in 2004

j)Inner Mongolia Autonomous Region achieved 100% bus access to Gacha Village in 2007

Fig. 11.1 (continued)

asphalt and concrete roads,' launching the nationwide Rural Highway Accessibility and Improvement Projects. That same year, the 'Western Mountainous Area Road Safety Project' began, with safety and disaster mitigation projects gradually extended to major national and provincial highways. In 2004, the 'Special Project for Rural Roads in Revolutionary Areas and Old Districts' was initiated. Following the 18th CPC National Congress, the 'Four Good Rural Roads' initiative was launched. Over two decades of relentless effort, China's physical geography remained unchanged, but its economic landscape has been dramatically transformed. The success of China's roads has become a microcosm of the Chinese dream.

In the fight against poverty, 'leaving no one behind' is not just a resonant slogan but a reality built upon countless challenging and complex projects, with many dedicated individuals in the highway sector even giving their lives for the cause. Among the most notable are the transformation of zip lines into bridges, the construction of the Dulongjiang road and Motuo highway, and the 'Village Access' project.

The Dulongjiang road and the Motuo highway exemplify the best of the new era's transportation sector's dedication to the spirit of 'two roads' in the construction of rural road and poverty-alleviation highway.

The Dulong ethnic group, numbering only a few thousand, predominantly resides in Dulongjiang Township within Gongshan Dulong and Nu Autonomous County, part of the Nujiang Lisu Autonomous Prefecture in Yunnan Province. Historically isolated due to the lack of road access, Dulongjiang Township was among the poorest areas in Yunnan and China. Since 1990, the Ministry of Transportation and the Yunnan Provincial Government prioritized Nujiang Prefecture for targeted poverty alleviation efforts. By 2002, 12 highway projects totaling 498 km were completed, transforming the transportation landscape of the prefecture from having no classified or asphalt roads. Notably, the 96.2 km Dulongjiang Highway, completed in 1999, ended the absence of highway access in China's last ethnic minority region. The subsequent Dulongjiang Highway Reconstruction Project, initiated in 2011, culminated in the opening of the 6,680-m Dulongjiang Tunnel through the Gaoligong Mountains in 2018, eliminating the annual closures due to heavy snowfall (Fig. 11.2).

Compared with the Dulongjiang Highway, the construction of the Xizang-Mutuo Highway (Zamo section) took longer, was more arduous and involved the most significant sacrifices in peacetime.

Motuo is home to over 11,000 people. From the early 1960s to 1994, the Motuo Highway in Xizang was attempted to be built four times. Despite these efforts, due to limitations in the economy, technology, and equipment of the time, the project was never fully completed. However, more than 200 highway constructors sacrificed their lives in the process, resting forever at the snowy mountains' base. The harsh conditions and the high cost in lives did not deter the highway workers; instead, it fueled their determination and resolve to overcome the 'forbidden zone'.

After 1995, the Xizang Autonomous Region's highway and transportation department annually allocated substantial funds for repairs to the damaged sections of the Motuo Highway during the dry season, managing to maintain temporary accessibility for agricultural vehicles with "south accessible during the dry season and north during the rainy season," yet the road was passable for less than three months annually. Constructing a year-round accessible highway in the challenging terrains of the Yarlung Zangbo River's great bend and the Yarlung Zangbo Grand Canyon

Fig. 11.2 Dulongjiang road and its tunnels

has not only placed immense pressure on Xizang's highway authorities but has also become a nationwide concern among road construction professionals.

In the 21st century, after several reconnaissance efforts, the reconstruction project for the Motuo Highway commenced on April 20, 2009. This marked the fifth construction effort in the history of the Motuo Highway. Originating at the intersection of National Highway 318 and the old Zhamo Highway in Pomi County, Zamu Town, the route passes through areas like Da'erqu, Bolonggong, Jinzhu Zangbu, Miri, Madi, and Ximo River, ending at Lianhua Square in Motuo County. The entire route spans 117.28 km, with an estimated project cost of 950 million yuan, fully funded by the state. By the end of 2010, the critical control project of the Motuo Highway—the Garongla Tunnel—was successfully completed, removing the largest obstacle for the comprehensive construction of the highway section from Garongla Mountain to Motuo County (Fig. 11.3). In 2013, the Motuo Highway was completed and opened to traffic, marking the first time that every county in China was connected by highway.

In April 2020, the rural passenger transportation shuttle buses in Motuo County were officially opened, passing through 5 townships, 12 administrative villages and 1 natural village in the county.

In May 2021, another highway to Motuo, the Pai Town to Motuo Highway in Miling County, Linzhi City, Xizang Autonomous Region, was fully opened.

Abuloha Village, the last in China to gain road access, derives its name from the Yi language, meaning 'a valley deep in the mountains' and 'a seldom visited place.' It falls under the jurisdiction of Wuyi Township in Butuo County, within the Liangshan Yi Autonomous Prefecture of Sichuan Province. Home to hundreds of villagers, Abuloha is nestled deep within the Jinsha River Gorge in the heart of the Greater Liangshan Mountains, bordered by mountains on three sides and a river on the fourth. Recognized as a village in severe poverty by the national government, its remote location and rugged terrain have historically left it isolated from road access.

In order to complete the task of poverty alleviation on time, in September 2018, the Sichuan Provincial Transportation Department launched a 4,000-m-long through-village highway project. This highway is the last through-village highway in Liangshan Yi Autonomous Prefecture, Sichuan Province, and its start of construction is not only an urgent requirement for poverty alleviation work in Sichuan Province, but also the ardent expectation of the poor people.

Situated on cliffs at an average altitude of around 1,500 m, the construction of this highway faced immense difficulties. By December 2019, after more than a year's effort, 3,000 m had been completed, with the remaining 1,000 m requiring passage through a large canyon of hard rock. To expedite construction, the local party committee and government resolved to advance the roadwork from both Weimu Village in Lago Township, along the cliff edges, and from the Abuloha Village Committee. Transporting the heavy machinery needed to reach Abuloha Village necessitated the use of large helicopters. In early December 2019, over the course of eight days, the Sichuan Highway and Bridge Construction Group used heavy-lift helicopters for 16 flights to deliver 50.1 tons of equipment and materials, including excavators, air compressors, generators, and barrels of diesel, to the site.

Fig. 11.3 Mutuo highway Garongla tunnel

On June 30, 2020, the opening of the Abuloha Village Highway, which cost more than ten million yuan per km, coincided with the launch of the 'Jintong Project' rural passenger lines. This 4,000-m-long road through the village has transformed the previously over 4-h trek along a goat path into a mere 10-min car journey, and travel to Butuan County town now takes just 2 h. The completion of this road has dramatically altered the village's fortunes, reducing the poverty rate, which stood at 71.94% at the end of 2019, and marking a successful culmination of the nation's village highway initiative.

References

1. China Highway Society. History of Chinese highways Book III [M]. Beijing: People's Transportation Press Co Ltd, 2017.
2. Ministry of Transport of the People's Republic of China. Statistical Bulletin on the Development of Highway and Waterway Transportation Industry [R]. Beijing, 2012.
3. Website of the Information Office of the State Council of the People's Republic of China. White paper on China's transportation development [EB/OL]. (2016-12-29) [2022-05-17]. http://www.scio.gov.cn/ztk/dtzt/34102/35746/35750/Document/1537406/1537406.htm.
4. Ministry of Transport of the People's Republic of China. Statistical Bulletin of Transportation Industry Development [R]. Beijing, 2021.
5. Ministry of Transportation and Communications. China road spectrum [M]. Beijing: People's Transportation Press, 2009.
6. Ministry of Transportation and Communications, China Highway and Transportation History Editorial Review Committee. China's highway history, second volume [M]. Beijing: People's Transportation Press, 1999.

Chapter 12
High-Speed Railway Racing: A Hundred Years of Dramatic Changes in China's Railways

In February 1911, an official of the Qing Dynasty made a journey from Beijing to the neighboring regions, shortly after the Beijing-Zhangjiakou Railway was inaugurated, enabling train travel from Beijing to Zhangjiakou. Departing from Xizhimen at 8:40 a.m., the official reached Zhangjiakou by 5:20 p.m. Staying in a comfortable inn and reminiscing about the journey's scenic views, including rolling hills and the Great Wall, he reflected on his predecessor, Zhang Penghe—a renowned minister from the Kangxi era who had traveled the same route to Russia, a trip that took six days. The arduous journey was fraught with 'hunger, thirst, wind, and sand.' Yet, traveling by train, he faced 'no hardship' despite the 'rain and snow' [4], feeling a stark contrast to the past.

More than a century has passed, and the profound transformation of China's railways continues to astonish. With the inauguration of the Beijing-Zhangjiakou high-speed railway on December 30, 2019, the journey from Beijing to Zhangjiakou has been reduced to just one hour by high-speed train. The evolution of the railways mirrors the nation's development. As General Secretary Xi Jinping aptly summarized, the Beijing-Zhangjiakou Railway was completed in 1909; a century later, in 2019, the high-speed counterpart was launched. This transition from the initial breakthrough of indigenous design and construction to achieving the world's most advanced levels, and from speeds of 35 km/h to 350 km/h, encapsulates the growth of China's railway system and a leap in the nation's comprehensive strength.

Over the past century, China's railways have evolved from non-existence to strength, from faltering beginnings and challenging expansions to a network of dense railways and high-speed trains. The shift from a piecemeal, "World's Fair" approach to unified standards and modern management, and from a loss of sovereignty and vulnerability to setting benchmarks and becoming a model, reflects a journey marked by hardship, innovation, and progress. The dramatic changes in China's railways over a hundred years highlight the relentless struggle and innovative spirit of the railway workforce under the leadership of the Communist Party of China. Reflecting on this

© China Communications Press Co., Ltd 2025

H. Lu (ed.), *A Brief History of Transportation in China*,
https://doi.org/10.1007/978-981-97-9851-3_12

century-long history offers insights and lessons for understanding the trajectory of China's transport development and advancing railway reform and development.

The World Enters the Age of Railroads

The railroad emerged from the convergence and interplay of various new technologies during the Industrial Revolution: extensive coal mining prepared the fuel for the railways; the invention of coke and the blast furnace propelled steelmaking technology forward; the development of the lathe enabled machine production; among other advancements. The breakthroughs in the coal mining, steelmaking, and machinery industries, all vital for railway construction, laid the groundwork for the eventual creation of the railroad.

Railroads is the integration of tracks and steam locomotives. Initially, these two technologies evolved independently. In the 16th century, some European mines began utilizing wooden tracks for ore transportation. Subsequently, engineers started reinforcing these wooden tracks with iron strips to extend their lifespan. In 1776, L-shaped cast iron tracks were introduced, replacing the wooden ones. By 1789, 'fish-belly' shaped iron rails and compatible vehicles with flanged wheels marked a significant stride towards modernization. In 1820, low-carbon steel tracks emerged, offering superior strength and flexibility compared to the earlier cast iron tracks. The advent of low-carbon steel rails paved the way for the later invention of steam locomotives.

At the end of the 18th century, James Watt's improvements to the steam engine facilitated the widespread application of steam power. This advancement led to the idea of steam-powered vehicles, prompting inventors in France, Britain, and the United States to experiment with the creation of steam automobiles that could navigate roads. Nonetheless, these ventures were unsuccessful. The failure was partly due to the nascent state of steam automobile technology and partly because the roads of the time, predominantly made of dirt, were inadequate. Even though gravel roads had been introduced in the early 18th century, they were still incapable of supporting the heavy weight of steam-powered vehicles.

Reflecting on the transformative changes in transportation technology at the dawn of the 19th century, it appears that land transportation was at a pivotal juncture, with various modes emerging like a combination of elements. The first option was steam-powered automobiles and roads. As previously mentioned, neither technology was sufficiently developed at that time, and it wasn't until the end of the 19th century, with the advent of internal combustion engines and more advanced asphalt paving techniques, that the automobile era truly began. Another alternative was the horse-drawn carriage on roads. Indeed, horse-drawn transportation was highly advanced in Europe at the time, particularly following the invention of macadam road construction, making carriages a common mode of transport. However, their limited capacity and high cost were significant drawbacks. With the advent of railroads, as Karl Marx observed in "Das Kapital," "On land, macadam roads were relegated to a secondary position by the railways [5]."

Another mode of transportation involved horses pulling vehicles along tracks, a method with a long historical lineage. This form of transport served humanity for centuries before the emergence of steam locomotives. Even subsequent to the steam train's invention, many countries—including France, Germany, Australia, Japan, and China—initially built their first railways utilizing horse-drawn traction. Within urban transportation, tramways powered by horses remained a pivotal means of conveyance in European and American cities throughout the 19th century.

Road steam cars, gravel roads, four-wheeled carriages... Pioneers in various fields were tirelessly exploring possibilities, making it difficult for those at the dawn of the 19th century to predict which mode of transportation would emerge dominant. History was on the verge of a remarkable convergence—the fusion of rail tracks with the steam locomotive. Rail tracks offered load-bearing and directional stability, addressing the then-prevailing issues of inadequate road support for heavy loads and the cumbersome management of steam locomotive direction; the steam locomotive, in turn, provided a transportation capacity far surpassing that of horse-drawn vehicles. A pivotal figure in this innovation was the British engineer Richard Trevithick. In 1804, Trevithick constructed the world's first steam locomotive to run on cast iron tracks. His significant contribution also included the development of high-pressure steam technology, a marked departure from James Watt's lower-powered steam engines, enabling greater locomotive power. Despite these advancements, Trevithick's inventions never achieved commercial success, leading him to eventually abandon his steam locomotive research.

The title of 'Father of Railways' is justly awarded to the British engineer George Stephenson, who, building upon Trevithick's earlier work, constructed his first steam locomotive in 1814. In 1825, under his guidance, the Stockton-Darlington Railway was inaugurated, marking the first time in human history that a steam locomotive was operational on a public railway. Stephenson is distinguished as the 'Father of Railways' rather than Trevithick due to his unwavering dedication and perseverance. He not only successfully commercialized railway technology but also transformed it into a global industry.

The inauguration of the Stockton-Darlington Railway heralded the official dawn of the global railway era, marking humanity's first departure from reliance on animal and wind power towards the age of mechanical propulsion. The advent of the railway immediately unveiled its remarkable allure. Within mere decades, the railway expanded from Britain to the rest of the world, exerting its influence across all spheres of human society. From shaping global politics and economies to affecting international and domestic conflicts, and from altering everyday life to impacting basic necessities such as food, clothing, shelter, and transportation, railways have profoundly transformed the course of history and catalyzed societal evolution. So significant were these changes that Max Weber, the renowned German sociologist and economist, heralded the railway as 'the most revolutionary tool in history [6].

See Table 12.1 for the inauguration of railroad by country.

Table 12.1 Inauguration of railroad by country [1]

Country	Earliest opening (year)	Country	Earliest opening (year)
United Kingdom of Great Britain and Northern Ireland	1825	Italy	1839
United States of America	1830	Switzerland	1844
French	1832	Spanish	1848
Belgium	1835	Brazilian	1851
German	1835	India	1853
Canadian	1836	Australia	1854
Russian	1837	Egypt	1855
Austrian	1838	Japanese	1872
The Netherlands	1839	Sino	1876

China's Railroads in the Late Qing Dynasty

The 19th century was the golden age of railroads, yet during this period, China, despite having recently basked in the glory of the Kangxi and Qianlong golden ages, was outwardly formidable but internally fragile and in decline. The nation was besieged by internal strife and external threats, financial resources were stretched thin, private capital was scarce, there was a severe shortage of diplomatic and financial expertise, the country was isolationist with closed borders, and prevailing thoughts and ideologies were conservative and regressive. All these factors combined to make the early development of railways in China an arduous journey, fraught with complexities and hardships.

In the early 19th century, knowledge of railroads was introduced to China by Western missionaries. Progressive Chinese figures like Lin Zexu, Xu Jiyu, and Hong Rengan also played a significant role in disseminating information about railroads. Following the Second Opium War, China's semi-colonial status intensified. To extend their influence into China's interior, Western powers continually pressed the Qing government to construct railroads. The Qing government, deeply apprehensive about the Western powers' encroachment and perceiving the railroad as a foreign novelty fraught with potential dangers, steadfastly resisted the introduction of railroads within the nation's borders.

China's first commercial railway, the Wusong Railway, was constructed without authorization by the British firm Jardine Matheson. It was completed in December 1876, only to be reclaimed and dismantled by the Qing government the following year. In the 1870s, amidst the Self-Strengthening Movement, some officials within the Qing government recognized the strategic and economic benefits of railroads and began advocating for their construction. In December 1874, Li Hongzhang, the Viceroy of Zhili and Northern Minister of Trade, first proposed the construction

Fig. 12.1 Tangxu railway (Tangshan-Xuguozhuang) coal trains

of railroads to the imperial court. This sparked intense debates within the Qing government about the merits of building railroads.

In November 1881, the Tangxu Railway, connecting Tangshan to Xugezhuang, was completed, marking the first railroad independently constructed by China. Initiated under the directives of Li Hongzhang, this coal transport railway was developed by the Kaiping Coal Mines. Spanning 9.7 km, the line was initially operated with horse-drawn vehicles before transitioning to steam locomotive traction (Fig. 12.1).

In 1887, under the approval and supervision of Liu Ming-chuan, the first governor of Taiwan, the construction of the Taiwan Railway commenced. The first segment, spanning over 28 km from Keelung to Taipei, was completed in 1891, followed by the completion of the Taipei to Hsinchu stretch, covering over 78 km, in 1893. Following the Sino-Japanese War, the Qing government was compelled to cede Taiwan to Japan, resulting in the transfer of Taiwan's railways into Japanese hands.

After the Sino-Japanese War, the significance of railroads increasingly came to the fore. In May 1889, the Qing government proclaimed a resolute commitment to developing railroads. At that time, facing internal and external challenges and with an empty treasury, the government had no choice but to resort to foreign loans for railway construction. In 1897, the Qing government established the Railway Corporation in Shanghai and appointed Sheng Xuanhuai as its director, tasking him with securing foreign loans for railway projects. In May 1897, the Qing government signed the 'Lu-Han Railway Loan Contract' with Belgium, marking the beginning of a surge in borrowing that became as unstoppable as a breach in the Yellow River's banks. Between 1897 and 1906, China entered into a total of 12 railway loan agreements with foreign entities. The influx of foreign capital, albeit relieving the financial constraints on railway construction, also precipitated the construction of major trunk lines such as the Beijing-Hankou, Beijing-Fengtian, Tianjin-Pukou, Shanghai-Nanjing, Zhengding-Taiyuan, and Bianluo Railways. Nevertheless, reliance on foreign loans led to significant concessions and losses, primarily in terms of management rights, hiring practices, procurement, and financial terms, including borrowing discounts, surplus profits, brokerage fees, discounts, and commissions.

The Beijing-Zhangjiakou Railway, completed by the Chinese themselves, was the first mainline railroad designed, constructed, and independently operated after its completion. In 1905, amidst the British and Russian rivalry, the Qing government resolved to undertake the construction of the Beijing-Zhangjiakou Railway on its own. The geographic conditions between Beijing and Zhangjiakou presented steep and complex challenges over a distance of more than 200 km, making construction extremely difficult, especially since the Chinese had no prior experience in independently building a railroad mainline. Consequently, Westerners were generally skeptical about the Chinese undertaking the project, with some foreigners even claiming that the Chinese capable of building the Beijing-Zhangjiakou Railway had yet to be born. Construction officially began on October 2, 1905, and the railway was completed in August 1909. Spanning 201.2 km, the railway traverses the Great Wall and the Yanshan Mountain Range, with its route characterized by undulating terrain and numerous sharp curves, presenting a formidable engineering challenge. Zhan Tianyou, the chief engineer of the Beijing-Zhangjiakou Railway, faced these difficulties with resilience, overcoming each engineering challenge as it arose. To navigate the Badaling Pass, Zhan Tianyou innovatively designed a 'Z' shaped switchback line (Fig. 12.2), reducing the length of the originally planned 1,800-m tunnel to 1,091 m, thereby lowering construction costs. The Juyongguan and Badaling tunnels were among the project's most challenging undertakings, with Zhan Tianyou personally overseeing their design and construction, implementing several technological innovations to expedite progress. The railway was inaugurated on September 24, 1909, and its completion was a sensational achievement both domestically and internationally. The significance of the Beijing-Zhangjiakou Railway extended far beyond the engineering feat itself; it bolstered the national pride and confidence of the Chinese people, showcasing their intellect and strength. This monumental achievement inspired generations of Chinese, with Premier Zhou Enlai lauding it as 'the glory of the Chinese people.

After the Sino-Japanese War, the imperialist powers took the way of dividing the spheres of influence and so on to build railroads in China directly, among which the three railroads of Dong-Sheng, Jiao-Ji and Yunnan-Vietnam were built and operated by Russia, Germany and France respectively by exerting pressure on China and investing in them directly.

In June 1896, aiming to 'ally with Russia to resist Japan,' the Qing government dispatched Li Hongzhang to sign the 'Sino-Russian Secret Treaty' with Russia, authorizing Russian investment in the construction of the Chinese Eastern Railway. The main line stretched from Manzhouli in the west, passing through Angangxi and Harbin, to Suifenhe in the east, with both termini connecting to the Trans-Siberian Railway. In August 1897, construction began simultaneously on two sections: from Manzhouli to Harbin and from Harbin to Suifenhe. In March 1898, Tsarist Russia

Fig. 12.2 Beijing-Zhangzhou railway Qinglongqiao station "Z" shape railroad

coerced the Qing government into signing the 'Lease of Port Arthur Treaty,' securing the rights to construct the South Manchuria branch line, which extended from Harbin through Changchun and Shenyang to Dalian. The Chinese Eastern Railway and its South Manchuria branch formed a 'T'-shaped network, placing the entire Northeast under Russian influence. Construction of the South Manchuria branch line commenced in June 1898. By 1901, the main and branch lines of the Chinese Eastern Railway were essentially completed, and by 1903, the entire network was operational, with the main line measuring 1,481 km and the branch line 944 km in length.

In 1895, leveraging its role in the 'Triple Intervention' as a pretext, France compelled the Qing Government to sign the 'Sino-French Agreement on Border and Commercial Issues,' securing the rights to extend the Vietnam Railway into Chinese territory. To this end, France established the Yunnan-Vietnam Railway Company, tasked with the construction and operation of the railway. The gauge of the Yunnan-Vietnam Railway was 1 m, stretching from Haiphong in Vietnam through Lao Cai to Yunnan-fu (now Kunming), covering a total of 855 km. The Yunnan segment started at Hekou on the Sino-Vietnamese border, passing through Bisezhai, Kaiyuan, Pingxiang, and Yiliang, before reaching Kunming, measuring 464.6 km in length. Construction commenced in 1903 and the railway became operational on April 1, 1910. Situated in mountainous terrain, the construction of the Yunnan segment of the Yunnan-Vietnam Railway was a formidable engineering challenge, featuring 425 bridges and 155 tunnels, with 80% of the route winding through rugged mountains.

After the Sino-Japanese War, Germany gradually expanded its influence in Shandong and, in March 1898, signed the 'Jiaozhou Bay Concession Treaty' with the Qing government, securing the construction rights for the Jiaoji Railway. The project was undertaken by the German Shandong Railway Company. The railway commenced at the port of Qingdao, passing through Jiaozhou, Gaomi, Weixian, Qingzhou, and Zibo, and terminated in Jinan, covering a distance of 394 km. Additional branch lines included a 39.2 km stretch from Zibo to Boshan and a 7.2 km segment from Zibo to

Hongshan, bringing the total length to 440.7 km. Construction of the Jiaoji Railway started in 1899 and it was officially opened for operation in 1904.

In February 1904, the Russo-Japanese War erupted, resulting in a significant defeat for the Russian forces. The subsequent Treaty of Portsmouth, signed by both countries, led Russia to agree to cede its leasehold in the Port Arthur territory (Lüshunkou) and the southern section of the South Manchurian branch line, starting from Kuanchengzi (now Changchun), including relevant branch lines, to Japan. In December 1905, Japan and the Qing government signed the 'Sino-Japanese Agreement Regarding Affairs in the Three Eastern Provinces,' forcing the Qing government to recognize all rights and interests transferred from Russia to Japan. Japan took over the segment of the South Manchurian branch line south of Changchun, renaming it the South Manchuria Railway, and established the South Manchuria Railway Company to manage and upgrade the railway. Furthermore, Japan constructed the Antung-Mukden Railway, stretching from Antung (present-day Dandong) to Mukden (present-day Shenyang), connecting with the South Manchuria Railway over 303.7 km. Additionally, Japan built the Xinmin-Fengtian Light Railway from Mukden to Xinmin in the west.

After occupying Taiwan, Japan undertook the renovation of the Keelung-Hsinchu Railway in 1899 and extended it southward to Kaohsiung. By 1908, this expansion was completed, resulting in the formation of the Keelung-Kaohsiung Western Railway, spanning a total of 408.5 km. In the eastern part of the island, the Taitung Railway was completed in 1926, stretching from Hualien to Taitung over a distance of 176 km.

The exploitation of China's railway rights by imperialist powers ignited the patriotic zeal of the Chinese people against imperialism, leading to the initiation of a movement to reclaim railway rights and the rise of commercially operated railways. In 1903, the people of Guangdong, Hunan, and Hubei provinces were among the first to call for the annulment of the 'Guangdong-Hankou (Yue-Han) Railway Loan Agreement' with the United States. Bowing to public pressure, the Qing government, in August 1905, signed an agreement with the U.S. to redeem the rights to the Guangdong-Hankou Railway. This act set a precedent, prompting demands from other provinces to nullify treaties with foreign powers and recover railway rights, propelling the movement to its peak. The momentum to regain railway rights also spurred the emergence of a wave of commercially managed railways. Between 1903 and 1910, 19 railway companies were established across 15 provinces, most of which were commercially operated. Amidst this surge, the Qing government, aiming to enhance railway management, established the Ministry of Posts and Communications in 1906 (the 32nd year of Guangxu's reign), which was responsible for overseeing shipping, railways, telecommunications, and postal services. This marked the establishment of China's first central administrative body dedicated to the transportation sector.

The reclamation of railroad rights and the initiative for domestically managed railroads unsettled the imperialist powers. Unwilling to relinquish their railroad interests, countries such as Britain, France, Germany, and the United States pressured the Qing government. Fearing escalation, the Qing government, citing the slow progress of commercially managed railroads and fundraising difficulties, announced in 1911 the nationalization of main railway lines and signed the 'Huguang Railways Loan Contract' with the banking consortia of the four countries. This decision sparked vehement opposition from the people of Hunan, Sichuan, Hubei, and Guangdong provinces, with Sichuan Province, in particular, witnessing a widespread movement to protect the railways. In response, the Qing government dispatched officials to Sichuan to address the railway issues, arresting leaders of the protection movement and killing dozens in what became known as the 'Chengdu Massacre.' The academic and business communities in Chengdu struck, and people across Sichuan resisted taxes and organized armed groups to march towards Chengdu. The Qing government's urgent redeployment of the Hubei New Army to quell the unrest in Sichuan left Wuchang vulnerable, allowing revolutionary forces to seize it on October 10, 1911. The railway protection movement significantly hastened the onset of the Xinhai Revolution.

Between 1876 and 1911, China constructed a total of 9,488 km of railroads. Spatially, the railroads of the late Qing Dynasty centered around Beijing, forming four main trunk lines: the Beijing-Fengtian (Jingfeng), Beijing-Hankou (Jinghan), Beijing-Zhangjiakou (Jingzhang), and Tianjin-Pukou (Jinpu) Railways, along with four connecting lines: the Zhengding-Taiyuan (Zhengtai), Tongqing, Jiaozhou-Jinan (Jiaoji), and Bianliang-Luoyang (Bianluo) Railways, which together laid the groundwork for the basic North China Railway Network. South of the Yangtze River, there were the Shanghai-Nanjing (Huning), Shanghai-Hangzhou (Huhang), Zhuzhou-Pingxiang (Zhuping), and Zhuzhou-Changsha (Zhuchang) lines. In South China, there were only a few short lines such as Guangzhou-Jiujiang (Guangjiu), Chaozhou-Shantou (Chaoshan), and Zhangzhou-Xiamen (Zhangxia), with no railroads in the northwest and only the Yunnan-Vietnam (Dianyue) line in the southwest region. Looking at the construction timeline, before the First Sino-Japanese War, the Qing government built only about 485 km of railroads, accounting for approximately 0.5% of the total length; the majority were constructed post-war. There were five gauges used initially: the standard gauge of 1.435 m covered 5,746 km (about 60.56%), the broad gauge of 1.52 m covered 2,556 km (about 26.94%), the narrow gauge of 1.067 m covered 107 km (about 1.13%), the meter-gauge of 1.0 m covered 768 km (about 8.09%), and the 0.762 m narrow gauge covered 311 km (about 3.28%).

China Railway in the Republican Period

January 1, 1912, Sun Yat-sen in Nanjing, sworn in as provisional president of the Republic of China. 1912 February Sun Yat-sen resigned as provisional president of the Republic of China, full planning for the national railroad. He put forward the important assertion that "transportation is the mother of industry, and railway is the mother of transportation", began to systematically study the overall planning of the national railroad construction, and began to raise funds, design railroad lines, etc. In August 1919, Sun Yat-sen published "Industrial Plan", centered on railroad construction, railroads and highways, inland waterways, telegraph and telephone, port construction, junction setup, and so on. With railway construction as the center, the plan was designed to combine railways with highways, inland waterways, telegraphs, telephones, ports, junctions, transportation industries and other related sectors, forming a systematic and complete idea of large-scale transportation. The plan was divided into six major railroad systems, constituting a "nationwide network of railroads, popular without stagnation", with the overall goal of making China a modern power with 100,000 miles (about 160,000 km) of railroads. Due to political turmoil and other reasons, Sun's plan was not successfully implemented.

During the Beiyang government era, the administration upheld the Qing dynasty's policies of nationalizing mainline railways and financing railway construction through foreign loans, while prohibiting the operation of commercially managed railways. A significant number of railway rights were auctioned off to service foreign debts. Between 1912 and 1915, the Beiyang government entered into nine railway loan agreements with various countries, relinquishing the rights and interests in 11 railways. From 1912 to 1927, approximately 3,900 km of railway lines were constructed, with around 2,100 km built within the provinces inside the Great Wall (Guannei) and about 1,800 km in the three northeastern provinces.

In October 1928, the Nanjing National Government established the Ministry of Railways to oversee the nation's railways. Initially, the Ministry attempted to implement 'unified management' and 'independent accounting.' However, due to the fragmentation caused by regional warlords and ineffective communication of government orders, these efforts yielded minimal results. After the Central Plains War in October 1930, which involved clashes between Chiang Kai-shek's forces and those of Yan Xishan and Feng Yuxiang, internal strife within the Kuomintang temporarily subsided, allowing for the consolidation of state-owned railway rights. In 1932, the Nanjing National Government issued China's first 'Railway Law,' establishing regulations for railway construction authority, technical standards, and funding sources. In 1933, the Ministry of Railways initiated a reorganization focused on commercialization, which included implementing through transport, improving passenger and freight services, and enhancing technical equipment. This reorganization led to significant increases in both the passenger and freight volumes and the revenue of state-owned railways.

Between 1928 and 1937, the National Government constructed approximately 3,600 km of railways within the provinces. The Zhuzhou to Shaoguan section of the Yuehan (Guangdong-Han) Railway was completed in 1936, marking the completion of the entire 1,096 km Yuehan Railway. The Longhai Railway was extended westward to Baoji, and the Jinpu (Tianjin-Pukou) Railway's Yangtze River ferry service commenced operations in 1933. Several provinces and districts also constructed a number of publicly operated railways, including the Zhegan (Zhejiang-Jiangxi), Tongpu (Tongliao-Pukou), and Huainan Railways. Mao Yisheng, a renowned Chinese bridge expert, oversaw the construction of the first dual-use highway and railway bridge designed by Chinese engineers—the Qiantang River Bridge (Fig. 12.3). The bridge, with a total length of 1,453 m and a main span of 1,072 m, began construction in August 1934 and the railway bridge section was opened to

Fig. 12.3 Public-railway bridge—Qiantang river bridge

traffic on September 26, 1937. With support from Zhang Xueliang, approximately 660 km of public railways were built in the Northeast region between 1928 and 1931. Additionally, there was some development of commercially operated private railways, such as the Jiangnan, Qike, and Gebishi Railways (Fig. 12.3).

After the September 18th Incident in 1931, Japan occupied the northeastern region of China and implemented a 'unified transportation' policy across railroads, highways, waterways, aviation, and communications. In October 1931, the railways were initially placed under the temporary administration of the pseudo-Northeast Transportation Committee by Japan, which later transferred control to the Transportation Ministry of the puppet state of Manchukuo, effectively under Japanese control. In February 1933, the South Manchuria Railway Company ('Mantetsu') and the puppet Manchukuo authorities signed a contract, allowing 'Mantetsu' to manage the region's railways under the guise of 'entrustment.' In March 1935, the Soviet Union unilaterally signed an agreement with Japan to sell the Chinese Eastern Railway to the Manchukuo government. This move completed Japan's 'unified' management of the northeastern railroads. Aiming to 'colonize the Northeast' and support its war of aggression against China, Japan undertook extensive new railway constructions based on the 'Development Plan for the Manchuria-Mongolia Railways.' From 1931 to 1945, Japan constructed approximately 5,700 km of railroads in the Northeast, with an additional 1,000 km underway. By the end of the War of Resistance against Japan, the total length of railways in the Northeast reached about 11,800 km, establishing an extensive network of railways in the region.

On July 7, 1937, the Second Sino-Japanese War fully erupted. The Japanese invasion forces prioritized the capture of railways as a military objective, resulting in the occupation of a large portion of the railways in North and East China by the Japanese army. By the end of the Anti-Japanese War, the Nationalist Government retained control of only approximately 1,400 km of railways, primarily in Yunnan and Shaanxi provinces. During the war, Japan exploited the human, material, and financial resources of the occupied territories to construct additional railways in North, Central, and South China to support its aggression and economic looting. In these occupied areas, Japan implemented colonial-style management and operations for the railways.

In December 1937, after the fall of Nanjing, the Nationalist Government relocated the capital to Chongqing, making the southwest and northwest regions the hinterland of the Second Sino-Japanese War effort. To enhance the transportation links between the front lines and the rear, as well as between domestic and international locations, and to facilitate the inflow of foreign aid and the export of agricultural and mineral products, the Nationalist Government employed various strategies. These included requisitioning civilian labor, repurposing old railway tracks, and reallocating materials from unfinished railway projects. As a result, approximately 1,900 km of railways were constructed in the southwest and northwest regions.

On August 15, 1945, Emperor Hirohito of Japan issued the Imperial Rescript on the Termination of the War via radio broadcast, leading to Japan's unconditional surrender. The Nationalist Government and the People's Governments in the liberated areas, led by the Communist Party of China, took control of the railways from Japanese and puppet forces. In the Northeast, the railways, including the Chilong (Central and Long) Railway, were jointly administered by both sides after being recovered. The railways in Taiwan, which had been under Japanese occupation for 50 years, were declared restored. The Chinese section of the Yunnan-Vietnam Railway regained its rights according to an agreement between China and France. In June 1946, the Nationalist authorities breached the truce agreement signed with the Communist Party, launching a full-scale assault on the liberated areas, plunging the railways once again into the turmoil of war. After the victory of the Second Sino-Japanese War, the Nationalist Government initiated the civil war. Apart from completing the restoration of the Qijiang Railway, no further railway construction was undertaken.

During the Republic of China era (1912–1949), a total of 17,103.1 km of railways were constructed across the country, averaging about 462.2 km of railway construction per year. From 1912 to 1927, due to the turmoil caused by warlord conflicts and political instability, railway construction relied heavily on foreign loans, resulting in a total of 3,946.7 km of railways built during these 15 years, with an annual average of 263 km. In the early 1930s, leading up to the full-scale outbreak of the Second Sino-Japanese War, there was a notable surge in railway construction. From 1928 to 1949, a total of 13,156.4 km of railways were constructed, including those built by Japan in China, averaging 626.5 km per year.

The construction of railroads before the establishment of New China was characterized by the following features: first, a late start and slow progress. The appearance of railroads in China was about half a century later than the appearance of the first railroad in the world. Secondly, a limited number and uneven distribution. By 1949, before New China's founding, a total of 21,800 km of national railroads (with some lines dismantled), were mostly built in the northeastern and coastal areas, with only over 1,600 km in the vast northwestern and southwestern regions. Third, heavy reliance on imported equipment and low efficiency. Railroad technical equipment heavily relied on imports, with inconsistent gauges and a variety of models, earning it the nickname "a museum of international railroads"; fragmented management led to low transportation efficiency. Fourth, subjugation to foreign control and loss of sovereignty over rail rights. Using rail property and operational income as collateral for loans and construction resulted in operational, management, employment, and procurement rights being held by foreign entities; with all rights to profits plundered by directly operated foreign railroads.

Creation of the People's Railway

After the founding of New China, the railroads returned to the people; on January 10, 1949, the Railway Department of the Chinese People's Revolutionary Military Committee was established in Shijiazhuang; in October 1949, it was reorganized into the Ministry of Railways of the Central People's Government, and the process of creating the people's railroads began.

Railroaders took the restoration of the national railroad network as their first priority, and in early 1949, most of the Northeast Railway had been restored. After the end of the Pingjin Campaign, the North China Railway was restored one after another, and in May, through passenger trains were opened between Beiping and Shenyang, and in July, the Jinpu Railway and Jiaozhi Railway were restored to operation, and the through passenger trains between Shanghai and Beiping, which had been interrupted for 12 years, were restored to operation. By the end of 1949, the Shanghai-Hangzhou, Huainan, Nanxun, Jingbao, Longhai, Zhegan, Jinghan, South Tongpu, Yuehan Railway and so on have been restored and opened to traffic. So far, the mainland's original major railroads are basically restored, and connected as a whole. 1950 years later, some bridges and lines were permanently restored. The restoration of the railroad network was of great significance in supporting the war of liberation and promoting economic recovery.

While the national railroad network was basically restored, the construction of new lines was also carried out one after another. At the early stage of the founding of New China, the construction of new lines was focused on the southwest and northwest regions, and the Chengdu-Chongqing Railway, Tianlan Railway and Xianggui Railway Laimu section were newly built. Among them, the Chengdu-Chongqing Railway was the first mainline railroad built after the founding of New China. During the period of the "First Five-Year Plan" (1953–1957), railroad lines such as Baocheng, Lanxin, Baolan, Fengsha and Yingxia were constructed. The construction of Wuhan Yangtze River Bridge started in 1955 and was opened to traffic in 1957, with a total length of 1,670 m. Wuhan Yangtze River Bridge is China's first bridge over the Yangtze River, after the opening of the Beijing-Han, Guangdong-Han two major trunk lines connected, jointly known as the Beijing-Guangzhou Railway, since then, "the moat into the road".

From 1958 to 1976, the construction of Baolan Line, Lanxin Line, Lanqing Line, Qiangui Line, Chengkun Line, Jingcheng Line, Taijiao Line, Waifu Line, Xiaoyong

Line, Sichuan-Guizhou Line, Guikun Line, Sichuan-Guizhou Line, Jiaozhi Line, Zhiliu Line, Xiangqian Line, Xiangyu Line, Jingyuan Line, Beijing-Tongshan Line, and Qinghai-Xizang Line from Xining to Golmud, and so on, was completed. Many new railroads and bridges and tunnels built during this period were rare in history for their complex geology, difficult engineering and high technical difficulties. The completion of these projects has raised the level of China's railroad engineering technology to a new height. Among them, the Chengdu-Kunming Railway from Chengdu to Kunming, a total length of 1,096 km, located in the Hengduan Mountains, through the size of the Liangshan Mountains, the total extension of bridges and tunnels accounted for 41.6% of the length of the line, there are 41 stations built on bridges or tunnels, along the route of the terrain of the steep, geologically complex, arduous engineering is rare in the world, is known as the "Museum of Geology". On July 1, 1970, the Chengkun Railway was opened to traffic. Nanjing Yangtze River Bridge began construction in 1960, was opened to traffic in 1968, the total length of 6788.55 m, was the world's longest two-lane dual-purpose public-railway bridge (Fig. 12.4). The bridge was designed and constructed by China itself, adopting a series of new technologies, which was a milestone in the history of bridge construction in China. At the same time of building new lines, the main technical transformation was also carried out through the construction of double lines and electrification. Duplex railroads grew from 860 km at the beginning of the founding of China to 7,630 km in 1978. Electrified railroad construction began to take off, and in 1961, the electrified railroad was built between Baoji and Fengzhou on the Bao Cheng line. By the end of 1978, the electrified railroad mileage of the whole country had reached 1,030 km.

In 1978, the operating mileage of national railroads reached 52,000 km, an increase of 30,000 km compared with that at the beginning of the founding of China. A number of new major trunk lines, the basic formation of the framework of the railroad network; network layout has been greatly improved, the southwest and northwest of the railroad length of the country's railroad length of the proportion of the beginning of the establishment of China from 6 to 24.5%. All provinces, autonomous regions and municipalities directly under the central government, except Xizang, have railroads, and Fujian, Ningxia, Qinghai and Xinjiang are connected by train for the first time.

Before the founding of New China, the railroad required locomotives, rolling stock and other equipment completely dependent on imports, a small number of factories can only play a maintenance task. After the founding of new China, in the face of foreign economic and technological embargo on China, the railroad people insisted on independence and self-reliance, and tenaciously carried out independent innovation, and gradually established the foundation of new China's railroad industry. 1952, Sifang Locomotive and Rolling Stock Factory copied the first steam locomotive, named liberation type, and then Sifang and Dalian Locomotive and Rolling Stock Factories made the trial of the Victory type, People's type, Forward type, and the construction type, which began the era of China's self-made steam locomotives. Then, Sifang Locomotive and Rolling Stock Factory and Dalian Locomotive and Rolling Stock Factory test-produced Shengli type, People's type, Forward type, Construction type and so on, which started the era of China's self-made steam locomotive. By

Fig. 12.4 Nanjing Yangtze river bridge

1979, the number of steam locomotives in the whole road reached a historical high of 7,899, of which domestic steam locomotives accounted for 86%. 1958 began to manufacture internal combustion locomotives and electric locomotives, and at the end of the 1960s, domestic internal combustion and electric locomotives were mass-produced and put into operation. By 1978, domestic internal combustion and electric locomotives accounted for 21% of the total number of locomotives. 1953, China began to design and manufacture its own buses, and successively mass-produced 21 and 22 types of buses, which covered hard-seat cars, hard sleeper cars, soft-seat cars, soft sleeper cars, dining cars and other types of buses. The manufacture of trucks has developed from 30-ton to 50-ton, and in 1956, it began to develop and produce 60-ton trucks; after the 1960s, a variety of 60-ton general-purpose trucks have been put into use. The comfort level of passenger cars has been continuously improved, and the load capacity of wagons has been continuously increased.

The People's Railway has always been under the banner of the Party, serving the overall situation of national development and playing an important role in various undertakings. After the outbreak of the war of liberation, the railroads in the liberated areas of northeast China immediately took up the important task of supporting the war of liberation, and set off a boom of "resurrecting the dead machine", "resurrecting the dead car", repairing and transporting, etc., and many heroes and models emerged. In the three wars, the majority of railroad workers to "the People's Liberation Army to fight where the railroad to where" as the slogan of action, braving the enemy's fire, completed a large number of railroad restoration and military transportation tasks, a strong support for the war of liberation. 1950 October, the Chinese People's Volunteer Army across the Yalu River, to carry out the war of resistance against the U.S. to aid North Korea. According to the instruction of the central government, a large number of railroad workers and railway corps members joined the volunteer army. Braving the bombing and strafing by enemy planes, they put themselves into the struggle of repairing the lines and bridges to ensure smooth railroad transportation, and built a steel transportation line that "can't be broken or bombed". More than 100,000 people participated in the battle, and more than 2,100 people sacrificed their lives to protect the "steel transportation line". In the rear, the national railroads also launched a vigorous mass movement to resist the U.S. and defend the country. In the course of nearly three years, the railroads, through the concerted efforts of the front and rear, transported more than 385,000 vehicles (weighing more than 8 million tons) of soldiers and military supplies for the Chinese People's Volunteers, making them the lifeline for the mobility of the troops and the supply of logistics.

After the founding of New China, the productivity of railroad transportation has been greatly improved, and the ability to provide transportation guarantee for the development of national economy has been greatly enhanced. Railroad passenger and freight traffic increased year by year, in 1952, the national railroad completed 130 million tons of freight traffic, passenger traffic 160 million passengers, respectively, than in 1949, an increase of 136.5 and 58.8%; "First Five-Year Plan" period, the railroad to improve the level of management, to explore the potential of the capacity, and strive to adapt to the growing demand for transportation. 1957, the completion of 270 million tons of freight traffic, passenger traffic 310 million passengers, the railway transport production efficiency has improved greatly, the ability to provide transport security for the development of the national economy has been greatly enhanced. In 1957, the completion of 270 million tons of freight, passenger traffic of 310 million passengers, completed 134.6 billion tons of cargo turnover, passenger turnover of 36.1 billion passenger km, respectively, than in 1952, an increase of 107.5, 91.2, 123.7 and 80.1%. In the following ten years, during the period of comprehensive construction of socialism, the passenger and freight transportation of the railroad maintained a substantial growth momentum.

The development of the new Chinese railroads has also experienced twists and turns, and the unrealistically high targets set during the "Great Leap Forward" created chaos in the railroad business. In the "Cultural Revolution", the railroads suffered a large impact, the main lines were almost paralyzed, the number of accidents increased dramatically, and the economic benefits fell dramatically. However, under the leadership of Deng Xiaoping and other leaders of the Party and the State, the railroad workers fought hard and still made important achievements in the railroad business. The situation of total paralysis of the national railroad did not happen after all, and the railroad construction has made important achievements, and the technical equipment has been greatly improved, and the completion of the Tanzanian Railway is a major achievement of foreign economic assistance. After the Tangshan earthquake, the railroad people rose up to fight against the earthquake and make significant contributions to the victory in the fight against the earthquake.

China's Railroads Since Reform and Opening Up

The reform and opening-up have spurred robust economic and social development in China, where the railroads once became significant bottlenecks, notably with the prominent issues of 'difficulty in buying tickets' and 'difficulty in transporting goods'. With the central government's strong emphasis and support, railway personnel have faced these challenges head-on, bravely innovating and accelerating development. This has resulted in a favorable trend of rapid growth and significant technological advancements over the past few decades.

A series of major railway construction campaigns were organized and implemented, significantly improving the scale and quality of the rail network. In the 1980s, the focus was on three key campaigns: the northern 'Battle of Daqin', the southern 'Attack on Hengguang', and securing 'East China', which accelerated the advancement of crucial railway corridors and achieved significant results. In the 1990s, the strategy emphasized 'Beijing-Kowloon' and 'Lanzhou-Xinjiang' lines, the 'Baozhong' and 'Houyue' quick campaigns, and further securing 'East China' and 'Southwest China', along with enhancing the Daqin line. This led to the construction of major capacity trunk lines including the Beijing-Kowloon, Lanzhou-Xinjiang, and Baozhong Railways, as well as the Guangzhou-Shenzhen high-speed rail and Beijing West Railway Station. Starting from 1998, the strategy aimed at a decisive effort in the Southwest, a strong focus on coal transport, high-speed railway construction, network expansion, and breaking through the 70,000 km mark, initiating a cross-century railway construction surge. Key achievements included the completion of the Baocheng and Zhuzhou-Changsha double lines, and the opening of new lines such as Neikun, Shuibei, Shuohuang, Xikang, and Shenyang, significantly enhancing major trunk line capacities. At the turn of the 21st century, China embarked on constructing several significant projects. The completion of the Daqin Railway (Fig. 12.5), with an annual freight volume reaching 450 million tons, marked China's railway heavy haul technology reaching world-class standards. The Beijing-Kowloon Railway (Fig. 12.6), linking Beijing and Hong Kong, significantly enhanced the north–south transportation capacity of China's railway network, playing a crucial role in the economic and social development of the regions along its route. The Qinghai-Xizang Railway marked significant achievements, ending Xizang's history of inaccessibility by rail, overcoming challenges such as permafrost, high altitude, and fragile ecology, making it the world's highest altitude railway with the longest route through permafrost areas. The Qingshui River Bridge, completed in 2003 on the Qinghai-Xizang Railway, became the world's longest plateau permafrost railway bridge (Fig. 12.7).

Implementing significant speed increases to boost railway productivity was a major focus in the 1980s when passenger trains' average speed was only around 50 km/h [7], and freight trains were even slower. State funding could only support the construction of a few hundred km of railway, insufficient for high-speed rail development. The railway sector innovated and explored ways to increase speed and efficiency on existing lines. In 1994, the speed improvement of the approximately

Fig. 12.5 Yongding River special bridge of Daqin (Datong-Qinhuangdao) railway. *Note* The Daqin Railway construction commenced in January 1985 and commenced operations in December 1992. The line extends from Hanjialing station on the North Tongpu railway in the west to Liucun station at the Qinhuangdao Hub in the east, traversing the four provinces (and municipalities directly under the central government) of Shanxi, Hebei, Beijing, and Tianjin, covering a total length of 653 km. The Daqin railway represents a first in Chinese railway history with its comprehensive planning approach integrating road, mine, port, and power aspects, and the synchronized construction of loading, transportation, and unloading facilities. It is New China's inaugural railway to operate 10,000-ton heavy-haul train combinations on a double-tracked, electrified, dedicated coal transport line. The image depicts a train crossing the Yongding River bridge on the Daqin railway

150 km long Guangzhou-Shenzhen line was completed, increasing the maximum speed from 100 to 160 km/h and serving as an effective model. Prioritizing speed increases as a key strategy, the railway sector vigorously pursued implementation. From 1997 onwards, China's railways underwent six major speed increases, covering a total of 22,000 km and reaching speeds of 140–250 km/h. These extensive speed increases significantly boosted railway transport capacity, efficiency, service quality, and competitiveness, leaving a profound impact across society. They also drove technological advancements in locomotive and rolling stock, safety management, and more, laying a solid foundation for high-speed railway development.

Advancing scientific research and technological innovation spurred the rapid development of high-speed railways. Preliminary research and systematic demonstrations for the Beijing-Shanghai high-speed railway were initiated in the 1990s, but the project faced delays in starting for various reasons. The Qinhuangdao-Shenyang passenger dedicated line (Fig. 12.8), started in 1999, served as China's 'test bed' for high-speed rail development. The indigenously developed 'China Star' train set achieved a test speed of 321.5 km/h. As the first high-speed railway in China, the completion and operation of the Qinhuangdao-Shenyang line provided valuable experience for large-scale high-speed railway construction. In 2004, the State Council approved the "Medium and Long-Term Railway Network Plan," marking the transition to focusing on high-speed rail (HSR) construction, supported by the strategic

Fig. 12.6 Beijing-Kowloon railway connects and passes through at Yongding River bridgehead in November 1995. *Note* The Beijing-Kowloon Railway was inaugurated in September 1996, extending a total of 2,398 km to Kowloon. The railway traverses nine provinces and municipalities, including Beijing, Tianjin, Hebei, Shandong, Henan, Anhui, Hubei, Jiangxi, and Guangdong, crossing five major river systems: the Haihe, Yellow, Huaihe, Yangtze, and Pearl rivers. It stands as another significant north–south trunk line between the Beijing-Guangzhou and Beijing-Shanghai railways, distinguished by its length, scale, investment, and the broad geographic areas it covers, setting records in Chinese railway construction at the time

Fig. 12.7 Qingshui River bridge of Qinghai-Xizang Railway. *Note* The first phase of the Qinghai-Xizang railway, from Xining to Golmud, spans 814 km. Construction began in 1958, and the line was opened to traffic in May 1984. The second phase, from Golmud to Lhasa, commenced in June 2001 and was fully operational by July 2006, covering a total length of 1,142 km, with 1,110 km of new tracks. This railway, connecting Xining to Lhasa, is the first to reach the hinterland of Xizang and is the world's highest-altitude high-altitude railway, featuring the longest route and traversing the most extensive permafrost areas

focus on major projects and the country's growing economic strength. High-speed railways such as Hefei-Jiujiang, Wuhan-Guangzhou, Beijing-Tianjin, Zhengzhou-Xi'an, Wenzhou-Fuzhou, and Fuzhou-Xiamen were successively inaugurated, eliciting significant attention both domestically and internationally. The Beijing-Tianjin Intercity Railway (Fig. 12.9), operating at 350 km/h, was China's fastest railway at the time. The 1,318 km Beijing-Shanghai high-speed railway (Fig. 12.10) is the world's highest capacity and standard high-speed railway line. By the end of 2012, China's railway operational mileage reached 98,000 km, with high-speed railways accounting for 9,356 km.

Adhering to independent innovation, the level of railway science and technology has significantly improved. Following the reform and opening up, to meet the rapid growth in railway demand, the rolling stock sector played a crucial role by intensifying efforts in independent innovation and technological research, thus advancing rolling stock manufacturing technology. In the 1980s, aiming to achieve a breakthrough in the rolling stock industry, efforts were made to enhance locomotive performance and diversify types. Series of Dongfeng-type internal combustion locomotives and Shaoshan-type electric locomotives were developed, marking the advent of the second-generation product series for new traction power. Additionally, 25-type

Fig. 12.8 "China Star" running on the Qin-Shenzhen passenger dedicated line. *Note* The Qin-Shen passenger dedicated line stretches from Qinhuangdao City in the west to Shenyang City in the east, covering a total length of 404.64 km, and was inaugurated in October 2003. The infrastructure was designed for speeds of 250 km/h, while the superstructure was planned for 200 km/h. In the 66.8 km segment from Shanhaiguan to Suizhong North, a comprehensive testing section was established, designed for speeds of up to 300 km/h. On November 27, 2002, the indigenously developed high-speed train "China Star" set a record on this line with a test speed of 321.5 km/h. This line has set several milestones in the technical R&D and construction of subgrades, bridges, overhead lines, signaling systems, and rolling stock. It is the first high-speed railway built in China, providing invaluable experience for the country's subsequent large-scale high-speed railway projects. The image depicts the "China Star" speeding along the Qin-Shen passenger dedicated line

Fig. 12.9 Beijing-Tianjin intercity railway (Photo by Luo Chunxiao). *Note* The Beijing-Tianjin intercity railway commenced construction in July 2005 and started operations in August 2008. Spanning from Beijing to Tianjin, the railway has a total operational length of 120 km, with the main line being 112.38 km long. It is China's first high-speed railway to achieve speeds of 350 km per hour

passenger cars and various freight wagons were developed to meet the booming needs of passenger and freight transport. In the 1990s, the industry vigorously implemented the 'Three Improvements' strategy, focusing on improving quality, improving level, and improving grade, leading to continuous breakthroughs in manufacturing capabilities and quality, achieving a systemic upgrade. More powerful and reliable Dongfeng and Shaoshan series locomotives were developed, along with 23-ton and 25-ton open wagons, high-grade new passenger cars, and double-decker passenger cars, significantly upgrading the main rolling stock products and markedly elevating their quality and standard.

Starting from the 1980s, China Railway began independently developing its own train sets, debuting with the KDZ1 electric train set in 1988. Throughout the 1990s, several diesel-powered train sets such as "Lushan," "Jiujiang," "North Asia," "New Dawn," "Shenzhou," and "Golden Wheel" were developed. At the same time, electric train sets like "Spring City," "Great White Shark," "Blue Arrow," "Central Plains Star," and "Changbai Mountain" were introduced. Specifically for the Qin-Shen Passenger Dedicated Line, high-speed train sets like "Pioneer" and "China Star" were developed. The "China Star" high-speed train set, leveraging domestic scientific research strengths, achieved groundbreaking results in areas such as bogie design, aluminum alloy body application, aerodynamics, traction braking, and train network testing, setting a record as "the fastest in China Railway" during its test runs on the Qin-Shen Passenger Dedicated Line.

Following sustained efforts in scientific research and independent innovation, by the end of the 20th century, China had established a comprehensive rolling stock industry with formidable manufacturing capabilities. The industry achieved standardization and serialization of rolling stock, securing the world's largest manufacturing scale in this sector. High-quality, cost-effective domestic rolling stock is continually

Fig. 12.10 Beijing-Shanghai high-speed railway Nanjing Dasheng Pass Yangtze River bridge. *Note* The Beijing-Shanghai high-speed railway commenced construction in April 2008 and was inaugurated in June 2011. Stretching from Beijing South Station to Shanghai Hongqiao Station, it spans a total of 1,318 km, features 24 stations, and is designed for speeds of up to 350 km per hour. This railway is part of the "eight vertical and eight horizontal" main high-speed railway corridors outlined in the 2016 revision of China's medium and long-term railway network Plan. The image shows the Dashengguan Yangtze River Bridge in Nanjing on the Beijing-Shanghai High-Speed Railway. This bridge serves as a shared crossing for the Beijing-Shanghai high-speed railway and the Shanghai-Wuhan-Chengdu railway, while also supporting a dual-line subway for Nanjing. With a total length of about 9,273 m, it is the world's first six-track railway bridge, with its double-span continuous arch being the largest in terms of span and design load among similar high-speed railway bridges globally

deployed in railway transportation, providing robust support for the sustained and rapid development of the national economy. Despite the dominance of foreign brands in the civil aviation and automotive sectors within the domestic market, Chinese railway rolling stock has uniquely maintained its stronghold, marking a significant achievement in national industrial development.

Since 2004, with the support of the government, China Railway has adopted a strategy of "introducing advanced technology, joint design and production, and creating Chinese brands," leading to the production of the Harmony series train sets by incorporating foreign advanced technology. By assimilating and innovating upon this technology, Chinese companies have further developed high-speed train sets such as the CRH380A and CRH380B, which have become key models for high-speed railways. The advancement in high-speed train set manufacturing technology has significantly bolstered the rapid development of high-speed railways.

During this period, China Railways vigorously promoted independent innovation and actively pursued key technological breakthroughs in railroad engineering, construction, communications, and signaling. As a result, the level of innovation in railway science and technology has significantly advanced.

Innovating railway service models to enhance passenger and freight service quality. Leveraging major speed increases and equipment improvements, railway operations were reformed by introducing 'overnight' and 'daytime' passenger trains, which were greatly welcomed by passengers, significantly enhancing passenger market competitiveness. Conduct and service improvements, along with strengthened marketing, expanding heavy haul transportation, and boosting freight efficiency, have contributed significantly to an early turnaround in transportation profitability.

Entering the era of high-speed railways, the speed of trains in China has steadily increased, railway capacity has significantly enhanced, and passenger service facilities and travel conditions have seen significant improvements. The China Railway Customer Service Center (12,306 website) initiated its trial operation on January 30, 2010. The Beijing-Tianjin Intercity Railway began trialing online ticketing on June 12, 2011, which was expanded to cover all train services by December 23, 2011. Starting January 1, 2012, a real-name ticketing system was implemented for all passenger trains nationwide. This effectively addressed the longstanding issues of ticket availability and accessibility, marking a significant advancement in passenger services. In freight transport, reforms centered around the actual cargo system were actively pursued, easing the receipt of goods and accommodating cargo owners, exploring the development of modern railway logistics, and contributing to the increase in freight volumes.

China Railway in the New Era

Since the 18th National Congress of the Communist Party of China (CPC), the railway sector has diligently executed the strategic directives of the CPC Central Committee and the crucial directives of General Secretary Xi Jinping, advancing into a phase of high-quality development within the framework of serving the national development agenda.

"Continued high-intensity investment has accelerated the construction of the high-speed railway network. Adapting to the needs of national economic development, the scale of investment in railway construction has reached the highest level in history. High-speed railway development has entered the fast lane, with an average of 3,500 km being commissioned annually. A number of high-speed railways, such as Harbin-Dalian (Ha-Da), Beijing-Guangzhou, Guiyang-Guangzhou, Nanning-Guangzhou, Hefei-Fuzhou, Lanzhou-Xinjiang, and others, have been completed and operational, accelerating the expansion from "four verticals and four horizontals" to "eight verticals and eight horizontals". China has successfully built the largest and most modern high-speed railway network in the world."

On December 1, 2012, the Harbin-Dalian High-speed Railway opened for operation. Construction started in August 2007, making it the world's first newly built high-speed railway in an alpine area to be put into operation. On December 26, 2012, the Beijing to Zhengzhou section of the Beijing-Guangzhou High-speed Railway was inaugurated, connecting with the previously completed Zhengzhou to Guangzhou section. This marked the full operation of the world's longest Beijing-Guangzhou High-speed Railway. On December 26, 2014, the entire Lanzhou-Xinjiang (Lanzhou-Urumqi) Passenger Dedicated Line was operational. With a total length of 1,786 km, it became the high-speed railway with the longest mileage constructed and opened to traffic in a single phase in the world. On December 30, 2015, the western section of the Hainan Island high-speed railway was opened, linking with the eastern section opened in December 2010, to form a loop around the entire island, marking the inauguration of the world's first island-loop high-speed railway. On December 28, 2016, the Guiyang-Kunming section of the Shanghai-Kunming High-speed Railway, stretching 2,252 km, was inaugurated, marking the complete operation of another major east–west high-speed railway corridor in China. On December 6, 2017, the Xi'an-Chengdu High-speed Railway, spanning 658 km, was opened, bridging the Qinling Mountains that have divided China's northwest and southwest for millennia, turning the once formidable 'Road to Shu' into history. On September 23, 2018, the Hong Kong section of the Guangzhou-Shenzhen-Hong Kong High-speed Railway commenced operation, integrating Hong Kong into the national high-speed railway network and achieving interconnectivity with the mainland's high-speed rail system. On December 30, 2019, the world's first intelligent high-speed railway capable of speeds up to 350 km/h—the Beijing-Zhangjiakou High-speed Railway—was inaugurated, playing a crucial role in serving the Beijing 2022 Winter Olympics and Paralympics (Fig. 12.11).

Fig. 12.11 Harbin-Dalian high-speed railroad (Photo by Luo Chunxiao). *Note* The construction of the Harbin-Dalian High-speed Railway started in August 2007 and it was opened for operation in December 2012. The line originates in Dalian City, Liaoning Province, and terminates in Harbin City, Heilongjiang Province, spanning across the three northeastern provinces with a total length of 904 km. This line is China's first high-speed railway designed for speeds of 350 km/h, constructed in the high-latitude cold regions. It is a crucial part of the Beijing-Harbin passenger dedicated line within the 'four vertical and four horizontal' framework, as outlined in the country's medium and long-term railway network plan

On October 10, 2018, General Secretary Xi Jinping presided over the third meeting of the Central Financial and Economic Commission and made a significant decision to fully initiate the planning and construction of the Sichuan-Xizang Railway. The meeting underscored the immense and enduring significance of the Sichuan-Xizang Railway's planning and construction for the nation's enduring peace and stability, as well as for the economic and social development of Xizang, and stressed the importance of its successful execution. The railway authorities earnestly studied and implemented General Secretary Xi Jinping's important directives on the Sichuan-Xizang Railway's planning and construction, advancing the project with a focus on high standards and quality from the outset. On November 8, 2020, construction formally began on the Sichuan-Xizang Railway's Ya'an-Linzhi section (Fig. 12.12).

Deepening independent innovation to establish a globally advanced high-speed railway technology system. By steadfastly pursuing independent innovation in railway science and technology, China's high-speed railway technology has continuously advanced, developing a globally advanced high-speed railway technology system with independent intellectual property rights. By the end of 2021, a comprehensive high-speed railway technology system encompassing high-speed railway construction, equipment manufacturing, and operational management was established, elevating the overall technology to global leadership, with certain areas

Fig. 12.12 Intelligent train set driving over Guanting reservoir bridge on the first day of Beijing-Zhanghai high-speed railway (Photo by Luo Chunxiao). *Note* On December 30, 2019, the Beijing-Zhangjiakou high-speed railway opened for operation. With a total length of 174 km and a maximum design speed of 350 km per hour, this line is the world's first high-speed railway to feature automated driving capabilities

achieving world-leading standards. This marks a pivotal transition from follower to leader in the field.

In the field of high-speed railway construction: Adapting to China's complex and diverse geological and climatic conditions, with a primary focus on original innovation, significant worldwide technical challenges have been overcome in areas such as high-speed railway roadbeds, tracks, long-span bridges, extensive tunnels, major passenger stations, and system integration. A comprehensive suite of technologies for constructing high-speed railways under various climatic conditions and geological settings has been systematically mastered.

In the field of high-speed railway technology and equipment, building upon the CRH trainsets introduced with advanced technology and joint design and production, Chinese railway authorities have furthered independent innovation and overcome technical challenges. This has led to the successful development of the "Fuxing" class trainsets, which boast complete independent intellectual property rights and world-leading technology. Notably, the "Fuxing" intelligent trainsets have achieved a global first by enabling automatic driving at speeds of up to 350 km/h. China has developed a range of "Fuxing" series products that span speed grades from 160 to 350 km/h, designed to operate across diverse environments, including plateaus, cold regions, and areas prone to wind and sand (Fig. 12.13).

In the field of high-speed railway operation and management: Full expertise has been developed in the comprehensive high-speed railway operation and management technologies for complex network conditions. Innovations in high-density, cross-line transport and scheduling technologies have been made for high-speed trains across various speed levels, addressing challenges related to diverse train compositions, speeds, extensive distances, and cross-line operations. This has enabled the efficient

Fig. 12.13 "Fuxing" China standard train g123 departing from Beijing south station

and high-frequency operation of trains on busy high-speed mainlines and intercity railways, achieving peak departure intervals of just 4–5 min.

In the field of high-speed railway safety production: Fully leveraging technology for safety assurance, over 2,700 monitoring points have been deployed on the intelligent 'Fuxing' trainsets. Technologies for in-transit monitoring of trains, including self-awareness, health management, and fault diagnosis, havc been developed, achieving comprehensive real-time monitoring of train operations. An online monitoring system comprising high-speed comprehensive inspection vehicles and trackside detection sensors among other equipment has been established. Utilizing big data analysis, this system has enabled precise monitoring of the operational status of high-speed railway infrastructure. Furthermore, systems for monitoring natural disasters such as wind, rain, and snow, along with foreign object detection alarms and earthquake early warning, have been developed, enabling comprehensive prevention and control of natural disasters and security risks.

By implementing three major initiatives—the passenger transportation quality improvement plan, the freight transportation incremental action, and the "Fuxing" brand strategy—railway authorities strive to meet the ever-increasing transportation needs of the people and enhance the competitiveness of the railway market. Daily schedule adjustments for high-speed railways will be fully implemented, optimizing and increasing the supply of high-speed passenger transport products. The number of high-speed train services continues to grow, with the network's reach continually expanding. By the end of 2021, more than 7,400 EMU trains were operated daily, accounting for 77% of all passenger train services. Initiatives such as the passenger station throughput project and improvements to restroom facilities, known as the 'toilet revolution,' aim to enhance the intelligent service levels at stations and on trains. Efforts include expanding the mutual recognition of security checks between railways and metros, standardizing and improving the quality of conventional rail services, and enhancing the passenger services of slower trains. The introduction of new, people-oriented measures such as facial recognition verification, online seat selection, QR code entry, onboard and station Wi-Fi services, online food ordering, and e-tickets has significantly improved the passenger travel experience and noticeably elevated the quality of passenger services. The in-depth implementation of the rail freight incremental action, improving and leveraging the western coal eastward and northern coal southward transport systems, and actively promoting the shift from road to rail, aim for a significant increase in rail freight volumes and market share.

Railway international cooperation and services for the 'Belt and Road' initiative have achieved new results. On December 3, 2021, the entire China-Laos Railway was inaugurated. Spanning 1,035 km from Kunming in Yunnan Province, China, to Vientiane, the capital of Laos, this railway is the first international line built and operated in cooperation between China and Laos, using Chinese standards and directly linked with China's railway network. Significant progress has been made in other international projects, including the Jakarta-Bandung High-Speed Railway and the Hungary-Serbia Railway. Since 2011, China Railway has successfully established the China–Europe Railway Express as a prominent international logistics service. This service operates regular freight trains between China and Europe, as well as countries along the 'Belt and Road,' primarily transporting goods in containers. On June 8, 2016, China Railway officially launched the unified branding for the China–Europe Railway Express. By early 2022, over 50,000 China–Europe freight trains had been operated, connecting 33 cities in China with 180 cities across 23 European countries.

Over the past 100 years, China's railroads have undergone radical changes.

From the perspective of railway construction, by the end of 2021, the total track length of China's railways increased from 21,800 km at the end of 1949 to 150,000 km, ranking second in the world. The double-tracking rate grew from 4.0 to 59.5%, the highest globally. The electrification rate surged from zero to 73.3%, placing China among the world leaders in this aspect. High-speed railways were developed from scratch, reaching a total length of 40,000 km and leading the world in this category. Today, China boasts an advanced and extensive railway system. The operational map of China's high-speed railway network is illustrated as shown (Fig. 12.14).

From the perspective of railway technological progress, China has mastered advanced railway construction technologies, capable of building lines, bridges, and tunnels in extremely complex geological conditions and with high construction challenges. A complete rolling stock industrial system has been established, enabling the production of some of the world's highest-quality rolling stock. Furthermore, advanced safety assurance technologies have been developed, significantly improving transportation safety levels. Consequently, China's railways have become highly reliable major arteries and exceedingly popular modes of transportation.

From the perspective of capacity and efficiency of railroad transportation, the number of passenger trains operated by China Railway has increased from more than

Fig. 12.14 On June 8, 2016, the inaugural ceremony of the China–Europe liner was held in Wuhan

300 in the early days of the founding of China to more than 10,000, the operating speed of passenger trains has increased from dozens of kilometers per hour to 350 km/h, and the annual number of passengers transported by rail has increased from 100 million to 3.66 billion, and the volume of goods transported by rail has increased from 56 million tons to 4.55 billion tons [2]. Passenger traffic, freight traffic, and cargo turnover and other major transportation economic indicators ranked first in the world [3]. China's railroad has developed from transportation limited by severe capacity constraints to adaptive transportation with relatively abundant capacity.

From the perspective of quality of railroad services, the conditions of service facilities have undergone dramatic changes, featuring both the most advanced high-speed trains and the high-quality, low-priced "green trains" to accommodate the needs of diverse passengers. The establishment of the 12,306 Railway Customer Service Center, along with the introduction of online ticketing, online meal ordering, electronic tickets, and facial recognition for station entry, has significantly enhanced the convenience, comfort, and safety of passengers' train travel.

China's high-speed railway has become a striking national emblem. On January 19, 2021, General Secretary Xi Jinping, during his inspection at the Taizi Cheng station on the Beijing-Zhangjiakou high-speed railway, emphasized that China's high-speed rail is a prime example of successful independent innovation. It has progressed from nonexistence to leading the world, evolving from technology introduction, digestion, and absorption to re-innovation and then to independent innovation. Over the decades, under the strong leadership of the Party, China's high-speed railway has undergone a historic transformation from catching up, to keeping pace, and finally to leading the way. This effort has resulted in the construction of the world's largest and most modern high-speed railway network, the development of a globally advanced high-speed railway technology system with independent intellectual property rights, and the establishment of the China high-speed railway brand, known for its world-class operational quality.

As an important national infrastructure and modernized means of transport, high-speed rail has had a wide and profound impact on all aspects of the economy and society. High-speed rail has enhanced the economic development attractiveness and impact of the areas along its route, strengthened the influence and leadership of central cities on surrounding regions, and reshaped China's economic geography. It has reduced travel time and distance, accelerated the pace of economic and social activities, promoted urban clusters and regional integration, and driven high-quality economic and social development. High-speed rail has improved public service levels, facilitated the flow of resources and factors, and supported the coordinated development of regional economies and societal prosperity. By leveraging its green and low-carbon advantages, high-speed rail has contributed to environmental protection, energy conservation, and emission reduction, playing a significant role in the advancement of ecological civilization. It has transformed people's lifestyles, offering safe, comfortable, and convenient travel options, significantly enhancing the public's sense of gain, happiness, and security. In the future, China's high-speed rail will continue to play a vital role.

"A century of transformations foreshadows the future. The railway in old China struggled to get started and faced difficulties in expanding under the national conditions of internal strife and weakness. Under the CPC's leadership, the new China Railway has achieved remarkable feats, transformed the nation, and captured global attention. Especially since the reform and opening, China's high-speed railroad construction has been a world-historical miracle, reshaping China's economic geography. Now, the blueprint for the new era has been set, outlining a mission that is both glorious and challenging. The railroad serves not only as a vital artery for rapid travel but also as the driving force behind the strategic development of land and the promotion of China's green growth."

References

1. Jin Shixuan, Xu Wenshu. History of the development of Chinese railroads (1876–1949) [M]. Beijing: China Railway Press, 1986.
2. Ministry of Railways Archives and History Center. China railway yearbook (1999) [M]. Beijing: China Railway Press, 1999.
3. State Railroad Administration. State Railway Administration releases Railway Statistics Bulletin 2020 [EB/OL]. (2021-04-19) [2022-05-08]. http://www.gov.cn/xinwen/2021-04/19/content_5600508.htm.
4. Anonymous. Qing dynasty Mongolian travelogue selection of thirty-four kinds [M]. Bi Aonan, collation. Beijing: Oriental Publishing House, 2015.
5. Marx. Capitalism [M]. Guo Dali, Wang Yannan, Translation. Shanghai: Shanghai Sanlian Bookstore, 2009.
6. Max Wiebel. A general history of the world economy [M]. Yao Zengjian, Translation. Shanghai: Shanghai Translation Publishing House, 1981.
7. Zeng Kunhua. History of Chinese railroads [M]. Beijing: Yanjing yinshuibu, 1924.

Chapter 13
Through Mountain and Cross Sea: The Outstanding Chinese Bridge and Channel

"In the process of transportation development, bridge and tunnel construction has always progressed alongside road development, inseparable and indispensable. 'Where there's a mountain, make a road; where there's water, build a bridge.' As key infrastructure for crossing treacherous rapids and towering mountains, the evolution of bridge and tunnel technology has allowed us to achieve many marvels and fulfill the dream of transforming insurmountable barriers into thoroughfares. Ancient bridges, whether stone or wood, each unique in structure, bear the marks of time and the mastery of bridge-building art. By appreciating these awe-inspiring bridge and tunnel creations, we not only connect with the wisdom of our ancestors but also cultivate our appreciation for humanistic and historical values, finding inspiration within them [1-54]."

Ancient Bridges Full of Wisdom

"Across the vast expanse of 9.6 million km^2, the east features vast plains with meandering waterways and a dense network of rivers, while the west is characterized by towering mountains, overlapping peaks, and countless ravines and cliffs. The north consists of endless grasslands and deserts. Bridges and tunnels, vital for traversing waterways and mountains, have naturally emerged as essential for survival and are commonly found. With their complex structures and challenging construction, these infrastructures often come with associated special buildings or architectural complexes. The scale and type of bridges and tunnels vary across different regions, reflecting diverse natural and social development conditions. Constructing these infrastructures demands significant human and material resources, as well as sophisticated organization and management."

The construction of bridges and tunnels relies on societal productivity and scientific and technological levels, and is subject to the requirements of politics,

© China Communications Press Co., Ltd 2025

H. Lu (ed.), *A Brief History of Transportation in China*,
https://doi.org/10.1007/978-981-97-9851-3_13

economics, the military, and more. It also has profound connections to people's lives, production, literature, art, and religion. Different eras, with their unique conditions, demands, technologies, and equipment, inevitably lead to varied processes and outcomes.

In the early stages of observing and understanding nature, humans encountered logs naturally spanning waterways and ditches, as well as stone slabs, beams, and arches formed by the earth's crust changes and erosion. These natural bridges enabled early humans to expand their activity range without being cut off from each other. Known as 'Born Bridges,' these natural formations were nature's gift and served as the earliest bridges for humankind. Inspired by these, early humans, using basic tools like stone, wood, and bone, began to craft rudimentary bridges, leading to the creation of various types we recognize today, including log bridges, wooden beam bridges, stone slab bridges, stone arches, brick arch bridges, and vine rope bridges.

To the Xia, Shang, Western Zhou period (21 BC–771 BC) of the Bronze Age and the Spring and Autumn and Warring States period (770 BC–221 BC) of the Iron Age, all walks of life in the society have been different degrees of development, commercial prosperity, transportation, trade and consequently developed, people's horizons are gradually broadened, the understanding of nature in the breadth and depth has been unprecedentedly improved; penetration between different scientific and technological, influence and effect, science and technology has been greatly advanced, the emergence of pig iron smelting and casting technology. Depth of the understanding of nature in the breadth and depth of the unprecedented improvement; different science and technology between the penetration, influence and role, science and technology has been a great progress, the emergence of pig iron smelting and casting technology. This period is the initial accumulation period of ancient Chinese science and technology.

As detailed in the 'Kao Gong Ji,' a notable work from the late Spring and Autumn period, the government's handicraft system encompassed 30 specialized production departments, reflecting a sophisticated division of labor. These included seven related to woodworking, six to metalworking, five each to leatherworking, dyeing, and painting, and others focused on the production of tools, weapons, musical instruments, containers, jade items, and construction, among various crafts. The 'Liuqi' section of 'Kao Gong Ji' offers a detailed account of the development and production processes of bronze ware and tools, marking a significant advancement in metallurgy. By manipulating the proportions of copper, tin, and lead, artisans could produce bronze alloys with varying properties, tailored for specific uses. This represents one of the earliest empirical scientific summaries of alloy compositions in the world.

The emergence of metal tools, such as bronze and iron, also contributed to the development of bridge construction techniques.

According to historical records, King Hui of Qin, during his reign in the Warring States period, constructed the 235 km long Baoshi Pass. This pass extended from Baocheng Baogu in Shaanxi Province to Xiegu in the ancient Meixian County (present-day Meixian County). The design featured palisades carved into the mountainside and combined cut walls with wooden beam structures, creating a form of post-and-beam bridges that were not common in Western engineering practices at the time. These innovative bridges comprised multiple wooden beams, serving as essential infrastructure in the region. Even today, similar wooden trestle bridges continue to facilitate passage for people and animals in remote mountainous areas where transportation options are limited.

In "The Book of Songs, Daya, Daming,' it is recorded that in 1184 BCE, King Wen of Zhou erected a pontoon bridge over the Weishui River for his marriage ceremony. This is the earliest recorded construction of a pontoon bridge, predating by more than 500 years the first such structure in the Western world, built over the Bosphorus Strait during King Darius I of the Persian Empire's invasion of Greece, as documented by the ancient Greek historian Herodotus.

During the Warring States period, the text 'Guanzi' includes the passage 'guide water flow…clear blockages, and carefully manage bridges and fords,' which implies the dredging of accumulated water and the clearing of obstructed waterways, with special attention to bridges at crossing points. This suggests that during the Warring States period, beam bridges spanning rivers or valleys were already present in the Yellow River basin and beyond. These beam bridges, characterized by their flat and straight appearance, were known in ancient times as 'flat bridges.' Constructed by placing wooden or stone beams (planks) on piers, these structures constituted the early form of beam bridges.

Li Daoyuan's 'Notes on the Water Classic,' written during the Northern Wei Dynasty, records a bridge on the Fen River, just one mile east of Yangqu County in the Taiyuan Prefecture of Shanxi Province. This bridge, built during the reign of Duke Ping of Jin from 557–532 BC, is a wooden beam bridge supported by thirty pillars. It is the earliest such bridge documented in ancient texts.

On the Ba River, 20 li northeast of Xi'an, stands the Ba Bridge, a bridge with stone piers and interchangeable wooden and stone beams. Originally built during the Western Han Dynasty, the bridge has been destroyed and reconstructed multiple times over the centuries. According to the Qing Dynasty's 'Wanliqiao Zhi,' the bridge typically underwent reconstruction every 60 years since the Song Dynasty. In 1834, 40 li southwest of Xi'an, the bridge was significantly rebuilt to its current form. Measuring approximately 354 m in length and 7 m in width, with 67 spans each about 6 m wide, the bridge is a simply-supported timber-beam structure with stone piers. Each pier is made up of six stone columns, each constructed from four layers of cylindrical stones, supported by a stone base, and reinforced by 11 cypress wood piles in a plum blossom arrangement (Fig. 13.1). This design makes the Ba Bridge one of the earliest examples of lightweight bridge piers in history and marks it as an enduring and still-in-use testament to ancient bridge construction techniques.

The small-scale water conservancy project to harness the Min River began on the Chengdu Plain around 320 BC, with the construction of Dujiangyan commencing a few decades later. At the Dujiangyan Fish Mouth, a multi-hole continuous bamboo cable bridge spanned the confluence of the inner and outer rivers. Originally named 'Zhupu Bridge' before the Song Dynasty, it was renamed 'Pingshi Bridge' after reconstruction, as illustrated in Fig. 13.2. The bridge was destroyed during the wars at the end of the Ming Dynasty. In 1803, during the eighth year of the Qing Dynasty's Jiaqing period, a bamboo bridge with multiple holes was rebuilt in the traditional

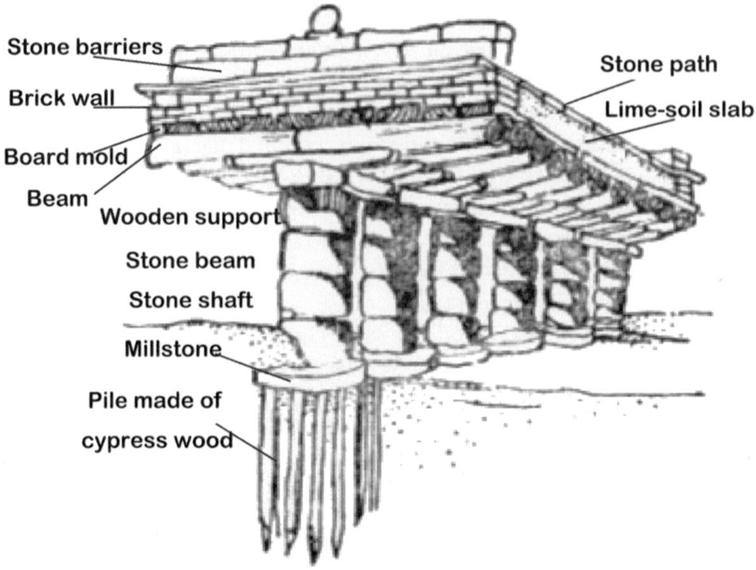

Fig. 13.1 Ancient Ba Bridge

Fig. 13.2 Zhupu Bridge, the earliest bamboo cable-stayed bridge in China

style, named Anlan Bridge, and remains in use. Due to the variable currents of the Minjiang River and consequent adjustments to Dujiangyan, the bamboo bridge also had to adapt, resulting in unique piers and spans. Measuring 340 m in length, with 8 spans and a maximum span of 61 m, the bridge is over 3 m wide and nearly 13 m high. It is supported by 8 wooden and 1 stone pier, each protected by anti-scour piles, as shown in Figs. 13.3 and 13.4. This bridge stands as an outstanding example of China's ancient bamboo bridges.

According to the 'Bamboo Annals,' in the 37th year of King Mu of Zhou's reign (circa 965 BC), the Zhou Dynasty launched a campaign against the state of Chu, deploying nine armies that reached the Jiujiang River, where they constructed bridges

Fig. 13.3 Bridge piers in the river

Fig. 13.4 Bridge deck

Fig. 13.5 Schematic of a cantilevered timber beam bridge

using structures that resembled the shapes of soft-shelled turtles and alligators as beams. This method involved stacking gravel or neatly arranged stones in the water to create a series of stepping stones, forming a causeway-style stone bridge, also known as a 'stepping stone bridge.' Such bridges, reminiscent of ancient construction techniques, can still be observed in some mountainous regions today.

At the beginning of the 3rd century, the Qiang people in the border area between Gansu and Xinjiang (Anxi—Turpan) innovated a new bridge structure—the cantilevered wooden beam bridge. This design employed round or square timbers, interlaced vertically and horizontally, with stacked stones on both sides serving as the foundation. The structure projected layer by layer towards the river, each layer extending about 1 m with the front end slightly inclined upwards to prevent downward deflection under load. When the span between the two ends is just three or four Zhang, simple wooden beams complete the construction, resulting in a cantilevered wooden beam bridge, as illustrated in Fig. 13.5.

In the 'Commentary on Waterways,' it is said: 'Further east, the river merges with Seven Mile Creek… Here lies a stone bridge, also known as the traveler's bridge… Comprising several arches, each meticulously constructed from stacked stones, the bridge stands tall and robust, a testament to exceptional craftsmanship. Despite the toll of time, it continues to serve travelers faithfully. Zhu Chaoshi mentioned in a letter to his brother that the bridge, located six or seven miles from the Luoyang Palace, is entirely built of large stones, rounded at the base to allow water flow and large enough for boats to pass beneath. An inscription on the bridge states that construction commenced in the eleventh month of the third year of Taikang (282 CE) during the Jin Dynasty, employing seventy-five thousand workers until its completion at the end of April…' This traveler's bridge at Seven Mile Creek near Luoyang is recognized as China's first documented stone arch bridge.

Ancient bridges in China are diverse in form, each showcasing a unique beauty that is a feast for the eyes. Leveraging natural materials like wood and stone, ancient China developed highly sophisticated bridge construction techniques. These efforts produced countless masterpieces, contributing to a rich architectural heritage that continues to inspire.

Zhaozhou Bridge (Fig. 13.6), also known as Anji Bridge and colloquially as Dashi Bridge, is situated in Dashiqiao Village, south of Zhao County in Shijiazhuang City, Hebei Province. Constructed between 591 and 599 during the Kaihuang era of the Sui Dynasty, it was designed by the renowned craftsman Li Chun. It is recognized as the world's first single-arch stone bridge with open spandrels, featuring a span

Fig. 13.6 Zhaozhou Bridge

of 37.02 m, an arch height of 7.23 m, and a net bridge width of about 9 m, accommodating pedestrians and vehicles. The bridge's construction showcases exceptional craftsmanship, making it the earliest preserved and largest span hollow-arch stone bridge in China, predating similar bridges in France by more than 1,000 years. In 1991, it was acknowledged by the American Society of Civil Engineers as the oldest 'civil engineering milestone in international history.' The bridge has withstood 10 floods, 8 wars, and numerous earthquakes. Since its completion, Zhaozhou Bridge has undergone eight major repairs and is listed as one of the first batch of national key cultural relics protected units.

Suzhou Baodai Bridge (Fig. 13.7), also known as Changqiao, was constructed between the 11th and 14th years of the Yuanhe period (816–819) of the Tang Dynasty. Situated 7.5 km southeast of Suzhou city, in the Wuzhong District of Jiangsu Province, the Baodai Bridge resembles a long dragon resting at the mouth of Tantai Lake, alongside the undulating canal. The bridge is renowned for its exquisite design, stretching approximately 316.8 m in length with 53 arches, and spans varying from 3.9 m to 6.0 m, the 15th arch boasting the largest span of 6.95 m. The sheer number of arches, the bridge's length, and its intricate design are exceptional in the history of both Chinese and international bridge construction. The bridge ingeniously adopts the 'bridge as a dike' concept to address lake drainage, featuring the inaugural use of continuous arches with both flexible and rigid piers. This approach skillfully incorporates staggered, small-span steep arches tailored to the landscape and flood control requirements.

In his masterpiece 'Qingming Riverside Scroll,' Northern Song Dynasty painter Zhang Zeduan exquisitely depicted the Kaifeng Rainbow Bridge, capturing the scene where 'the bustling dream fades at the two bridges, leaving only the lingering Bian River flowing east.' The Kaifeng Rainbow Bridge is a single-span equilateral folded-line wooden arch bridge (Fig. 13.8), with a span of approximately 25 m and a clear span of about 20 m. The arch rises about 5 m above a clearance of 5.5–6 m from the water surface, and the bridge width is an impressive 8 m, with a rise-to-span ratio

Fig. 13.7 Suzhou Baodai Bridge

Fig. 13.8 Kaifeng Rainbow Bridge in Qingming riverside scene

of about 1/5. It was the first bridge to employ longitudinal and transverse wooden members to form a self-stabilizing wooden arch structure. The Kaifeng Rainbow Bridge is meticulously designed and intricately constructed, marrying practicality with beauty in a testament to outstanding architectural skill and artistic finesse.

Anping Bridge, constructed in 1138 and completed in 1152 during the Southern Song Dynasty, spans from Jinjiang to Nan'an in Fujian. The bridge measures 811 zhang (approximately 2,623 m) with 362 arches, extending over 5 miles and is colloquially referred to as the 'Five-Mile Bridge.' For an extended period, it held the distinction of being not only China's longest bridge but also the longest stone bridge in the world, earning the reputation 'no bridge in the world surpasses this one' (Fig. 13.9). It has been designated as one of the first batch of national key cultural relics protected units.

Lugou Bridge is situated 13 km west of Beijing, just outside the Guang'anmen Gate, spanning the Yongding River (formerly known as Lugou River). Construction began in 1189 and was completed in 1192 during the Jin Dynasty's Mingchang era. This large eleven-arch stone bridge measures 212.2 m in length, and including the abutments at both ends, extends 266.5 m, making it the oldest and longest ancient bridge still standing in northern China (Fig. 13.10). Marco Polo, the renowned Italian traveler who visited during the Yuan Dynasty, described the Lugou Bridge as 'the finest and most unique bridge in the world.' It has been designated as one of the first batch of national key cultural relics protected units.

Fig. 13.9 Fujian Anping Bridge

Fig. 13.10 North view of Lugou Bridge

Jihong Bridge, constructed during the Chenghua period of the Ming Dynasty (1465–1487), is located in Yongping County, Dali Bai Autonomous Prefecture, Yunnan Province, spanning the Lancang River. The bridge boasts a clear span of 57.3 m and a total length of 113.4 m. It is comprised of 18 iron cables, including 16 base cables and 12 load-bearing cables grouped in sets of 4, with an additional handrail cable on each side. Nestled between two towering mountains, the suspended rope bridge presents an extraordinary challenge and is one of the oldest surviving cable bridges in the world.

Luding Bridge, also known as Imperial Bridge, is located in Luding Town, Luding County, Ganzi Xizang Autonomous Prefecture, Sichuan Province, spanning the Dadu River. Construction commenced in the 44th year of the Kangxi Emperor's reign (1705) and was completed the next year, with Kangxi Emperor himself inscribing 'Luding Bridge' and erecting a royal stele at the bridgehead. The bridge, with a span of 103 m, ranks among the earliest and largest suspension bridges globally. It is 3 m wide and over 10 m high, built from 13 iron chains. Small chains link the main chains every 5 m, and wooden planks are laid over the 9 base chains to form the deck, with two chains on each side serving as railings. The abutments are constructed from dressed stone, and the pavilions are wooden, characterized by their elegantly high eaves and stately simplicity. The bridge gained international fame on May 29, 1935, when the Chinese Workers' and Peasants' Red Army crossed it during the Long March, in the celebrated 'seizing of the Luding Bridge.

The First Industrial Revolution, initiated in mid-18th century England, heralded the era of machine replacement of hand tools and saw the steam engine become widely used as a power source. Following these developments, the period after 1870 marked the onset of the Second Industrial Revolution, characterized by the widespread application of electricity, the invention of the internal combustion engine and new modes of transportation such as automobiles, trains, and airplanes, and the creation of novel communication methods. Despite these global advancements, China remained under feudal rule during the Industrial Revolutions, predominantly reliant on a small peasant economy. The development of bridge construction technology was constrained, continuing to rely on traditional wood and stone materials. However, notable innovations were achieved in the construction of garden bridges.

The Seventeen-hole Bridge, spanning the east embankment of Beijing's Summer Palace, is a beautifully constructed joint-arch stone bridge (Fig. 13.11). Built during the Qing Dynasty's Qianlong period (1736–1795), it connects South Lake Island to the west with Langru Pavilion to the east. Measuring 150 m in length and comprising 17 arches, the bridge is gracefully elevated like a crescent moon. Its balustrades are intricately adorned with uniquely carved small lions. The bridge's design features a harmonious succession of curves, with each arch decreasing in size, creating a rhythm that echoes a grand chorus and infuses the structure with graceful dynamism. Inscriptions on the bridge, 'Lingwei Yanyue' on the north and 'Xiudong Lingbo' on the south, encapsulate its profound cultural significance and its remarkable aesthetic, academic, and practical values.

The Six Bridges of the Western Causeway (Fig. 13.12) in Beijing's Summer Palace are inspired by the Su Causeway of Hangzhou's West Lake, featuring six distinct bridges named Jiehu, Binfeng, Yudai, Jing, Lian, and Liu, arranged from north to south. Between the Liu and Lian Bridges is Jingming Lou, named after a line from Fan Zhongyan's 'Essay on Yueyang Tower' that extols the tranquil beauty of spring. The embankment is densely planted with peach and willow trees, which blossom into a vibrant spectacle of green and red in spring, endowing the area with the charm of 'Jiangnan of the North.

Fig. 13.11 Seventeen Hole Bridge in the summer palace, Beijing

a) Liu Bridge

b) Lian Bridge

c) Mirror Bridge

d) Jade belt bridge

e) Binfeng bridge

e) Jiehu lake bridge

Fig. 13.12 Six Bridges on the west embankment of the summer palace, Beijing

Yangzhou Wuting Bridge, also known as the Lotus Bridge, is located outside the North Gate of the Lianxing Temple (formerly known as the Fahai Temple) in Yangzhou City, Jiangsu Province, about 3 km from the city; spanning the Slender West Lake, it connects the south to He Garden and the north to Shou'an Temple, serving as a necessary passage for those visiting Guanyin Mountain and Pingshan Hall. The bridge was first constructed in the twenty-second year of the Qianlong era of the Qing Dynasty (1757), and is a uniquely designed stone arch bridge. Its elegant structure features yellow tiles and vermilion pillars, complemented by white balustrades and colorful painted caisson ceilings inside the pavilions, creating a grand and luxurious appearance. Inspired by the Five-Dragon Pavilions in Beijing's Beihai Park and lacking a natural waterway like that of Beihai, the artisans innovatively integrated pavilions and bridge to form a 'pavilion bridge,' consisting of five pavilions grouped together over the bridge structure. The bridge's body is constructed in an arched configuration, comprising three types and a total of 15 different arch openings, thereby establishing a series of ethereal arches atop the robust foundation of the bridge. Beneath the bridge lie four wings with 15 interconnected arch openings on the main and side facets (Fig. 13.13).

Shanghai Yuyuan Jiuqu Bridge was built during the Jiajing era of the Ming Dynasty (1522–1566), and is a renowned scenic bridge within Yuyuan Garden in Shanghai. Its structure is ingeniously designed with nine winding turns, allowing visitors to meander through, experiencing multiple visual shifts. This design significantly enhances the scenic beauty, making it an overwhelmingly enchanting sight (Fig. 13.14).

Fig. 13.13 Yangzhou Wuting Bridge

Fig. 13.14 The Nine Curve Bridge in Shanghai Yuyuan Garden

Fig. 13.15 Chengyang Bridge, Guangxi

Chengyang Bridge in Guangxi, also known as Yongji Bridge, was initiated in 1912, took 12 years to construct, and was completed in 1924. Located in Sanjiang Dong Autonomous County, Liuzhou City, in the Guangxi Zhuang Autonomous Region, it spans the Linxi River and showcases distinctive ethnic architectural elements as a stone pier and wooden beam bridge. The bridge features four spans and five pier platforms (built with stone masonry), extending a total length of 64.4 m. The bridge's wooden beams are comprised of eight rows of fir trees, each with a diameter of 1.6 feet, stacked in three layers to form the bridge deck. This is topped with planks, vertical posts, and tiles, creating a covered promenade-style walkway, hence known as a 'covered bridge' or 'wind and rain bridge' (as depicted in Fig. 13.15). Five pavilions stand on the bridge; the central one is a four-story hexagonal tower pavilion, flanked by two four-story square tower pavilions on each side, and two are hall-shaped pavilions. At first glance, the intricate arrangement of pavilions and towers presents a majestic and imposing sight.

Ancient Chinese bridges are an integral part of ancient Chinese civilization. "They embody the threefold value of cultural relics in terms of history, art, and science, with their scientific and technological aspects even more pronounced than those of magnificent palaces, altars, temples, and monasteries. Ancient bridges showcase the great achievements in science, technology, cultural arts created by humanity throughout the course of history, representing a highly valuable aspect of both tangible and intangible cultural heritage."

"Creating new lyrics with the tenderest rhymes, I softly sing as my flute intertwines. As the melody ends, along the Songling path I tread, looking back at the misty waves and fourteen bridges spread." The beauty of a bridge encompasses the beauty of scenery, form, craftsmanship, harmonious contrast, and the fusion of engineering and art, offering an intuitive beauty that resonates with the soul. A bridge is an amalgamation of practicality and artistry, infused with a wealth of cultural and emotional elements, showcasing diverse and splendid vistas that bring emotional pleasure and enjoyment to people. Joseph Needham, a renowned British historian of science and technology, once said, "No Chinese bridge lacks beauty, and many possess unique charms."

By the middle and late 19th century, with the introduction of railroads into China, many steel railroad bridges were built, with more than 3,000 bridges in preliminary statistics, mainly in the form of steel plate girder bridges, steel truss girder bridges, steel plate and truss hybrid bridges, steel and wood hybrid bridges, etc. Typical bridges include the Jiyun Canal Bridge (a 62-m span under-supported single-line steel truss bridge) on the Tang(Shan)-Xu(Ge Zhuang) Railway, built in 1888; the Songhua River Railway Bridge on the Harbin to Manzhouli Railway, completed in 1901 (1 hole 33.54-m simply-supported upper-bearing steel truss bridge + 8 holes 76.804-m simply-supported curved chord under-supported steel truss bridge + 10 holes 33.54-m simply-supported upper-bearing steel truss bridge); the Huailai Bridge on the (Beijing) Zhang(Zhangjiakou) Railway located between Kangzhuang and Zhangjiakou (7 holes 30.5-m simply-supported upper-bearing steel truss bridge); and the Zhengzhou Yellow River Bridge (Old Bridge) on the Pinghan Railway, built in 1905 (26 holes 31.5-m under-supported steel truss girder + 52 holes 21.5-m upper-supported steel plate girder + 24 holes 31.5-m under-supported steel truss girder).

Modern Bridges that Thrive

China's modern bridges can be recognized as having been built at the beginning of the 20th century. In 1901, the first automobile was imported into Shanghai, and in 1888, the first steel truss bridge was constructed in Tianjin. With its own bridge companies and engineering talents in bridge construction, China saw rapid development in the construction of railway bridges.

Shanghai Waibaidu Bridge, completed in 1907, was designed by the Cleveland Company of England. Located at the confluence of Suzhou Creek and the Huangpu River in downtown Shanghai, it is a lower-bearing, simply-supported riveted steel truss bridge (Fig. 13.16), with both spans measuring 52.178 m and a width of 18.4 m. Between 2008 and 2009, the bridge underwent a comprehensive relocation and overhaul, during which 63,000 rivets were replaced, accounting for about 40% of the original structure's total rivets. Additionally, the timber pile foundation was replaced with concrete piles, while maintaining the original appearance of the parts of the bridge piers above the river. Waibaidu Bridge, in conjunction with the Shanghai Tower, forms a harmonious and iconic reflection of Shanghai's architectural heritage spanning over a century.

The Zhongshan Bridge over the Yellow River in Lanzhou, Gansu (Fig. 13.17), was designed by Americans, constructed by Germans, and involved Chinese participation. It took three years to complete and was opened to traffic in the first year of the Xuantong era of the Qing Dynasty (1909). The bridge's superstructure consists of a through-type steel truss, standing 5.1 m high, with a total of 5 spans, each spanning 45.9 m and a total width of 8.46 m. The abutments on both the north and south sides

Fig. 13.16 Shanghai Waibaidu Bridge

Fig. 13.17 Zhongshan Yellow River Bridge, Lanzhou, Gansu, China

are made of cement mortar masonry, while the middle four piers are constructed from high-strength, rapid-setting cement in a gravity-style structure, with caisson foundations excavated down to bedrock. All materials and equipment required for the bridge's construction, from steam engines and cement steel bars to nuts and bolts, were imported from Germany. The bridge underwent comprehensive repair and reinforcement in 1954, including the addition of arched steel beams on top of the original parallel chords. To promote Yellow River tourism, the bridge was further reinforced in 2004, which included jacking up the entire superstructure, strengthening the piers by enlarging their cross-section, and replacing the bridge deck. As a result of raising the bridge deck, it now serves as a pedestrian bridge.

Fig. 13.18 Qiantang River Bridge

Qiantang River Bridge in Hangzhou, Zhejiang Province (Fig. 13.18) is the first double-decker bridge in China, designed and constructed indigenously, serving both highway and railway purposes with a steel truss girder structure. The lower deck accommodates the railway, stretching 1,322.1 m for single-line traffic, while the upper deck serves as a highway bridge, extending 1,453 m. The bridge has a total width of 9.94 m, with the roadway measuring 6.5 m wide. The Qiantang River Bridge stands as a milestone in Chinese bridge engineering history, being the first modern steel bridge masterminded, designed, and built by Chinese engineers, namely Mr. Mao Yisheng, the Director of the Engineering Department, and Mr. Luo Ying, the Chief Engineer, marking a monumental achievement in China's bridge construction history.

In the 1950s, with industries poised for development, China's transportation infrastructure saw dramatic advances, laying a solid foundation for the nation's sustainable economic growth. Leveraging Soviet technology, numerous small to medium-span concrete and stone arch bridges, as well as culverts, were constructed. The highway sector produced hundreds of sets of standard designs for prefabricated reinforced concrete plate girders and T-beams, as well as arch culverts. Designs and manufacturing for various types of bridge-building machines under 80 tons were undertaken, and specialized bridge construction teams were established. Significant bridges built during this period include the Yongjiang No. 1 Bridge in Nanning, Guangxi (1964, a 7 × 55 m reinforced concrete thin-walled box cantilever bridge), the Dan River Arch Bridge on the Taijiao Line (1959, with an 88 m span hollow-ribbed arch bridge), and the Hongshui River Bridge in Laibin, Guangxi (1978, a 105 m + 2 × 90 m reinforced concrete thin-walled box arch bridge). In 1957, with assistance from Soviet experts, the Wuhan Yangtze River Bridge was constructed, marking China's first dual-purpose bridge for both road and rail across the Yangtze River. By 1968, leveraging domestic expertise, the Nanjing Yangtze River Bridge was completed. Additionally, from the 1960s to the 1980s, prestressed concrete T-girder bridges, continuous beam bridges, truss arch bridges, box arch bridges, and rigid frame arch bridges were developed to varying extents. In 1964, a new bridge structure, the hyperbolic arch bridge, was introduced, with notable examples being the Xiangjiang River Bridge in Changsha, Hunan, and the highway approach bridge of the Nanjing Yangtze River Bridge, leaving a dazzling legacy in China's bridge engineering history.

Wuhan Yangtze River Bridge is a dual-purpose bridge for both railroad and highway use, with a total length of 1,670 m, including a 1,156-m main bridge, a 303-m north bank approach bridge, and a 211-m south bank approach bridge (Fig. 13.19). The upper deck serves as a highway bridge, featuring a two-way, four-lane design, 18 m in width, with pedestrian walkways 2.25 m wide on each side. The lower deck accommodates a double-track railway bridge, 14.5 m wide. The main bridge structure consists of three connected continuous segments, each comprising three spans (making a total of nine spans) with each span 128 m long, supported by eight piers. The carbon steel truss girders feature a diamond-shaped web and H-shaped cross-section. A pioneering method of pipe column drilling was used for the foundation construction. The Wuhan Yangtze River Bridge was one of the 156 projects aided by the Soviet Union in China, with construction commencing on September 1, 1955, and the bridge opening to traffic on October 15, 1957. As a key achievement of New China's First Five-Year Plan, the bridge's design was featured on the third set of Renminbi notes issued in April 1962, becoming a significant symbol of New China's national construction.

Nanjing Yangtze River Bridge, situated between the Gulou District and the Pukou District in Nanjing, is the first bridge across the Yangtze River that was independently designed and constructed by China. It serves both railroad and highway purposes

Fig. 13.19 Wuhan Yangtze River Bridge

and is a large double-deck steel truss bridge. Construction began in January 1960, with the railroad section opening to traffic on September 30, 1968, and the highway section following on December 29, 1968. The main bridge consists of 9 piers and 10 spans, stretching 1,576 m long with the largest span reaching 160 m. The upper deck accommodates a 15-m-wide, four-lane highway extending 4,589 m in length, while the lower deck features a 14-m-wide, double-track railway spanning 6,772 m (Fig. 13.20). The steel main girders of the bridge were constructed using domestically developed 16Mn low-alloy bridge steel, incorporating four types of foundations including pipe-column and caisson foundations. From October 2016 to December 2018, the bridge underwent a comprehensive maintenance and renovation while closed to traffic. The Nanjing Yangtze River Bridge holds significant importance in both the history of Chinese and world bridges, symbolizing a major achievement in China's economic construction and marking a critical milestone in Chinese bridge engineering.

The Changsha Xiangjiang Bridge in Hunan, now known as the "Orange Island Bridge," was completed in 1972. Spanning the Xiangjiang River and crossing over Orange Island in Changsha City, Hunan Province, it stands as China's largest highway double-arch bridge (Fig. 13.21). The total length of the bridge is 1,532 m, with the main span comprising 8 × 76 m plus 9 × 50 m. The east bank approach bridge consists of four spans ranging from 18 to 27 m, making the main bridge 1,250 m long and 20 m wide. A branch bridge extends from the middle of the main bridge towards the south side of Orange Island, with spans of 6 × 30 m plus 4 × 20 m, and is 282 m long and 8 m wide. In 1981, the design of the Changsha Xiangjiang Bridge was honored with the National Excellent Design Project Award for the 1970s.

In the 1970s, China embraced modern construction technologies for cable-stayed bridges, suspension bridges, arch bridges, and the use of prestressing techniques. In February 1975, China's first experimental cable-stayed bridge, the Yunyang Cable-Stayed Bridge in Sichuan (a double-tower concrete cable-stayed bridge with spans of 34.91 m + 75.84 m + 34.91 m), was completed. However, due to the construction

Fig. 13.20 Nanjing Yangtze River Bridge

Fig. 13.21 Hunan Changsha Xiangjiang Bridge

of the Three Gorges Dam project, this bridge was demolished in October 2006. In 1977, the Minghe Bridge in Hebei, the first highway prestressed concrete continuous girder bridge with a span of 45 m, was completed. 1978 saw the completion of the country's first prestressed concrete bridge designed to withstand high seismic intensity—the Tangshan Xinluanhe Bridge (a series of six 4×40 m prestressed concrete T-shaped continuous girders). In 1982, Shanghai saw the construction of its first double-tower, double-cable-plane prestressed concrete cable-stayed bridge—the Shanghai Mao'gang Bridge with spans of 85 m + 200 m + 85 m, which was demolished in 2020.

The 1980s marked a renaissance in bridge construction in China. Guided by the insistence, suggestions, and support of Professors Li Guohao and Xiang Haifan from Tongji University, Shanghai led the way with the construction of the Shanghai Nanpu Bridge, a cable-stayed bridge independently designed and constructed by Chinese engineers. This was followed by the construction of the Yangpu Bridge, signaling the beginning of an era in which China independently designed and built mega-span bridges over major rivers. Several cable-stayed bridges with spans exceeding 600 m were constructed, including the Qingzhou Minjiang Bridge with a main span of 605 m featuring stacked beam diamond-shaped towers (completed and accepted on December 25, 2002), the Wuhan Second Yangtze River Bridge with a main span of 618 m and mixed beam diamond-shaped towers (opened in June 1995), and the Nanjing Second Yangtze River Bridge (now known as the Nanjing Baguazhou Yangtze River Bridge) with a main span of 628 m and steel box girder diamond-shaped towers (opened on March 26, 2001). Concurrently, developments in concrete cable-stayed bridges like the Jinan Yellow River Bridge propelled the rapid advancement of China's long-span cable-stayed bridge construction into the forefront of the global stage.

Nanpu Bridge, located in the Dongjiadu area of Huangpu District, Shanghai, commenced construction on December 25, 1988, and opened to traffic on December 1, 1991. Spanning a total length of 8,364 m, the bridge serves as a two-way six-lane urban expressway. The main bridge is 846 m long with a span configuration of 40.5 m + 76.5 m + 94.5 m + 423 m + 94.5 m + 76.5 m + 40.5 m, and a width of 30.35 m. It stands as a double-tower, double cable-plane, stacked beam cable-stayed bridge independently designed and constructed by China. The towers of the main bridge are in a folded H-shaped reinforced concrete structure, supported by high-strength galvanized steel wire stay cables. The east and west approach bridges measure 3,746 m and 3,754 m in length, respectively (Fig. 13.22).

Yangpu Bridge serves as a vital artery in Shanghai, linking Yangpu District with Pudong District. Construction commenced on May 1, 1991, and the bridge was inaugurated on October 23, 1993, featuring a two-way, six-lane urban expressway. The bridge spans a total length of 8,354 m, with the main bridge extending 1,172 m. It is a double-tower, double cable-plane, composite girder cable-stayed bridge, with span arrangements of 40 m (Puxi transition span) + 99 m + 144 m + 602 m + 144 m + 99 m + 44 m (Pudong transition span), and the main bridge deck is 30.35 m wide. Upon completion, the Yangpu Bridge (Fig. 13.23) boasted the world's longest span for a composite girder cable-stayed bridge at that time.

Jinan Yellow River Bridge was officially started in December 1978 and opened to traffic in July 1982. The total length of the bridge is 2022.8 m, with a width of 19.5 m, of which the traffic lane is 15 m. The main bridge consists of a 488-m, 5-span continuous prestressed concrete box girder system with a double-tower, double-cable-plane cable-stayed bridge. The span arrangement is 40 + 94 + 220 + 94 + 40 m, featuring a closed twin-cell box girder, A-shaped portal towers, and a

Fig. 13.22 Nanpu Bridge

Fig. 13.23 Yangpu Bridge

fan-shaped cable arrangement, constructed using the hanging basket method. The approach bridge totals 1534.8 m, with the south approach comprising 24 spans and the north approach 27 spans, both utilizing pre-tensioned 30-m span prestressed composite girder bridges. At that time, it was the longest-span prestressed concrete cable-stayed bridge in Asia (Fig. 13.24).

Ting Kau Bridge, Hong Kong spans the Rambler Channel and is a 3-tower, 4-span large cable-stayed bridge with a span configuration of 127 m + 448 m + 475 m + 127 m. Designed by German engineer J. Schlaich, it was completed and opened to traffic in May 1998. The bridge deck is uniquely divided into two sections, each situated on either side of the single-column towers, and is balanced by the lower stay cables. Stay cables connect the top of the main tower to the base of the adjacent towers, giving the bridge a more slender appearance that also contributes to superior wind resistance during strong storms. It is a masterpiece of combined structural engineering and represents a perfect union of art and technology. The bridge has been recognized as one of the most beautiful bridges of the 20th century.

Prestressed concrete girder bridges have been a fundamental bridge structure in China. Built in 1971, the Fuzhou Wulong River Bridge is a large-span prestressed concrete T-shaped rigid frame with hanging beams in the mid-span. The span configuration is 58 m + 3 × 144 m + 58 m, with a total length of 552.22 m and a total width of 12 m. The superstructure was constructed using two methods: cantilever assembly and cantilever casting. The Shayang Han River Bridge in Hubei, completed and opened to traffic on July 1, 1985, features a 62.4 m + 6 × 111 m + 62.4 m 8-span continuous prestressed concrete girder bridge, spanning 792.7 m in length and 12 m in width. The superstructure comprises a variable cross-section single-box single-cell girder, constructed with longitudinal and vertical bi-directional prestressing reinforcement, using cantilever casting. In 1988, China's first large-span prestressed concrete continuous rigid frame bridge, the Guangdong Panyu Luoxi Bridge (Fig. 13.25), was completed, marking the entry of China's prestressed concrete bridge construction into the international advanced ranks. In 1997, the auxiliary channel bridge of the Humen Bridge in Guangdong was completed with a span arrangement of 150 m + 270 m + 150 m. As a 3-span prestressed concrete continuous rigid frame bridge with a single-box, single-cell cross-section and dual-column hollow thin-walled piers, it was recognized as the world's largest span continuous rigid frame bridge at the time of completion.

Traditional stone bridges have also seen enhancements and developments. In July 2000, the Dan River Bridge (Fig. 13.26), the largest span stone arch bridge in China

Fig. 13.24 Ting Kau Bridge, Hong Kong

Fig. 13.25 Rock Creek Bridge

Fig. 13.26 Shanxi Danhe Bridge

at the time, was completed in Jincheng City, Shanxi Province. The bridge features a full hollow-section, variable cross-section, catenary curve, hingeless stone arch structure, with spans of 2 × 30 m + 146 m + 5 × 30 m. The main span's clear span is 146 m, with a clear rise of 32.444 m, yielding a rise-to-span ratio of 1/4.5. The bridge has a deck width of 24.8 m and a height of 80.6 m. The balustrade is adorned with over 200 stone carvings depicting the history and culture of Jincheng City and nearly 300 traditional stone lions, symbolizing a harmonious blend of modernity and traditional culture.

Steel pipe concrete arch bridges, originating from Japan and other countries, have been developed in China due to their small construction load, ease of cantilever assembly, and cost-effectiveness. In April 1991, the first steel pipe concrete tie arch bridge with a main span of 115 m, featuring an equal cross-section and a dumbbell-type catenary curve, was constructed in Wangcang, Guangyuan, Sichuan. This was followed by the construction of hundreds of steel pipe concrete arch bridges, including the notable Guangzhou Yabunsha Bridge. This 3-span continuous, self-anchored, center-supported truss arch bridge, with a main span of 360 m, was completed in 2000 and became a prominent example. Additionally, reinforced concrete box arch bridges employing rigid frameworks were developed. In June 1997, the Chongqing Wanzhou Yangtze River Bridge, formerly known as the Wanxian Yangtze River Bridge (Fig. 13.27), was completed. It spans 856 m in length and 23 m in width, with a four-lane configuration in both directions. The main bridge is a reinforced concrete box arch bridge, crossing the river in a single span of 420 m without underwater foundations. The main arch, with a 1/5 rise-to-span ratio and a catenary arch axis, features a steel pipe concrete-concrete composite structure with a three-chamber box section. It measures 7 m in height and 16 m in width, with a width-to-span ratio of 1/26.25 and a height-to-span ratio of 1/60. Upon completion, it held the record for the longest span among similar bridge types worldwide.

China's innovative prestressed concrete truss-arch composite bridge was validated with the construction of the Jiangjie River Bridge (Fig. 13.28), spanning the Wujiang River at Weng'an County, Qiannan Buyi and Miao Autonomous Prefecture, Guizhou Province. The bridge's span configuration is 20 m + 25 m + 30 m + 330 m + 30 m + 20 m, with a total length of 461 m, a width of 13.40 m, and a height of 263 m. The main span of 330 m was, at that time, the world's largest concrete truss bridge. The Jiangjie River Bridge was completed and opened to traffic in June 1995.

Jiangxi Jiujiang Yangtze River Bridge is located over the Yangtze River, spanning between Xunyang District in Jiujiang City, Jiangxi Province, and Huangmei County in Huanggang City, Hubei Province. It is a double-deck, dual-line bridge serving both railroad and highway purposes. The highway bridge was completed and opened to

Fig. 13.27 Wanzhou Yangtze River Bridge

Fig. 13.28 Jiangjie River Bridge

traffic in January 1993, and the railroad bridge formally opened for operation in June 1995. Upon completion, it was the largest dual-use road-rail bridge on the Yangtze River. The upper level is a four-lane highway, 14 m wide, with sidewalks 2 m wide on both sides, and a total length of 4460 m. The lower level accommodates a double-track railroad, with a total length of 7675.4 m. The main bridge comprises a 4-span 11-section steel truss girder, arranged from north to south as follows: 3 × 162 m continuous steel truss girders, a flexible arch structure of 180 m + 216 m + 180 m steel truss girders, and 2 × 162 m continuous steel truss girders.

Guangdong Humen Bridge. Humen Bridge is the first modern suspension bridge in mainland China. Construction of the Humen Bridge began on October 28, 1992, and it was opened to traffic on June 9, 1997. The main channel bridge of the Humen Bridge (Fig. 13.29) features a single-span 888 m suspension bridge design, with a main cable sag-to-span ratio of 1/10.5. The main cable has a diameter of 687.2 mm and a center-to-center distance of 33 m. The stiffening girders are made of flat, closed, streamlined steel box sections, with the box girder having a total width of 35.6 m and a height of 3.012 m at the center of the bridge axis.

Jiangyin Yangtze River Highway Bridge. The Jiangyin Yangtze River Highway Bridge (Fig. 13.30) is a critical river crossing project that connects the national trunk lines from Tongjiang to Sanya and Beijing to Shanghai. Construction began on November 22, 1994, and the bridge was opened to traffic on September 28, 1999. The main bridge consists of a 336.5 m + 1385 m + 309.34 m orthogonal anisotropic plate steel box girder suspension bridge, with a sag-to-span ratio of 1/10.5. The box girder has a height of 3 m and a width of 32.5 m, with the total width including the cantilevered fairing reaching 36.9 m. The cable towers stand approximately 190 m tall, and the main cable is composed of 169 strands, each containing 127 parallel high-strength galvanized steel wires with a diameter of 5.35 mm and a tensile strength of 1600 MPa. The approach bridges feature spans of 30 and 50 m using prestressed

Fig. 13.29 Humen Bridge main channel bridge

Fig. 13.30 Jiangyin Yangtze River highway bridge

concrete simply supported beams. The Jiangyin Yangtze River Highway Bridge is China's first mega-scale steel box girder suspension bridge with a span exceeding 1,000 m, marking a significant milestone.

Shantou Bay Bridge, Guangdong. The Shantou Bay Bridge (Fig. 13.31) was completed in December 1995. The main bridge features a main span of 452 m with a pre-stressed concrete box girder design. The bridge deck has a clear width of 23.8 m. The main towers, standing at 95.1 m, are constructed from reinforced concrete and utilize gravity anchorages. The main cable has a sag-to-span ratio of 1/10 and is assembled using the Prefabricated Parallel Wire Strand (PWS) method. The hangers are made from twisted galvanized steel wires, attached with cold-cast anchors, and are spaced 6 m apart.

Tsing Ma Bridge, Hong Kong. The Tsing Ma Bridge (Fig. 13.32) opened in May 1997, spanning the strait between Tsing Yi and Ma Wan and connecting Hong

Fig. 13.31 Shantou Bay Bridge

Fig. 13.32 Tsing Ma Bridge, Hong Kong

Kong's Lantau International Airport to the urban areas. The total length of the bridge is 2160 m, with the main span being 1377 m. It was the world's longest dual-purpose suspension bridge for both road and rail traffic at the time of its completion. The stiffening girders are a hybrid structure of steel trusses and steel box girders, with a width of 41 m. The upper deck accommodates six road lanes, while the lower level within the steel box girder is designed for railway traffic, including two emergency lanes for use during typhoons. The bridge towers reach a height of 206 m, and the navigational clearance below the bridge is 79 m.

Entering the 21st century, China initiated another peak in bridge construction. While continuing to build bridges across major rivers and mountain ranges, the focus has gradually shifted towards sea-crossing bridges. The opening of the Hangzhou Bay Bridge in 2008 and the Hong Kong-Zhuhai-Macao Bridge in 2018 marked new pinnacles in China's bridge construction achievements. Simultaneously, the design and building technologies for various other bridge types have seen comprehensive enhancements. The full-scale development and application of new materials, technologies, techniques, and equipment have further elevated the quality of bridge construction, positioning China among the world's leading nations in bridge engineering.

Donghai Bridge. Construction of the Donghai Bridge commenced on June 26, 2002, and it was inaugurated for traffic on December 10, 2005. It stands as China's inaugural sea-crossing endeavor built over the vast offshore sea area. The bridge adheres to the standards of a two-way, six-lane highway, featuring an emergency stopping lane, measuring 31.5 m in width, and designed for speeds up to 80 km/h. The main channel bridge is characterized by a single-plane cable system supported by two inverted Y-shaped towers, each 140.2 m tall, with a principal span of 420 m. Additionally, the Zhushan section of Donghai Bridge is a dual-tower, dual-plane cable-stayed bridge with a steel–concrete composite beam, having a main span of 332 m.

Hangzhou Bay Bridge. Construction of the Hangzhou Bay Bridge began on June 8, 2003, and it was officially opened on May 1, 2008. Spanning from Zhengjiadai in Haiyan, Jiaxing City, Zhejiang Province, to Shuiluwan in Cixi, Ningbo City, the bridge stretches over 36 km. It comprises the north and south channel bridges along with the approach bridges. The north channel bridge features a double-tower steel box cable-stayed design with spans of 70 m + 160 m + 448 m + 160 m + 70 m, whereas the south channel bridge is a single-tower steel box girder cable-stayed structure with spans of 100 m + 160 m + 318 m. The sea approach bridges have spans of 70 m each, totaling 18.27 km in length, and the southern tidal flat approach bridges have 50 m spans, extending 10.1 km. Both the 50 and 70 m beams are factory-made, prestressed concrete box girders, initially simply supported and subsequently made continuous. Continuous box girders with spans ranging from 30 to 80 m are used on the landsides and the northern tidal flats.

Hong Kong-Zhuhai-Macao Bridge. Spanning the Lingdingyang area at the Pearl River's mouth, the Hong Kong-Zhuhai-Macao Bridge (Fig. 13.33) officially opened on October 24, 2018. The comprehensive bridge-tunnel system stretches 55 km, comprising 22.9 km of bridge structures and a 6.7 km undersea tunnel. Key features include a double-tower cable-stayed bridge with a 458 m main span over the Qingzhou Channel, a double-tower cable-stayed bridge with a 268 m main span over the Jiuzhou Channel, and a triple-tower cable-stayed bridge with a 258 m main span for the River-Sea Direct Shipping Channel. The approach bridges consist of a 15 km steel box girder with 110 m spans and a 5 km section with 85 m spans. Designed to highway standards, the bridge accommodates six lanes in each direction and measures 33.1 m in width.

Sutong Yangtze River Highway Bridge. Situated between Nantong City and Suzhou (specifically Changshu City) in Jiangsu Province, the Sutong Yangtze River Highway Bridge (Fig. 13.34) spans the Yangtze River and was inaugurated in 2008. The bridge extends over 8206 m, adhering to the dual-carriageway, six-lane highway standard. Its main navigation span features a double-tower, double-plane cable-stayed bridge with a steel box girder, spanning 2088 m with segments of 100 m + 100 m +

Fig. 13.33 Hong Kong-Zhuhai-Macao Bridge

Fig. 13.34 Sutong Yangtze River Highway Bridge

300 m + 1088 m + 300 m + 100 m + 100 m. The towers rise 300.4 m, providing a navigational clearance of 62 m, supported by 131 large-diameter, ultra-deep drilled piles and a foundation slab covering 5471.375 m^2 (113.75 m × 48.1 m). The auxiliary bridge comprises a 268 m pre-stressed concrete continuous rigid frame. The Sutong Yangtze River Highway Bridge is distinguished as the world's first cable-stayed bridge to exceed a 1000-m main span.

Shanghai Lupu Bridge (Fig. 13.35). Located over the Huangpu River, between the former Luwan District (now part of Huangpu District) and the Pudong New Area in Shanghai, construction of the Lupu Bridge began in October 2000 and it was opened to traffic on June 28, 2003. The bridge serves as a two-way, six-lane urban expressway. The main span of the Lupu Bridge is designed as a steel arch bridge with a basket-handle tied-arch configuration, featuring spans of 100 m + 550 m + 100 m. This single-span structure crosses the river with a sag-to-span ratio of 1/5.5. The bridge's deck follows a vertical curve with a radius of 9,000 m. The arch ribs measure 5 m wide at the top and 3 m at the base, with the total bridge width being 29.8 m.

Wuhu Yangtze River Bridge. Completed in September 2000, the Wuhu Yangtze River Bridge (Fig. 13.36) features a railway bridge that stretches 10,624.4 m and a highway bridge that spans 6,078.4 m, with a combined rail-road section extending 2,668.4 m and the main bridge's steel girders measuring 2,193.7 m in length. The primary channel is spanned by a cable-stayed bridge with towers of reduced height, featuring spans of 180 m + 312 m + 180 m, while the secondary channel is crossed by a continuous truss girder bridge consisting of four spans, each 144 m long. The main span of 312 m on the short-tower cable-stayed bridge surpasses the 300-m span milestone for heavy-load railway bridges in China. In its category and under similar loads, this plate-truss composite short-tower cable-stayed bridge holds the world record for span length.

Fig. 13.35 Shanghai Lupu Bridge

Fig. 13.36 Wuhu Yangtze River Bridge

Nanjing Yangtze River Second Bridge, now known as Nanjing Baguazhou Yangtze River Bridge (Fig. 13.37), was completed in March 2001, located 11 km downstream from the existing Nanjing Yangtze River Bridge. The southern branch features a 5-span continuous steel box girder cable-stayed bridge, extending 1,238 m with spans of 58.5 m + 246.5 m + 628 m + 246.5 m + 58.5 m. At the time, it boasted the largest span among China's cable-stayed bridges, with a deck width of 32 m, excluding the cable-stayed anchorage zones. The northern branch comprises a 5-span continuous beam bridge with a three-way prestressed box girder system, featuring spans of 90 m + 3 × 165 m + 90 m.

The Chongqing Shibanpo Yangtze River Bridge, completed in 2007, features a main span of 330 m, incorporating a hybrid continuous rigid-frame and combination structure system of steel and concrete. To reduce the structural weight and improve the span capability of the continuous rigid-frame bridge, a 108-m section in the center of the mid-span utilizes steel box girders. The bridge's design philosophy and construction techniques are recognized for their innovation.

Fig. 13.37 Nanjing Yangtze River Second Bridge

The Chongqing Chaotianmen Yangtze River Bridge features a main bridge designed as a three-span continuous center-bearing steel truss arch bridge, with spans of 190 m + 552 m + 190 m. This bridge incorporates a double-decker layout catering to both vehicular and rail transport. The upper deck accommodates a two-way, six-lane urban roadway flanked by pedestrian sidewalks, spanning a total width of 36.5 m. The lower deck centrally houses a dual-track urban rail system and was officially opened on April 29, 2009.

The Nanjing Dasheng Pass Yangtze River Bridge features a main bridge consisting of a six-span continuous box girder-arch combination, with dual main spans of 336 m each. This makes it the bridge with the largest span among its type worldwide, with a width of 41.6 m. It accommodates two high-speed rail lines, two Class I rail lines, and two Nanjing subway lines, totaling six lines, making it the highest load standard high-speed railway bridge in the world. The bridge was officially opened in January 2011.

The Zhoushan Xihoumen Bridge, part of the Zhoushan archipelago connection project, is a two-span continuous, split-type, fully floating steel box suspension bridge. It features a main span of 1,650 m and a total length of the steel box girders of 2,220.8 m, with the girders standing 3.26 m high and spanning 37.08 m in width. The bridge incorporates a central slot of 6.0 m and a novel separation of the double-box section, a design first for suspension bridges worldwide. The bridge commenced operations in December 2009.

The main spans of both the Jiangsu Taizhou Yangtze River Highway Bridge, opened on November 25, 2012, and the Anhui Maanshan Yangtze River Highway Bridge, which commenced operations on December 31, 2013, are designed as 1,080-m, three-tower, two-span steel box girder suspension bridges.

The Jiangsu Wufengshan Yangtze River Bridge, which opened to traffic on June 30, 2021, has a main span of 1,428 m. The upper level accommodates a dual-carriageway, eight-lane expressway with the bridge deck's total width at 46 m, and the lower level features a railroad line with a 4.6-m track spacing. The bridge's design includes a steel truss main girder suspension structure with spans of 84 m + 84 m + 1092 m + 84 m + 84 m. The stiffening trusses have a center width of 30 m, a height of 16 m, and a sectional spacing of 14 m.

The Shanghai-Suzhou-Tongzhou Yangtze River Bridge, which was inaugurated on July 1, 2020, spans a total length of 11.072 km, including a combined public-railway bridge segment of 6,989 m. The upper deck serves a two-way, six-lane highway (part of the Tongxi Expressway), while the lower deck accommodates a two-way, four-track railway. The main navigational span features a dual-purpose steel truss girder double-tower cable-stayed bridge with spans of 140 m + 462 m + 1092 m + 462 m + 140 m and main towers soaring to a height of 330 m.

On December 28, 2020, the Guangxi Pingnan Third Bridge (Fig. 13.38), holding the title for the world's largest span steel tube concrete arch bridge, was officially inaugurated. The bridge stretches a total of 1,035 m, featuring a main span of 575 m and a deck width of 36.5 m, accommodating two-way, four-lane traffic. Several construction techniques employed in building this bridge have set new global bench-marks for arch bridge construction, marking a significant leap from China matching to leading in bridge construction technologies.

The construction of these high-quality bridges, which reflect the highest level of bridge construction in the world today, signifies that the overall construction level of Chinese bridges has crossed into the top ranks of the world.

Fig. 13.38 Three bridges in Pingnan, Guangxi

Tunneling Projects Throughout the Country

China's landscape is characterized by its ruggedness, with mountains, plateaus, and hills making up approximately 67% of its land area, alongside numerous mountain ranges, rivers, and an expansive maritime territory. This complex and diverse topography has presented significant challenges to the development of transportation infrastructure. In "The Difficulty of the Shu Road," the poet Li Bai famously likened the treacherous paths through the ancient Shu territory to the daunting task of ascending to the heavens. Faced with towering and precipitous mountains in ancient times, people were left with no choice but to traverse the peaks or construct plank roads along the mountainsides for passage.

Humans have known for a long time to utilize natural caves for shelter. When society developed to the point of being able to make digging tools, there were tunnels dug by hand, but the exact date could not be traced. Constrained by economic and technological means, the construction conditions of tunnels were extremely harsh, safety issues were prominent, labor intensity was enormous and time-consuming, so the number of tunnels built in ancient times was very small.

The earliest documented instance of an underground man-made structure in China dates back to the early Eastern Zhou Dynasty (around 700 BC), as described in "Zuo Zhuan." The text recounts the famous anecdote where Duke Zhuang of Zheng dug through the earth to springs, creating a tunnel to meet his mother. The backstory involves Duke Zhuang's mother, Lady Jiang, who tolerated the rebellion of Duke Zhuang's brother, Gong Shu Duan. After Gong Shu Duan's unsuccessful coup and subsequent flight to the state of Gong, Duke Zhuang, in a moment of anger, exiled his mother to Ying City (now northwest of Linying County, Henan Province) and vowed never to see her again "unless in the Yellow Springs" (an expression referring to the afterlife). However, Duke Zhuang later regretted this decision and longed to see his mother. A clever official named Ying Kao Shu devised a solution that allowed Duke Zhuang to keep his vow while still reuniting with his mother. He dug a tunnel more than ten feet deep until water sprung forth, then constructed a wooden room beside the spring to symbolize the Yellow Springs, allowing the mother and son to meet "in the afterlife." Overjoyed by the reunion, Duke Zhuang famously exclaimed, "In this great tunnel, such joy abounds!" He then brought his mother back to the palace. The record suggests that this structure was an underground building made of wood, and it was well-constructed.

Many imperial tombs and accompanying burial sites throughout China's history were constructed as underground projects. The term "tunnel" in ancient Chinese initially referred to a tomb passage, a sloped path leading from the surface to the burial chamber. In ancient texts, this was known as "sui", also referred to as "enviable path" or "tunnel". Given the size and weight of the ancient emperors' and nobility's coffins, a sloped access was necessary for their transportation into the tombs. Tomb passages exposed on the surface without earthen cover were known as enviable paths, while those covered with earth were termed tunnels. Historical records, such as in "Nan Shi" (Southern History), recount instances emphasizing the use of these tomb passages.

For example, Wang Yi's biography mentions an emperor humorously noting his passage through such a tunnel as if entering a family tomb. Wang Anshi, in his epitaph for Ma Gong, speaks of setting inscriptions in a tomb passage to ensure their enduring legacy. From the Shang Dynasty through the Qing Dynasty, a vast number of tombs and tomb passages were constructed, including the globally renowned Terracotta Army pits of the Qin Shi Huang Mausoleum, which were created using the open-cut method. These pits, with a total length of approximately 2.5 km and a cross-section of around 3.7 m by 3.2 m, employed a wooden half-frame structure without mid-supports, characterized by large logs and mats on the beam tops, and without backing behind the columns. The construction of these imperial tombs represents the pinnacle of ancient underground engineering, serving as a precious heritage of underground engineering technology.

In the ancient art of grottoes, a significant number of tunnels were constructed, with architectural feats such as the Dunhuang Mogao Grottoes, Longmen Grottoes, and Yungang Grottoes achieving high levels of accomplishment. The Karez system represents an ancient marvel of hydraulic engineering, consisting of underground artificial water channels found throughout northwest and north China. Its design involves excavating horizontal tunnels from a water source to guide the water to the surface, primarily for the irrigation of farmland.

Before the invention of gunpowder, ancient Chinese primarily employed basic tools such as hammers, chisels, and wedges for rock excavation. Subsequently, leveraging the physical properties of rocks, they developed techniques like the fire-setting method, where rocks were heated and then cooled abruptly with water or vinegar to induce thermal shock and cracking, facilitating their removal. This involved using highly flammable pine or cypress as fuel to burn the rock surfaces for 4–5 h, followed by quenching the heated surfaces with water or vinegar, causing them to cool suddenly and crack naturally. The cracked pieces were then pried away using iron tools. This fire-quenching technique, also known as thermal shock fracturing, was a commonly used method in China's history, notably by Li Bing, the governor of Shu County during King Zhao of Qin's reign, for major water conservancy projects in Sichuan.

The Shimen Tunnel, located on the ancient Baoxie Road, is recognized as China's earliest tunnel constructed for transportation purposes. Situated beneath Qipan Mountain (also known as Jitouguan) at the Baogukou of Hanzhong in Shaanxi Province, it was completed in the ninth year of Emperor Ming of the Eastern Han Dynasty's Yongping reign (66 AD). The tunnel's construction is commemorated by stone inscriptions, "Ode to the Stone Gate" and "Inscription on the Stone Gate," detailing its history.

The Baogukou in Hanzhong, a critical and perilous passage on the Baoxie Road, is characterized by a steep, towering rock wall on the Baoshi River's southern bank, known as "Baoping." The rapid river currents made it challenging to construct plank roads in this area. Travelers, hindered by the Qipan Mountain's obstacles, were forced to detour, navigating around the mountain. In response, Emperor Ming of the Han Dynasty, Liu Zhuang, issued an imperial edict in the fourth year of Yongping (61 AD) to carve a tunnel through the mountain obstructing the plank road's path. The project, led by Chu Jun, the prefect of Hanzhong County, began with extensive preparations.

Over two thousand workers were mobilized in the sixth year of Yongping (63 AD), and the tunnel was completed by the ninth year (66 AD). Historical records reveal that this undertaking involved the labor of 2,690 individuals from Guanghan, Shu County, and Ba County, aimed at facilitating the Baoxie Road. The Shimen Tunnel runs north to south, parallel to the Baogu River, aligning the tunnel floor with the plank road level. It features varying heights with a difference of 30–50 cm from south to north. The east wall measures 16.5 m in length, while the west wall extends 15 m; the north entrance is 3.75 m high and 4.1 m wide, compared to the south entrance, which is 3.45 m high and 4.2 m wide. The tunnel's inner walls exhibit no tool marks, indicating a smooth, well-finished rock surface. According to the "Compendium of Ancient Monuments of Baogu Valley," which includes verses from the Qing Dynasty, the Shimen was constructed using the fire-setting technique. This method involved utilizing abundant firewood from the Baoshi River's banks to heat and fracture the stone, followed by manual removal with hammers. As the earliest transportation tunnel in China with documented use of the "fire-setting method," the Shimen Tunnel also stands as the oldest known manually excavated transportation tunnel worldwide that accommodated vehicular traffic (Fig. 13.39).

At that time, the term 'tunnel' was not used, so it was metaphorically referred to as a 'stone gate', hence the name Stone Gate of the Trestle Trail. In the second year of the Yongping era (509 AD) during the Northern Wei Dynasty, Wang Yuan's 'Stone Gate Inscription' records: "This gate was indeed carved out during the Yongping period of the Han Dynasty." According to Li Daoyuan's 'Commentary on the Water Classic' from the Northern Wei Dynasty, "The Baoshui River further flows southeast through the smaller Stone Gate, with the passage being over six zhang (a unit of length) through the mountain," referring to the Shimen Tunnel. Shimen Tunnel is a remnant of the valuable ancient architecture; this great project is the first man-made tunnel through the mountain in the history of mankind. Baoxie Path's Stone Gate along with its cliffside stone inscriptions was designated in 1961 as one of the first batch of national key cultural relics protection units. Unfortunately, during the construction of the Shimen Reservoir from 1969 to 1971, the tunnel was submerged. To protect the cultural relics, some of the exquisite cliff stone carvings were carefully removed and are now displayed in the Hanzhong Museum.

Since its opening during the Yongping era of the Eastern Han Dynasty, Shimen has witnessed generations of officials, merchants, scholars, and poets who, inspired by the majestic views, recorded events, composed poetry, and expressed their thoughts by engraving them on the cliffs both inside and outside of Shimen. This tradition has given rise to the magnificent spectacle of Shimen cliff carvings. According to a cultural relics survey conducted in 1960, there are 104 types of inscriptions at the Shimen site, with 34 types found on the inner walls alone, spanning from the Han and Wei dynasties to the Ming and Qing dynasties. The 13 types from the Han and Wei periods are particularly renowned, collectively known as the "Han-Wei Thirteen Inscriptions" or "Shimen Thirteen Inscriptions" (Figs. 13.40), and are highly esteemed by scholars and calligraphers across the ages.

Fig. 13.39 Shimen Tunnel

Fig. 13.40 Thirteen pieces of Shimen

The invention of gunpowder revolutionized tunneling engineering, making the drilling and blasting method the primary approach. China's ancients made significant contributions to tunnel construction. Before gunpowder's advent, ancient techniques involved using hammers, steel chisels, and primitive methods like fire-setting and quenching with water for mining and tunneling. Around the 7th century, Tang Dynasty pharmacologist Sun Simiao documented the production of black powder. After 1225, China's method of producing gunpowder spread to Islamic countries through India and reached Europe in the late 13th century. In 1627, Austrian industrialists first used gunpowder for mining; by 1666, France employed it in excavating the Languedoc Canal tunnel, possibly the first public tunnel excavated with gunpowder. Until the 1860s, tunnel construction worldwide predominantly involved manual drilling and blasting with black powder. The invention of the rock drill in 1851 marked the onset of modern construction methods, facilitating the construction of numerous tunnels. In 1861, during the construction of the Simplon Railway Tunnel through the Alps, pneumatic rock drills were first used, replacing manual drilling. In 1867, during the construction of the Hoosac Railway Tunnel in the USA, nitroglycerin explosives were introduced, replacing black powder. In 1866, Swedish inventor Alfred Nobel developed dynamite, significantly aiding the excavation of hard rock. In 1872, dynamite was first used in constructing the St. Gotthard Tunnel in south-central Switzerland, one of the world's most renowned tunnels. This potent yellow explosive replaced black powder, easing the excavation of hard rock and becoming widely used in tunnel engineering, thereby advancing tunnel construction technology and efficiency. After nearly a century of effort, China has developed today's high-efficiency, large-scale multi-head jumbo drills.

From 1874, when China began constructing its first narrow-gauge railroad from Shanghai to Wusong, until the overthrow of the Qing Dynasty in 1911, China had built approximately 9,100 km of railroads over more than 30 years. In this period, on the 10 main railroad lines constructed, totaling about 4,600 km, there were more than 230 tunnels with a combined length of about 42 km. The first railway tunnel in Chinese history, the Shiqiuling Tunnel, is located in Keelung, Taiwan. Construction started in the fourteenth year of the Guangxu Emperor (1888), and it was completed two years later in 1890, spanning a total length of 261 m. Between 1898 and 1904, the Xing'anling Tunnel was constructed, stretching 3,078 m, and was the longest broad-gauge railway tunnel in Asia at that time. The most significant tunnel project of this era was the Badaling Tunnel on the Beijing-Zhangjiakou Railway, personally designed and overseen by the eminent engineer Zhan Tianyou. It measures 1,091 m in length, was constructed in just 18 months, and was completed in 1908. This tunnel also marked China's first successful construction of a trans-mountain railway tunnel.

From October 1911 to October 1949, China constructed over 370 railway tunnels totaling about 100 km in length on more than 40 main and branch railway lines, covering approximately 7,000 km. Among these, the Ducao Tunnel on the second line of the Bin-Sui Railway was the longest in China at the time, measuring 3,840 m. Constructed between 1939 and 1941, this tunnel passed through granite strata and was built using the top-down guide pit method, with concrete lining for support.

Since the 1990s, tunnel construction in China has seen a phase of rapid development. A significant milestone was the construction of the Qinling Mega Tunnel on the Xikang Railway, marking the first use of Full-Face Tunnel Boring Machine (TBM) technology in the country. In 2017, China's first intelligent three-arm rock drilling rig with full independent intellectual property rights rolled off the production line at the China Railway Engineering Equipment Group's Tunnel Equipment Manufacturing subsidiary, significantly advancing the domestication of specialized tunnel construction equipment. In 2008, China developed its first composite Earth Pressure Balance Shield Machine with independent intellectual property rights. By October 2020, the China Railway Engineering Equipment Group had independently developed and manufactured over 1,000 shield machines, exporting them to more than 20 countries and regions. To tackle the challenges of constructing tunnels through quicksand, the "curved pipe curtain + freeze" construction technique was adopted for the Gongbei Tunnel of the Zhuhai Link of the Hong Kong-Zhuhai-Macao Bridge. For high-gas tunnels, hydraulic fracturing gas extraction technology was employed. The ultra-wide sections were excavated in 8 steps using the double sidewall guide pit method, while weak surrounding rock was addressed through stepwise construction or by leaving a core of soil intact. For sudden mud and water inflows, a dual-liquid grouting method using cement and sodium silicate was used for reinforcement and water stoppage. China's tunnel construction technology has made significant leaps, equipping the country with the capability to construct tunnels through the Himalayas, the Hengduan Mountains, and even undersea tunnels.

From manual excavation with steel drills and sledgehammers, to the use of hand-held pneumatic drills, advancing to hydraulic rock drilling rigs, and eventually adopting full-section Tunnel Boring Machines (TBM) and shield construction technology, tunnel construction in China has experienced a three-stage leap. The development of tunnel and underground engineering construction methods in China has evolved from manual tools, through small machinery and semi-mechanization, to large-scale integrated mechanization and finally to the factory-like efficiency of full-section TBMs. This progression has enabled comprehensive operations including excavation, debris removal, anchoring, support, and lining, significantly enhancing both personal safety and productivity. The development of tunnel and underground engineering construction technology in China has entered a new era, marking the integration of traditional and modern techniques. This new phase is characterized by a blend of drilling and blasting methods supported by large machinery, combined with modern shield, immersed tube, and fully mechanized TBM construction, heralding a new paradigm in the field.

Since the 1980s, with the rapid development of the national economy, the pioneering roles of highways and railways across the country have become increasingly significant. The level of tunnel survey, design, and construction has continuously improved, incorporating many advanced technologies from abroad. Notable projects include the Wutongshan Tunnel in Shenzhen, the Banzhangshan Tunnel in Zhuhai, the Gushan and Mawei Tunnels in Fujian, and the Qidaoliang Tunnel in Gansu, among others. By the end of 1990, China had constructed more than 10 tunnels exceeding one km in length. The Gushan Tunnel in Fujian was particularly noteworthy, being equipped with lighting, sound absorption, moisture control, communication, and fire prevention systems, along with closed-circuit television surveillance and radar speed measurement systems, making it China's first modern highway tunnel.

By the end of 2000, China had constructed 1,684 highway tunnels with a combined length of 628 km. In the 1990s, several extra-long dual-tunnel expressways were completed, including the Jinyun Mountain and Zhongliang Mountain Tunnels (3,165 m) on the Chengdu-Chongqing Expressway, and the Daxiling-Huwuling Tunnel (4,116 m) on the Zhejiang Yongtaiwen Expressway, extending the maximum length of individual highway tunnels in China to over 3,000 m. These projects marked significant advancements in handling ventilation, landslides, gas emissions, groundwater, as well as in operational management and traffic monitoring technologies. In the late 1990s, the construction of the Erlangshan Tunnel (4,160 m) on the Sichuan-Xizang Highway, the Huaying Mountain Tunnel (4,634 m) in the Guang'an area of Sichuan, and the Jiudingshan Tunnel (3,204 m) on the Yunnan Chudai Expressway set new records for long mountain tunnel construction in China. The completion and opening of the underwater highway tunnel beneath the Pearl River in Guangzhou and the tunnel crossing beneath the Huangpu River in Shanghai (both exceeding 3,000 m in length) signaled a new level of proficiency in constructing underwater immersed tunnels in China. During the Ninth Five-Year Plan period, 504 new tunnels totaling 278,000 linear meters were constructed.

By the end of 2010, China had constructed 7,384 highway tunnels totaling 5,122.6 km in length. Among these, there were 265 extra-long tunnels spanning 1,138 km and 1,218 long tunnels covering 2,020.8 km. Notably, nearly 20 tunnels ranging from 5 to 10 km in length were completed, including the Longtan Tunnel in Hubei, the Fangduoshan Tunnel in Chongqing, and the Cangling Tunnel in Zhejiang. Additionally, five ultra-long tunnels over 10 km were constructed, such as the 18 km Qinling Zhongnanshan Highway Tunnel, the Maiji Mountain Tunnel in Gansu, and the Mud Mountain Tunnel in Sichuan. This period saw the emergence of numerous long mountain tunnels and deep-water submarine tunnels, with significant advancements in construction, operational management, and service technologies. Notable among them are about 20 underwater tunnels, including the approximately 8.9 km Shanghai Chongming Yangtze River Tunnel and the 8.7 km Xiamen Xiang'an Submarine Tunnel. In mountainous areas, a considerable number of continuous arch tunnels, narrow clearance tunnels, half-tunnels, and covered tunnels were constructed, while in urban vicinities, five wide eight-lane tunnels and underground interchange tunnels were built. The Xiamen Xiang'an Submarine Tunnel marked a new breakthrough in undersea tunnel construction, while the shield diameter of the Shanghai Cross-River Tunnel reached an impressive 15.43 m. The Erlang Mountain Tunnel in Sichuan, with a length of 4.2 km and an entrance elevation of 2,200 m, is among the deepest in China's highway tunnel network, with a burial depth of 830 m and maximum ground stress reaching 50 MPa. It faced numerous geological challenges such as rock bursts, large deformations, dark rivers, and rich groundwater, with pressurized heads in exploration holes reaching 115.4 m. The Huaying Mountain Tunnel in Sichuan encountered coal seams, karst formations, faults, and high-stress cores, along with various toxic and hazardous gases like gas, natural gas, petroleum gas, and hydrogen sulfide. The 5.4 km long Yanmen Pass Tunnel in Shanxi traversed 27 faults, showcasing the complex and advanced level of tunnel construction technology in China.

By the end of 2020, China had constructed 21,316 highway tunnels, totaling 21,999.3 km in length. This includes 1,394 extra-long tunnels spanning 6,235.5 km and 5,541 long tunnels covering 9,633.2 km. Between 2016 and 2020, the number of highway tunnels in China increased at an annual rate exceeding 1,000 tunnels and 1,200 km in length, indicating a rapid pace of development. By the end of 2020, there were 18 extra-long highway tunnels over 10 km in length, 9 of which were completed in the last five years, with 22 more currently under construction. On September 30, 2021, the Qinling Tiantai Mountain Tunnel, featuring six lanes in each direction and extending 15.56 km, was opened to traffic and is currently the largest highway tunnel in the world. The Tianshan Shengli Tunnel, which is under construction and extends 22.13 km, is set to become the world's longest highway tunnel.

Simultaneously, China has constructed a significant number of complex highway tunnel projects in areas characterized by loess soil, extreme cold, high altitude, high geostress in soft rock, water-rich karst formations, high geothermal temperatures, severe rockbursts, and high gas levels. Among these, the Huangyan Expansion Project in Shaanxi Province, completed and opened in 2016, includes 26 single-bore, three-lane loess tunnels (20 of which are purely in loess), featuring the world's largest excavation span of 17.85 m and height of 12.18 m, with a maximum cross-sectional area of 177.1 m^2, making it the largest loess highway tunnel in terms of excavation span and cross-section. Additionally, the Mila Mountain Tunnel on the Lalin Highway, completed and opened in 2019, stands as the world's highest-altitude highway tunnel, with its entrance at 4,752 m and exit at 4,774 m above sea level, spanning a total length of 5,727 m.

In recent years, China has successfully completed nearly a hundred underwater highway tunnels utilizing techniques such as the immersed tube method, drill and blast method, and shield tunneling. Notably, the Hong Kong-Zhuhai-Macao Bridge features the world's longest highway immersed tube tunnel, which is also unique for being a deep-buried immersed tube and holding the record for the largest cross-sectional area of any highway tunnel globally. The Garongla Tunnel in Xizang is recognized as the plateau tunnel with the most complex geological conditions worldwide, achieving four unparalleled aspects: the most complex meteorological conditions, the steepest gradient, the highest seismic intensity, and the most active geological structure. The Qinghai Kunlun Mountain Tunnel is the longest permafrost tunnel globally. Its construction technique, which involves spraying wet concrete in permafrost regions, has filled a gap in both domestic and international tunnel construction practices. The Hangu Pass Tunnel, the world's longest wet loess tunnel, spans 7,851 m. Its completion signifies a significant breakthrough in the construction of long, large, and high-risk tunnels in China, elevating the country to a leading position in the field.

The Qinling Zhongnanshan Highway Tunnel, stretching 18.02 km, is the world's longest dual-bore highway tunnel and the longest highway tunnel in China. The Shantou Bay Tunnel, 6.68 km in length and featuring six lanes in both directions, is the first in China to be situated in an VIII-degree earthquake zone and traverse complex seabed geology, with a massive shield tunnel diameter of 15.03 m.

An analysis of historical data reveals that between 1980 and 2020, China constructed a total of 12,412 railroad tunnels, spanning approximately 17,621 km, which accounts for 90% of the total length of China's railroad tunnels. Notably, during the "11th Five-Year Plan" to "13th Five-Year Plan" period, railroad tunnel development in China experienced rapid growth, with 9,260 tunnels completed, extending about 15,316 km and representing 78% of the total railroad tunnel length in China (Fig. 13.41). Specifically, during the "11th Five-Year Plan" period (2006–2010), 2,262 railroad tunnels were constructed, totaling approximately 2,686 km. In the "12th Five-Year Plan" period (2011–2015), 3,611 tunnels were added, covering about 6,038 km. Finally, during the "13th Five-Year Plan" period (2016–2020), an additional 3,387 tunnels were completed, extending roughly 6,592 km.

By the end of 2021, China's operational railway network spanned approximately 151,000 km, featuring 17,532 operational railroad tunnels totaling around 21,055 km. Additionally, 2,418 tunnels were under construction, extending approximately 6,414 km, and 6,226 tunnels were in the planning stages, with a projected total length

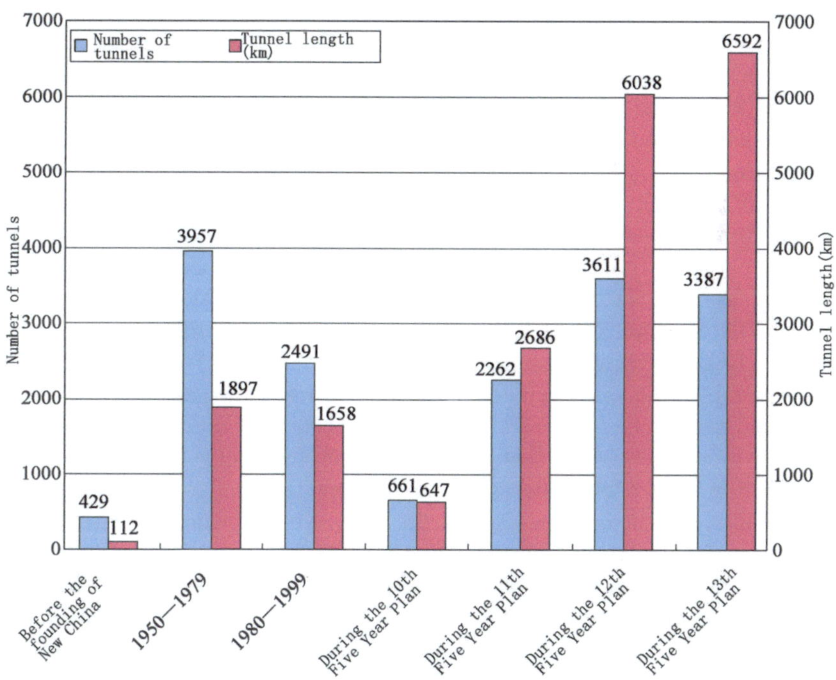

Fig. 13.41 China's railway tunnel construction achievements

of about 15,266 km. Among the operational tunnels, 235 were extra-long, covering about 3,152 km in total. Notably, 11 of these extra-long tunnels exceeded 20 km in length, cumulatively spanning about 262 km. By 2022, the longest operational tunnel in China was the New Guanjiao Tunnel in the Xigaze section of the Qinghai-Xizang Railway, stretching over 32 km. Other notable long tunnels included the West Qinling Tunnel on the Lanzhou-Chongqing Line and the Taihang Mountain Tunnel on the Shitai Passenger Line, both exceeding 25 km in length. The longest tunnel under construction was the Eastern Ring Tunnel of the Guangzhou-Foshan Eastern Ring Line, measuring 45.04 km, with the Sichuan-Xizang Railway's Sedhi La Tunnel and the Dali-Rui Railway's Gaoligong Mountain Tunnel also surpassing 30 km. The longest planned tunnel was the Shenzhen Tunnel of the Guangzhou-Dongguan-Shenzhen Intercity Railway, reaching 58.86 km, with the Yigong Tunnel of the Sichuan-Xizang Railway and the Yangcheng Lake to Suzhou New District Tunnel of the Suzhou-Wuxi-Shanghai Intercity Railway also exceeding 40 km.

High-speed rail serves as a prominent showcase of China's advanced technology, characterized by tunnels with large cross-sections and long distances, where construction and operational standards significantly surpass those of conventional rail tunnels. As of the end of 2021, China's high-speed rail network spanned approximately 40,900 km, with tunnels totaling around 6,473 km, representing about 15.8% of the network. Of these, extra-long tunnels accounted for roughly 1,141 km, making up about 2.8%. There were also 3,409 km of high-speed rail tunnels under construction, with extra-long tunnels comprising about 981 km, or approximately 28.8% of this construction. Additionally, planned high-speed rail tunnels extended approximately 6,678 km, with extra-long tunnels reaching about 1,425 km, constituting around 21.3% of the planned construction. Through these achievements, China has firmly established itself as a leading nation in high-speed rail and mega-tunnel construction.

As of 2022, the longest railway tunnel is the Qinghai-Xizang Railway's New Guanjiao Tunnel, measuring 32.456 km in length, and is already operational. The No.1 Jiwan Dashan Tunnel on the Guizhou to Nanning high-speed railway stretches 17,012 m and was successfully pierced in March 2022. The Ya'an to Linzhi segment of the Sichuan-Xizang Railway plans to construct several extra-long tunnels exceeding 10 km each, with tunnels accounting for 84% of the route and totaling 843 km in length. The longest among these, the Yigong Tunnel, extends 42.5 km and commenced construction at the end of 2021. The Hangzhou-Shaoxing-Taiwan high-speed railway's Xiaobeishan Tunnel, consisting of No.1 and No.2 tunnels, spans a total of 596 m. It is China's first single-bore, four-track, large-section high-speed railway tunnel, designed for speeds up to 350 km/h with an excavation cross-section of 350 m^2, and opened to traffic on January 8, 2022.

During the 13th Five-Year Plan, China added 3,329.3 km to its municipal subway tunnel network, averaging an annual increase of 665.9 km and an average annual growth rate of 21%. This five-year expansion surpassed the total accumulated length of subway tunnels prior to this period. By the end of 2020, the total length of municipal subway tunnels in China reached 5,422.3 km, marking an increase of approximately 2,858.3 km since 2016. With 5,481.9 km of subway lines under construction, urban rail transit network plans in 67 cities have been approved, encompassing over 7,000 km of planned lines. Of these, subway lines account for 5,426.8 km, representing 76.58% of the total planned expansion.

According to statistics from China's Urban Rail Transit Annual Report, by the end of 2021, 50 cities in mainland China had operational urban rail transit lines totaling 9,192.62 km. Out of this, subways accounted for 7,253.73 km, making up 78.9% of thc total and securing China's position as the global leader in this area. Specifically, Shanghai's subway system extended 795.63 km, representing 10.97% of China's total subway length. Beijing's subway reached 708.9 km, comprising 9.77% of the total, while Guangzhou's system spanned 564.0 km, accounting for 7.78%. Shanghai, Beijing, and Guangzhou rank as the top three cities in China in terms of subway mileage.

Future Prospects for Bridge and Tunnel Development

Bridge and tunnel construction, initially reliant on natural materials like wood, stone, and rattan, underwent significant transformations during the first and second Industrial Revolutions. These periods introduced steam engines, electricity, and iron, leading to the development of railroads and automobiles, and consequently, the rapid growth of the transportation sector. The ongoing innovation in foundational theories, combined with the advent of novel construction techniques and equipment, significantly accelerated the advancement of bridge and tunnel construction technologies worldwide in the latter half of the 19th century.

The Chinese nation has always been characterized by a spirit of innovation, particularly evident in the construction technology of ancient bridges and tunnels. Techniques used in arch and garden bridges were far more advanced than those in Western countries. During the 18th and 19th centuries, as the Western Industrial Revolution surged ahead, China, under Qing Dynasty rule, did not keep pace with global industrial advancements. In the first half of the 20th century, the construction of numerous bridges and tunnels in China had to rely on importing Western building materials, construction techniques, and equipment, resulting in the country's bridge and tunnel construction technology lagging overall behind Western standards.

The 1980s marked a renaissance in China's bridge and tunnel construction, fueled by the reform and opening-up policy. This era saw the widespread adoption of prestressing technology, the proliferation of electronic computing, advances in structural theory, and the maturation of finite element analysis. Coupled with continuous improvements in construction techniques and equipment, this period led to the creation of several world-class mega bridges and tunnels. Consequently, China's overall technical prowess in bridge and tunnel construction ascended to a globally advanced standing.

Entering the 21st century, the application of refined structural calculation and analysis techniques, ongoing advancements in high-strength steels, the increasing adoption of ultra-high-performance concrete, the implementation of whole-component assembly and short-line precasting methods, the mature use of cable-suspended construction techniques, and the widespread use of Tunnel Boring Machine (TBM) technology have collectively elevated the construction standards of bridges and tunnels. As a result, China's bridge and tunnel construction has comprehensively advanced into the ranks of the world's leading nations.

At the close of the 20th century, a new economic revolution quietly arose. Over 200 years following the first Industrial Revolution, a revolution centered on information and knowledge industries ushered humanity into a new era of the knowledge economy.

The knowledge economy era is fundamentally characterized by intelligence and high efficiency. The synergy between human intellect and computer networks has elevated knowledge innovation to the most valuable asset, becoming both the backbone of the economy and the cornerstone of various industries.

During the planning and design phases of bridges, highly advanced computer-aided methods are employed for effective and swift optimization and simulation analyses. The application of Virtual Reality (VR) technology allows stakeholders to realistically preview the bridge's appearance and functionality before construction. It also enables the simulation of the structure's response to earthquakes and typhoons, its environmental impact, and its aesthetic appeal during different times of the day, thereby aiding in the decision-making process.

During the bridge's manufacturing and erection stages, intelligent manufacturing systems will be employed to fabricate components in the factory. Subsequently, satellite positioning and remote control technologies will enable the management and oversight of the bridge's construction from headquarters located thousands of miles from the construction site.

Upon completion and commissioning, bridges will be safeguarded and maintained operational through automated monitoring and management systems. Should any faults or damage occur, health diagnostic and expert systems will automatically pinpoint the affected areas and suggest maintenance strategies. In essence, bridge projects in the knowledge economy era will embody characteristics of intelligence, informatization, and remote control, akin to advancements in other sectors. Intelligent construction robots, governed by computer software, will execute underwater and aerial tasks under the guidance of central headquarters. These operations will be carried out with precision, adhering strictly to the project plans, presenting a grand vision of bridge engineering in the 21st century.

Data indicates that by the end of 2020, China was home to 38,114 (16,798 highway and 21,316 railway) tunnels, spanning a combined length of 41,000 km (19,630 for highways and 21,999.3 for railways). According to statistics from the International Tunneling and Underground Space Association (ITA), approximately 50% of the world's tunnels under construction are in China. This positions China as a leading tunnel power with the highest number of highway and railway tunnels, the largest scale of construction projects, and the fastest development pace in the world. China's tunnel construction and operational technologies are advancing to the forefront globally.

References

1. Du Shiran, Fan Chuyu, Chen Meidong, et al. The history of science and technology in China (first and second volumes) [M]. Beijing: Science Press, 1982.
2. Han Bolin. History of the development of bridges in the world [M]. Shanghai: Knowledge Publishing House, 1987.
3. Tang Huancheng. Bridge [M]. Beijing: China Railway Press, 1981.
4. Mao Yisheng. History of Chinese ancient bridge technology [M]. Beijing: Beijing Publishing House, 1986.
5. Writing Group for the History of Bridges. Bridge history [M]. Shanghai: Shanghai Science and Technology Press, 1979.
6. Zhou Li [M]. Chen Shuguo, edited by Chen Shuguo. Changsha: Yuelu Shusha, 1989.
7. Poetry [M]. Wang Xiumei, translation and commentary. Beijing: China bookstore, 2015.
8. Guan Guan [M]. Fang Xuanling, note. Liu Ji, Supplementary note. Liu Xiaoyi, proofreading. Shanghai: Shanghai Ancient Books Publishing House, 2015.
9. Li Daoyuan. Shui Jing Zhu [M]. Chen Qiaoyi, note. Hangzhou: Zhejiang Ancient Books Publishing House, 2001.
10. Anonymous. Ancient Bamboo Book Chronicle [M]. Jinan: Qilu Shusha, 2010.
11. Yinghe, et al. Shiqu Baoji Sanji [M]. Shanghai: Shanghai Bookstore Press, 1988.
12. Marco Polo. Marco Polo's travels [M]. Beijing: China Literature and History Press,2008.
13. Li Hequn. Selected literature on ancient Chinese bridges [M]. Wuhan: Huazhong University of Science and Technology Press, 2008.
14. Jiang Kui. Bai Shi Dao Ren Poems [M]. Shanghai: Shanghai Bookstore Press, 1987.
15. Joseph Lee. Technology and civilization in China [M]. Taipei: Commercial Press, Taiwan, 1980.
16. Zuo Qiu Ming. Zuo Zhuan [M]. Jiang Jizhang, punctuation. Changsha: Yuelu Shusha, 1988.
17. Li Yanshou. Nan shi [M]. Beijing: Zhonghua shubu, 1975.
18. Wang Anshi. The Collected Writings of Mr. Linchuan [M]. Beijing: Beijing Library Press, 2004.
19. Xue Fengfei. Bogu Cliffs [M]. Wuhan: Hubei People's Publishing House, 1999.
20. Wang Chang. Jinshi jinshi zhi zhu [M]. Beijing: China Bookstore,1985.
21. Hong Yixuan. Pingjin Read Tablet Records [M]. Hu Zhengwu, Xu Sanmei, punctuation. Shanghai: Shanghai Ancient Books Publishing House, 2017.
22. Luo Y. Chinese stone bridges [M]. Beijing: People's Transportation Press, 1959.
23. Teng J. J., Shen P.. Modern bridge architectural design [M]. Beijing:People's Transportation Press,2008.
24. Fritz Leonhardt. Art and Modeling of Bridge Architecture [M]. Xu Xingyu, Gao Yanjie, Jiang Weilong, translation. Beijing: People's Transportation Press, 1988.
25. Pan Hongxuan. Ancient bridges [M]. Beijing: Zhonghua shubu, 1982.
26. Tang Huancheng. Ancient bridges of China [M]. Beijing: Cultural Relics Press, 1957.
27. Qiao Hong. Ancient bridges of China [M]. Hefei: Huangshan Shusha, 2014.
28. Fan Fan. Bridge aesthetics [M]. Beijing: People's Transportation Press, 1987.
29. Luo Zhewen, Liu Wenyuan, Liu Chunying. Chinese famous bridges [M]. Tianjin: Hundred Flowers Literature and Art Publishing House, 2006.
30. Matthew Wells. World Famous Bridge Design [M]. Zhang Hui, Li Nan, Translation. Beijing: China Construction Industry Press,2003.
31. Editorial Committee of China Railway Bridge History. History of Chinese Railway Bridges [M]. Beijing: China Railway Press,1987.
32. T.Y. Lin, S.D. Stotesbury. Structural concepts and systems [M]. Wang Chuanzhi, et al, Translation. Beijing: China Construction Industry Press,1985.
33. Wang B. H. Structural development of cable-stayed bridges and Chinese experience [M]. Beijing: People's Transportation Press, 2003.
34. China's Bridges Writing Group. China's bridges [M]. Beijing: People's Transportation Press, 2003.

35. Chinese Society of Civil Engineering, Bridges and Structural Engineering Branch. Excellent bridges in China [M]. Beijing: People's Transportation Press, 2006.
36. Writing Group for Three Combinations of Reinforced Concrete Truss Arch Bridges. Reinforced concrete truss arch bridge [M]. Beijing: People's Transportation Press, 1977.
37. Xiang Haifan, Prospect of World Bridge Engineering in the 21st Century [J]. Journal of Civil Engineering, 2000(3):1-6.
38. Xiang Haifan. Major technological innovations in the development of world bridges [J]. Guangxi transportation science and technology, 2003(5):1-7.
39. Information Archives of Highway Planning and Design Institute, Ministry of Transportation and Communications. Analysis of the technical level of highway bridges at home and abroad in the 1990s [R]. Beijing: Ministry of Transportation and Communications, 1993.
40. WU Zhongwei. Development trend and problems of high performance concrete (HPC) [J]. Construction Technology, 1998, 29(1):8-13.
41. Department of Transportation. Construction of double-curved arch bridge [R]. Beijing: Ministry of Transportation and Communications, 1968.
42. Feng Naiqian. Development and application of high-performance concrete and ultra-high-performance concrete [J]. Construction Technology, 2009, 38(4):1-6.
43. YUAN Xue-zhan, JIN Taili, LI Lezhou. Stability analysis of river bridges [C] Proceedings of the 1994 Bridge Symposium of the Chinese Highway Society of Bridge and Structural Engineering. Beijing: People's Transportation Press, 1994.
44. Huangshi Yangtze River Highway Bridge Administration. Huangshi Yangtze River Highway Bridge [M]. Wuhan: Wuhan University Press, 1998.
45. Qian Dongsheng. Independent development of China's large-span suspension bridges [C] Proceedings of the 1995 bridge symposium of the Chinese highway society bridge and structural engineering society. Beijing: People's Transportation Press, 1995.
46. Chen Derong, Xu Fengyun. Prospects for the development of steel-reinforced concrete arch bridges from the Yibin Bridge [C]. National Bridge Symposium. 1991.
47. Wang Yicai. Development and history of tunnels [J]. Journal of Xi'an Highway College, 1985, 1(03):1.
48. Wu Y. Development history of China's tunnels [J]. Construction enterprise management, 2018(11):1.
49. Yan Jinxiu. 40 years of development of tunnel engineering technology in China [J]. Tunnel Construction (in Chinese and English), 2019, 39(4):8.
50. Jiang Shuping. Current situation and prospect of highway tunnel construction technology in China [J]. Transportation World, 2003(2):6.
51. Journalist. China is already the country with the most road tunnels and the fastest development in the world--Speech of Jiang Shuping, Chairman of Tunnel Engineering Branch of China Highway Society, at the opening ceremony of the 2013 National Academic Exchange Conference on Highway Tunnels (excerpt) [J]. Tunnel Construction, 2013, 33(10):1.
52. China Journal of Highway, Editorial Board. Review of Academic Research on Transportation Tunnel Engineering in China-2022 [J]. China Highway Journal, 2022, 35(4):40.
53. Chen Jianxun, et al. Topical report on road tunnel development research [R]. Xi'an: Chang'an University, 2019.
54. TIAN Siming, WANG Wei, GONG Jiangfeng. Development and Prospect of Railway Tunnels in China(with Statistics of Railway Tunnels in China as of the End of 2020)[J]. Tunnel Construction,2021(2):1.

Chapter 14
Flight Dreams: From Ancient Aspirations to Contemporary Realization

As one of the quintessential modes of transportation, China's air transport and aerospace engineering, despite their late start, have seen rapid development. Currently, China stands as a major player in the aerospace field, steadily advancing towards becoming a leading aerospace power. Air transportation, with its unique features and advantages, plays an indispensable and often irreplaceable role in transportation, facilitating not only long-haul transoceanic travel but also fast and efficient courier logistics. The high-speed nature of air transport has made global travel far more accessible, and with ongoing advancements in space technology, space transportation and travel are poised to become part of everyday life. The ancient Chinese dream of flight is gradually becoming a reality. The remarkable accomplishments of China's aviation and space endeavors showcase the ingenuity, perseverance, and relentless effort of the Chinese people, creating inspiring tales and leaving behind enduring legacies of resilience and triumph over challenges.

Since its establishment, PRC's civil aviation industry has undergone a remarkable journey of development from its nascent stages to becoming a major force, particularly since the reforms and opening-up policies were implemented. By 2021, the number of civil aviation airports expanded from 78 in 1978 to 248 (excluding Hong Kong, Macao, and Taiwan). The national airports' passenger throughput reached 907 million, cargo and postal throughput hit 17.828 million tons, and aircraft movements tallied at 9.777 million. These figures represent increases of 390, 283, and 285 times, respectively, compared to 1978. The share of passenger turnover within the comprehensive transportation system surged from 2 to 33.1%. For 16 consecutive years, China's civil aviation passenger volume has maintained the second highest globally, effectively meeting the increasing demand for air transport among the population, significantly supporting the country's strategy for globalization, and aligning well with the civil aviation demands posed by economic and social development.

Simultaneously, China's space industry has transitioned from humble beginnings to remarkable achievements. Since 2016, the sector has accelerated into an era of innovative development. Space infrastructure has seen steady progress,

© China Communications Press Co., Ltd 2025
H. Lu (ed.), *A Brief History of Transportation in China*,
https://doi.org/10.1007/978-981-97-9851-3_14

with the completion and operationalization of the BeiDou Global Satellite Navigation System, the near-completion of the high-resolution Earth Observation System, and the steady enhancement of satellite communication and broadcasting capabilities. The lunar exploration project's "three-step" strategy concluded successfully, marking the commencement of the China Space Station construction. The "Tianwen-1" mission extended China's reach from the Earth-Moon system to interplanetary exploration, achieving globally recognized milestones.

This chapter traces the millennia-old dream of flight within the Chinese nation, from the challenging inception of China's aviation and aerospace endeavors to their present-day burgeoning success. It revisits the pioneers' formidable journey of innovation, reflects on the indomitable spirit of research and perseverance among aerospace professionals, and captures the exhilaration and emotion evoked as the ancient dream of flight gracefully transitions into the reality of contemporary China.

The Millennium Dream

The sky ignites boundless imagination. Ancient Chinese, gazing upward, were mesmerized by the expansive celestial river, weaving stars into mythological tales like Chang'e in the cold Moon Palace and the star-crossed lovers, the Cowherd and the Weaving Maiden, divided by the Milky Way. "From atop a perilous tower a hundred feet high, one feels they could pluck the stars from the heavens." The aspiration to soar into the azure skies, to traverse the infinite cosmos, to embark on space voyages, and to pioneer space transportation has perennially been a cherished dream of the Chinese people.

Qu Yuan once expressed his longing for the cosmos with the verses "I once dreamed of ascending to the heavens, my soul wandering the midway without anchor… Wishing to step forth and climb to the skies, yet still embodying the essence of times past" from "The Nine Chapters," and "Embarking with my spirit to ascend through the mists, veiling myself with drifting clouds as I journey on high" from "Distant Travel" [1]. Similarly, Su Shi's "Ode to the Red Cliff" features the renowned line "Accompanying flying immortals on journeys, embracing the bright moon for eternity" [2]. Although ancient thinkers varied in their celestial imaginings, they universally shared an aspiration to soar beyond the earthly bounds.

The aspiration to soar through the skies spurred ancient individuals into practical efforts to explore the realm above. As recounted in "Mozi-Lu Wen," during the Spring and Autumn period, the forefather of carpentry, Lu Ban, crafted a bird from bamboo, named "Wooden Magpie." This creation was capable of remaining airborne for three days and nights, marking it as one of the earliest flying machines designed by the Chinese (Fig. 14.1) [3]. Subsequent inventors, such as Zhang Heng in the Eastern Han Dynasty and Han Zhihe in the Tang Dynasty, also developed similar rudimentary flying devices.

Over 600 years ago, an individual named Wandu (born Tao Chengdao, but commonly referred to as "Wandu" following an honor bestowed by Zhu Yuanzhang)

Fig. 14.1 Luban's "wooden magpie"

endeavored to realize humanity's centuries-old dream of flight. He embarked on this venture by sitting in a chair attached to 47 rockets, holding a kite, aiming to ascend to the skies. Unfortunately, the rockets exploded, leading to Wandu's tragic demise. Though Wandu did not achieve flight, his exploratory spirit continues to inspire generations. Revered as the "first person in aerospace history," a statue commemorating him is prominently displayed in the theme park at the Xichang Satellite Launch Center (Fig. 14.2). In memory of Wandu, the International Astronomical Organization (IAO) has recognized his contributions by named a crater on the moon after him.

After the founding of PRC, and especially since the reform and opening up of China, China's aerospace and aviation technology and endeavors have made great strides, opening the way for the Chinese people to travel the skies and explore the universe.

It is these seemingly distant fragments of the reality of human spaceflight that, after millennia of profound accumulation, have together cultivated the sacred dream of flight within the Chinese nation. This dream has continuously fueled the aspirations of China's modern aerospace and aviation sectors, turning them into tangible realities.

Fig. 14.2 Manto statue

The Rough Beginnings of Chinese Aviation [4]

On September 21, 1909, Feng Ru (Fig. 14.3), a Chinese expatriate in America, piloted his self-built aircraft, Feng Ru No. 1, soaring over 800 m above Oakland, California. His successful and safe landing turned a new leaf in the annals of Chinese aviation history.

In March 1919, China's first civil aviation institution, the Aviation Affairs Office, was established by the Beiyang Government. Subsequently, in May 1920, China inaugurated its first civilian airline route, the Beijing-Tianjin section of the Beijing-Shanghai line. This marked the beginning of the gradual development of civil aviation in early 20th-century China.

In the late 1920s and 1930s, as developed nations started to form a global aviation network, some companies initiated flights to China. In August 1930, China National Aviation Corporation (CNAC), a joint venture between China and the US, was established with substantial support from the Nationalist Government, positioning it ahead of its competitors in technology and operations. Following World War II, both CNAC

Fig. 14.3 Father of Chinese aviation–Feng Ru

and Central Airlines (restructured from the Sino-German joint Eurasian Airways) saw significant advancements in technology and services. Concurrently, on December 20, 1946, the Nationalist Government and the United States signed the Sino-US Air Transport Agreement, paving the way for air transport development between the two countries. This agreement designated three locations for each nation's airlines to operate passenger, cargo, and mail services, and established three major routes: the North Pacific, Pacific, and Atlantic.

In December 1946, the National Government authorized Air France to operate two monthly flights from Saigon (present-day Ho Chi Minh City) to Shanghai, which was changed to a weekly service in 1947. In the same year, Air France conducted a trial flight on the Paris to Shanghai route using a DC4 aircraft. By May 1948, an agreement between the French Government and the National Government allowed both China and France to designate one or several air transport entities for commercial operations between Kunming and Hanoi. Subsequently, Air France launched a cargo route from Saigon, Vietnam to Kunming, China, passing through Hanoi, Vietnam, using DC3 aircraft. Air China also established a route from Shanghai, China, to Hanoi, Vietnam, via Kunming, China.

After the end of the Second World War, civil aviation transportation in Europe and the United States ushered in the climax of development. In the 1950s and 1960s, in the process of restoring and developing civil aviation in Europe and the United States, in addition to the development and application of new aircraft types, the airlines were constantly being improved and strengthened and new routes were being opened up.

Rapid Development of Aviation in New China

After the founding of PRC, China's aviation industry entered a new stage of development and gradually regained control of the country's civil aviation from Western countries.

Fig. 14.4 Aircraft of the "two airlines" uprising

On November 2, 1949, the Political Bureau of the Central Committee of the Communist Party of China decided to establish a Civil Aviation Bureau under the People's Revolutionary Military Committee, with guidance from the Air Force Command. On November 9 of the same year, Liu Jingyi and Chen Zhuolin, the general managers of China Airlines and Central Airlines, led an honorable uprising of the staff from both airlines in Hong Kong, known historically as the "Two Airlines Uprising." They returned to Beijing and Tianjin with 12 aircraft, providing vital material and technical resources for the nascent civil aviation sector in New China (Fig. 14.4).

In 1950, there were only 30 small aircraft in China, the annual passenger traffic was only 10,000, and the total cargo turnover was only 1.57 million tons-kilometers.

To accelerate the development of the aviation sector, the central government initiated reforms in the civil aviation management system. On February 27, 1958, the State Council announced that the Civil Aviation Administration of China (CAAC) would henceforth be under the leadership of the Ministry of Transportation. On November 7, 1960, following discussions and principle approval by the State Council's Organization Committee, it was decided to rename the CAAC to the "General Administration of Civil Aviation of the Ministry of Transportation." This entity became a comprehensive general administration within the ministry, responsible for managing and operating transport and specialized aviation, and directly overseeing the work of regional civil aviation administrations. On April 13, 1962, the 53rd meeting of the Standing Committee of the Second National People's Congress resolved to rename the General Administration of Civil Aviation of the Ministry of Transportation to the "General Administration of Civil Aviation of China." Subsequently, on April 15, 1962, the central government decided to reclassify the General Administration of

Civil Aviation from a bureau under the Ministry of Transportation to one directly under the State Council, with the Air Force directly managing its operational, party and government affairs, and personnel.

In the 1950s, the inauguration of the Beijing-Irkutsk air route by China and the Soviet Union not only facilitated smoother connections between China's capital and the Soviet Union along with Eastern European nations but also enabled further expansion to various cities in Europe and several countries across North and South America through intermodal transport. During the 1960s, the establishment of air links with Pakistan opened China's western gateway, allowing civil aviation flights from China to reach Europe and Africa directly via Pakistan.

In January 1964, China and France established diplomatic ties. By June 1966, the two countries had signed an air transport agreement, paving the way for Air France to inaugurate flights to Shanghai in September of the same year. This made Air France the first Western airline to operate flights to the People's Republic of China. On September 20, the "Chateau de Chevigny" Boeing 707 made its maiden landing on Chinese soil, inaugurating a regular service between Paris and Shanghai. The establishment of this inaugural route between China and France helped New China's civil aviation break through the Western blockade, propelling the nation's connections with Western countries into the "air age."

In 1980, China's civil aviation sector was restructured to separate from military governance, transitioning the General Administration of Civil Aviation of China (GACAC) from Air Force affiliation to a direct entity under the State Council, adopting a more corporate management approach. At that time, the entire civil aviation fleet comprised only 140 transport aircraft, most with a capacity of 20–40 passengers. There were only 17 medium to large aircraft capable of carrying more than 100 passengers and just 79 airports. By 1987, further reforms differentiated government roles from business operations. The civil aviation industry underwent a significant reform, characterized by the establishment of six national flagship airlines (Air China, China Eastern Airlines, China Southern Airlines, China Southwest Airlines, China Northwest Airlines, and China Northern Airlines), and six regional administrations (North China, East China, Central South, Southwest, Northwest, and Northeast) based on the existing administrative framework. These changes marked the ascent of New China's air transport industry.

The final two decades of the 20th century marked a period of significant reform and opening-up for China's civil aviation, aligning with both the era's developmental trends and the nation's strategic needs. Early into the reform era, China's civil aviation sector initiated systemic reforms. Throughout the early 1980s, it acquired state-of-the-art aircraft such as the Boeing 747SP, Boeing 747–200, Boeing 737, and MD-80, progressively narrowing the gap with global civil aviation standards. From 1987, a structural reorganization separated regional administrations, airlines, and airports, leading to the establishment of several key and local airlines, with airports operating as independent entities. Concurrently, the expansion of domestic and international routes advanced. To accommodate WTO accession and deepen systemic reforms, civil aviation intensified its international engagement by broadening foreign investment, liberalizing investment ratios, expanding bilateral and multilateral relations with other countries' aviation sectors, gradually liberalizing air traffic rights on a reciprocal basis, and actively participating in global aviation matters.

To adapt to the demands of marketization and globalization, the State Council introduced the "Civil Aviation System Reform Program" in 2002. This led to the restructuring of nine state-owned airlines under the Civil Aviation Administration into three major groups: Air China, China Eastern Airlines Group Corporation, and China Southern Airlines Group Corporation. Additionally, three major aviation service support groups were reorganized: China Civil Aviation Information Group Corporation, China Aviation Fuel Group Corporation, and China Aviation Equipment Import and Export Group Corporation. By 2003, the reform of the administrative management system of the General Administration of Civil Aviation of China (CAAC) was completed, transitioning from a three-tier to a two-tier management structure, comprising CAAC and regional administrations. The reform of airport localization was substantially completed by 2004 (excluding Beijing Capital International Airport and airports within the Xizang Autonomous Region). In the same year, reforms to establish a market-driven pricing mechanism were initiated following the guidelines set out in the "Notice on Issues Related to Domestic Airfare Management." In 2010, CAAC released the "Strategic Concept for Building a Strong Civil Aviation Nation," defining a strong civil aviation nation as one whose civil aviation industry is among the global leaders in comprehensive strength, plays a strategic role in national economic and social development, and possesses significant international competitiveness, influence, and innovation capability."

Development of Airports in China

Beijing Nanyuan Airport, established in 1910, holds the distinction of being China's earliest airport. Initially used by the royal family during the Yuan Dynasty, Nanyuan's low-lying, fertile terrain made it a favorite royal hunting ground through the Yuan, Ming, and Qing Dynasties, and later, a military drill field for the Qing army. In 1904, France, aiming to market its nascent aviation technology to China, brought two small planes to Beijing for demonstrations. The open and flat expanse of Nanyuan was chosen for these flight performances. By 1910, the Qing government acquired a Farman biplane from France and established China's first aircraft repair facility on the parade ground of the Yi Jun (the Qing government's main army, named "Yi Yong Batulu" for the bravery of its leader, Song Qing). The facility, initiated by Liu Zuocheng and Li Baojian, who had studied in Japan, also included a basic runway, marking the advent of China's first aircraft and airport [12].

"At the founding of New China, there were only 36 small-scale and rudimentary airports in the country (Fig. 14.5 is the Tianjin Air Terminal established in the 1920s). By 1978, the number of airports had increased to 78, reaching 175 by 2010, and 248 by 2021, over three times the number at the start of the reform and opening-up period. The historical change in the number of airports is shown in Fig. 14.6."

Fig. 14.5 Tianjin air terminal established on the Shanghai-Peiping route in the 1920s

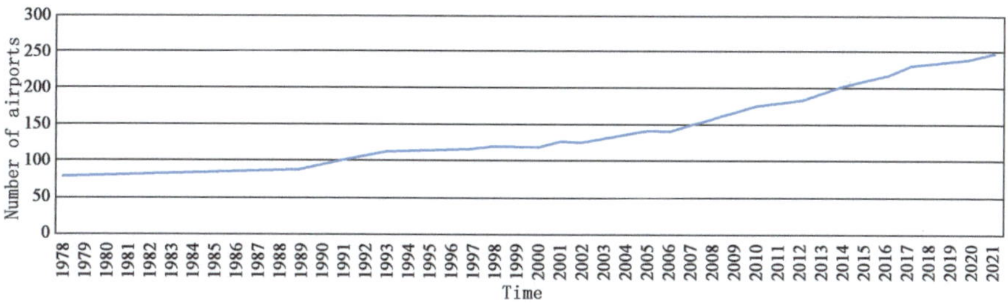

Fig. 14.6 Changes in the number of transportation airports in China (Excluding Hong Kong, Macao, and Taiwan), 1978–2021

"Immediately after the founding of New China, the Civil Aviation Administration under the Central Military Commission immediately commenced airport construction work. Especially during the 'Great Leap Forward' campaign that began in 1957, provinces, autonomous regions, and municipalities initiated airport construction booms in their capitals and key cities, building a number of airports [5]."

In the 1960s, to open international routes and accommodate the technical demands of large jet aircraft, China rapidly rebuilt and expanded Shanghai Hongqiao and Guangzhou Baiyun airports, turning them into international airports. Subsequently, China also constructed, renovated, and expanded several airports, such as Taiyuan Wuxu Airport, Hangzhou Jianqiao Airport, Lanzhou Zhongchuan Airport, Urumqi Diwopu Airport, Hefei Luogang Airport, Tianjin Zhangguizhuang Airport, and Harbin Yanjiagang Airport, among others. Since air transportation during this period served only a relatively small number of public officials, the demand for airports was at the 'aircraft-centric' stage.

The significant leap in China's airport construction began with the reform and opening-up. With the implementation of the reform and opening-up policy, the significance of civil aviation airports has increasingly become apparent [6]. At the same time, China has successively introduced large-scale, medium- and long-range, wide-body jet airplanes, promoting enhancements in airport construction standards, size, and safety measures.

After 1984, provincial capitals and large and medium-sized cities ignited a construction boom in civil aviation airports, a rarity in civil aviation history.

The sustained, rapid growth of China's national economy and civil aviation has necessitated the construction of larger-scale modern airports, with several coming into operation since the 1990s. Many small and medium-sized airports have also undergone new construction, remodeling, and expansion.

The guiding principle for airport construction during this period was to 'focus on the construction of key airports', while gradually expanding financing channels and increasing investment. During the 'Eighth Five-Year Plan' period (1991–1995), infrastructure investment in civil aviation reached 12.2 billion yuan, with technology upgrade investments at 6.09 billion yuan. In the 'Ninth Five-Year Plan' period (1996–2000), infrastructure investment soared to 68 billion yuan, and technology upgrade investments to 12.6 billion yuan, which were 5.57 times and 2.07 times the amounts of the 'Eighth Five-Year Plan' period, respectively. During the 'Tenth Five-Year Plan' period (2001–2005), investment in airport construction continued to grow, with the industry's total fixed asset investment reaching 94.7 billion yuan. In the 'Eleventh Five-Year Plan' period (2006–2010), this investment increased to 140 billion yuan, and during the 'Twelfth Five-Year Plan' period (2011–2015), it reached 200 billion yuan. In the new development stage of airports, the technology of China's airport construction and operation has reached the world's leading level, and the construction and operation of Beijing Daxing International Airport (Fig. 14.7) is a representative case of China's first-class airport construction [7, 8].

Beijing Daxing International Airport, also known as Beijing's Second International Airport, is a mega international aviation hub located between Yufa Town

Fig. 14.7 Overhead view of Beijing Daxing international airport

and Lixian Town in Beijing's Daxing District, and Guangyang District in Langfang City, Hebei Province. It is classified as a 4F-class international airport, a world-class aviation hub, and a new driver of national development. As of February 2021, the terminal area spans 780,000 m^2; the civil aviation apron accommodates 223 aircraft stands, including 76 close stands and 147 remote stands. There are four operational runways: the east, north, and west runways are 60 m wide and measure 3,400 m, 3,800 m, and 3,800 m in length, respectively, while the second west runway is 3,800 m long and 45 m wide. Additionally, there's a 3,800-m-long military runway. The airport is designed to handle 72 million passengers, 2 million tons of cargo and mail, and 620,000 aircraft movements annually by 2025.

The construction of Beijing Daxing International Airport was completed in less than 5 years. At the end of 2014, the National Development and Reform Commission issued a notice approving the feasibility study report for the new Beijing airport project, with a total investment of 79.98 billion yuan in the airport project. Construction of the terminal area officially began in September 2015. Subsequently, the airport was constructed at 'China speed,' a term denoting rapid progress. By June 2017, the terminal's steel structure was topped out, and by December 2018, all four runways were completed. In June 2019, the main construction project was officially accepted, and by the end of August 2019, Beijing Daxing International Airport had completed operational acceptance and received its final use permit. The airport held its commissioning ceremony on September 25, 2019.

Achieved at 'China Speed,' Beijing Daxing International Airport has set many records: the world's largest single-structure airport terminal, the world's largest earthquake-resistant terminal with seismic isolation, the first airport terminal in the world to feature double-decker departures and arrivals, the world's first airport with a high-speed railway running beneath it, and the world's largest seamless terminal without structural joints.

Beijing Daxing International Airport serves as a multi-modal, three-dimensional transportation hub, featuring an integrated transport network structured as 'five verticals and two horizontals.' With 16 rail tracks running directly beneath the terminal building, it enables seamless intermodal transfers among airplanes, high-speed trains, and highways. The 'five verticals' include the Beijing-Kaifeng Expressway, Beijing Daxing International Airport Expressway, Beijing-Taiwan Expressway, Jingxiong Railway, and Daxing Airport Express Line. The 'two horizontals' refer to the Beijing Daxing International Airport North Line Expressway and the Langfang-Zhuozhou Intercity Railway.

Aircraft Manufacturing and the Birth of the Great Aircraft

After decades of growth since the founding of New China, the country's aeronautical manufacturing industry expanded from small beginnings to extensive manufacturing capabilities, evolving from repairs and imitation to indigenous development. This industry, born amidst the Korean War, saw the formal establishment of the Aviation

Industry Management Committee on April 17, 1951, following the promulgation of the Decision on the Construction of the Aviation Industry. Supported by significant state funding, the aviation sector entered a phase of key large-scale development. With technical assistance from the Soviet Union, several aviation higher education institutions were established, 13 key backbone enterprises commenced operations, and the industry swiftly transitioned from maintenance to manufacturing. China's first aircraft, the CJ-5, its first jet fighter, the J-5, and its first helicopter were successfully tested in 1954, 1956, and 1958, respectively. In 1959, the successful test of the J-6, China's first supersonic jet fighter, positioned the country among the few globally capable of mass-producing jet fighters at that time [13, 14].

In the 1960s and 1970s, China's aviation manufacturing industry embarked on a period of independent construction and development, entirely reliant on its own capabilities, and persevered through numerous challenges and external disruptions. In 1965, the design of the indigenously developed Q-5 ground attack aircraft was finalized. By 1966, the supersonic J-7 fighter had been successfully prototyped. The maiden flight of China's first self-designed and manufactured high-altitude, high-speed J-8 fighter in 1969 marked a significant advancement in the country's fighter aircraft development. During this period, with an eye on long-term, comprehensive planning, efforts were made to enhance the scientific research capabilities of the aviation industry, leading to the establishment of the Aviation Research Institute and 22 specialized design and research institutes. Accompanied by the expansive third-line construction project aimed at diversifying industrial infrastructure for strategic reasons, the layout of China's aviation manufacturing industry saw considerable improvement. By the late 1970s, the industry had expanded beyond its traditional bases in the Northeast, North, and East China to establish significant research and production capacities in the third-line areas of South-Central, Southwest, and Northwest China.

Since the reform and opening up, China's aviation manufacturing industry has advanced significantly, promoting large-scale 'military-to-civilian and domestic-to-international' strategic initiatives. It has persevered through exploration, adapting and evolving amidst change, and maturing through market competition. To reverse its long-standing developmental delays, the civil aviation sector has undertaken the development and enhancement of over 20 models. Internationally, it has made a historic transition from basic international collaborations and technology importation to exporting crucial aeronautical products and technologies, and engaging as an equal partner in major aviation projects. Particularly since the 21st century, the industry has faced significant developmental opportunities, entering a new era of accelerated growth. This has positioned China among the nations with advanced, specialized, and systematized aviation industry capabilities. Notably, China's fifth-generation fighters, such as the J-20, are set to undertake future air and maritime sovereignty missions for the Chinese Air Force. Moreover, China has acquired the technology for the R&D and production of turbofan regional jets and advanced helicopters [15].

In April 2002, China's large aircraft project was initiated, with the ARJ21 regional passenger aircraft project officially launched. The C919 project was formally established in 2007, and its engineering prototype was delivered in 2009. The first C919 aircraft was assembled and rolled off the production line in November 2015, and it completed its maiden flight in Shanghai in May 2017. Reflecting the national will, the large aircraft project holds a pivotal status and employs numerous new and key technologies in airframe design and manufacturing to ensure the large aircraft's competitiveness [16].

Space Transportation and Space Travel

Ten thousand years is too long; seize the day. Driven by intense patriotic passion, unwavering dedication, and a spirit of innovation in pursuit of truth, China's space professionals have dedicated their intellect, youth, and entire lives without reservation to the cause of space exploration, contributing to the resurgence and revitalization of the Chinese nation, and serving humanity as a whole.

Under the steadfast leadership of older generation central leaders such as Mao Zedong, Zhou Enlai, Deng Xiaoping, and with the high emphasis and directive of General Secretary Xi Jinping, China's space endeavor, symbolized by the 'two bombs and one satellite' initiative, has made historic strides. Esteemed scientists like Qian Xuesen, Wang Daheng, Wang Xiji, Sun Jiadong, Ren Xinmin, Chen Fangyun, Yang Jiachi, Tu Shou'e, Huang Weilu, Zhao Jiuzhang, Yao Tongbin, Qian Ji, Guo Yonghuai, and Xie Guangxuan (the pioneering chief designer of the Long March 3 rocket and the 'combined bombs' technology) have propelled China from nascent beginnings to becoming a global aerospace power. The saga of space research and application is marked by brilliance, carved out of relentless dedication and the prime years of these luminaries, making every moment of this historical journey profoundly significant and meticulously chronicled.

Unforgettable Years of Space Development

This section focuses only on space developments related to transportation, specifically those achievements pivotal to the construction of space transportation systems. This includes rockets and missiles, where missile technology served as a foundational precursor to rocket development, and satellites, which underpin the design of transportation spacecraft. Also covered are the launching and landing systems, notably within the framework of Project 714 [17].

In February 1956, renowned scientist Qian Xuesen submitted his 'Opinions on Establishing China's National Defense Aviation Industry' to the central authorities. The following month, the State Council drafted the '1956–1967 Outline for the Long-Term Plan for Scientific and Technological Development,' aiming to achieve independent development of China's jet and rocket technologies within 12 years.

In April 1956, the Aviation Industry Committee of the People's Republic of China was established to unify and lead China's aviation and rocketry programs.

Nie Rongzhen was appointed as the director, with Huang Kecheng and Zhao Erlu serving as deputy directors.

On May 10, 1956, Vice Premier Nie Rongzhen submitted 'Preliminary Opinions on Establishing Missile Research in China' to the Central Committee. This proposal was approved at the Central Military Commission meeting chaired by Premier Zhou Enlai on May 26. Premier Zhou tasked the Aviation Industry Commission with organizing the missile management and research entities.

In January 1958, the Ministry of Defense formulated the Outline of the Ten-Year (1958–1967) Development Plan for Jet and Rocket Technology.

Following the launch of the Soviet Union's first artificial Earth satellite, several prominent Chinese scientists, including Qian Xuesen and Zhao Jiuzhang, suggested initiating research into a Chinese satellite project. Concurrently, various universities commenced related scientific studies. The Chinese Academy of Sciences was tasked with creating a preliminary plan for developing artificial satellites. This initiative, code-named the 'Five-Eight-One' mission, led to the formation of the "Five-Eight-One Group".

In April 1958, the construction of China's first carrier rocket launch site began. On May 17 of the same year, Chairman Mao Zedong, at the second meeting of the Eighth National Congress of the CPC, called for the country to develop artificial satellites. On February 19, 1960, the first launch of China's self-designed and manufactured experimental liquid-fueled sounding rocket was successful. In September of the same year, the sounding rocket was successfully launched again. On November 5 of the same year, the first launch test of China's copy of the Soviet 'P-2' missile was successful, and on March 21, 1962, the launch test of China's first independently developed medium- and short-range missile failed.

In January 1963, the Chinese Academy of Sciences established the Interstellar Travel Committee, led by Zhu Kezhen, Pei Lisheng, Qian Xuesen, Zhao Jiuzhang, and others, to study and formulate long-term plans for interstellar travel. On April 29, 1964, the National Defense Science and Technology Commission (NDSC) reported to the Central Committee, envisioning the launch of China's first satellite in 1970 or 1971. On June 29 of the same year, China's self-developed medium- and short-range missile was successfully tested again. On July 19 of the same year, the first biological rocket was successfully launched.

In 1965, the Central Special Committee approved the launch vehicle development plan formulated by the Seventh Ministry of Machinery Industry for the period from 1965 to 1972, and tasked the Chinese Academy of Sciences with formulating the satellite series development plan; on 30 June 1966, Premier Zhou Enlai inspected the Jiuquan launch base, observed the launch test of a medium- and short-range rocket, congratulating on its successful launch. On October 27 of the same year, the missile nuclear weapon launch test was successful, the warhead hit the target accurately, and the nuclear explosion was realized. In November of the same year, the development of the Long March 1 carrier rocket and the "Dongfanghong 1" artificial satellite began. On December 26 of the same year, China's developed medium-range missile's first flight test was largely successful; in 1967, the *Peace II* solid-fueled weather rocket was successfully test-fired; on January 30, 1970, the first flight test of a medium-

and long-range missile was successful. At 21:35 on April 24 of the same year, the *Dongfanghong No. 1* artificial satellite was successfully launched. This was the first artificial satellite launched by China. On May Day, Chairman Mao Zedong and other leaders hosted a reception at Tiananmen Square for the developers of the satellite and launch vehicle.

In 1971, the manned spaceflight project ("Project 714") was initiated, but the project was discontinued in 1975 (due to economic and technical difficulties). On September 10, 1971, the first flight test of the Dongfeng V intercontinental missile was largely successful. On November 26, 1975, China launched a recoverable satellite, which successfully returned to Earth on the 29th as planned. On January 7, 1979, a new launch method for long-range missiles was successfully tested. On May 18, 1980, China successfully launched a long-range carrier rocket into the predetermined area of the Pacific Ocean. The CPC Central Committee, the State Council, and the Central Military Commission sent congratulatory messages. On June 10 of the same year, a celebratory meeting was held in Beijing's Great Hall of the People, attended by leaders such as Deng Xiaoping, Hu Yaobang, Li Xiannian, Chen Yun, Peng Zhen, and Xu Xiangqian, with Hu Yaobang delivering a key speech. On April 8, 1984, China's first geostationary orbital experimental communication satellite was successfully launched and was accurately positioned over the equator at 125 degrees east longitude on the 16th. On April 7, 1990, the Long March 3 rocket successfully placed the Asia-1 satellite into its intended orbit on its seventh launch. This marked the first time China launched a foreign satellite, signifying China's formal entry into the international commercial launch service market. The Long March 3 rocket became one of the main rockets used by China for launching light satellites into synchronous transfer orbit for the international market during that period.

China's Manned Space Program and Space Transportation and Space Travel

Under the strong leadership of Deng Xiaoping, Jiang Zemin, Hu Jintao, Xi Jinping, and other successive central leadership teams, and the effective organization and command by Ding Henggao, the first commander-in-chief of China's manned spaceflight project, along with his deputies Shen Rongjun and Liu Jiyuan, the second commander-in-chief Cao Gangchuan, and his deputies Hu Shixiang, Luan Enjie, Wang Liheng, as well as subsequent commanders-in-chief such as Zhang Youxia, Li Jinai, Chang Wanquan, Zhang Qingwei, Ma Xingrui, Xu Dazhe, and Yuan Jiajun, under the technical guidance of the pioneers of the 'Two Bombs, One Satellite' project, and through the heroic efforts of numerous scientists including Wang Yongzhi (the first chief designer of China's manned spaceflight project), Qi Faren (the chief designer of the Shenzhou spacecraft), and Zhou Jiping (chief designer), China's manned spaceflight development has made a historic leap from zero to becoming one of the world's top three powers in manned space exploration.

In March 1986, pioneers of the “Two Bombs, One Star” project, including Wang Daheng and Yang Jiachi, the initiators of the National 863 High-Tech Program, suggested including the “manned spaceflight project” in the National 863 Program, which was adopted by the Central Authorities. However, the Central Authorities required the project to be established as a separate initiative due to the significant investment required, leading to a delay until 1992. In January 1992, Liu Jiyuan, then Vice Minister of the Ministry of Aerospace Industry and concurrently General Manager of the China National Space Administration, proposed to the Central Authorities to formally initiate the manned spaceflight project as soon as possible, which was accepted. On September 21, 1992, the Standing Committee of the Political Bureau of the CPC Central Committee approved the implementation of the manned spaceflight project (code-named “Project 921”) and established a “three-step” development strategy for its execution.

The first step involves launching a manned spacecraft, constructing a preliminary supporting experimental manned spacecraft project, and conducting space application experiments. In the second step, following the first manned spacecraft’s successful launch, achieve breakthroughs in the rendezvous and docking technology between manned spacecraft and other space vehicles, and use manned spacecraft technology to retrofit and launch a space laboratory, addressing short-term, manned space applications of a certain scale. In the third step, construct a manned space station to address larger-scale, long-term manned space application challenges.

On September 25, 2010, the Standing Committee of the Political Bureau of the CPC Central Committee approved **the Manned Space Station Project Implementation Plan**, and the Manned Space Station Project was officially initiated. China’s manned spaceflight project consists of eight major components: the astronaut system project, space application system project, manned spacecraft system project, launch vehicle system project, launch site system project, **telemetry, tracking, and command (TT&C) system project**, landing site system project, and the space laboratory (China Space Station) system project. Today, China’s manned spaceflight project **has completed all its planned objectives, and the Chinese Space Station is now in its on-orbit construction phase.**

Construction of China’s Autonomous Space Transportation System

The successful implementation of China’s manned spaceflight project has enabled the preliminary construction of an independent space transportation system. China has successively developed the Long March series of launch vehicles (Long March 1 to Long March 11), which can meet all of China’s space transportation and travel needs. As of May 10, 2022, the Long March series has completed 420 flights, achieving a reliability that ranks among the world’s best. China has also developed the Shenzhou series of manned spacecraft (Shenzhou-1 to Shenzhou-14) and the Tianzhou series of cargo spacecraft (Tianzhou-1 to Tianzhou-4), with the Shenzhou series having completed 13 successful flights and the Tianzhou series 4. These spacecraft have effectively supported China's space exploration and logistics missions. Additionally, China has established a manned spacecraft launch system centered on the Jiuquan Launch Site and a cargo spacecraft launch system centered on the Wenchang Launch Site [18].

In addition, China has also established multiple landing systems, including the Dongfeng landing site, as well as an emergency landing system. Furthermore, a comprehensive global tracking and control system covering land, sea, air, and space has been established, along with the Beijing Aerospace Command Center, which serves as the main measurement and control command center.

Chinese Space Transportation System Major Transportation Achievements

The successful implementation of China's manned spaceflight project has not only led to the initial construction of China's autonomous space transportation system but also to the successful completion of its space transportation missions, including astronaut missions from Shenzhou I to Shenzhou XIII. In November 1999, Shenzhou I conducted the project's first unmanned flight test. This was followed by Shenzhou II in January 2001, Shenzhou III in March 2002, and Shenzhou IV in December 2002, each performing subsequent unmanned flight tests for the project.

On October 15, 2003, Yang Liwei aboard the Shenzhou V manned spacecraft successfully completed China's first manned space flight, orbiting the Earth 14 times in space and lasting 21 h and 23 min.

From October 12 to 17, 2005, astronauts Fei Junlong and Nie Haisheng aboard the Shenzhou VI manned spacecraft orbited the Earth 76 times, completing a mission that lasted 4 days, 19 h, and 33 min, achieving China's first multi-day manned space flight.

On September 25, 2008, the Shenzhou VII manned spacecraft was launched into space by a Long March 2F carrier rocket from China's Jiuquan Satellite Launch Center, carrying astronauts Zhai Zhigang, Liu Boming, and Jing Haipeng. At 16:34 on September 27, during its 29th orbit, the Ground Flight Control Center issued an extravehicular activity (EVA) command. Astronauts Zhai Zhigang and Liu Boming, donning China's indigenously developed 'Feitian' and the Russian-imported 'Seahawk' extravehicular space suits, respectively, successfully conducted China's first spacewalk.

At 05:58 on November 1, 2011, the Shenzhou VIII spacecraft, tasked with the first rendezvous and docking mission, was launched from the Jiuquan Satellite Launch Center in China. At 01:45 on November 3, the Tiangong-1 target vehicle and Shenzhou VIII successfully completed their first rendezvous and docking. On November

16 at 18:30, Tiangong-1 successfully executed its second orbital adjustment and entered a long-term management orbit.

On June 11, 2013, astronauts Nie Haisheng, Zhang Xiaoguang and Wang Yaping successfully entered space on board the Shenzhou-10 manned spacecraft.

On September 15, 2016, the Tiangong-2 space laboratory was successfully launched from the Jiuquan Satellite Launch Center. On October 17, 2016, the Shenzhou-11 manned spacecraft was successfully sent into space using the Long March 2F Y11 launch vehicle from the Jiuquan Satellite Launch Center. On October 19, 2016, the Shenzhou-11 manned spacecraft successfully completed an automatic rendezvous and docking with Tiangong-2. Astronauts Jing Haipeng and Chen Dong then entered the Tiangong-2 space lab.

At 09:22 on June 17, 2021, the Long March 2F Y12 launch vehicle, carrying the Shenzhou-12 manned spacecraft, was launched from the Jiuquan Satellite Launch Center, successfully sending astronauts Nie Haisheng, Liu Boming, and Tang Hongbo into space. At 15:54 on the same day, the Shenzhou-12 spacecraft, after entering orbit, successfully completed its orbital configuration and, along with the previously docked Tianzhou-2 cargo spacecraft, formed a three-module combination. On August 20, following approximately six hours of extravehicular activity, the Shenzhou-12 crew successfully completed their second spacewalk at 14:33. On September 16, the Shenzhou-12 spacecraft undocked from the space station complex. The following day, at approximately 13:30 on September 17, the return capsule of the Shenzhou-12 safely landed at the Dongfeng Landing Site.

At 00:23 h on October 16, 2021, the Long March 2F Y13 rocket, carrying the Shenzhou 13 manned spacecraft, successfully launched astronauts Zhai Zhigang, Wang Yaping, and Ye Guangfu into space from the Jiuquan Satellite Launch Center. The flight crew was in excellent condition, and the launch was a complete success.

At 18:51 on November 7, 2021, female astronaut Wang Yaping, donned in China's new-generation 'Feitian' extravehicular space suit, along with Zhai Zhigang, successfully exited from the node module of the Tianhe core module. She became the first Chinese female astronaut to perform an extravehicular activity (EVA), marking a milestone in extravehicular spacewalks by Chinese women.

At 9:56 a.m. on April 16, 2022, the return capsule of the Shenzhou XIII manned spacecraft successfully landed at the Dongfeng landing site, marking the complete success of the Shenzhou XIII manned mission. At about 10:00 a.m., astronauts Zhai Zhigang, Wang Yaping, and Ye Guangfu safely returned and were in good health. The flight crew spent 183 days working and living in the Tianhe core module of the Chinese space station, setting a new record for the longest continuous flight by Chinese astronauts in orbit. With this, the key technology verification phase of the Chinese space station is complete, moving next to the on-orbit assembly and construction phase.

At 10:44 a.m. on June 5, 2022, the Long March 2F Y14 rocket, carrying the Shenzhou XIV manned spacecraft, was ignited and launched at the Jiuquan Satellite Launch Center. About 577 s later, the Shenzhou XIV manned spacecraft successfully separated from the rocket and entered the predetermined orbit. The flight crew was in good condition, and the launch was a complete success. At **0:33 a.m. on September**

2, after approximately 6 h of extravehicular activities, astronauts Chen Dong, Liu Yang, and Cai Xuzhe worked closely to complete all the scheduled tasks outside the spacecraft. Chen Dong and Liu Yang have safely returned to the **Wentian laboratory module**, marking the extravehicular activities a complete success.

With the continuous development of China's human spaceflight and commercial spaceflight, space travel will also be realized by some members of the public in the future.

Cargo Ship Transportation in Space (Chinese Space Station in Orbit)

At 19:51 on April 20, 2017, China's first cargo spacecraft, Tianzhou-1, was successfully launched aboard the Long March 7 Y2 rocket from China's Wenchang Space Launch Complex. This launch marked the culmination of the second phase of China's manned spaceflight project's 'three-step' development strategy, signifying the dawn of China's space station era [19]. Tianzhou-1, designed specifically for space station construction and operation tasks, measures 10.6 m in length and 3.35 m in maximum diameter, comprising a cargo module and a propulsion module. At its maximum load, the spacecraft weighs 13.5 tons, with a payload capacity of 6.5 tons, making it one of the largest in China's spacecraft fleet.

At 20:55 on May 29, 2021, the Long March 7 Y3 rocket carrying the Tianzhou-2 cargo spacecraft was launched on schedule at China's Wenchang Space Launch Complex. Approximately 604 s later, the spacecraft successfully separated from the rocket and precisely entered its planned orbit. At 21:17, the solar arrays deployed successfully, marking the launch as a complete success [20]. At 5:01 on May 30, Tianzhou-2 completed a rapid rendezvous and docking with the Tianhe core module. Tianzhou-2 became the first spacecraft to dock with the space station's core module. This docking utilized fast automatic rendezvous and docking technology, notable for its speed, completing the entire process in 8 h without the need for ground intervention.

At 15:10 on September 20, 2021, the Long March 7 Y4 rocket, carrying the Tianzhou-3 cargo spacecraft, was launched at the Wenchang Space Launch Complex in China [21]. About 597 s later, the spacecraft successfully separated from the rocket and precisely entered its predetermined orbit. With this launch, China's manned space program achieved 20 successful missions out of 20 attempts, marking the delivery of the second shipment to the core module of the space station.

Tianzhou-3 was launched from the coastal region, while Tianzhou-2 is still in orbit. The two spacecraft are set to fly together until they are 'directly received and confirmed' by the Shenzhou-13 astronauts. According to the manned spaceflight project plan, both cargo ships will dock simultaneously at both ends of the Tianhe core module, forming a linear configuration in preparation for the arrival of the

Shenzhou-13 manned spacecraft. This marks the first time in the history of China's manned spaceflight that two Chinese cargo spacecraft are simultaneously in orbit.

At 1:56 a.m. on May 10, 2022, the Long March 7 Y5 rocket, carrying the Tianzhou-4 cargo spacecraft, was launched at the Wenchang Space Launch Complex in China. About 10 min later, the spacecraft successfully separated from the rocket and entered its predefined orbit. At 2:23 a.m., the spacecraft's solar panels deployed successfully, marking the launch a complete success. Subsequently, the Tianzhou-4 cargo spacecraft is set to rendezvous and dock with the orbiting space station complex. This launch marks the 22nd mission of China's manned spaceflight program, the first launch transitioning from the key technology verification stage to the in-orbit construction phase of the space station, and the 420th flight of the Long March rocket series. After entering orbit, Tianzhou-4 completed its status setup and, at 8:54 Beijing time on May 10, 2022, successfully docked with the aft port of the Tianhe core module using an autonomous rapid rendezvous and docking mode. Tianzhou-4 carried consumables for the three astronauts of the Shenzhou XIV manned spacecraft for a six-month in-orbit stay, propellants, application experiment devices and sample materials, spare parts, and some of the payload. After the docking, it transitioned to the combined flight phase.

China's Moon Project and Deep Space Transportation and Interplanetary Travel

Under the effective leadership of the central leaderships under Hu Jintao and Xi Jinping, with the foundational guidance of the first commander-in-chief Luan Enjie and chief designer Sun Jiadong, as well as the strong leadership of subsequent commanders-in-chief and chief designers, coupled with the heroic efforts of countless scientists, China's moon exploration project has achieved initial success.

In 1993, Liu Jiyuan, then General Manager of the China National Aerospace Industry Corporation (CNACI) and Director of the CNSA, proposed a lunar exploration project to the central government, which received support from Song Jian, the State Councilor in charge. Song Jian personally led the evaluation, but due to economic conditions and systemic reasons at the time, a consensus was not reached, and the project was not immediately approved. In 1999, the newly established National Defense Science and Industry Commission (NDSIC) formally started planning and reviewing the lunar exploration project, initiating preliminary scientific and technological efforts.

On January 23, 2004, Premier Wen Jiabao approved the lunar exploration project. The first meeting of the lunar exploration project's leading group was held on February 25, 2004, where the 'General Requirements for the Development of the Lunar Exploration Project' were passed. On March 15, 2004, the National Defense Science and Industry Commission appointed the chief commanders and chief designers for the five major systems. By November 2004, the formulation of the project's overall technical plan was completed.

In February 2006, the State Council promulgated the 'Outline of the National Medium- and Long-Term Plan for Scientific and Technological Development (2006–2020)', which explicitly listed 'manned spaceflight and the moon exploration project' among the country's 16 major scientific and technological initiatives. The CNAIC had already conducted studies on the necessity and feasibility of the moon exploration project as early as 1994, completed a study on the technical plan for the moon exploration satellite in 1996, and concluded a study on the key technologies of the moon exploration satellite in 1998, followed by further in-depth evaluations.

The Moon Exploration Project is divided into four phases, with the first phase aimed at achieving lunar orbit exploration. The Chang'e-1 satellite was launched on October 24, 2007, and conducted effective exploration for 16 months. On November 12, 2008, China released its first complete lunar map, created from Chang'e-1's data, which was the most comprehensive lunar image available globally at that time. In March 2009, Chang'e-1 was intentionally crashed into the Moon, marking the successful insertion of China's self-developed satellite into lunar orbit and the acquisition of a complete lunar map.

The second phase of the Moon Exploration Project aimed to achieve a soft landing on the lunar surface and conduct automatic inspections and surveys. Chang'e-2, launched on October 1, 2010, served as a precursor mission, conducting several technical verifications and extended experiments for the phase, successfully fulfilling and surpassing all predetermined tasks [9]. Chang'e-3 was launched on December 2, 2013, and successfully achieved a soft landing in the Moon's Mare Imbrium on December 14. The following day, China's first lunar rover, 'Jade Rabbit,' deployed from Chang'e-3, began its lunar surface survey, acquiring vast amounts of engineering and scientific data. As of September 1, 2020, Chang'e-3 had been on the Moon for 2,453 days. It is currently in a long-term management phase, with some lander scientific payloads still operational, despite being officially retired.

The third phase of the Moon Exploration Project aimed to achieve unmanned sample return. On October 24, 2014, China conducted the third phase's re-entry and return flight test, validating key technologies for the return capsule's re-entry to Earth at near the second cosmic speed. The test was a complete success, with the service module separating from the return capsule on November 1, and the return capsule smoothly landing in the designated area. Afterwards, the service module proceeded with additional experiments, including tests in a large elliptical orbit with an apogee of 540,000 km and a perigee of 600 km, exploration around the Earth-Moon L2 point, and returning to lunar orbit to conduct experiments related to the Chang'e-5 mission. The service module is set to continue with further extended experiments.

At 2:23 a.m. on December 8, 2018, China successfully launched the Chang'e-4 probe aboard a Long March 3B rocket from the Xichang Satellite Launch Center, marking the beginning of a new chapter in lunar exploration. The Chang'e-4 mission achieved two unprecedented feats: the first-ever soft landing of a probe on the far side of the Moon; and the first implementation of relay communications between a spacecraft at the Earth-Moon L2 point and both the Earth and Moon. In addition, Chang'e-4 has yielded numerous significant original scientific findings and has provided valuable experience for future innovation in deep space exploration.

At 4:30 a.m. on November 24, 2020, China successfully launched the Chang'e-5 probe aboard the Long March 5 Y5 rocket from the Wenchang Space Launch Complex, smoothly placing it into the designated orbit. This marked the commencement of China's first mission to return samples from an extraterrestrial body [22]. After entering the Earth-Moon transfer orbit, Chang'e-5 conducted two orbit corrections and two lunar braking maneuvers to successfully enter circumlunar orbit. Following this, the probe underwent separation, descent, and powered landing, with the lander and ascender module touching down in a predetermined area on the lunar surface on December 1 to begin sampling. On December 3, the ascender ignited and launched into orbit accurately, completing a rendezvous and docking with the orbiter and return module on December 6, where the samples were transferred. Afterward, it executed a controlled lunar landing as planned. Between December 12 and 16, after two Earth-Moon transfers and two orbit corrections, the return capsule separated from the orbiter and re-entered Earth on December 17, successfully landing in the designated area in Siziwangqi, Inner Mongolia. With this mission, China became the third country in the world, after the United States and the Soviet Union, to achieve lunar sample return.

On April 24, 2022, the third phase of China's Moon Exploration Project came to a successful conclusion, and the fourth phase has already launched, with China's space industry embarking on a new journey of interstellar exploration. China's space industry will adhere to the forefront of global space development and the nation's major strategic space needs, launching Chang'e-6, Chang'e-7, and Chang'e-8 probes, and carrying out critical mission technology research and the construction of international lunar research stations. Notably, Chang'e-6 aims to sample the Moon's far side and explore the development of a constellation of communication and navigation satellites around the Moon.

Preliminary Construction of China's Autonomous Deep Space Transportation System

The successful implementation of China's Moon Exploration Project means that China has initially succeeded in building an independent deep space transportation system [10, 11]. China has used the Long March 3 and 5 launch vehicles to successfully complete the lunar missions for the Moon Exploration Project. As of May 10, 2022, China's Long March series of launch vehicles have made 420 flights, and their reliability has reached the world's advanced level. In addition, China has successively developed the Chang'e-1 to Chang'e-5 lunar probes, and has successfully completed the lunar missions of China's Moon Exploration Project. In the future, China aims to develop spacecraft for deep space transportation and potentially for interstellar travel, and to successfully realize China's manned lunar landing and Chinese astronauts' lunar exploration. China has already built a deep space transportation and lunar mission launch system, primarily based at the Xichang Launch Site.

In addition, China has established several landing and emergency landing systems in Inner Mongolia and other regions, as well as a comprehensive deep-space tracking and control system spanning land, sea, air, and space, and thc Beijing Aerospace Command and Control Center, and has successfully carried out deep-space missions and interplanetary journeys for Mars exploration.

References

1. Liu Xiang. Chu Ci [M]. Wang Yi, note. Hong Xingzu, Supplementary note. Sun Xuexiao, proofreading. Shanghai: Shanghai Ancient Books Publishing House, 2015.
2. Su Shi. Dongpo ji [M]. Xi'an: Sanqin Publishing House, 2022.
3. Mozi. Mozi [M]. Bi Yuan, ed. Wu, Xumin, proofreader. Shanghai: Shanghai Ancient Books Publishing House, 2014.
4. WANG Y, SUN X. China in the history of world aviation [J]. Civil Aviation Today, 2016(04):172–181.
5. SHI Bao-Hua, XU Wei, WANG Wen-Jie, et al. Current situation and development of airport construction in China [J]. Road construction machinery and construction mechanization, 2010, 27(02):14–19,8.
6. Xu Gang. On the impact of China's airport construction on China's civil aviation development [J]. Civil Aviation Management, 2017(12):47–49.
7. Chen Nanping. Forging a new power source for national development with high-quality Daxing airport construction [J]. Transportation Construction and Management, 2019(06):44.
8. ZHAO Jianhua, ZHANG Xiaofeng, WANG Lubing, et al. Green Operation Practice and Prospect of Beijing Daxing International Airport [J]. Environmental Protection, 2021, 49(11):18–21.
9. Hao Zhe, Cai Jinman, Liu Ningzhe. Another Major Leap in China's Spaceflight History--An Interview with Wu Weiren, Chief Designer of the Moon Exploration Project, and Wu Zhijian, Deputy Commander-in-Chief [J]. National Defense Science and Technology Industry, 2011(09):16-17.
10. Information Office of the State Council of the People's Republic of China. China's Spaceflight in 2021 [M]. Beijing: People's Publishing House, 2022.
11. LIU Zhusheng, SUN Lingyi. Development and prospect of space transportation system [J]. Science in China:Technology Science, 2012, 42(05):493–504.
12. Chen Weisuan. From Nanyuan airport to Daxing international airport the epitome of rapid development of new Chinese airports [J]. China economic weekly, 2019(17):16–17.
13. Haiyin Tou. Research on the Target Orientation and Development Path of China's Civil Aviation Manufacturing Industry [D]. Beijing: Beijing Jiaotong University, 2015.
14. Lu Xiaopeng. History and Prospect of the Development of Large Civil Aircraft and China's Strategy for Developing Large Civil Aircraft [J]. World Science and Technology Research and Development, 1995(05):68–69.
15. LI Zhiqiang, LI Jing, HAN Ye. Key manufacturing technologies of large aircraft engines [J]. Defense Manufacturing Technology, 2011(05):13–16.
16. Xie Xipeng. Exploration of key technologies of large aircraft manufacturing [J]. Science and Technology Innovation Herald, 2013(34):34.
17. Liu Jiyuan. Sixty years of Chinese spaceflight [M]. 2 ed. Beijing: Peking University Press, 2016.

18. ZHU Yu, WANG Zhi, QI Zhihong. China's spaceflight-from a space power to a space power [J]. Science and Technology Herald, 2007(23):17–22.
19. Zhao Jincai. China's first cargo spacecraft Tianzhou-1 successfully docked with Tiangong-2 space laboratory [J]. Science, 2017, 69(03):31.
20. Lai Yun, Zhang Miluo. The launch mission of Tianzhou-2 cargo spacecraft achieves complete success [J]. Science and Technology Communication, 2021, 13(11):8.
21. Xinhua News Agency. Tianzhou-3 cargo spacecraft a complete success [J]. Enterprise Observer, 2021(09):8.
22. Chang'e 5 set off, strive to realize China's space history 5 "first" [J]. Infrared, 2020, 41(12):50.

Chapter 15
Urban Transport: Supporting and Leading the Way for Rapid Urbanisation and Motorisation in China

With the rapid advancement of socio-economic and transportation technologies, our cities continue to develop and evolve. Nowadays, considering both the quantity and quality of urban development, China's cities hold a significant place in the annals of global urban history, spanning from ancient cities that signify the evolution of civilization to modern cities guided by eco-development principles, serving as leaders and exemplars.

High population size and density are China's cities' most notable characteristics. This drives China's commitment to green development and the construction of compact cities. Consequently, China's urban transportation development aims to create a comprehensive system, prioritizing rail as the backbone and green modes as the main components. This chapter's core objective is to outline the defining characteristics of Chinese cities and the consequent goals for a rational transportation structure; it reviews the evolution of urban transportation, analyzes the primary challenges and their solutions, and encapsulates the urban and transportation development insights since the reform and opening up. It also elucidates the new urbanization direction through metropolitan areas and city clusters, along with the future of deeply integrated comprehensive urban transportation.

City Size and Urban Motorization

With the advancement of socio-economic factors and scientific technology, China's cities have experienced significant transformations in both size and structure. Grasping this historical evolution is crucial for a comprehensive understanding of urban transportation's current state and underlying causes, addressing urban transportation issues, and navigating the challenges of modern transportation.

© China Communications Press Co., Ltd 2025
H. Lu (ed.), *A Brief History of Transportation in China*,
https://doi.org/10.1007/978-981-97-9851-3_15

City Classification

Since the founding of PRC, urban construction has developed rapidly, especially since the reform and opening up. The built-up area of China's cities has grown from 5,000 km^2 in 1978 to 62,000 km^2 in 2021, an 11.4-fold increase, as shown in Fig. 15.1 (Note: the statistics in this chapter do not include Hong Kong, Macao, and Taiwan).

Since the founding of New China, China has made numerous adjustments to the criteria for classifying city sizes to align urban construction with urban development realities and facilitate categorization, guidance, and policy-making.

In 1955, the State Construction Committee first proposed criteria for categorizing cities as large, medium, and small in its Report on the Current Situation of Urban Construction Work and Several Issues. The criteria defined large cities as those with a population of 500,000 or more, medium-sized cities as those with populations between 200,000 and 500,000, and small cities as those with less than 200,000 people. In 1980, the Committee revised the Interim Provisions on Urban Planning Quotas and Indicators, introducing a new category for mega-cities with populations exceeding 1 million. In 1984, the State Council's Regulations on Urban Planning reinstated the 1955 standards. The 1989 Urban Planning Law of the People's Republic of China, building on the 1984 standards, specified that city sizes should be determined based on the non-agricultural population in urban and suburban areas. In 2014, the State Council issued the Circular on Adjusting the Standard for Classifying the Size of Cities (Guofa [2014] No. 51), using the resident population in urban areas as the statistical basis to divide cities into five categories and seven grades, as detailed in Table 15.1.

According to the data from the seventh national census, there are currently seven super-large cities in China: Shanghai, Beijing, Shenzhen, Chongqing, Guangzhou, Chengdu, and Tianjin. These cities constitute 1.02% of the total number of cities, and account for 10.76% of the population and 18.1% of the regional GDP, respectively. Additionally, there are 14 large metropolitan cities: Wuhan, Dongguan, Xi'an, Hangzhou, Foshan, Nanjing, Shenyang, Qingdao, Jinan, Changsha, Harbin, Zhengzhou, Kunming, and Dalian, which make up 2.04% of the total number of cities, with their populations and regional GDP shares being 10.25 and 14.66%, respectively, as detailed in Table 15.2.

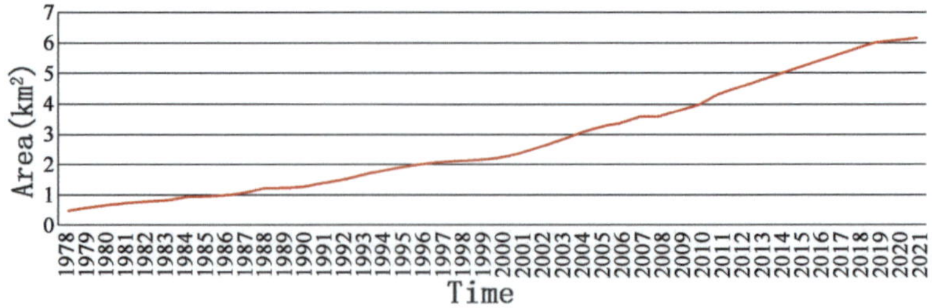

Fig. 15.1 Change curve of China's urban built-up area (1978–2021) [1]

Table 15.1 Of the five categories and seven grades of urban classification standards issued by the state council

Classification		Urban resident population (10,000)
Megacity		≥1000
Megacity		[500, 1000)
Large cities	Type I big city	[300, 500]
	Type II large cities	[100, 300)
Medium-sized city		[50, 100)
Small city	I type small city	[20, 50)
	Type II small cities	<20

Table 15.2 Overview of the country's mega-mega-cities in 2022

Classification	City		Population		Gross regional product (GDP)	
	Quantities (pcs)	Percentage (%)	Quantities (billions of people)	Percentage (%)	Quantities (Trillions of dollars)	Percentage (%)
Megacity	7	1.02	1.5	10.76	18.4	18.1
Megacity	14	2.04	1.43	10.25	14.9	14.66

Status and Development Trend of Motorization

Since the reform and opening up, the number of motor vehicles owned in China has risen from 1.59 million in 1978 to 395 million in 2021, **increasing by approximately 247.4 times**, with an average annual growth rate of 14.7%. The number of privately owned cars (specifically passenger cars) grew from **0.259 million** in 1978 to 262 million in 2021, **increasing by about 1010.6 times**, as illustrated in Fig. 15.2.

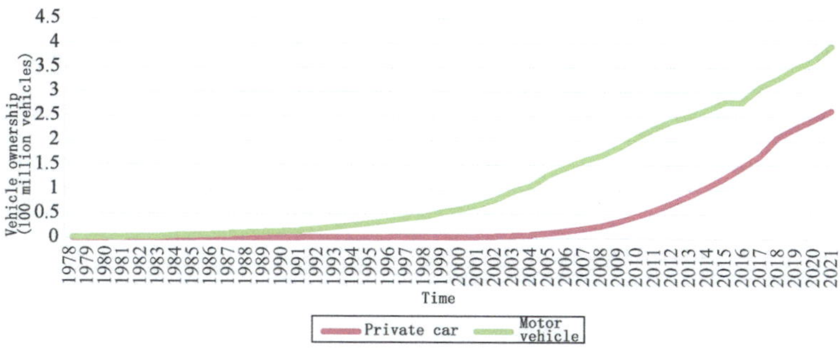

Fig. 15.2 Curve of change in motor vehicle and private automobile ownership in the country (1978–2021)

The report of the 19th CPC National Congress proposes: "Construct an urban and rural development pattern that facilitates coordinated development among large, medium, and small cities, as well as small towns, with city clusters as the core."

In the "Key Tasks for New Urbanization and Integrated Urban–Rural Development in 2021," the National Development and Reform Commission proposes to "strengthen the capacity of urban agglomerations and metropolitan areas, shift the development approach of super-large and large cities, enhance the level of modernization in urban construction and governance, advance urbanization with county towns as key platforms, and hasten the progress of integrated urban–rural development" [2].

The Outline of the Fourteenth Five-Year Plan for the National Economic and Social Development of the People's Republic of China and the Visionary Goals for the Year 2035 proposes: "Adhere to the path of new-type urbanization with Chinese characteristics, deeply advance the people-oriented new urbanization strategy, and foster the coordinated and distinctive development of large, medium, and small cities, as well as small towns, supported by city clusters and metropolitan areas, to offer a higher quality of urban life to more people."

This indicates that since the reform and opening up, China's urbanization and motorization processes have been rapidly advancing. Going forward, the integrated urban–rural development led by urban agglomerations will guide China's urban growth, fueling the ongoing momentum for urbanization and motorization.

Transportation Development and Urban Prosperity

Urban Development Goals

The Athens Charter, issued in 1933, summarizes key stages and insights in urban development. It notably clarifies the city's basic functions: residence, work, recreation, and transportation. Among these, the first three are specific activities tied to fixed locations, while transportation is essential for facilitating them. Without effective transportation, these functions cannot be fully realized. Thus, the quality of urban transportation significantly influences a city's quality of life, productivity, happiness index, and sustainability.

The development goals of modern cities are green, smart, humanistic, livable, innovative, and resilient. Research has shown that commuting time is the primary factor affecting urban residents' well-being. Therefore, prioritizing green development, employing smart technologies, and aiming for an improved urban life are the focus in building modern cities.

City Size and Transportation Modes [3]

A city is a settlement formed through human commercial, cultural, political, religious, and administrative activities. Its size, spatial layout, and residential patterns largely depend on the distance and range of people's activities within a relatively short time frame (usually within a day), which in turn are influenced by the available modes of transportation and technological advancements in transit.

The evolution of transportation modes is a crucial factor in altering the spatial structure and land-use patterns of cities. Based on transportation advancements, city development can be broadly categorized into distinct eras: the walking and horse-drawn carriage era, the tram era, the automobile era, and the era of integrated transportation. As depicted in Fig. 15.3, different transportation methods significantly affect people's mobility and accessible areas, resulting in varied urban structures and land-use patterns.

In the era of walking and horse-drawn carriages, people's activity radius was very limited, with cities typically featuring a concentric circle layout with dense settlements around the CBD. These circles had a radius generally not exceeding walkable distances, leading to smaller cities with high population densities. For example, in 1850, Boston's maximum travel distance from the city center was 3.2 to 4 km; by 1860, Philadelphia's population density reached up to 36,400 people per km^2. Many European and American cities before the 19th century were characterized by high population densities and relied primarily on walking for transportation. This

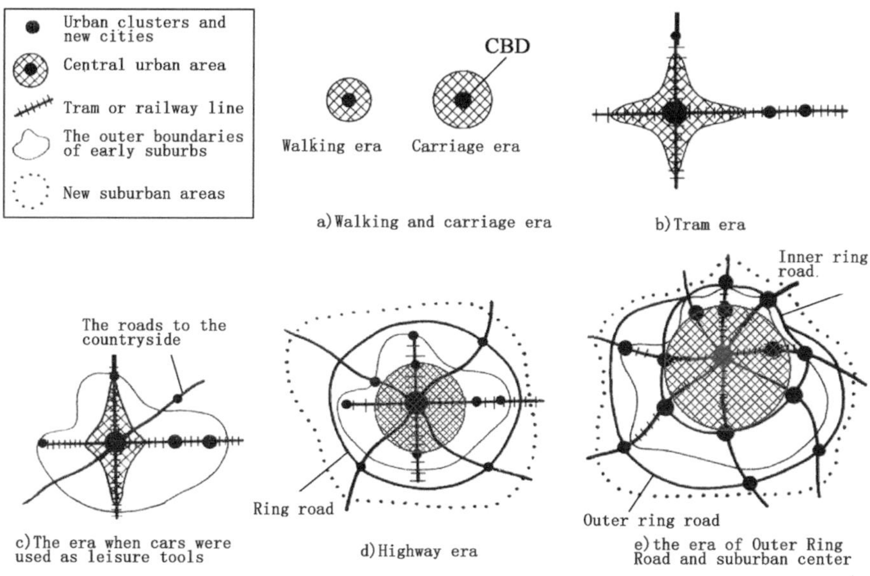

Fig. 15.3 Urban layout patterns supported by different transportation modes

pattern changed with the advent of the railroad era in the early 20th century. By 1905, Philadelphia's population density had fallen to 12,900 people per km^2.

From the end of the 18th century, the emergence of capitalist industry in Europe led to an increased concentration of populations in cities, making urban commuting a pressing issue. Trams, as the earliest extensively used motorized urban transport, became widespread in Europe and America by the late 19th century, playing a pivotal role in the development of early industrial and commercial cities. Between 1889 and 1892, tram systems were adopted in all major U.S. cities. By the early 20th century, the U.S. had 32,180 km of tram lines, making trams a primary commuting method. The advent of streetcars allowed people to reach the city center quickly and with less effort, greatly enhancing travel convenience. The proximity to the city center ceased to be the sole determinant of residential choice; instead, the distance to tram stations became crucial. Thus, the introduction of tram transport significantly spurred the development of city outskirts, with residential construction and land development activities near tram lines becoming highly active. Due to the high investment and long construction period, an extensive tram network was initially unfeasible, leading to concentrated residential areas along the lines. Consequently, urban facilities like hotels and commercial buildings were also built along these lines for convenient access. From the original concentric circle layout, the city's shape evolved into a star-like form, with rail lines branching out, causing the city to expand primarily along these lines, resulting in a more dispersed urban form.

In the early 20th century, the United States led the way into the automobile era. By the 1920s and 1930s, the US had nearly 30 million automobiles. The advent of automobiles significantly enhanced travel flexibility and convenience, improving connections between tram corridors and suburbs, and stimulating land development activities between railway branches. Original narrow transport corridors along railway lines were widened, breaking away from strip development and star-shaped urban forms. With the construction of urban freeways and highways and the widespread adoption of cars, connectivity between urban and suburban areas improved further. Around the city center, multiple sub-centers formed, leading to the emergence of numerous satellite cities. The city gradually evolved into a form that expanded outward more evenly, supported by a backbone of railway lines and expressways. Entering the automobile era, differences emerged between the US and Europe in transportation choices and urban development, influenced by land constraints and national policies.

In the early 20th century, the United States led the transition into the automobile era, with nearly 30 million automobiles by the 1920s and 1930s. The advent of automobiles significantly enhanced travel flexibility and convenience, enhancing transportation between tram trunk lines and between suburbs. The originally narrow corridors along the tram lines were widened, disrupting the strip and star-shaped development patterns. With the advent of urban expressways and highways, connectivity between urban and suburban areas improved, leading to the emergence of multiple sub-centers around city centers and numerous satellite cities around urban peripheries. The city gradually evolved into a more balanced outward expansion, structured around both rail and automobile expressways. The shift to the automobile era marked a divergence in transportation mode choices and urban development patterns between the United States and Europe, influenced by land area constraints and national policies.

Entering the 21st century, as economic growth and urban expansion in developing countries have continued, developing rail transportation has emerged as an effective solution to urban transportation challenges. By 2021, China had opened 269 rail transit lines across 50 citics, totaling a length of 9,193 km.

Evolutionary Patterns of Urban Development

With advances in transportation technology and urban socio-economic development, urban scale and form have evolved, resulting in four stages of urban development: monocentric city, polycentric city, metropolitan area, and urban agglomeration. Throughout these stages, transportation changes have played a crucial role in expanding urban space and shaping urban structure. The evolution process of urban scale and form is depicted in Fig. 15.4 [3, 4].

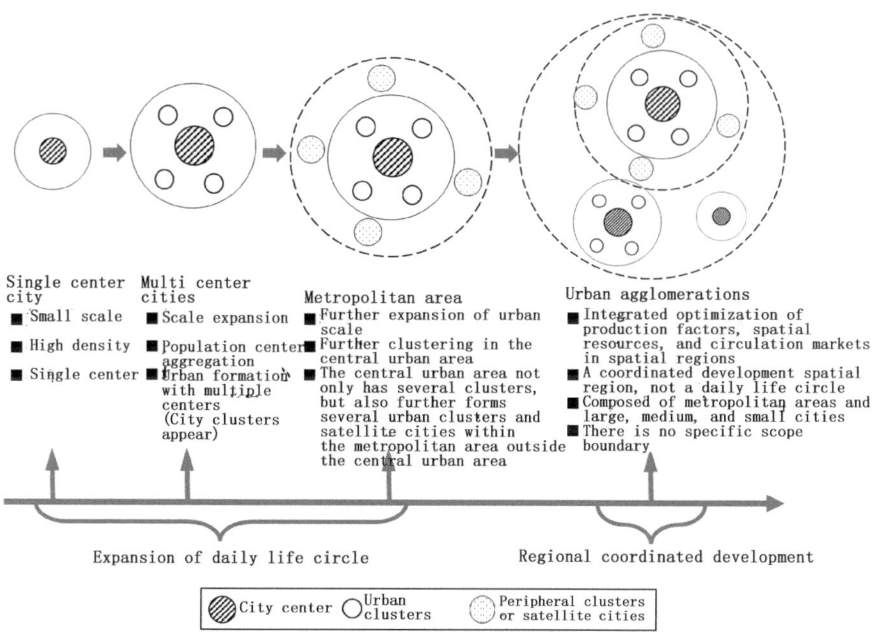

Fig. 15.4 Schematic diagram of the evolutionary process and form of urban scale and form development

The Development of Urban Transportation in China

Urban Rail Transit

The Beginning Stage of Urban Rail Transit—The Era of Streetcars

The earliest urban rail transit in China was the tram. In the 25th year of the Guangxu period of the Qing Dynasty (1899), Beijing took the lead by introducing several streetcars from Siemens, a German company, and constructed a line from Majiapu outside Yongdingmen to Chongwenmen. This line was inaugurated the following year and spans 7.5 km (Fig. 15.5) [1].

In the 32nd year of the Guangxu Emperor's reign during the Qing Dynasty (1906), Tianjin launched China's first officially operational tram line, which spanned 5.16 km (Fig. 15.6).

Since then, Shanghai, Shenyang, Harbin, Dalian, Changchun, and other cities have built tram line.

High-Capacity Urban Rail Transit in Infancy

China's first large-capacity urban rail transit is Beijing Subway Line 1, which was planned starting from 1953. The first phase of the Beijing Subway project officially commenced on July 1, 1965 (Fig. 15.7), was completed and began operations on

Fig. 15.5 The first streetcars introduced in China [5]

Fig. 15.6 China's first officially operated streetcar line [6]

October 1, 1969, spanning 23.2 km, and officially opened to the public on September 11, 1981 (Fig. 15.8). This marked the beginning of the development of China's subway systems, with Tianjin Metro Line 1 officially commencing operations on December 28, 1984. In the early 1990s, construction began on Shanghai's subway system, culminating in the opening of its first line in 1993.

Fig. 15.7 Groundbreaking ceremony for the country's first subway, 1965 [7]

Fig. 15.8 Beijing metro phase I [8]

Steady Development Phase of Urban Rail Transit

After the 1990s, China intensified its investment in urban transportation infrastructure and consciously began to vigorously develop rail transit. Beyond Beijing and Shanghai, cities such as Guangzhou, Shenzhen, and Dalian also initiated the construction of urban rail transit systems.

Before 2000, Beijing, Hong Kong, Tianjin, Shanghai, and Guangzhou had subways.

The first straddle-type urban rail transit line was completed and opened in Chongqing in 2004, the first urban rail transit line using linear motor trains was completed and put into operation in Guangzhou in 2005, and the first magnetic levitation train line was put into operation in Shanghai in 2006.

As of 2009, 10 cities in China's mainland have opened and operated rail transit, with a total operating mileage of 952.8 km.

Rapid Development Phase of Urban Rail Transit

By the end of 2015, 26 cities in China have opened 116 urban rail transit lines, with an operating mileage of 3,618 km, of which 2,658 km is subway, presenting a trend of more new operating lines, fast growth of passenger flow and diversification of system modes (Fig. 15.9).

By the end of 2021, a total of 50 cities in China had opened and were operating 269 urban rail transit lines, with an operating mileage of 9,193 km, and carrying a total of 23.71 billion trips, including subways, light rails, monorails, urban express rails, trams, and magnetic levitation transportation [9, 10].

By the end of 2021, the top ten cities with the longest operational urban rail transit mileage were Shanghai, Beijing, Chengdu, Guangzhou, Wuhan, Nanjing,

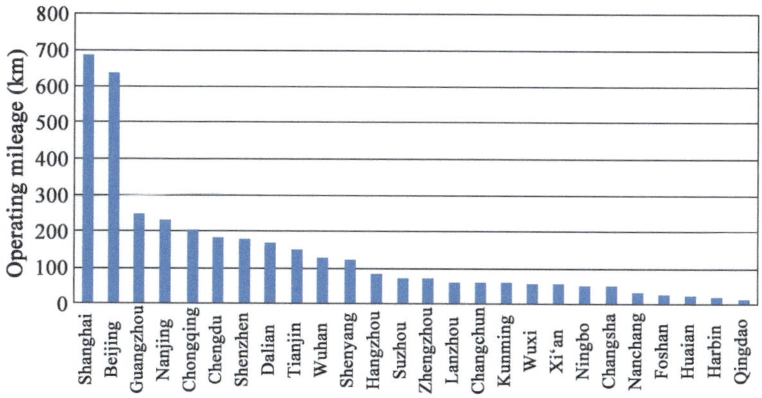

Fig. 15.9 Status of urban rail transit development in China, 2015

Shenzhen, Chongqing, Hangzhou, and Qingdao. Among them, Shanghai and Beijing had operational rail transit mileages of more than 900 km and 800 km, respectively (Fig. 15.10).

To sum up, the development of urban rail transit in China has taken a considerable amount of time. Beijing Subway Line 1, the first line, was completed on October 1, 1969. It was not until 1984 that China's second urban rail transit line, Tianjin Subway

Fig. 15.10 China's urban rail transit operating mileage at the end of 2021 Top 10 cities in length [1]. *Note* The orange color in the graph shows the additional length in 2021

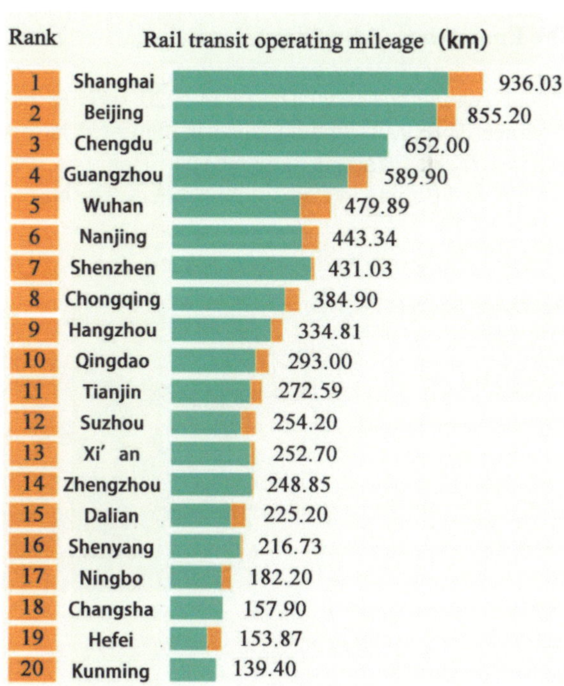

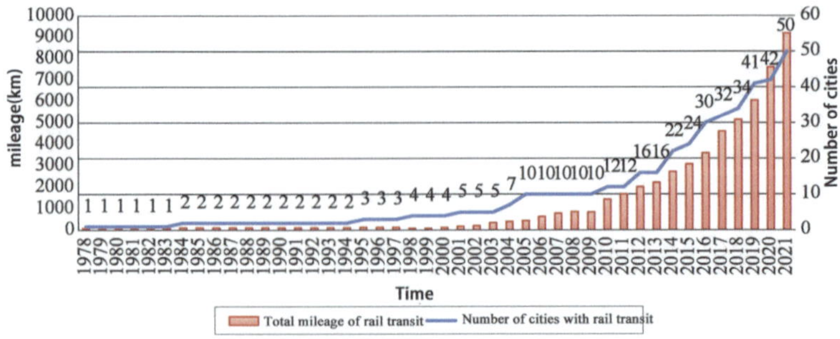

Fig. 15.11 Changes in the number of cities with rail transit and mileage in China's Mainland, 1978–2021. *Note* Plotted based on data published by China Urban Construction Statistics Yearbook of the Ministry of Housing and Construction and China Railway Transportation Association

Line 1, was completed. By 1995, urban rail transit lines had been opened in 3 cities, totaling 63 km; by 2000, this had expanded to 4 cities with a total of 117 km. By 2021, 50 cities had opened urban rail transit lines, totaling 9,193 km, as illustrated in Fig. 15.11.

Regular Urban Public Transportation

The Emergence of Bus Routes

China's regular bus service began in 1920, serving the suburbs of Nantong in Jiangsu Province. By 1922, Shanghai had inaugurated its first officially operated bus line, followed by Hangzhou in the same year, Tianjin in 1924, Shenyang in 1925, Chengdu in 1926, and Beijing in 1935, among others. The buses in Beijing in 1935 are depicted in Fig. 15.12.

Fig. 15.12 Buses in Beijing, 1935 [11]

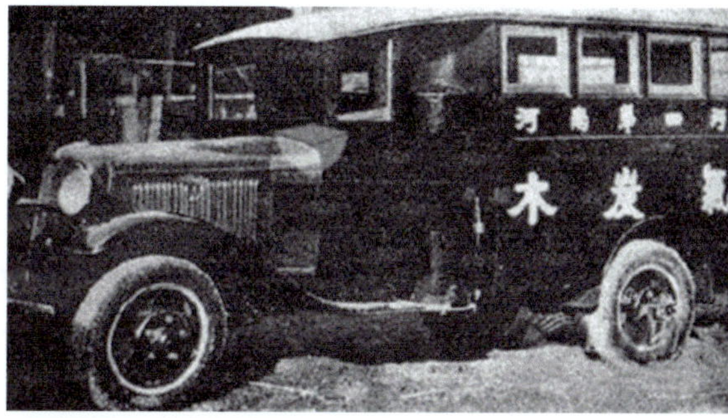

Fig. 15.13 Buses in Guangzhou at the beginning of the liberation period [12]

Fig. 15.14 Beijing trolley buses in the 1960s [13]

Initial Development of Conventional Public Transportation

At the early stage of the founding of New China, more than 20 cities already had bus services. By the beginning of the reform and opening-up period in 1978, China's urban bus fleets comprised a total of 25,839 vehicles, with a total route length of 46,384 km, and an annual passenger volume reaching 13.2 billion. The buses in Guangzhou at the early post-liberation period are depicted in Fig. 15.13, while the trolley buses in Beijing during the 1960s are shown in Fig. 15.14.

Steady Development of Conventional Public Transportation

Conventional public transportation has developed steadily since the reform and opening up. By the end of 2010, the country had 3,726 km of bus-only lanes, a parking area of 3,689.7 million m^2 for public buses and trams, maintenance yards totaling 805.9 million m^3, 563 urban passenger ferry terminals, and 420,500 vehicles for public bus and tram operations nationwide. There were 33,672 operational routes for buses and trams, with a total route length of 633,700 km, and an annual passenger volume reaching 67.012 billion trips [10].

Table 15.3 Basic information on public transportation development

Particular year (years)	Number of public trams (million vehicles)	Bus-only lane Total length (kilometers)	Public transportation lines total length (10,000 km)	Public transport passenger volume (billions of visits)
2015	56.18	8569	89.43	765.40
2019	69.33	14,952	133.6	691.76
2020	70.44	16,552	148.21	442.36

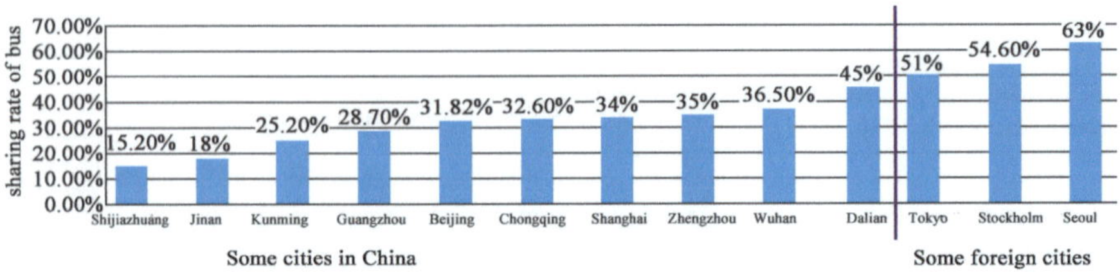

Fig. 15.15 Comparison of transit share rates in selected cities. *Note* Data are from city statistics or conversions, and represent the proportion of all modes of public transportation

Rapid Development of Conventional Public Transportation

Conventional public transportation entered a period of rapid development by 2020, with a total of 704,400 buses in operation and the total length of public transportation lines reaching 1,482,100 km (Table 15.3). These figures represent increases of 1.68 times and 2.33 times, respectively, compared to 2010.

At the same time, while the public transport modal share has increased, cities with higher shares like Dalian (45%), Wuhan (36.5%), and Shanghai (34%) still lag significantly behind international cities such as Seoul (63%), Stockholm (54.6%), and Tokyo (51%). Furthermore, the modal share in many Chinese cities falls below 20%, for instance, Jinan at 18%, as detailed in Fig. 15.15.

The Emergence and Development of BRT

Since 2005, the Bus Rapid Transit (BRT) system has rapidly developed in several cities across the country and has played an important role in improving public transport services and facilitating medium to long-distance bus journeys.

China's first BRT line commenced operations in 2005, situated along Beijing's South Central Axis Road, spanning 15.7 km with 17 stations, utilizing dedicated lanes and median-aligned stations. In 2010, Guangzhou launched its first BRT corridor (Fig. 15.16), extending 22.9 km with 26 stations. This corridor integrates 31 BRT routes and permits 30 additional bus lines to use sections of the BRT lanes, offering

Fig. 15.16 Guangzhou BRT Corridor [14]

free transfers within the corridor. The Guangzhou BRT corridor introduced innovative operational and organizational practices for BRT systems. At its inception, the corridor achieved a peak hourly flow of 29,900 passengers in one direction, earning the "Special Award for Sustainable Urban Transportation" from the World Wide Fund for Nature (WWF) in 2011 and was recognized as a "United Nations Lighthouse Project for Climate Change" in 2012.

Prioritizing Public Transportation

Entering the 21st century, urban areas have faced intensified challenges with traffic congestion and pollution, making the transformation and urgent development of urban transportation a critical necessity. Prioritizing public transportation has emerged as a crucial strategy for the sustainable development of cities and their transport systems.

In 2005, the General Office of the State Council sent the directive on prioritizing urban public transportation development to various ministries including the Ministry of Construction, National Development and Reform Commission, Ministry of Science and Technology, Ministry of Public Security, Ministry of Finance, and Ministry of Land and Resources, leading regions and departments to prioritize public transportation in urban planning.

In 2008, as part of the State Council's ministerial system reform, the responsibility for overseeing urban passenger transportation was transferred to the Ministry of Transport, unifying the management of intercity, urban–rural, and urban passenger transportation across the country.

In 2011, the Ministry of Transport initiated the National Public Transit-Oriented Cities Construction Demonstration Project. Subsequently, in 2012, 2013, and 2016, the initiative was expanded to 87 cities nationwide, enhancing their public transit capabilities, infrastructure, equipment, and service quality. This period also saw significant advancements in the reforms of urban public transport development priorities, financial and tax support systems, and the application of theoretical and engineering innovations.

In 2012, to hasten the adoption of the urban public transportation priority strategy, the State Council issued the 'Guiding Opinions on Prioritizing Urban Public Transport Development,' elevating public transport to a primary focus within urban transport development and marking a new phase in prioritizing urban and public transport.

In 2016, the Ministry of Transport published China's inaugural national-level public transportation plan, the '13th Five-Year Development Outline for Urban Public Transport.' This plan advocated for leveraging the construction of transit-oriented cities to advance the development of cities led by green transportation.

In December 2017, the Ministry of Transport formally designated Shanghai and Nanjing in Jiangsu Province as national 'Public Transport Cities' demonstration cities, presenting them with official plaques. This marked a significant milestone in the priority development of urban public transportation.

In 2017, four national ministries and commissions launched the 'Action Plan for the Enhancement of Civilization and Smoothness of Urban Road Traffic' across the country, strongly advocating for green transportation and prioritizing public transit.

The 'Outline for Building a Powerful Transportation Country' released by the CPC Central Committee and the State Council in 2019, and the 'Outline for the National Comprehensive Three-dimensional Transportation Network Planning' released in 2021, both emphasized the direction for green and public transit-focused development.

In 2021, General Secretary Xi Jinping, in his keynote at the Second United Nations Global Sustainable Transport Conference titled 'Intersecting with the World, Connecting with the Times, and Striding Forward on the Path of Sustainable Development,' highlighted the necessity of prioritizing ecological considerations, embracing green and low-carbon progress, and fostering an economic system geared towards green, low-carbon development to ensure a comprehensive green transformation of economic and social advancement as a sustainable strategy.

In March 2022, the Ministry of Transport published the 'Management Measures for the National Public Transport-Oriented City Construction Demonstration Project,' setting standard requirements for the construction, application, implementation, evaluation, and dynamic assessment of the project. During the '14th Five-Year' period, the development of public transit-oriented cities will adhere to these guidelines, with a significant update being the optimization of indicators—shifting from the original 20 assessment indicators and 10 creation indicators, along with several characteristic indicators, to 19 indicators and one characteristic indicator..

City Bike

Bicycle into China

In 1817, the German Karl Drais invented the 'two-wheeled carriage,' which is considered the world's earliest prototype of the bicycle [15]. By the 1860s, after being first mass-produced in France, bicycles were introduced to China by Western merchants and missionaries [16]. The earliest documented introduction of bicycles in China comes from an article in the Shanghai Xinbao, dated November 24, 1868: 'It is observed that there are several bicycles in Shanghai, ridden by individuals seated atop the vehicle with one wheel before and one behind. Riders use their toes to propel themselves forward. Another type involves a person pedaling, akin to operating a seesaw, with both front and rear wheels spinning swiftly, allowing for an energy-efficient means of travel. This spectacle was not witnessed by just one person, but likely by many.' [17] In the eleventh year of the Guangxu Emperor's reign (1885), British, French, and German companies began importing bicycles and their components into Shanghai, categorizing them as 'hardware and sundries.' Initially, bicycles were rare in China, with the first riders being Westerners in cities with significant foreign populations like Beijing, Tianjin, Hankou, Guangzhou, and Yantai. The next groups to adopt cycling were students who had studied abroad and the children of high-ranking officials and wealthy merchants. The last emperor, Puyi, even went as far as to remove over thirty thresholds that had been part of the Forbidden City's gates for centuries to facilitate bicycle riding (Fig. 15.17).

Fig. 15.17 Thresholds lying at the forbidden city's palace gates over the centuries sawed off for the ease of riding a bicycle [18]

Fig. 15.18 "Flying Pigeon" brand bicycles [19]

By the 1940s, China had established its own bicycle manufacturing companies, significantly increasing the number of users within the bicycle transportation system.

In 1950, 'Flying Pigeon,' the first entirely domestically-produced bicycle brand of New China, was launched in Tianjin (Fig. 15.18).

Bicycle Kingdom

In the 1970s, bicycles had already secured an important place in people's lives, and traveling by bicycle was highly enviable.

By 1982, approximately 35 million bicycles were present across 232 cities in China, averaging one bicycle for every 2.8 people.

By the 1990s, as productivity increased, bicycle transportation became the predominant mode of transportation, making China home to the world's largest number of bicycle users and earning it the title of 'the Kingdom of Bicycles' (Fig. 15.19) [20].

The ups and downs of bicycle development

With the booming domestic automobile industry, the growth of bicycle transportation has decelerated, and its influence has waned.

In the 21st century, the concept of green transportation has increasingly gained prominence, and coupled with rapid economic growth, domestic bicycle transportation has entered a new phase of development, facing many new opportunities and challenges.

As of the end of 2020, China had approximately 400 million bicycles, the highest number globally.

Despite the gradual decline in China's urban cycling environment and a downward trend in its modal share, bicycles, known for their health, environmental, economic, and convenience benefits, have received increasing attention in recent years, indicating a trend towards revival.

Fig. 15.19 Bicycle transportation [21]

Electric bicycle

At the end of the 20th century, with the development of electric drive technology, electric bicycles emerged, featuring pure electric drive systems and safe speed limits (Fig. 15.20), facilitating travel for both urban and rural residents in China. By the end of 2021, annual e-bike sales in China had exceeded 30 million units, with an estimated 300 million units in circulation [20].

Fig. 15.20 Electric bicycle transportation [22]

Rental bikes

The rental bike market in China has gone through three stages of development.

In the first phase, from 2007 to 2010, the public bicycle model that emerged abroad was introduced to China. Governed primarily at the municipal level, this phase was characterized by docked public bicycles.

The second phase, from 2010 to 2014, saw the emergence of companies specializing in the public bicycle market, with a continued focus on docked public bicycles.

The third phase, from 2014 to the present, has been marked by the rapid development of the mobile Internet, leading to the emergence of Internet bike sharing and the more convenient dockless public bicycles, which have begun to replace docked ones.

Since 2007, public bicycle systems with intelligent operation and management of practical value have been formally introduced in China, and has started trial operation in Beijing, Hangzhou, Wuhan and other cities, and gradually expanded to other cities. 2016, the rapid development of shared bicycle, of which "Mobay" is a typical representative. Since 2016, shared bicycles have rapidly developed, with 'Mobike' being a notable example. This led to an overlap in functions with the government-invested public bicycle systems, slowing their development (Figs. 15.21).

Currently, bike sharing is in a phase of refined operations, necessitating both innovation and standardization.

Fig. 15.21 Shared versus public bicycles [23, 24]

Transit-Oriented Development Model for Deep Integration of Transportation and Land Use

The high population density of Chinese cities means that rail transportation plays a dominant role in mega-cities. Additionally, due to national characteristics like resource constraints and 'big city diseases' such as traffic congestion, China's cities aim to achieve land conservation, efficient travel, and sustainable transportation. A primary strategy for achieving these goals is integrating urban transportation with land use, promote mixed land use, ensure work-life balance, and develop a multi-centric urban structure with a corresponding eco-friendly transportation system. In particular, the Transit-Oriented Development (TOD) model, which deeply integrates transportation with land use, is key.

Since the reform and opening up, Chinese cities have extensively researched and explored the deep integration of transportation and land use, gaining valuable experience. Cities like Chongqing, Chengdu, Shanghai, Shenzhen, Hangzhou, Tianjin, and Guiyang have undertaken theoretical and practical explorations. Notably, the Chongqing Shapingba TOD project stands out as a highly valuable case study [25].

In December 2010, the Chongqing Municipal Government and the Ministry of Railways signed the minutes of a meeting, designating Shapingba station as a pilot project for the comprehensive development of national railroad stations, and specified that the development plan for the station's overground portion would be led and formulated by the Chongqing Municipality, with implementation organized by a project ownership company established by the Municipality.

On December 30, 2020, the first high-speed railway TOD project in the country within a shopping district, the Chongqing Shapingba Station Railway Comprehensive Transportation Hub TOD project (Chongqing Jinsha Tianjie Shopping District), was fully operational (Fig. 15.22).

The project spans 330 mu (approximately 220,000 m), with a total construction area of about 750,000 m^2, including 480,000 m^2 allocated for commercial development, and 270,000 m^2 dedicated to the comprehensive transportation hub and other public buildings, with a total investment of 10.5 billion yuan. The project

Fig. 15.22 Planning and design of Chongqing Shapingba station railway comprehensive transportation hub TOD project

features over 500 domestic and foreign brands, spanning various sectors including viral dining, unique cultural and creative products, boutique home furnishings, trendy sports, digital gadgets, and parent–child education. Within three days of its opening, the shopping district saw over 900,000 visitors and generated more than 100 million yuan in revenue, making it one of the more popular shopping centers in Chongqing.

The project merges transportation and urban functionalities, aligning with the construction needs of the city's core area, and is pioneering in proposing the use of high-speed railway station overbuilds for urban space utilization. The project thoroughly considers the multi-level intersection layout with the area's urban roads, ensuring the high-speed railway station is seamlessly integrated into the entire urban framework, linking the Three Gorges Plaza on the north and Shapingba Park on the south side. This connection fosters a modern urban center that integrates external transportation with shopping, entertainment, leisure, and health facilities.

The comprehensive transportation hub system spans seven levels, all situated underground: Level B1 is the bus station, Level B2 houses the taxi stands and high-speed railway platforms, Levels B3 and B5 are designated for public parking, Level B4 includes the transfer hall and dedicated parking for shared vehicles, and Levels B6 and B7 serve as the rail transit station concourses. Each level features distinct transportation functions that are seamlessly connected, integrating various facilities such as the Chengdu-Chongqing high-speed railway, four urban rail lines, bus stations (with 27 bus routes), taxi stands, and public parking into a vast underground space. Clear signage complements this setup, creating an efficient, multi-level transit interchange system that enables quick transfers between different modes of transportation within a 100-m walking distance. This hub exemplifies a truly people-centered approach to 'zero-distance interchanges'.

The project has realized a win–win situation for all parties. It successfully completed the construction of a comprehensive transportation hub, bridged the divide in the city caused by the railway, enhanced the cityscape, boasts an annual passenger flow of 180 million and total retail sales of consumer goods amounting to 5 billion yuan, provides employment for 40,000 people, and achieves annual tax revenue of nearly 600 million yuan.

The project's success factors include forming a consensus on TOD development concepts, establishing a cooperative win–win mechanism for all stakeholders, exploring and making breakthroughs in land policy to ensure success (such as the stratification and rights confirmation of three-dimensional land space), and adopting unified planning, design, and construction..

As an early project, the Chongqing Shapingba TOD project also has its regrets: the absence of TOD planning and integrated planning and design from the very beginning, the uncontrollable negotiation process for stakeholders' benefit sharing, the difficulty in overcoming technical specifications such as fire protection, and the lack of legal support for land layering.

TOD has emerged as a crucial breakthrough for achieving high-quality urban development and a vital approach to constructing future ideal cities (Fig. 15.23). It is recognized globally, proven by practice, capable of introducing new dynamics to

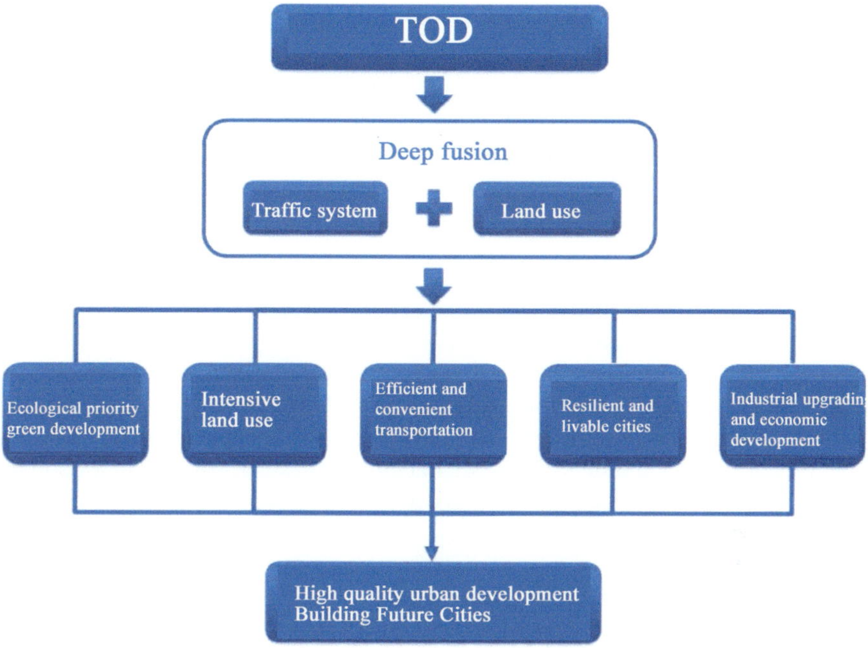

Fig. 15.23 TOD as an important way for high-quality urban development

urban development, reshaping cities, and yielding significant socio-economic benefits. It serves as the cornerstone for developing 'cities on tracks' and urban clusters around hub nodes.

The Ideal Human-Centered City and Its Transportation

Ideal City of the Future

The process of urbanization has been accompanied by the advancement of human civilization. From primitive, agricultural, and industrial stages to ecological civilization, our understanding of ideal cities has continuously deepened. Today, we recognize that urban needs comprise two main categories: basic needs and aspirations for betterment, aiming for cities that are healthier, safer, and more livable.

In summary, the ideal future city should embody the following main characteristics: green, smart, humanistic, livable, innovative, and resilient [4].

A green city respects natural laws and fosters harmony between humans and their environment. It is a composite green ecosystem encompassing nature, society, and the environment, characterized by nature's harmonious coexistence, social equity and peace, and a low-carbon, energy-efficient environment. This concept of greenness permeates all aspects, including green-led transportation, energy-efficient and low-carbon construction, clean and environmentally friendly industries, and behaviors marked by harmony, thrift, and respect for nature.

The green city concept carries the crucial mission of guiding the healthy development of Chinese cities, representing a deep interpretation and succinct encapsulation of China's urban development aspirations.

Wisdom

Smart cities are the inevitable outcome of technological advancement and the pursuit of a better quality of life, and serve as a vital tool for tackling challenges and achieving high-quality development. Building a smart city hinges on strategic top-level design, emphasizing real-time and precise analysis of the current state, unveiling and leveraging underlying mechanisms and principles, alongside the self-evolution capability of systems and subsystems. Characteristics of smart cities include comprehensive sensing, digitalization, intelligence, and sophisticated management and decision-making processes.

Humanities

A humanistic city represents the pinnacle of urban development and aesthetic aspirations, central to why people gravitate towards urban living for a fulfilling life. It is a crucial component of sustainable development. Urban characteristics and humanistic atmospheres evolve gradually through the city's economic and social history and constitute the essence of a city's enduring appeal.

Future city development should harmonize the city's natural beauty, cultural heritage, landscape, industrial framework, and functional identity, preserving its urban fabric and enhancing its cultural richness, culminating in a people-centered, culturally rich, and distinctive humanistic city.

Liveable

The future development of cities will foster more livable environments, characterized by cleanliness, blue skies, efficient transportation, convenience, and comfort. Residents will no longer endure the city's filth and chaos or worry about shopping, healthcare, housing, education, and other concerns.

Moreover, the moral quality of urban residents will continually enhance. Every city dweller will contribute to crafting a civilized urban space. Through their actions, they will educate and influence their peers and descendants, fostering interactions that align with the ethos of a healthy city, thereby nurturing a living environment that is safe, healthy, clean, comfortable, and convenient.

Blaze new trails

Innovation capacity is essential not only for high-quality development but also as a fundamental prerequisite and driving force for addressing urban challenges and constructing better cities. The innovative prowess of a nation or city fuels its creativity and dynamism, acting as a crucial mechanism for maintaining a leading edge. It forms the bedrock for cities to sustain development, foster high-value industries, and unravel complex urban development issues.

Cities of the future should serve as platforms that continuously spark innovative thoughts, acting as hubs for the aggregation, dissemination, and integrated application of knowledge. They should be incubators for new products and breeding grounds for fresh ideas, thus also becoming venues for lifelong learning and talent development.

Resilience

The Fourteenth Five-Year Plan and Vision 2035 of the People's Republic of China advocate for the development of 'resilient cities'. Resilient cities are characterized by the capacity to anticipate abnormal events, exhibit robust system resilience during such events, and possess the ability to recover afterward.

Urban Transportation Development Under the Goal of Building a Strong Transportation Nation

The Outline for Building a Powerful Transportation State sets an ambitious goal to create a safe, convenient, efficient, green, and economical modern transportation system, along with first-class facilities, technology, management, and services. This aims to satisfy the people, ensure robust security, and position the nation as a leader in global transportation. Specific development objectives for urban transportation include.

① Significantly enhance intelligent, safe, green, and shared transportation, alleviate urban traffic congestion, and substantially improve the barrier-free travel service system;

② Prioritize urban public transport development; enhance the construction of urban public transport facilities, foster integration with other transport modes, encourage green public transit usage, guide individual motorized travel wisely, and upgrade the urban pedestrian and non-motorized transport system. This also involves enhancing the quality of walking, cycling, and other forms of travel, and improving accessibility features;

③ Respect urban development laws by promoting urban unity, system integrity, and growth, coordinating urban functions and land use, and strategically planning comprehensive urban transportation systems;

④ Refine the urban road network with a rational structure of expressways, primary and secondary roads, and branch roads; enhance micro-circulation and road accessibility; and significantly advance the intelligence of urban transportation infrastructure;

⑤ Systematically plan and develop urban parking facilities and augment infrastructure for charging, hydrogen refueling, gas refueling, and public transit stations.

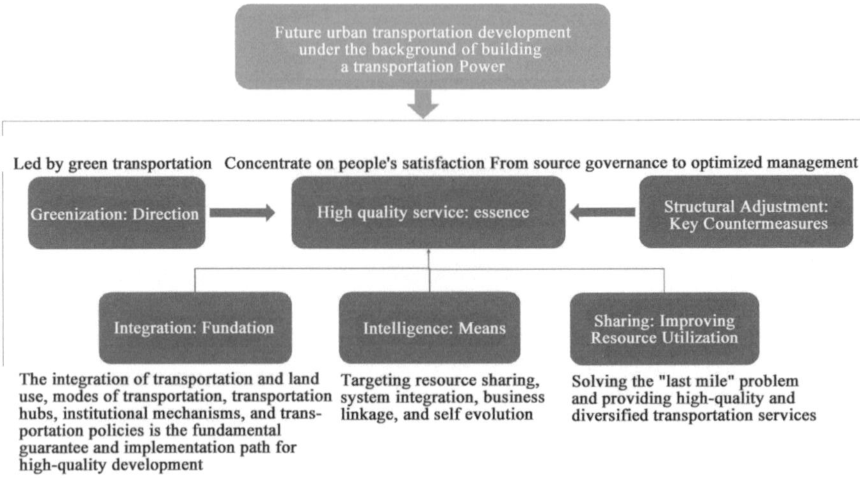

Fig. 15.24 Requirements for urban transportation development in the construction of a strong transportation nation

High-quality service is our fundamental goal. Integration is key to achieving high-quality development; structural adjustment is essential for addressing urban transport challenges; greening guides the evolution of urban transport; intelligence provides solutions to congestion and safety concerns; and sharing is an effective strategy for managing resource scarcity (Figs. 15.24).

① High-quality service aims to establish a world-class transportation system that meets the people's aspirations for a better life, addressing multi-level demands and achieving widespread satisfaction.
② Integration spans transportation and land use, diverse transportation modes, comprehensive transit hubs, management and organization, big data sharing, institutional mechanisms, and cohesive policies. Crucially, the synergy between transportation and land use is key to overcoming 'big city disease,' serving as a foundational strategy for addressing transportation issues at their source.
③ Structural adjustment is crucial for addressing urban transportation challenges systematically, focusing on urban layout, traffic configurations, road network designs, and right-of-way allocations.
④ Green transportation focuses on developing an urban system dominated by sustainable modes like public transit, walking, and biking, aiming to increase the share of eco-friendly travel..
⑤ Intelligent transportation is pivotal for enhancing traffic efficiency, safety, and service quality. By employing advanced technologies such as holographic sensing, information sharing, in-depth analysis, scientific decision-making, meticulous management, precise services, proactive safety measures, and vehicle–road synergy, it establishes a comprehensive transportation system. This

approach facilitates resource sharing, system integration, operational coordination, and self-evolution, significantly elevating the intelligence and modernity of urban transportation infrastructure.

⑥ Shared transportation enhances the efficiency of resource utilization and addresses the gap between transportation supply and demand, particularly when resources are scarce. It not only facilitates the resolution of transportation challenges but also significantly contributes to the enhancement and quality of end-point transit services. As a vital component of sustainable transportation development, shared transportation emphasizes the importance of making transportation facilities (such as parking and charging stations), vehicles (including non-motorized vehicles, cars, and trucks), data, and services widely accessible. This approach aims to optimize resource use and effectively tackle urban mobility issues.

References

1. Ministry of Housing and Urban-Rural Development of the People's Republic of China. China urban construction statistics yearbook 2020 [M]. Beijing: China Statistics Press, 2021.
2. National Development and Reform Commission of the People's Republic of China. Key Tasks for New Urbanization and Integrated Urban and Rural Development in 2021 [R]. Beijing: National Development and Reform Commission of the People's Republic of China, 2021.
3. Lu Huapu. Analysis of urban transportation [M]. Beijing: China Water Resources and Hydropower Press, 2001.
4. Institute of Transportation Studies, Tsinghua University. Research on rational scale structure of metropolitan area based on transportation perspective [R]. Beijing: Tsinghua University, 2022.
5. Sohu. Traffic expands dramatically as old and new tools compete [EB/OL]. [2022-03-18]. https://m.sohu.com/n/256811527/?v=3&_trans_=000014_baidu_ss.
6. Heze Bus Company. Why are buses called "several roads" instead of "several numbers"? [EB/OL]. (2017-11-05) [2022-03-15]. https://www.sohu.com/a/202470595_705673.
7. China Quality Miles. History returns to Huadiwan: The glory and dream of a mega-city [EB/OL]. (2020-08-04) [2022-04-01]. https://baijiahao.baidu.com/s?id=1674095766817061624&wfr=spider&for=pc.
8. Ping'an Pucheng. The reason China built its first subway, actually this ......[EB/OL]. (2019-06-22) [2022-05-15]. https://weibo.com/ttarticle/p/show?id=2309404385936178815154.
9. China Railway Transportation Association [EB/OL]. [2022-03-18]. https://www.camet.org.cn.
10. Ministry of Transportation of the People's Republic of China. Statistics [EB/OL]. [2022-03-18]. https://www.mot.gov.cn/tongjishuju.
11. Liu, Meili. Through the history of time and space, Beijing public transportation 5 road dragon ridge line on the bus experience [EB/OL]. (2019-08-16) [2022-03-18]. https://www.chinabuses.com/buses/2019/0816/article_90794.html.
12. Sohu. Conform to the trend of the times China cab change [EB/OL]. (2019-06-22) [2022-05-15]. https://www.sohu.com/a/71651894_364493.
13. Zulonghan. A brief history of my bus [EB/OL]. (2018-11-21) [2022-03-18]. https://www.163.com/dy/article/E14802BB0521D1G8.html.
14. Public Transportation Information. Liu Shaokun: Chinese experience in sustainable transportation development [EB/OL]. (2019-12-21) [2022-04-18]. https://www.sohu.com/a/361825304_99949100.

15. YIN Zhaowei,LI Zhiping. The Return of the Bicycle: An Interpretation of 1817-2050[J]. Urban Transportation,2022,20(2):127–129.
16. Xu Tao, Duan Wenjuan. The centuries-old origins of bicycles and China–Interview with Xu Tao, Assistant Researcher, Institute of History, Shanghai Academy of Social Sciences [J]. Earth, 2017(1):34–36.
17. Tian Bolan. Experts described the bicycle kingdom built history from the strange skills to civilization wonders [N/OL]. Oriental Morning Post, 2012-07-31 [2022-04-30]. http://news.sohu.com/20120731/n349455771.shtml.
18. Sohu. Puyi sawed off how many thresholds for cycling? |Anecdotes [EB/OL]. (2016-12-28) [2022-04-30]. https://www.sohu.com/a/122789787_162197.
19. CCTV News. [New China's first] The first domestic bicycle - Flying Pigeon Bicycle [EB/OL]. [2022-03-18]. http://news.cctv.com/2019/07/19/ARTINKmdB32Xk102Clz2j7zC190719.shtml.
20. China Cycling Association [EB/OL]. [2022-03-18]. http://www.china-bicycle.com.
21. People's Daily. Trip card takes off star, bikes sold out [EB/OL]. (2022-06-30) [2022-06-30]. https://baijiahao.baidu.com/s?id=1737031230350392554&wfr=spider&for=pc.
22. Netease. The city with the most electric cars in China, the former "city of motorcycles" has just turned into "the city of electric cars" [EB/OL]. (2018-04-18) [2022-06-30]. https://www.163.com/dy/article/DFMA8URC0524F8OV.html.
23. Pickleball Insights. Rule of law and civilization reflections induced by bike sharing [EB/OL]. (2020-09-07) [2022-06-30]. https://baijiahao.baidu.com/s?id=1677098539639591675.
24. Yang Jianyong. Comment: Where is the future of bike sharing? [EB/OL]. (2017-11-16) [2022-06-30]. https://www.sohu.com/a/204667061_220528.
25. Yi B. The practice of TOD in Chongqing [R]. Chongqing: Chongqing Urban Comprehensive Transportation Hub (Group) Co.

Chapter 16
Transportation Towards the Globe: Great Upsurge of Ports and Water Transportation

In the long history of China, water transport has always assumed the important mission of prospering the economy, exchanging culture, and promoting international economic and trade exchanges, and has played a decisive role. At present, comprehensively deepening reform is a breakthrough and leading key link in the strategic layout of the "Four Comprehensives". The realization of the international passenger flow and logistics development goals of the "Outline for Building a Powerful Transportation Nation" depends on strong aviation and maritime transport. In particular, the establishment of reliable, economical and efficient global transportation services and a complete supply chain system is an indispensable and important component in achieving the goal of a modern and powerful socialist country. Historical experience shows that when maritime transport is prosperous, the economy will be strong. The port is the support point and base of maritime transportation, and it is the support and destination of water transportation. Therefore, this chapter chooses ports and water transportation as the subject, reviews the historical context, and provides an outlook to development to support the dream of becoming a strong country.

China's Earliest Port and Its Development Background

In ancient China, handicrafts were highly developed, and to meet the global exchange demands of handicraft goods, the shipbuilding industry flourished, giving rise to maritime routes and prosperous coastal ports. The Records of the Grand Historian (Shi Ji) in the chapter on "Trade and Commerce" notes, "Panyu is also one of these bustling metropolises" [1]. This indicates that with the vitality of coastal ports, seaports gradually developed into coastal cities, playing a crucial role in China's economic development and overseas exchanges.

© China Communications Press Co., Ltd 2025
H. Lu (ed.), *A Brief History of Transportation in China*,
https://doi.org/10.1007/978-981-97-9851-3_16

The Birth of the Ancient Port

In the primitive society, our ancestors engaged in maritime activities using simple watercraft such as Gourd Floater and rafts (see Fig. 16.1). To facilitate rest and docking during their journeys, they also created rudimentary harbor points for fishing boats, known as semi-natural and semi-social human-formed harbors. During this period, the initial form of ports emerged. In the Xia, Shang, and Zhou dynasties, with the advent of plank boats and sail technology, watercraft continued to evolve, leading to larger-scale maritime activities and the emergence of rudimentary fishing and military ports. During this period, people were able to navigate along the coastal areas to regions including today's Korean Peninsula, the Japanese archipelago, and the Indochinese Peninsula.

During the Shang Dynasty, Panyu was the earliest seaport formed. Panyu is taken from "Yu of Fanshan", which is located in the Yuexiu District of Guangzhou Municipality, which is today's Guangzhou Port. Panyu is located in the lower reaches of the three rivers in the west, north and east. In ancient times, it was located on the coast of the bay, where people see "rising sea to the south" (the ancient name of the South China Sea), with excellent water transportation conditions. During the Zhou Dynasty, according to the "Book of Zhou" cited in "Yiwen Leiju", "The People of Yue paid tribute with boat to King Cheng of Zhou Dynasty" [3], the route of the boat tribute was via the East China Sea, across the Yellow Sea, across the Bohai Sea, into the Yellow River, and into the Wei River, indicating that existence of sea ports in Shaoxing, Zhejiang today.

Fig. 16.1 Ancient gourd floater for water crossing [2]

The Beginning of the Port During the Spring and Autumn Period and the Warring States Period

During the Spring and Autumn Period and the Warring States Period, maritime traffic between the Yangtze River Delta, the ancient Yellow River Delta, the Liaodong Peninsula, the Korean Peninsula and Southeast Asian countries began to increase, and a number of seaports of a certain scale were formed. Some of these ports have become ruins, and some are still important ports in China.

Jieshi Port is one of the famous ports in the Spring and Autumn Period and the Warring States Period. It is a famous northern port and the gateway to the sea of the Yan Kingdom. It is named after Jieshi Mountain. Its geographical location generally refers to the harbors and boat docking places along the coast of Qinhuangdao in ancient times. It provided connection between Yandu Ji and Liaodong, and was also an important coastal place connecting central Hebei, Shandong, Henan and Northeast China. According to the Records of the Great Wall of China, Jieshi Port first emerged in the Spring and Autumn Period. By the middle and late Warring States Period, as the Yan Kingdom became stronger and its territory expanded, Jieshi Port became more prominent and maritime traffic further expanded. Despite the ups and downs, it still played an important role until the end of the Qing Dynasty [4]. Jieshi, Zhuanfu (now Zhifu District, Yantai, Shandong), Langya (now Huangdao District, Qingdao, Shandong), Juzhang (now Ningbo, Zhejiang), and Kuaiji(now Shaoxing, Zhejiang) were known as the five seaports in China at that time. The geographical locations of Jieshi Port and Langya Port are shown in Fig. 16.2.

Langya Port is located at the mouth of the ancient Yellow River (today's Jiaonan, Shandong). It is named after its proximity to Langya Mountain. There is a saying that "For those who carried historical significance as a seaport, Liangya marks the beginning" [5]. Liangya Port was established no later than the Spring and Autumn Period and had developed into a renowned major port by the Warring States Period. It prospered in fishing and salt industries and served as a cultural hub. In the first year of King Ding of Zhou (468 BCE), King Goujian of Yue relocated his capital to Liangya, constructed Liangya City, and "dominated the east of the pass, observing the East Sea from Liangya, seven miles around, to gaze at the East Sea" [6]. Liangya also became the political center of the later Yue state.

The Rapid Development of Ports During the Qin and Han Dynasties

During the Qin and Han Dynasties, unified national power promoted the development of the navigation industry and also led to the vigorous development of ports. During this period, the Maritime Silk Road emerged. Panyu Port (today's Guangzhou Port), Hepu Port, and Xuwen Port were all the starting points of maritime foreign trade. The time when the ports in the three places emerged and when they came into play were

Fig. 16.2 The geographical location of Jieshi Port and Langya Port

different. Panyu emerged as the earliest foreign trade port; after Emperor Wu of the Han Dynasty pacified the Nanyue Kingdom, Panyu's status was replaced by Xuwen and Hepu, and the latter two became important foreign trade ports in the middle and late Western Han Dynasty; later, due to the turmoil in Xuwen and Hepu areas, Panyu slowly restored its important position as a foreign trade port in the Eastern Han Dynasty.

Panyu Port is located at the mouth of the Pearl River. It is not only an important inland port city, but also a metropolis for overseas trade. "Historical Records" records that overseas treasures such as "pearls, rhinos, and tortoiseshells" and specialties of Sichuan, Yunnan, and Guizhou such as "fruits and cloth" were distributed in Panyu, which soon became the largest commercial center and port city in the south of the Five Ridges.

The development of water transport trade in Panyu Port can also be revealed from the unearthed cultural relics. In the 1980s, the tomb of Nanyue King Zhao Mei (also known as Zhao Hu) was excavated in Guangzhou. Among the unearthed cultural relics was a petal-shaped white silver box. Judging from the craftsmanship such its shape, decoration, and gilt outer casing on the rim, it had obvious Persian style; in addition, African ivory and other overseas products were also unearthed in the tomb, indicating that during the Western Han Dynasty, Nanyue, one of the vassal states, had carried out maritime trade with Persia, Africa and other places through Panyu Port (Fig. 16.3).

In the sixth year of Yuanding (111 BC), Emperor Wu of the Han Dynasty pacified South Vietnam and redivided South Vietnam into nine counties. Hepu became a must-pass port for tribute merchant ships. At the same time, Hepu Port is also the starting point of the Maritime Silk Road. In the 'Book of Han—Treatise on Geography,' it is recorded: 'The voyage from Xuwen to Hapu takes about five months, where the state of Du Yuan is situated. Further, a journey of about four months by boat leads to the state of Yilumo. Additionally, after a boat journey of over twenty days, one arrives at the state of Chenli. On foot, a journey of more than ten days leads to the state of Fugandulu.' [8]. Hepu Port is located on the northeastern coast of Beibu Gulf in today's Guangxi Zhuang Autonomous Region. One could leave the country through the Indochina Peninsula to the south, or pass through Xuwen and transfer to the north via Dongye. It was a maritime transportation hub and a main trade port in the Han Dynasty a (Fig. 16.4).

Xuwen is now Xuwen County, Zhanjiang City, Guangdong Province. It is located at the southern end of Leizhou Peninsula and faces the Qiongzhou Strait to the south. It is the gateway from Guangdong to Hainan Island. "Yu Di Ji Sheng" records: "In the seven miles south of Xuwen County, goods are accumulated here for those who need, which is beneficial to transactions." [9]. Merchant ships anchored in Xuwen can obtain supplies, fresh water, food and goods, etc. At the same time, Xuwen was located in the center of Hepu County at that time, which made it convenient for the rulers to administer Leizhou Peninsula and Hainan Island. Before Emperor Wu of

Fig. 16.3 Exotic cultural relics unearthed from the tomb of the Nanyue King [7]

Fig. 16.4 The route of the maritime silk road in the Han Dynasty

the Han Dynasty opened the ocean route from the South China Sea to the Indian Ocean, Hepu and Xuwen had already become civil commercial ports owing to their geographical advantages. After Emperor Wu of the Han Dynasty opened up the ocean route, Xuwen became the official port of departure for overseas trade and foreign exchanges.

Tang and Song Dynasties When Port Transportation Was Most Prosperous

The Tang and Song Dynasties were the first peak of the development of China's navigation enterprise. Maritime trade was very developed and shipping activities were extremely active. Against this background, seaports responsible for cargo handling and distribution, as well as berthing, supply and repair of sea-going ships developed rapidly. During the Tang Dynasty, the economy in the south of the Yangtze River developed rapidly, and southeastern ports such as Guangzhou Port, Quanzhou Port, Fuzhou Port, Mingzhou Port (today's Ningbo Port), Yangzhou Port, and Dengzhou Port were increasingly developed. The Northern Song Dynasty implemented an opening-up policy and committed to the development of overseas trade, which promoted the development of ports, especially outward-serving ports, and enabled the ports of the Song Dynasty to further develop after Tang Dynasty. In the Northern Song Dynasty, municipal shipping departments were established in Guangzhou, Hangzhou, Mingzhou (now Ningbo), Quanzhou, and Mizhou (now Jiaozhou, Qingdao), dedicated to the administration of ships, trade, taxation and other matters. The above-mentioned areas were also the location of important ports, so important ports such as Guangzhou Port, Hangzhou Port, Mingzhou Port, and Quanzhou Port appeared. The above four ports are called the four major ports of the Song Dynasty.

Yangzhou Port was the largest inland river port in the Tang Dynasty and an important overseas trade port. It was located at the intersection of the Yangtze River and the Sui-Tang Grand Canal. It was a distribution center for goods and materials from the north and south of the country and a water and land transportation hub. "Baoqing Siming Zhi" records: "Within the territories of Han's Yangzhou and Jiaozhou, bordering the southeastern sea, encompassing diverse foreign nations overseas. During the era of changing winds and tides, when merchant ships from various countries arrived, the Tang Dynasty appointed an official to oversee and manage their affairs at the market and harbor." [10]. It can be seen that the vast southeastern region, including Yangzhou, was a gathering place for foreign merchants at that time. Many Persians and Arabs settled here, engaging in trade activities such as ceramics, jewelry, and silk. Ships departing from the port of Yangzhou could cross the East Sea directly to Japan, attracting numerous Japanese and Korean Monks and laymen studying abroad to travel in and out of Yangzhou. With the flow of people and the expansion of commercial activities, the port of Yangzhou reached unprecedented prosperity, giving rise to the saying 'Yangzhou, rich and prosperous, surpassing all under heaven.' [11].

During the early Tang Dynasty, Hangzhou Port served as a major inland transportation hub. Speaking of Hangzhou Port, one cannot ignore the Qiantang River. Originating from the Yellow Mountains, the Qiantang River flows into the East China Sea around the Zhoushan Islands. Hangzhou Port, owing to Qiantang River that was conveniently connecting to the open sea, has historically been a crucial dock for river and maritime transportation. Vessels departing from Hangzhou Port could travel northward through the Grand Canal, reaching the capital city of Bianliang. Heading east through Hangzhou Bay, ships could sail to the South Seas and even as far as the Persian Gulf. During the Northern Song Dynasty, the Maritime Transport Office for the east and west Zhejiang was established in Hangzhou, ships departing from there had to go through procedures and taxation at the office, demonstrating its crucial role in maritime trade. Especially during the Southern Song Dynasty, as Hangzhou became the de facto capital and the national economic center shifted southward, the development of Hangzhou Port became even more prosperous. In addition to domestic vessels from various regions, a considerable number of Japanese, Arab, and Southeast Asian merchant ships constantly frequented the port. However, due to the fierce tides at the mouth of the Qiantang River, the natural navigation conditions were not as favorable as those at the Mingzhou Port at the mouth of the Yong River, leading to an accelerated development pace for Mingzhou Port in subsequent periods.

Mingzhou Port is a natural deep-water port. It is located in the eastern part of Zhejiang Province, facing the Zhoushan Islands. It is connected to the Beijing-Hangzhou Grand Canal across the Cao'e River and Qiantang River, connects the Yangtze River and Huaihe River to the north, and connects to overseas countries in the east. It is a connection hub for overseas transportation and inland shipping. In the third year of the Yuanfeng era of Emperor Shen Zong during the Song Dynasty (1080 AD), the Ministry of Central Affairs issued an order stating, 'For all Chinese merchants traveling to Goryeo and the various feudal states of Japan coming to China, only Qingyuan [Mingzhou] is allowed to receive and dispatch them.' [10]. This decree solidified the crucial role of Mingzhou Port in trade with Goryeo and Japan during the Song Dynasty. In comparison to the Tang Dynasty, the goods exported from Mingzhou Port during the Song Dynasty included not only silk and ceramic products but also copper coins, books, and other items, playing a significant role in cultural dissemination.

Quanzhou Port is located on the southeast coast of Fujian, rising during the Tang Dynasty, flourishing during the Five Dynasties and Ten Kingdoms, and reaching its peak development in the Song Dynasty. It was the largest port for foreign trade during the Song Dynasty and was jointly recognized with the Alexandria Port in Egypt as the world's two major superports [53]. Quanzhou Port experienced rapid growth during the Song Dynasty, primarily attributed to two factors: first, its political significance—after the Song Dynasty moved its capital to Hangzhou, Quanzhou became an outer port for the capital, handling a substantial amount of tribute (Fig. 16.5) and maritime trade activities; second, its economic advantages—the exquisite ceramics produced in the kilns of Fujian, Longquan kilns in Zhejiang, and the silk products from Jiangnan turned Quanzhou Port into a prosperous region, promoting its foreign trade development. Situated at the mouth of the Jinjiang River, Quanzhou Port facilitated overseas traffic to both the East and West Oceans. It extended eastward to Japan, southward to the countries in the South Sea, and westward to Persia, Arabia, and East Africa. The main imported goods were spices and medicinal items, while exports mainly consisted of silk and ceramics.

Guangzhou Port has been in existence since the Warring States period. After the Northern Song Dynasty pacified the Southern Han regime in Lingnan, in order to develop trade between Guangzhou Port and Southeast Asia and the Arabian region, they reconstructed the Maritime Transport Office in Guangzhou. At that time, the Northern Song Dynasty did not yet include the regions of the Two Zhejiangs and Fujian, and Guangzhou Port was the sole foreign trade port for the Northern Song Dynasty. The Northern Song government dispatched honest officials to govern Guangzhou, demonstrating their emphasis on the importance of Guangzhou Port. In this context, Guangzhou Port flourished, giving rise to the scene described as 'A thousand gates illuminated by the sun, a city of jade with ten thousand roofs emitting smoke' [9]. In the early Southern Song Dynasty, Guangzhou remained the largest port in the country, trading goods such as ivory, frankincense, and medicinal items. However, towards the end of the Southern Song Dynasty, as the Mongol armies descended southward, Guangzhou Port suffered from the calamity of the Yuan forces, destroyed in the flames of war, and began to lag behind the emerging port of Quanzhou.

Tributary Trade (Tribute Ship)

Fig. 16.5 Busy tribute scene [12]

The Development of Port and Maritime Activities Before the Policy of Closed-Door Isolation

Yuan Dynasty: The Golden Age of Seaports

During the Yuan Dynasty, the development of China's seaports reached its zenith. The maritime industry of the Yuan Dynasty was in a transitional period, often referred to as the last golden age of Chinese ports before the modern era. The vast territory of the Yuan Dynasty and the rulers' emphasis on maritime activities contributed to this development. When the Yuan Dynasty was established, Emperor Shizu Kublai Khan nearly conquered all the land on the Chinese map. Gradually shifting his focus to the boundless oceans, he regarded them as a new conquest target. At the same time, Kublai Khan engaged in maritime exploration for trade, to consolidate political power, and to showcase national prestige.

Considering the relatively underdeveloped economy in the northern regions, the rulers of the Yuan Dynasty primarily established open ports for foreign trade in coastal areas south of the Yangtze River, setting up market offices in seven seaports. After the Yuan forces marched from north to south, seizing Jiang, Zhe, and Min, Emperor Shizu Kublai Khan established market offices in Qingyuan (modern-day Ningbo), Shanpu, Quanzhou, and Shanghai in the 14th year of the Zhiyuan era (1277). Later, offices were added in Hangzhou, Wenzhou, and Guangzhou. By the 30th year of the Yuan era (1293), market offices in Wenzhou and Hangzhou were suspended. In the second year of the Dade era (1298), Shanghai and Shanpu were incorporated into

Fig. 16.6 The prosperous Quanzhou Port during the Yuan Dynasty [14]

Qingyuan. Thus, Guangzhou, Quanzhou, and Qingyuan became the three key ports of the Yuan Dynasty.

Quanzhou Port continued its development from the Song Dynasty and reached its peak during the Yuan Dynasty, expanding both the scope of trade activities and the number of trading countries (Fig. 16.6). Kublai Khan issued a decree stating, "Those foreign merchants who understand my intentions are welcome to declare them. If they are willing to visit, I will treat them with favor. Mutual market transactions are to be conducted freely according to individual preferences." [13]. This decree allowed Quanzhou Port's overseas trade to extend beyond tribute purposes, promoting investment and commerce. This decree received responses from numerous overseas countries, with more than ten countries sending envoys and merchants for visits in year 1279 in the unprecedented quantity.

The Buddhist temple Chengtiansi in Quanzhou preserves a Yuan Dynasty stone stove with an inscription that reads: "In front of the Sun family mansion in Quanzhou, the faithful Pu Limo, accompanied by his wife, Li Erniang, together with devotees from all directions, sincerely seek to achieve Buddhist enlightenment. Joyfully donating a celestial furnace to the Lingying Chan Temple, it shall be perpetually offered. Praying for peace and security in this world and the fruition of blessings in the next life." [15] In this short passage, it reflects a blend of Buddhist and Islamic characteristics, showcasing the collision and integration of various cultures and religions. During the Yuan Dynasty, Quanzhou Port exported ceramics and imported spices, earning the maritime Silk Road during this period the moniker "Spice and Ceramics Route." When Marco Polo visited China in 1275, he was astonished by the prosperity of Quanzhou Port, praising it in his travelogue as "one of the largest ports in the world, with a multitude of merchants gathered here and goods piled up like mountains [16].

After the Yuan Dynasty conquered the Southern Song, they renamed Mingzhou Port to Qingyuan Port [17]. Yuan Dynasty poet Zhang Zhu wrote in the poem "Seeing off Huang Zhongyu to Qingyuan Maritime Office": "Controlling the islands, gathering merchant ships, Pearls and fragrances mixed with ivory and elephants, The tax revenue is so abundant." [18] This poem vividly depicts the thriving foreign trade situation at Qingyuan Port during that time. According to the "Continuation Records of Siming" from the (Zhizheng) era, there were over 220 types of goods transported to Qingyuan by the end of the Yuan Dynasty, a significant increase from the 160 types during the Song Dynasty [19], indicating the development of maritime trade during the Yuan Dynasty. The Zhoushan Archipelago around Qingyuan served as a natural bridge for voyages to and from the Korean Peninsula and the Japanese archipelago, providing excellent natural navigation conditions to Korea and Japan. Qingyuan Port was the hub for China's trade with Goryeo (modern-day Korean Peninsula). During the Yuan Dynasty, the maritime route from China to Goryeo was well-established. Goods imported to Qingyuan Port from Goryeo included ginseng, pine nuts, safflower, Goryeo bronze wares, almonds, and white turmeric. Exports from Qingyuan included raw silk, ceramics, tea, books, and other products from Jiangnan. After the death of Emperor Shizu Kublai Khan, Yuan Dynasty's plans for Japan were temporarily suspended, but Yuan-Japan trade began to develop. Qingyuan Port became the preferred destination for Japanese merchant ships in trade with the Yuan Dynasty. The trade was primarily dominated by Japanese ships coming to the Yuan Dynasty, importing copper coins, ceramics, Buddhist materials, and timber from Japan [14].

In comparison to Quanzhou Port, which was spared from warfare, Guangzhou Port faced multiple sackings by the Yuan forces towards the end of the Southern Song. As a result, Guangzhou Port lagged behind in terms of economy, politics, military conditions, and the overall development of the maritime industry, being overshadowed by Quanzhou Port. In the 23rd year of the Yuan era (1286), Kublai Khan sent envoys from Guangzhou Port to Southeast Asia to attract countries to trade with the Yuan Dynasty. Guangzhou Port was then tasked with managing trade affairs between the Yuan Dynasty and countries like Nanyang (present-day ASEAN countries), leading to its resurgence. According to the "Records of South Seas in the Dade era," 'it was a gathering place for foreign ships, with treasures and valuables abundant, making it a foreign trade hub.' [20]. During the Yuan Dynasty, Guangzhou Port mainly conducted trade with countries like Zhenla and Champa, importing exotic treasures such as ivory, rhinoceros horns, pearls, scented wood, and precious fabrics for the enjoyment of royalty and nobility. While Guangzhou Port's position was inferior to Quanzhou Port during the Yuan Dynasty, it remained an important overseas trade port. Yuan Dynasty's advancements in maritime technology and transportation capabilities, compared to the Song Dynasty, laid the foundation for foreign merchants to conduct goods and material exchanges through seaports. This made the Yuan Dynasty the most prosperous era for maritime foreign trade in Chinese history.

Maritime Prohibition Policy During the Ming and Qing Dynasties

As feudalism gradually became conservative and rigid, the Ming and Qing dynasties implemented a "maritime prohibition" policy, severely hindering the further development of China's maritime industry and the continuous progress of maritime science and technology. China's maritime trade entered a period of decline from prosperity.

The maritime prohibition, also known as "closing the country to the outside world," was a policy rigorously enforced by the Ming and Qing governments to restrict foreign communication and trade. The connotations of maritime prohibition differed between the Ming and Qing dynasties. During the Ming Dynasty, maritime prohibition only banned private overseas trade by civilians, not foreign countries coming to trade. It was a policy where the government monopolized overseas trade. In the early Ming Dynasty, the main ports for foreign trade were Quanzhou, Ningbo, and Guangzhou, each with a Maritime Transport Office. During the Yongle era (1403–1424), the maritime prohibition policy relaxed, and Liujiagang became the starting port for Zheng He's voyages to the Western Seas, developing due to its advantageous geographical location near the mouth of the Yangtze River. By the Jiajing era (1522–1566), coastal areas were often harassed by Japanese Wokou pirates. Emperor Jiajing Zhu Houzhao closed the ports of Quanzhou and Ningbo, leaving only Guangzhou Port. At the same time, Liujiagang, gradually declining due to Wokou pirate invasions (see Fig. 16.7). In the first year of the Longqing era (1567), the Ming Dynasty opened a legal private trade port—Yuegang (modern-day Zhangzhou Port), marking the end of the maritime prohibition policy. "Dong Xi Yang Kao" records: "Permission to trade in the Eastern and Western Oceans." [21].Ships departing from Yuegang could sail to countries including Japan.

In the early Qing Dynasty, the implementation of maritime prohibition was primarily due to the significant threat posed by the Zheng family in Taiwan to the Qing rule. In order to combat and prevent such anti-Qing political forces, Emperor Shunzhi ordered the enforcement of maritime prohibition. After the reign of Emperor Qianlong, to prevent Western colonial invasions, the Qing Dynasty began to implement strict maritime prohibition and a policy of closing the country to the outside world. Simultaneously, the Qing Dynasty considered itself the "Celestial Empire," believing that it was rich in resources and self-sufficient. By keeping the country closed for an extended period, they strictly restricted the interactions of their people with the outside world.

In terms of the timeline, during the Shunzhi period (1644–1661), the Qing Dynasty officially issued the "Maritime Prohibition Edict." During the Kangxi era (1662–1722), after successfully recovering Taiwan, the Qing Dynasty announced the lifting of maritime prohibition and the opening of ports such as Xiamen, Ningbo, Shanghai, and Tianjin. They imported gold and silver minerals, abalones, shark fins, and other marine products from Japan and imported grains from Nanyang to address domestic food shortages. Simultaneously, they developed trade between the north and south, leading to the resurgence of ports and prosperity. In the twenty-second year of the

Fig. 16.7 The scene of the Ming Dynasty army fighting Japanese pirates [22]

Qianlong reign (1757), the Qing government ordered the cessation of all foreign trade, except in Guangzhou, known as the "Single-Port Trade" policy. From then on, the Qing Empire maintained trade only through Guangzhou Port with the West. The maritime Silk Road trade through Guangzhou experienced greater development than during the Tang and Song dynasties, forming an unprecedented global trade cycle. This trade continued and remained stable until the eve of the Opium Wars.

Imperialism's Gunboat Diplomacy and Concession Territories

In the late Qing Dynasty, the corruption of the Qing government intensified, imperialism began to invade China, and China gradually lost its territorial water sovereignty. Western powers have gradually seized our country's inland waterway shipping rights. Ships of foreign capitalist companies have been rampant in our country's inland rivers, grabbing a lot of benefits.

Ports Completely Controlled by Foreigners

In 1840, the Opium War broke out. Britain compelled the Qing government to sign the Treaty of Nanjing, stipulating that "the tariffs on imports and exports, as well as other charges, shall be fixed after due consultation and agreement" [23]. China transitioned from autonomous tariff determination to negotiated tariffs. Simultaneously, the ports of Guangzhou, Fuzhou, Xiamen, Ningbo, and Shanghai were gradually opened as commercial ports (see Fig. 16.8). The era of the Thirteen Hongs in Guangzhou, which monopolized foreign trade for over a century, came to an abrupt end. China's port development entered the period of the "Five Treaty Ports," marking a significant shift in its trade policies.

From this period onwards, Chinese ports began to fall under the control of foreign powers, becoming "bridgeheads" for imperialist aggression against China's resources

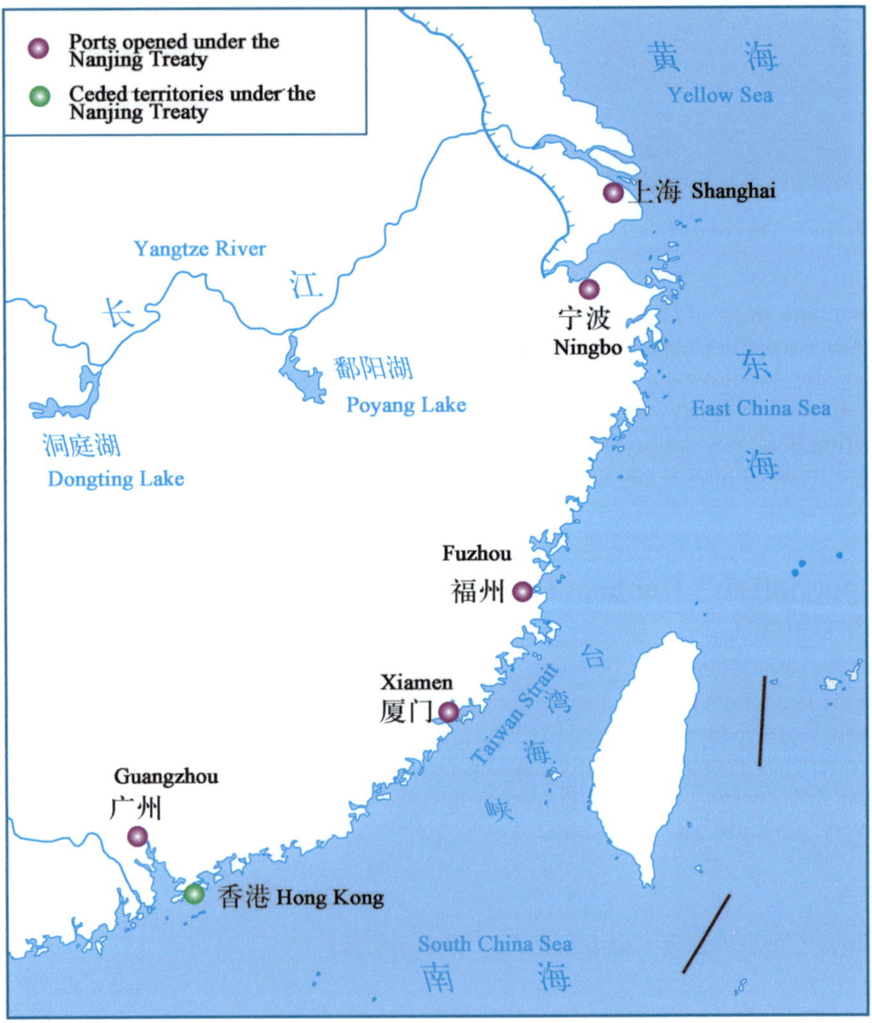

Fig. 16.8 Ports and ceded territories opened by the treaty of Nanjing [24]

[54]. Imperialism started flooding China with goods and plundering Chinese raw materials.

Guangzhou Port was the only port open to foreign trade during the Qing Dynasty and was demanded by Britain to remain open. Ningbo Port and Shanghai Port, due to their strategic locations in the Yangtze River Delta and at the mouth of the Yangtze River, also became ports demanded by Britain. The forced opening of Fuzhou Port and Xiamen Port by Britain was primarily for the tea trade. During the negotiations of the Treaty of Nanjing, the British side claimed, "For the sale of tea, Fuzhou is the most convenient, and we strive for permission to trade" [25], insisting on the opening of Fuzhou. From 1854 until the 1880s, the volume of tea trade between China and Britain remained consistently high.

In 1856, the Second Opium War erupted. Britain and France forced the Qing government to sign a series of unequal treaties. The Treaties of Tianjin and Beijing were both signed during this period, compelling the Qing government to open commercial ports such as Qiongzhou, Chaozhou (later changed to Shantou), Danshui, Taiwan (later changed to Tainan), Zhenjiang, Jiangning (Nanjing), Dengzhou (later changed to Yantai), Tianjin, Niuzhuang (later changed to Yingkou), Hankou, Jiujiang, and others. They constructed wharves within their respective concession areas, seizing not only the right to build ports but also the management of the ports in China. Under the political, economic, and military oppression of imperialism, the Qing government began to lose administrative sovereignty over customs, allowing Western powers to control Chinese customs.

In 1894, the First Sino-Japanese War broke out. After the Qing government's defeat, they were forced to sign the Treaty of Shimonoseki, opening ports in Chongqing, Shashi, Suzhou, Hangzhou (see Fig. 16.9). This change illustrates the deepening extent of imperialist aggression, extending from the coastal areas to the inland, intensifying economic exploitation, and deepening the semi-colonial and semi-feudal processes in China. With the loss of trade rights along the coast and the increase in commercial ports, China's coastal trade became completely monopolized by foreign powers.

Seizure of Inland River Navigation Privileges by Foreign Powers in China

The process of China losing its inland river navigation rights began roughly during the period of the Second Opium War. After the Second Opium War, foreign powers seized sovereignty over coastal, river, and inland river navigation through unequal treaties such as the Treaty of Tianjin, Treaty of Aigun, Treaty of Yantai, Treaty of Shimonoseki, and a series of Aftermath Treaties on Commercial Regulations [26].

"Global History" records, "Western powers not only controlled coastal ports but also some major inland river channels." The Treaty of Tianjin also included a specific provision on Yangtze River navigation rights, stating, "Various ports along

Fig. 16.9 The opening of China's ports during the period of imperialist aggression

the Yangtze River are open for British merchant vessels," and "British vessels are allowed to navigate from Hankou upstream to various destinations, with no more than three ports, as the area for the transportation of goods and trade by British vessels." [24]. Yangtze River inland river navigation rights began to diminish. In the same year, the Sino-Russian Treaty of Aigun stipulated: "From the Black Dragon River, Songhua River, Ussuri River, henceforth only China and Russia are allowed navigation; no other foreign vessels are allowed to navigate these rivers." [27]. With this, Russia gained navigation rights on the Black Dragon River, Songhua River, and Ussuri River. In 1876, citing the Margary Affair, Britain forced the Qing government to sign the Treaty of Yantai, allowing foreign steamships to travel from Hankou upstream to Yichang. In 1895, the Sino-Japanese Treaty of Shimonoseki was signed, permitting Japanese steamships to enter Chongqing, carrying passengers and cargo. Thus, foreign steamships obtained privileges for navigating the upper reaches of the Yangtze River.

In addition to the Yangtze River inland river navigation rights, foreign powers also gained navigation privileges on other inland rivers and their tributaries [28]. According to Article 6 of the Sino-Japanese Treaty of Shimonoseki, foreign ships' navigation on the Yangtze River extended from the mainstream to its tributaries. In 1902, imperialist powers further expanded their inland river and harbor navigation rights through the "Continuation of Negotiations on Trade and Navigation Treaty between China and Britain," designating places such as Changsha in Hunan as commercial ports. Ten locations in Guangdong, including Baitu Port, Luoding Port, and Ducheng, were set aside as temporary stops for boarding and disembarking passengers and cargo, following the regulations for mooring on the Yangtze River. At this time, foreign ships were allowed to use places such as Rongqi, Maning, Jiujiang, Gulao, Yong'an, Houlai, Lubu, Yuecheng, Ludu, and Fengchuan as locations for boarding and disembarking passengers [29].

During this period, the trend of inland river steamship navigation spread like mercury, infiltrating every possible route. The navigation rights on several inland rivers in Guangdong and Guangxi were almost entirely seized, and the navigation rights of several southern rivers, including the Yangtze and Pearl rivers, were also encroached upon.

Inland river navigation rights were crucial. From the aftermath of the Opium War to the fall of the Qing government, the navigation rights of various inland river routes had effectively been undermined by the invaders. Imperialism monopolized China's inland river shipping industry throughout this period. The provisions regarding specific rights in unequal treaties merely formalized existing facts, "legalizing" illegal activities.

As Li Yumin pointed out in "The Treaty System of Modern China," every loss of China's inland river navigation rights was accompanied by armed aggression and threats of war from the foreign powers. The late Qing government was powerless to resist, allowing foreign vessels to navigate China's coastal, inland river, and harbor areas, greatly facilitating their economic aggression, which caused significant harm to China's politics, national defense, and economy [30].

The Ports and Water Transportation in the Early Years of the People's Republic of China

Gradual Recovery and Development of Ports

Before the establishment of the People's Republic of China, ports were long subjected to foreign forces, becoming strongholds for imperialist nations to plunder China's resources and wealth. Due to prolonged national division and conflicts that resulted in economic backwardness, the construction of ports faced numerous challenges. Ports were not only limited in number and scale but also characterized by chaotic and irrational layouts. They were mainly concentrated in the trading ports controlled by imperialist powers. The development of ports during that period was distorted, serving as bases for Western powers to exploit China's resources and market their goods.

At the time of the establishment of the People's Republic of China, the ports were extremely backward. The country had over 100 berths, with only five ports—Dalian Port, Qinhuangdao Port, Qingdao Port, Shanghai Port, and Bayuquan Port—having deep-water berths capable of handling ten thousand-ton vessels. The total length of port terminals was just over 20,000 m, and the total cargo throughput was less than ten million tons. Many terminals were severely damaged due to long-term neglect, and waterways were heavily silted. Most terminals handled general cargo with simple operations and rudimentary equipment, having only around 200 sets of loading and unloading machinery, with manual labor being the primary method.

After the establishment of the People's Republic of China, the ports experienced a rebirth. Due to imperialist maritime blockades and the focus on economic development in the interior, transportation primarily relied on railways, resulting in slow growth in maritime activities. During this phase, port development primarily focused on technological upgrades and restoration. To quickly change this backward situation, the country initiated the restoration and reconstruction of ports, repairing dilapidated seaports and renovating major coastal ports such as Dalian, Tianjin, Shanghai, and Qinhuangdao. Several new coastal ports, including Zhangjiagang Port, Zhanjiang Port, and Fangchenggang Port, were also constructed. Notably, the planning for the reconstruction of the Tanggu New Port in Tianjin began in 1949 and became the first deep-water seaport completed after the establishment of the People's Republic of China. Figure 16.10 shows a postage stamp issued in 1954 for the Tanggu New Port in Tianjin. Zhanjiang Port, completed in 1956, was the first comprehensive seaport independently designed and constructed after the establishment of the People's Republic of China. Specialized mechanized coal and ore terminals were built in ports like Wuhan and Nanjing, with the Yuxi Kou Coal Terminal, initiated in 1955, being the first specialized coal terminal built after the establishment of the People's Republic of China.

From 1949 to 1973, through reconstruction, expansion, and new construction, the number of port berths in China increased to over 600, including over 90 deep-water berths capable of handling ten thousand-ton vessels. The total length of port terminals

Fig. 16.10 Tanggu New Port stamps in Tianjin

increased to over 50,000 m, and the total cargo throughput reached over one hundred million tons."

The First Peak of Inland Waterway Construction

The first peak of inland waterway construction occurred in the mid to late 1950s. During the First Five-Year Plan period, the focus was on the comprehensive improvement of the Yangtze River and Jialing River, achieving 24-h navigation. This turned the treacherous waterways into accessible routes, reducing the shipping time from 10 days to 4–6 days. From 1958 to 1961, the large-scale improvement of inland waterways began with the expansion of the Grand Canal project. The navigable length of the Grand Canal was 1044 km, with waterways deeper than 1 m covering 970 km. It was divided into the northern section (SuBei) and the southern section (SuNan), with the northern section being 461 km long and the southern Sect. 583 km long. The focus was on the improvement of the northern section from Xuzhou to Yangzhou, covering a river segment of 404 km. The project required a total investment of 200 million yuan and involved a massive workforce of 1.23 million people. The scale of investment and the number of participants were unprecedented in the first 20 years of waterway construction in New China, surpassing anything seen in the previous 50 years of other construction projects.

The First Peak of Port Construction

With changes in the international environment and China's foreign relations, especially the restoration of China's legitimate seat in the United Nations and the normalization of Sino-American relations and Sino-Japanese diplomatic relations, there was a surge in establishing diplomatic relations between China and relevant countries. In the early 1970s, as China's foreign relations developed, the number of countries establishing diplomatic relations with China exceeded 90. Foreign trade rapidly expanded, leading to a substantial increase in trade volume and cargo transportation. Against this backdrop, the contradiction of inadequate capacity for handling coastal port cargo became increasingly prominent, resulting in the "three pressures" issue (pressure on ships, cargo, and vehicles). Faced with this dilemma, on February 27, 1973, Premier Zhou Enlai instructed that the situation should be transformed within three years to alleviate the challenges faced by the ports.

Against this backdrop, the nation witnessed the onset of the first wave of port construction, later known as the "Three-Year Major Port Construction." Crude oil export terminals were constructed in ports such as Dalian, Qinhuangdao, and Qingdao. Simultaneously, unloading terminals were established in ports like Shanghai, Nanjing, and Guangzhou, meeting the transportation needs of oil fields such as Daqing, Panjin, and Shengli. New port areas were developed in Qinhuangdao, Tianjin, Yantai, Lianyungang, Shanghai, Ningbo, Fuzhou, Zhanjiang, and Fangchenggang, alleviating the urgent requirements for miscellaneous cargo transportation.

From 1973 to 1982, a total of 51 deepwater berths were constructed nationwide, adding a handling capacity of 120 million tons. During this period, China independently designed and built a 50,000/100,000-ton crude oil export terminal in Dalian, marking a significant milestone [31]. The development and construction during this period helped train and cultivate China's port construction teams, laying a solid foundation for future port development. Several model workers emerged during this time, leaving behind numerous outstanding stories. For example, during the construction of Dalian Nianyuwan Oil Port and Tianjin New Port, the construction team, in the face of harsh weather conditions, demonstrated heroic spirit and mutual encouragement, overcoming difficulties through continuous day and night operations. This period also saw the emergence of advanced collectives like the "March 8th Team" and the "Youth Assault Team" (see Figs. 16.11 and 16.12) [32].

The 'Three-Year Big Port Construction' effectively alleviated the backwardness of port and shipping infrastructure at that time, achieving remarkable benefits. During

Fig. 16.11 During the construction period of Dalian Neiyyuwan Oil Port, the China Communications First Navigation Bureau had a distinguished group of female workers known as the "March 8th Team" [32]

Fig. 16.12 July 1, 1973, Youth Assault Team at the construction site of the new wharf in Tianjin Port [32]

this period, Qinhuangdao Port faced challenges due to the mismatch between port capacity and demand, leading to issues such as vessel congestion, cargo pressure, and port blockage. Over 200 foreign and ocean-going vessels were in the port every day, but only one-third could berth for operations, while two-thirds waited at anchor for loading and unloading. In 1972, the average port stay for each foreign vessel

Fig. 16.13 Qinhuangdao Port crude oil terminal [32]

reached 10.9 days, with more than 20 vessels often having a port stay of around one month. After three years of construction, Qinhuangdao Port underwent significant transformation.

From 1973 to 1975, a total investment of 1.6 billion yuan was made, with the construction of four new berths and the extension of the quay line from 1,297 to 2,593 m. The East Port area was developed, and the first crude oil pipeline transport terminal in China was built (see Fig. 16.13). The port's throughput doubled, increasing from 5.12 million tons in 1972 to 15.57 million tons in 1975, rising from the sixth to the third among coastal ports in the country [33].

Development of Transport Ships

During this period, alongside the significant expansion of ports, there was a corresponding development in transport vessels. In the 1950s, wooden sailing ships were the primary means of transportation. In the eve of the founding of the People's Republic of China, motorized vessels (with a total capacity of over one million tons) were coerced by the Nationalist forces to leave the mainland, leaving behind wooden sailing ships for inland water transport. Throughout the entire 1950s, the tonnage of wooden sailing ships remained above 3 million tons, becoming the backbone of waterway transportation. It wasn't until the mid-to-late 1950s that coastal transport saw the addition of some passenger and cargo ships, while inland water transport welcomed a batch of thousand-ton barges and high-powered towboats [34].

In the 1960s and 1970s, cement ships and "second-hand ships" experienced rapid development. In the early 1960s, cement ships began operations, and by the 1970s, their numbers surged. The application of cement vessels expanded from cement sailing ships to cement barges and further to motorized cement ships, covering both inland and coastal transport. To strengthen China's ocean shipping fleet, break the long-term passive situation of leasing a large number of foreign vessels, the Ministry of Communications, in 1963, began utilizing China Bank's international funds to purchase "second-hand ships" through loans. In accordance with Premier Zhou's instructions, nearly one million tons of vessels were purchased on average each year, with 100,000-ton vessels being domestically constructed [34].

Restoration and Development of Shipping

The construction of ports supported the recovery and development of shipping. In April 1949, the People's Liberation Army issued orders to blockade the Yangtze River, leading to the interruption of passenger and cargo transportation along the river. By June 1st, cities such as Nanjing and Wuhan were liberated, and the blockades and martial law along the Yangtze were lifted [34]. The first vessels to resume operations were passenger and cargo liners serving the Shanghai-Hankou route. On June 3rd, the "Jiangling" liner made its inaugural voyage from Hankou, becoming the first passenger and cargo liner to resume service on the Yangtze River and in Shanghai [34]. Simultaneously, the "Hailiao" liner (as shown in Fig. 16.14) broke through the heavy blockades imposed by the Nationalist government at sea. It arrived at Dalian Port in the liberated area just before the founding of the People's Republic of China, becoming the first sea liner to reach the liberated area and the first to hoist the Five-Star Red Flag of the People's Republic of China after its establishment [35].

Fig. 16.14 The 5-cent banknote with the "Hailiao" liner printed on it [35]. The rapid development of ports and water transportation since the reform and opening up

The Second Peak of Port Construction

During the "Sixth Five-Year Plan" and "Seventh Five-Year Plan", the port entered the second construction climax, and the construction of coastal ports entered a stage of rapid development. The country listed ports as a strategic focus of national economic construction in the "Sixth Five—Year Plan " (1981–1985). During the "Sixth Five-Year Plan" period, a total of 54 deep-water berths were built, with an additional handling capacity of 100 million tons. After five years of construction, China's ports with 10,000-ton berths increased from 11 in 1980 to 15 in 1985, with a throughput of 317 million tons in 1985 [34].

The "Seventh Five-Year Plan" (1986–1990) witnessed the fastest development in the 40-year history of coastal port construction, with 186 berths constructed and an additional throughput capacity of 150 million tons. Ninety-six deepwater berths were completed, surpassing the total built in the first 30 years of the People's Republic of China. The newly constructed ports included 18 coal berths, three container terminals, and numerous large-scale loading and unloading berths for commodities like ore and fertilizer, reaching a level that ranked among the world's best. The number of ports with deepwater berths increased to over 20, and nine ports achieved an annual throughput exceeding 10 million tons [34]. The early 1980s saw the completion and opening of the Shekou Wutong Wharf, as depicted in Fig. 16.15.

Fig. 16.15 Shekou Wuwan Jetty Wharf was completed and opened in the early 1980s [36]

The Third Wave of Port Construction

To adapt to the further development of the socialist market economy, a third wave of port construction emerged. With the deep implementation of the reform and opening-up policy and new changes in the international shipping market, China began to focus on the deepening and specialization of berths. In particular, during the Fourth Session of the Seventh National People's Congress, the "Ten-Year Plan for National Economic and Social Development and the Outline of the Eighth Five-Year Plan" was deliberated and approved. It clearly stated that transportation is a fundamental industry. The key focus of port construction was on the hub ports located on the main maritime channels of China and the terminals for three major transportation systems: coal, containers, and roll-on/roll-off ships for passengers and cargo [37].

The country constructed a number of large-scale, technologically advanced modern terminals and berths in the key hub port areas. It established transportation systems for coal, containers, crude oil, and imported iron ore, effectively alleviating the bottleneck of infrastructure and capacity contradictions in the early stages of reform and opening up. A hierarchical layout framework was basically formed, with 20 main hub ports such as Dalian, Qinhuangdao, Tianjin, Qingdao, Shanghai, and Shenzhen, supplemented by regionally important ports, and appropriate development of small and medium-sized ports. By the end of 1997, coastal ports in the country had a total of 1,446 berths at intermediate level and above, including 553 deep-water berths, with a throughput capacity of 9.58 billion tons. This was four times the capacity at the beginning of the reform and opening up, and the throughput volume increased from 317 million tons in 1980 to 9.68 billion tons in 1997 [37]. The scene of the construction of Yantian Port in 1988 is shown in Fig. 16.16.

Fig. 16.16 The first pile was laid for the start-up project of Yantian Port in 1988 [38]

The Second Wave of Inland Waterway Construction

The second peak of inland waterway construction occurred in the 1980s and beyond. The first ten years were a stage of rehabilitative improvement, with major projects including the Jiangsu North Section of the Grand Canal navigation project. From 1984 to 1998, with an investment of over 600 million yuan, the 404-km section from Xuzhou to Yangzhou was constructed according to the third-level navigation standards, allowing thousand-ton vessels to travel directly from Xuzhou to Yangzhou. Other significant projects during this period included the first-phase project of the 575-km Qianton-level waterway from Xijiang Guigang to Guangzhou [34]. The subsequent ten years marked a stage of planned improvement with intensified efforts. Key projects included the 532-km 500-ton level waterway project from Xiangfan to Hankou on the Han River, the 303-km Qianton-level waterway project on the Lanxiu section of the Yangtze River, and the 20,000-ton level outbound waterway project at the mouth of the Minjiang River. Particularly during the "Ninth Five-Year Plan" period (1996–2000), the number of major projects significantly increased. These projects included the 20-billion-yuan second-phase project of the Xijiang waterway, enabling the 847-km waterway from Guangzhou to Nanning to meet the thousand-ton level standards. There was also an 86.2-million-yuan project for the improvement of the Jiepai waterway in the middle reaches of the Yangtze River, a 120-km Qianton-level waterway project from Dayu Crossing to Zhuzhou on the Xiangjiang River, achieving seamless navigation between the river and the sea, and a 15-billion-yuan project for the continued construction of the Shandong section of the Beijing-Hangzhou Grand Canal. After completion, a fleet of 2,000-ton vessels could travel directly to Jining, ensuring unimpeded water transport in the Su-Lu region [34]. The waterway improvement project of the southern section of the Grand Canal from 1992 to 1997, with an investment of 2.7 billion yuan, was the largest, highest-standard, most challenging, and most significant waterway improvement project in the history of the southern canal. The deep-water channel governance project at the mouth of the Yangtze River, which began at the end of 1997 (Fig. 16.17), alleviated the deep-water channel issues hindering the development of Shanghai as an international shipping center [34].

A New Round of Port Building Upsurge

After entering the 21st century, the process of economic globalization accelerated, technological revolution rapidly advanced, and industrial structure continuously optimized and upgraded. The competition for comprehensive national strength has intensified. To adapt to the changing international situation and the rapid development of the national economy, major ports across the country have actively conducted strategic research on port development, developed and constructed port information systems, and invested substantial funds in the construction of large deep-water

Fig. 16.17 Yangtze estuary deep waterway improvement project [39]

and specialized berths, ushering in another wave of port construction climax. As of the end of 2003, there were a total of 4,274 productive berths in coastal ports nationwide, including approximately 748 berths with a capacity of over 10,000 tons. The comprehensive throughput capacity was 1.67 billion tons, and the total cargo throughput reached 2.064 billion tons [34].

Development of Transport Ships

In the 1980s, specialized ships experienced rapid development. The reform and opening up brought a refreshing change to the appearance of the fleet, shifting the structure of cargo fleets from single-purpose general cargo ships to multi-purpose ships, bulk carriers, and specialized vessels. The rapid development of specialized ships, especially container ships, since the purchase of the first semi-container ship "Pingxiangcheng" by China Ocean Shipping Company in 1978, improved the variety and structure of transportation vessels in China. By 1990, various types of container ships reached 99, with more than 100,000 standard containers. After the 1990s, while the net tonnage of the national water transportation fleet slowed down or even decreased, the number of vessels began to decrease significantly. Adjusting the tonnage structure of the fleet and enhancing its competitiveness became a major task [34].

Overall, in the 50 years since the founding of the People's Republic of China, the composition of ships has undergone a transformation from wooden ships to cement ships and then to steel ships. The types of ships have evolved from general cargo ships and passenger-cargo ships to various types of vessels. The power of ships has passed through four stages: human-powered wooden ships, wind-powered sailing ships, coal-fired steamships, and oil-fired diesel ships. In terms of business operations, the main entities engaged in ship operations have shifted from state-owned shipping enterprises towards a diversified direction involving state-owned, collective, and private entities. This has formed a development pattern where ocean-going, coastal, and inland fleets advance together, meeting the needs of national economic development and foreign trade [34].

Ports and Water Transport Today and Tomorrow

China's Status as a Major Water Transport Country Has Been Established

Entering the twenty-first century, especially since the 18th National Congress of the Communist Party of China, the country has established five major coastal port clusters in the Bohai Rim, Yangtze River Delta, Southeast Coast, Pearl River Delta, and Southwest Coast regions. It has basically formed a multi-level development pattern led by major ports such as Shanghai Port, Dalian Port, Tianjin Port, Qingdao Port, Ningbo-Zhoushan Port, Shenzhen Port, and Guangzhou Port, with other ports developing together [40]. The service capacity and infrastructure level of ports have continuously and significantly improved. Among the top 10 ports globally in terms of cargo throughput and container throughput, China holds 8 and 7 positions, respectively (Tables 16.1 and 16.2). In 2020, China's ports achieved a cargo throughput of 14.55 billion tons (Fig. 16.18) and a container throughput of 260 million TEUs, maintaining the world's top position [41].

China's water transportation accounts for approximately 90% of the country's foreign trade cargo transport and a significant volume of inter-regional material transportation tasks, making it an indispensable mode of transport within China's comprehensive transportation system. In 2020, the proportion of national waterway cargo turnover in the overall transportation cargo turnover was 53.9%, an increase of 4 percentage points compared to 2018. The intermodal transportation volume of containers through ports by rail and water reached 6.87 million standard containers, representing a 53% growth compared to 2017. Water transportation has played a pioneering role in the development of the Yangtze River Economic Belt, coordinated development in the Beijing-Tianjin-Hebei region, integrated development in the Yangtze River Delta, and the construction of the Guangdong-Hong Kong-Macao Greater Bay Area, aligning with major national strategic initiatives. The changes in the foreign trade cargo throughput and container throughput at national ports since 2011 are illustrated in Figs. 16.19 and 16.20, respectively.

International maritime shipping has achieved rapid development. Maritime transportation handles approximately 90% of China's foreign trade cargo, playing a crucial role in promoting foreign trade development and safeguarding national economic security. In 2020, Chinese ports handled 1.4 billion tons of imported metallic ores, 660 million tons of imported petroleum and products, 240 million tons of imported coal, and 140 million tons of imported grains, accounting for around 90% of the respective total foreign trade import volumes in these categories.

Table 16.1 World port cargo throughput ranking

Port 2021 ranking (2020 ranking)	Country and area	Port name	2021_cargo throughput (10,000 tons)	2020_cargo throughput (10,000 tons)	Year-on-year growth rate (%)
1 (1)	China	Ningbo-Zhoushan	122,405	117,240	4.4
2 (2)	China	Shanghai	76,970	71,104	8.2
3 (3)	China	Tangshan	72,240	70,260	2.8
4 (5)	China	Qingdao	63,029	60,459	4.3
5 (4)	China	Guangzhou	62,367	61,239	1.8
6 (6)	Singapore	Singapore	59,964	59,074	1.5
7 (7)	China	Suzhou	56,590	55,408	2.1
8 (8)	Australia	Hedland	55,327	54,705	1.1
9 (10)	China	Rizhao	54,117	49,615	9.1
10 (9)	China	Tianjin	52,954	50,290	5.3
11 (11)	Netherlands	Rotterdam	46,871	43,681	7.3
12 (12)	South Korea	Busan	44,252	41,120	7.6
13 (13)	China	Yantai	42,337	39,935	6.0
14 (18)	China	Taizhou	35,291	30,111	17.2
15 (24)	China	Jiangyin	33,757	24,705	36.6
16 (15)	China	Dalian	31,553	33,401	−5.5
17 (17)	China	Huanghua	31,134	30,125	3.3
18 (16)	China	Nantong	30,851	31,014	−0.5
19 (19)	South Korea	Gwangyang	29,206	27,332	6.9
20 (20)	China	Shenzhen Port	27,838	26,506	5.0

The advantages of inland waterway transportation have been further leveraged. On May 20, 2019, the second phase of the 12.5-m deepwater channel project below Nanjing on the Yangtze River was successfully completed and accepted, officially entering the operational phase, enabling 50,000-ton ocean vessels to directly reach Nanjing Port. Since the 13th Five-Year Plan, China has continuously strengthened the development of inland waterway transportation infrastructure, forming a high-grade waterway network centered around the Yangtze River trunk line, the Xijiang River shipping trunk line, the Grand Canal, and high-grade waterways in the Yangtze River Delta and the Pearl River Delta. As of the end of 2020, the total navigable mileage of the national inland waterways reached 127,700 km, ranking first globally, with 16,100 km of high-grade waterways. In 2020, the national inland water transport freight volume reached 3.82 billion tons, consistently maintaining its position as the world's top performer for multiple years.

The construction of the Shanghai International Shipping Center has achieved positive results. Shanghai Port has become the world's most densely connected and widely radiating port for container shipping. In 2020, the container throughput at

Table 16.2 World port container throughput ranking

Port 2021 ranking (2020 ranking)	Country and area	Port name	2021_ container throughput (10,000 TEU)	2020_ container throughput (10,000 TEU)	Year-on-year growth rate (%)
1 (1)	China	Shanghai	4703	4350	8.1
2 (2)	Singapore	Singapore	3747	3687	1.6
3 (3)	China	Ningbo-Zhoushan	3108	2872	8.2
4 (4)	China	Shenzhen	2877	2655	8.4
5 (5)	China	Guangzhou	2418	2317	4.4
6 (6)	China	Qingdao	2371	2201	7.7
7 (7)	South Korea	Busan	2269	2181	4.0
8 (8)	China	Tianjin	2027	1835	10.4
9 (9)	China Hong Kong	Hong Kong	1780	1797	−0.9
10 (10)	Netherlands	Rotterdam	1530	1435	6.6
11 (11)	United Arab Emirates	Dubai	1374	1349	1.9
12 (12)	Malaysia	Klang	1374	1324	3.7
13 (14)	China	Xiamen	1205	1141	5.6
14 (13)	Belgium	Antwerp	1202	1203	−0.1
15 (15)	Malaysia	Tanjong Pagar Puerto La Paz	1120	980	14.3
16 (17)	USA	Los Angeles	1068	921	15.9
17 (16)	Taiwan, China	Kaohsiung	986	962	2.5
18 (19)	USA	Long beach	938	811	15.7
19 (21)	USA	New York-New jersey	899	759	18.5
20 (18)	Germany	Hamburg	870	853	2.0

Shanghai Port reached 43.5 million standard containers, maintaining its position as the world's top port for the 11th consecutive year. Six of the world's top ten ship management organizations and ten formally registered members of the International Association of Classification Societies have established regional headquarters or branches in Shanghai. Shanghai's maritime insurance and cargo insurance premium income rank second only to London and Singapore. The China Export Container Shipping Index (CCFI) published by the Shanghai Shipping Exchange has become one of the three major global shipping market indices. According to the "2021 Xinhua-Baltic International Shipping Center Development Index Report," the top 10 global shipping center cities in comprehensive strength in 2021 are Singapore, London, Shanghai, Hong Kong, Dubai, Rotterdam, Hamburg, Athens/Piraeus,

Fig. 16.18 Changes in China's port cargo throughput since 2011. *Note* Data comes from the Ministry of Transport

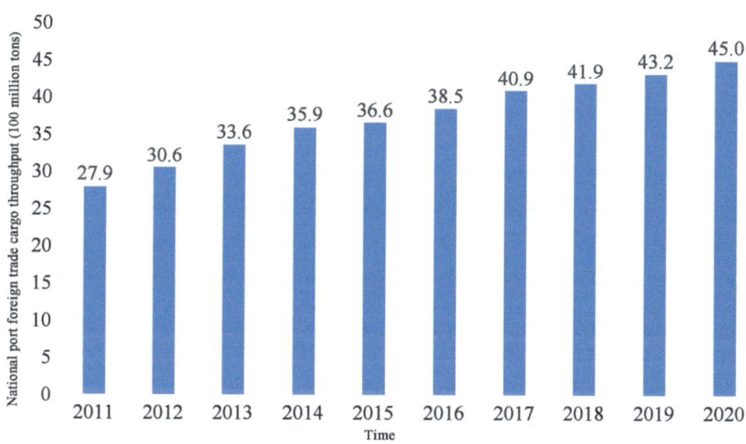

Fig. 16.19 Changes in foreign trade cargo throughput at national ports since 2011. *Note* Data comes from the Ministry of Transport

New York/New Jersey, and Ningbo-Zhoushan. Shanghai, leveraging hub construction, dual-drive development of the service industry, mechanisms such as the Free Trade Pilot Zone, technological innovation empowerment, and coordinated development in the Yangtze River Delta, continues to enhance its global resource allocation capabilities and maintains a comprehensive ranking of third place. Ningbo-Zhoushan Port, maintaining the world's highest cargo throughput for 12 consecutive years, has entered the top ten in comprehensive strength for the first time.

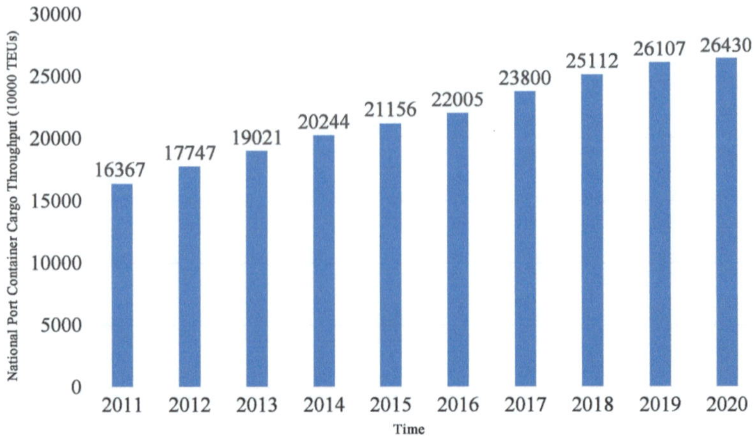

Fig. 16.20 Changes in container throughput of national ports since 2011. *Note* Data comes from the Ministry of Transport

The construction of automated container terminals is at a leading global level. Major container ports are accelerating the promotion of unmanned and intelligent terminal operations. Automated terminals in Qingdao Port, Shanghai Port, and others have been completed and put into operation, reducing operating costs and improving efficiency. Tianjin Port has built a new automated container terminal, actively working towards creating a "zero-carbon terminal." Port enterprises are promoting the enhancement of operational efficiency through truck identity card verification, appointment systems, and advancing the construction of port cloud data centers and electronic documentation. Many ports are accelerating the application of new technologies such as the Internet of Things, big data, Beidou Navigation, blockchain, 5G, and artificial intelligence. The large-scale promotion and application of the Yangtze Electronic Navigational Chart is illustrated in Fig. 16.21 [42, [43].

The pace of construction for green ports and ecological waterways is accelerating. The government has continuously introduced supportive policies, and significant progress has been made in certain areas of green port construction [45]. The utilization efficiency of port shorelines continues to rise, and substantial progress has been made in the rectification of illegal docks, effectively addressing issues related to ship and port pollution. The proportion of ships using shore power when berthed at ports is steadily increasing. According to preliminary survey statistics, by the end of 2019, there were over 5,400 completed port shore power facilities nationwide, covering more than 7,000 berths (including water service areas) [46]. The construction of ecological waterways has shown remarkable results, achieving harmonious development between water transportation and the environment (Fig. 16.22). The safety development level is continually improving, and equipment facilities are

Fig. 16.21 Yangtze River waterway map app demonstration [44]

Fig. 16.22 Yangtze River ecological waterway construction [47]

continuously being enhanced. A maritime supervision and service system covering port areas and nearshore waters has been basically established, with the continuous improvement of capabilities in water safety supervision and navigation security.

Comprehensive Commencement of Building a Strong Nation in Water Transport

In November 2018, during his inspection in Shanghai, General Secretary Xi Jinping connected via video to the automated Phase IV terminal of Yangshan Port. He received a briefing on the construction and operation of the terminal. General Secretary Xi Jinping emphasized that for a strong economy, it is imperative to be a maritime and shipping power.[1] On January 17, 2019, during his inspection of Tianjin Port, General Secretary Xi Jinping stressed that to develop the economy and strengthen the country, especially in transportation, maritime shipping must take the lead. Efforts should be made to build world-class smart and green ports to better serve the coordinated development of the Beijing-Tianjin-Hebei region and the Belt and Road Initiative.[2]

The 19th Party Congress Report made a significant judgment that socialism with Chinese characteristics has entered a new era. China's economic development has also entered a new era, characterized by the transition from high-speed growth to high-quality development. In the future, the overall strategic approach for the development of waterway transportation is guided by Xi Jinping Thought on Socialism with Chinese Characteristics for a New Era. It adheres to the new development philosophy, focuses on the changes in China's major social contradictions, follows the requirements of high-quality development, and revolves around the overall layout of the "Five-Sphere Integrated Plan" and the "Four-Pronged Comprehensive Strategy." The main line is to promote qualitative, efficient, and dynamic changes in water transportation development. Key strategic objectives include the construction of a global service network, an international shipping center, and the development of the Golden Waterway. The focus is on the construction of high-level waterways with the Yangtze River as the core, the improvement of the integrated transportation system focusing on rail-water transport, the establishment of an international shipping center with institutional innovation as the main breakthrough, the construction of a global shipping service network with the Belt and Road Initiative as the focus, and the enhancement of international influence through the standardization of international rules and conventions. These strategic tasks effectively support the implementation of major national strategies, the construction of a transportation power, a manufacturing power, a maritime power, a trade power, and highlight the role of water transportation as a crucial support for China's economic and trade development, national security, major diplomatic strategies, and the construction of a community with a shared future for humanity [48].

---

[1] See "Xi Jinping: Firm Confidence and Determination in Reform and Opening Up—Accelerating the Enhancement of Urban Capacity and Core Competitiveness", People's Daily, November 8, 2018, Page 01.

[2] See Xi Jinping: Steady Progress, Daring Responsibility, Innovation, and Achievements—Driving New and Greater Advances in Coordinated Development of Beijing-Tianjin-Hebei", People's Daily, January 19, 2019, Page 01.

After 2020, China's strategy to become a water transport power is mainly set in two stages, with goals set five years ahead for each stage [48].

By 2030, the goal is to build a water transport power, and China's overall water transport development level will generally be among the world's best. This includes the comprehensive completion of the global shipping service network, the Chinese characteristic international shipping center, and the Golden Waterway across the entire basin. An efficient two-way sea-land transportation logistics service system adapted to China's economic and trade status will be fully established. While further perfecting the traditional shipping service network for trade with other countries, a global shipping service network with a focus on serving countries along the Belt and Road will be basically completed. The inclusive development concept of water transport will be continuously highlighted. The world-class Chinese characteristic international shipping center will be basically completed, and the development level of some modern shipping service industries will rank among the best in the world. The ability to allocate shipping resources to countries along the Belt and Road, with a focus on service, will be significantly improved, becoming a leader in the world maritime governance system. China's solution to the development of the world maritime industry will be increasingly recognized, contributing to the development of the world maritime industry and enhancing its contribution to world trade. It effectively supports the improvement of China's comprehensive national strength and international influence [48].

By 2045, the overall goal is for China's water transport development to be at the world's leading level. This includes a substantial increase in international influence and discourse power in various aspects of the international maritime governance system, the Chinese characteristic international shipping center, and the construction of the Golden Waterway. The comprehensive completion of the global shipping logistics service network will be achieved. The global competitiveness of the Chinese characteristic international shipping center will be further enhanced, and the development level of several sectors of the modern shipping service industry will be among the best globally. It will have strong influence and discourse power, becoming a major contributor to the construction of the global maritime governance system. China's water transport development philosophy will be widely accepted by most countries and regions, and its solutions will be widely adopted by the international community. Water transport will become a key support for China's economic and trade development, national security, major diplomatic strategies, and the construction of a community with a shared future for humanity. It further highlights the role of water transport as a leader in the all-around opening-up of the new pattern and strongly supports the goal of building China into a modern socialist country with Chinese characteristics [48].

Overseas Port Construction and International Logistics Supply Chain Improvement

China's involvement in overseas port construction started early. In 2003, China COSCO Shipping Corporation began cooperation with the Singaporean PSA International to operate container terminals. With the introduction of the "Belt and Road" initiative, Chinese enterprises have accelerated their pace of going global, continuously improving the layout of overseas port networks, and promoting infrastructure connectivity, making them pioneers in the construction of the "Belt and Road." Among them, the Port of Piraeus serves as an exemplary collaboration between China and Greece (Fig. 16.23).

Enterprises participating in the construction of overseas ports not only involve infrastructure development but, more importantly, encompass the post-terminal operation and management. Some also include the development of industrial zones near the port, the construction of integrated transportation systems, and the improvement of shipping service networks. Collaborative port construction not only promotes the enhancement of shipping service networks but also contributes to the creation of a global port and shipping logistics service network system, supporting the prosperous development of international trade for both China and the world. This holds significant importance and plays a crucial role in China's international capacity cooperation, modern supply chain establishment, and the joint construction of a community with a shared future for humanity.

In recent years, China's overseas port cooperation and construction have deepened and solidified, with continuous improvement in the overseas port network and

Fig. 16.23 Exemplary collaboration between China and Greece—Port of Piraeus [49]

notable construction achievements. According to incomplete statistics, as of 2019, China has been involved in the cooperative construction of 58 ports worldwide, spanning 38 countries and regions across all six continents. Particularly since the introduction of the "Belt and Road" initiative, the number of ports with Chinese involvement has grown rapidly, from 9 in 2012 to 58 in 2019, averaging an annual increase of 7 ports (Fig. 16.24). Geographically, the majority are distributed in Asia, Europe, and Africa, with 20 in Asia (35%), 17 in Europe (29%), 10 in Africa (17%), and 19% in other regions (Fig. 16.25). Container terminals constitute the majority of cooperative construction, accounting for 74%, while others mainly include terminals for bulk cargo, oil products, and other goods (Fig. 16.26). In terms of whether the terminal is newly built, the majority involve investment and acquisition of existing terminals (76%), with only 24% being newly constructed, primarily in developing economies. Regarding the degree of participation in terminal operation and management, the majority are joint ventures, with local governments or enterprises holding controlling interests, accounting for 71%. The cooperative periods for terminals vary, with some as short as 20–30 years and others extending up to 99 years. In terms of functionality, the terminals primarily serve the economic and social development of the local regions, while also serving the hinterland and peripheral maritime logistics networks.

At present, the enterprises participating in the construction of overseas port networks are mainly dominated by shipping and port enterprises. Among them, China COSCO Shipping Corporation's COSCO Shipping Ports Limited and China Merchants Group's China Merchants Port Group Co., Ltd. are the main representatives. The number of port terminals jointly constructed by these two companies accounts for 24% and 54% of China's total overseas port cooperation, respectively (Fig. 16.27). The participating enterprises also include large state-owned port enterprises, major investment groups, private enterprises, and others.

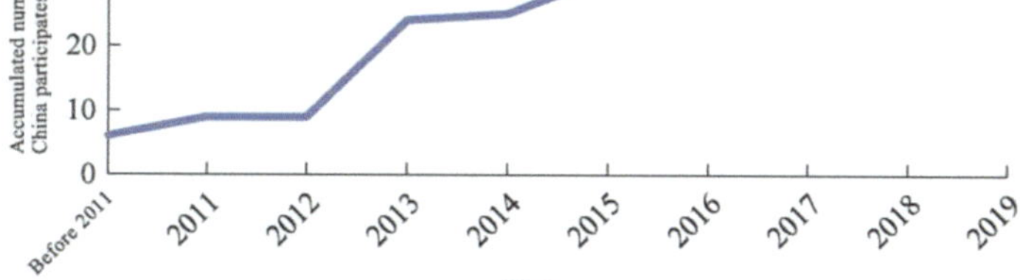

Fig. 16.24 Changes in the cumulative number of China's overseas cooperative ports

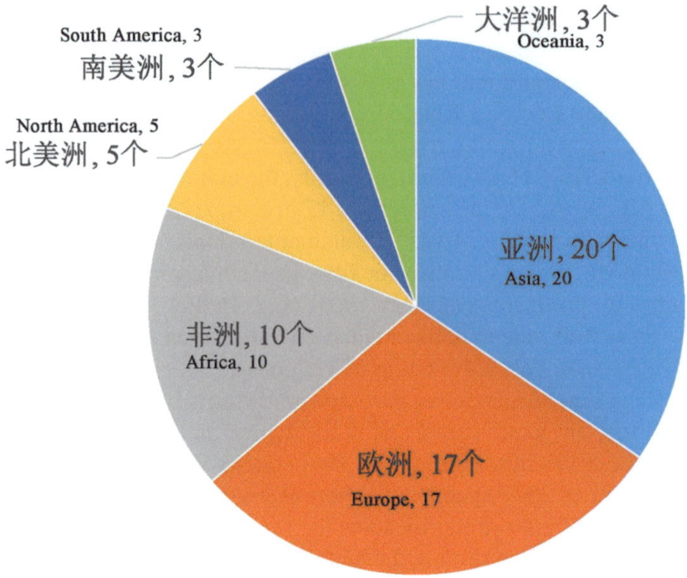

Fig. 16.25 Regional distribution, number, and proportion of China's overseas cooperative ports

Fig. 16.26 The port type structure of China's overseas cooperative ports

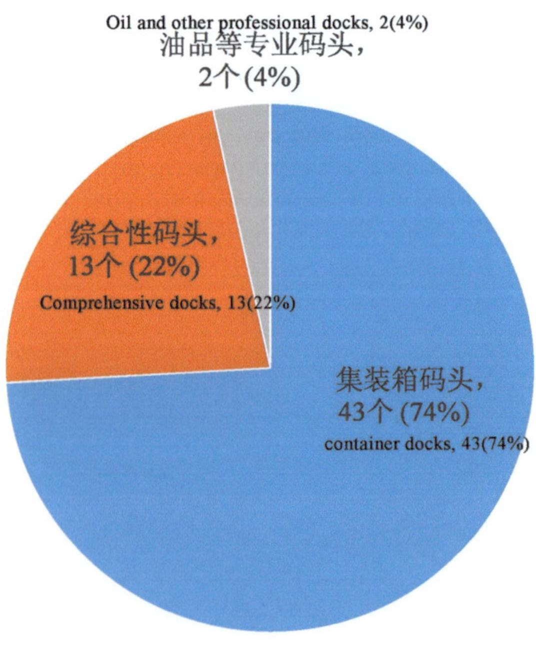

Fig. 16.27 The corporate structure of China's overseas cooperative ports

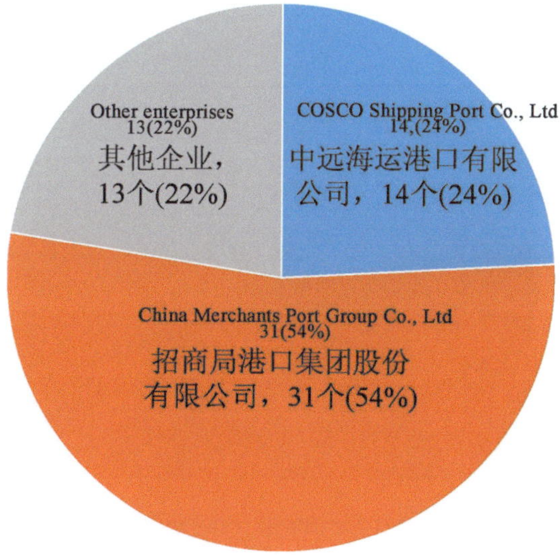

Collaborative construction of overseas ports and the enhancement of the global port and shipping service network have improved China's logistics network service capabilities on a global scale, providing solid support for the prosperity and development of world trade. For instance, China COSCO Shipping Corporation's ocean routes cover over 160 countries and regions with more than 1,500 ports, and approximately 53% of the group's total assets are overseas, contributing to around 63% of the total profits from overseas operations [50]. Through the collaborative construction of ports and industrial parks, China's advantageous capacities are rapidly expanding globally. Collaborating with local enterprises, it further expands the consumer market and promotes the continuous improvement of China's industrial competitiveness. For example, Guangxi Beibu Gulf International Port Group Co., Ltd. participates in the construction of the Malaysia-China Kuantan Industrial Park (Fig. 16.28), bringing China's advantageous industries to Malaysia and advancing international capacity cooperation to promote the transformation and upgrading of China-Malaysia cooperation [51].

Fig. 16.28 Kuantan Port jointly built by China and Malaysia [52]

References

1. Sima Qian. *Shi Ji*. Beijing: Zhonghua Book Company, 1959.
2. Ifeng News. Ingenious indeed, the ancients used this as a "lifebuoy" when crossing rivers [EB/OL]. (2017-08-25) [2022-05-04]. https://www.jiemian.com/article/1577805.html.
3. Ouyang Xun. Yi Wen Lei Ju. Shanghai: Shanghai Ancient Books Publishing House, 1982.
4. Zhang Yukun. Chronicle of the Great Wall of China. Nanjing: Jiangsu Phoenix Science and Technology Publishing House, 2016.
5. Fujihachi Fujita. A Series of Researches on Ancient Maritime Communications in the South China Sea. Translated by He Jianmin. Beijing: The Commercial Press, 1936.
6. Gu Donggao. Chronicle of Major Events in the Spring and Autumn Annals. Beijing: Zhonghua Book Company, 1993.
7. Baidu Wenku. The History of Ancient Chinese Maritime Navigation [EB/OL]. (2016-08-15) [2022-05-04]. https://wenku.baidu.com/view/84ab50a86629647d27284b73f242336c1eb93092?pcf=2&bfetype=new&bfetype=new.
8. Ban Gu. History of the Han Dynasty. Annotated by Yan Shigu. Beijing: Zhonghua Book Company, 1997.
9. Wang Xiangzhi. Yu Di Ji Sheng. Chengdu: Sichuan University Press, 2005.
10. Luo Jun, et al. Baoqing Simingzhi (Local Chronicle of the Southern Song Dynasty). Ningbo: Ningbo Publishing House, 2011.
11. Sima Guang. Zizhi Tongjian. Beijing: Zhonghua Book Company, 2012.
12. China National History. Deception and Manipulation in the Tributary System: Was the "Myriad Nations Coming to Court" Really as Glorious as We Imagine? [EB/OL]. (2018-08-11) [2022-05-04]. https://www.sohu.com/a/246597640_486911.
13. Song Lian, et al. History of the Yuan Dynasty. Beijing: Zhonghua Book Company, 1976.
14. China Art News. The Artistic Impression of China: Ancient Civilization and Contemporary Society - The Opening of the "Belt and Road" Artistic Investigation Special Research Exhibition by the Oil Painting Academy of China National Academy of Painting at the National Art Museum of China [EB/OL]. (2018-01-12) [2022-05-02]. https://www.sohu.com/a/216331913_534797.
15. Wu Youxiong. Quanzhou Religious Culture. Fuzhou: Fujian People's Publishing House, 1998.
16. Marco Polo. The Travels of Marco Polo. Translated by Liang Shengzhi. Beijing: China Literature and History Publishing House, 2008.

17. Liu Hengwu, Ma Min. The Status of Qingyuan Port in Foreign Trade During the Yuan Dynasty [J]. China Port, 2014(9): 16–18, 61.
18. Chen Zhuo. Compilation of Poetry from the Song and Yuan Dynasties. Haikou: Hainan Publishing House, 2000.
19. Wang Yuangong. (Zhizheng) Siming Xuzhi (Local Chronicle of the Siming Region in Zhejiang Province, China). Ningbo: Ningbo Publishing House, 2011.
20. Chen Dazhen. Local Chronicles of the South China Sea During the Dade Period. Beijing: Zhonghua Book Company, 1990.
21. Zhang Xie. Notes on the Investigation of the East and West Indies. Beijing: Zhonghua Book Company, 1981.
22. Lengyan Author Team - Memories of the Ming Dynasty. Did the Maritime Ban in the Ming Dynasty Force Common People to Join the Japanese Pirates? Clearly an Ancient Trade War Tactic, but Demonized as Isolationism [EB/OL]. (2021-06-21) [2022-05-04]. https://user.guancha.cn/main/content?id=523637&page=1.
23. Yu Nengmo, et al. Compilation of Chinese and Foreign Treaties. Beijing: The Commercial Press, 1936.
24. Xinhua News Agency Client. Outlook | Fuzhou Accelerates the Construction of a Modern International City [EB/OL]. (2021-06-15) [2022-05-04]. https://baijiahao.baidu.com/s?id=1702623390213833851&wfr=spider&for=pc.
25. Wen Qing, et al. The Beginnings and Ends of Foreign Affairs (A Compilation of Official Documents on Foreign Relations in the Qing Dynasty). Taipei: Wen Hai Publishing Co., Ltd., 1990.
26. Cui Lewen. On the Loss of China's Customs Sovereignty Since the Opium War [J]. Journal of the Party School, 1990(03): 34–37.
27. Liu Jinzao. Continued Documentary History of the Qing Dynasty. Hangzhou: Zhejiang Ancient Books Publishing House, 1988.
28. Zhu Jie. A Discussion on the Invasion and Influence of Late Qing Powers on the Navigational Rights of the Yangtze River [D]. Hefei: Anhui University, 2015.
29. Wang Xianqian and Zhu Shoupeng. Donghua Lu and Donghua Xulu. Shanghai: Shanghai Ancient Books Publishing House, 2007.
30. Li Yumin. The Treaty System in Modern China. Changsha: Hunan Normal University Press, 1995.
31. China Communications Construction Company First Harbor Engineering Co., Ltd. Tianjin Port! China's First Self-Designed and Constructed Container Terminal [EB/OL]. (2021-04-30) [2022-05-05]. https://m.thepaper.cn/baijiahao_12495758.
32. SOHU. "Illustrating the History of CCCC Through Images" - Tenacious Builders of New China (3): Three Years of Port Construction [EB/OL]. (2020-09-28) [2022-05-05]. https://www.sohu.com/a/421481146_692431.
33. Qinhuangdao News. Qinhuangdao Port in the "Three-Year Transformation of Port Appearance" [EB/OL]. (2021-08-30) [2022-05-01]. https://baijiahao.baidu.com/s?id=1709523989739192644&wfr=spider&for=pc.
34. Comprehensive Planning Department of the Ministry of Transport. A Compilation of Statistical Data on Transportation in New China for 50 Years (1949-1999). Beijing: China Communications Press, 2000.
35. Yizang Network. A Five-Cent Bill Tells the Story of a Red Uprising from Darkness to Light [EB/OL]. (2013-01-10) [2022-05-05]. http://www.ecangw.com/bbs/portal.php?mod=view&aid=212.
36. Inviting the Moon to Talk About History with Three People. Exhibition Review (III) | Reform and Opening-Up Bring Prosperity [EB/OL]. (2021-08-17) [2022-05-05]. https://www.sohu.com/a/484027762_121124389.
37. Shi Chunlin. 1949-2019: 70 Years of China's Port Construction and Layout Changes (Part II) [J]. China Maritime, 2019(11): 25-28.
38. Guangming Net. The "Pioneer Ox" Spirit in the Photographs of Shenzhen Port [EB/OL]. (2020-10-16) [2022-05-05]. https://www.sohu.com/a/425143114_162758.

39. State-owned Assets Supervision and Administration Commission of the State Council. CCCC: Accelerating the Upgrade and Replacement of Ports [EB/OL]. (2017-08-04) [2022-05-05]. http://old.sasac.gov.cn/n4470048/n4470081/n7361946/n7361977/n7361996/n7514761/c7514911/content.html.
40. Ministry of Transport of the People's Republic of China. Ministry of Transport Announces the "National Coastal Port Layout Plan" (Full Text) [EB/OL]. (2007-07-20) [2022-02-03]. http://www.gov.cn/gzdt/2007-07/20/content_691642.htm.
41. Maritime Information Network. "Global Port Development Report (2021)" Officially Released [EB/OL]. (2022-04-08) [2022-05-05]. https://weibo.com/ttarticle/p/show?id=2310474755952959357144.
42. Yin Lin. Practice and Development Thoughts on the Construction of Smart Ports in China [J]. Port Technology, 2019(08): 1–3.
43. Lin Rong. Situation and Prospects for the Construction of Smart Ports during the "14th Five-Year Plan" Period [J]. Port Technology, 2020(10): 1–3.
44. Xinhua Net. Phased Achievements in the Integration of Main and Branch Services of the Yangtze River Electronic Navigational Chart [EB/OL]. (2021-11-24) [2022-05-05]. http://www.xinhuanet.com/2021-11/24/c_1128094795.htm.
45. Guangming Net. The Time is Right for Green Port Construction [EB/OL]. (2018-08-16) [2022-04-16]. https://m.gmw.cn/baijia/2018-08/16/30560903.html.
46. Water Transport Bureau of the Ministry of Transport. Construction and Utilization of Shore Power at Chinese Ports [EB/OL]. (2020-04-28) [2022-03-01]. http://tugboat.chinaports.org/News_info.htm?id=205399&c=1.
47. Yangtze River Waterway. Supporting the "Backbone" of the Yangtze River – A Chronicle of the Construction of the 6-Meter Deepwater Channel Renovation Project from Wuhan to Anqing on the Yangtze River Trunk Line [EB/OL]. (2020-10-26) [2022-05-05]. https://www.sohu.com/a/427492410_807041.
48. Liu Changjian. Strategic Considerations for the Development of China's Water Transport under the Background of Building a Powerful Transportation Nation [J]. China Water Transport, 2019, (12): 14–15.
49. COSCO SHIPPING. Maritime Silk Road Port Tour (I)|When Piraeus in Greece Joins Hands with COSCO SHIPPING [EB/OL]. (2018-08-13) [2022-05-05]. https://www.sohu.com/a/247058246_672776.
50. China Water Transport Network. Total Overseas Assets of COSCO SHIPPING Group Exceed 400 Billion [EB/OL]. (2018-11-02) [2022-03-01]. http://www.zgsyb.com/news.html?aid=35443.
51. State-owned Assets Supervision and Administration Commission of the State Council. From Guangxi Beibu Gulf to Malaysia's Kuantan Port - Mutual Benefit and Development through the "Belt and Road" Initiative [EB/OL]. (2017-06-22) [2022-03-01]. http://www.sasac.gov.cn/n2588025/n2588129/c4626498/content.html.
52. Sohu. Malaysia-China Kuantan Industrial Park: Integration of Port and Industrial Park Creates a Model for Sino-Malaysian Production Capacity Cooperation [EB/OL]. (2018-01-09) [2022-05-05]. https://www.sohu.com/a/215710156_732289.
53. Anonymous. Quanzhou Houzhu Port Was the Largest Harbor in Ancient China [J]. Waterway Transport Digest, 2004(11): 68.
54. Stavrianos, L. S. A Global History: Since 1500. Translated by Wu Xiangying and Liang Chimin. Shanghai: Shanghai Academy of Social Sciences Press, 1999.
55. Ministry of Transport of the People's Republic of China. Ministry of Transport Releases the Full Text of the "National Inland Waterway and Port Layout Plan" [EB/OL]. (2007-07-20) [2022-02-03]. http://www.gov.cn/gzdt/2007-07/20/content_691664.htm.

Chapter 17
Pipeline Transportation: Energy Arteries Spanning Thousands of Miles

Pipeline transportation is one of the five major modern transportation methods, and modern pipeline transportation has a relatively short history of only 80 years. However, the history of the development and utilization of oil and natural gas, important subjects of pipeline transportation, in China is long-standing. The scale, layout, role, and impact of pipeline transportation in contemporary China are significant. As the energy arteries, oil and gas pipelines span from north to south and from west to east, supporting the country's development and modernization, leaving behind an unforgettable history.

Go Deep into History to Explore the Origin of Pipelines

In the long history of Chinese civilization, the discovery and utilization of oil and natural gas stand out as significant achievements. China is the earliest country in the world to discover and utilize oil and natural gas.

The earliest records of China's discovery and utilization of oil can be traced back to the years 54–92 during the Eastern Han Dynasty, as documented in the "Records of the Grand Historian · Treatise on Geography" written by Ban Gu (32–92). It mentions that in Shangyun Gaonu[1] County, there was "flammable water" [1]. The "Tai Ping Guang Ji," compiled by Li Fang and others in the Northern Song Dynasty, quotes the Tang Dynasty's "Youyang Za Zu," stating: "Gaonu has stone fat water, which is greasy, floating on the water like lacquer. It is extracted for cart axle lubrication, and lamps, which is extremely bright" [2]. Shen Kuo, in his work "Meng Xi Bi Tan" from the Northern Song Dynasty, mentioned that "there is a great amount of oil, born in the endless depths of the earth," and predicted that "this substance will

[1] "Gao Nu" is located in the area around present-day Yan'an in Shaanxi, and "Wei Shui" refers to a tributary of the Yan River, known as Yanhe today.

© China Communications Press Co., Ltd 2025

H. Lu (ed.), *A Brief History of Transportation in China*,
https://doi.org/10.1007/978-981-97-9851-3_17

surely become prevalent in the world" [3]. Throughout history, oil has been used for lighting, cooking, medicine, and even as a tax payment [4].

According to the " Records of Hua Yang " written by Chang Qu in the Eastern Jin Dynasty, the discovery, utilization, and pipeline transportation of natural gas can be traced back to the Qin and Han periods more than 200 years BC. The ancient people of Linqiong in Sichuan used gas for salt boiling, connecting bamboo segments to form pipelines for transporting natural gas and brine, known respectively as "fire bamboo pipes" and "water bamboo pipes" [5]. By the Ming and Qing Dynasties, natural gas in Sichuan had already been developed and utilized on a certain scale. The "The Exploitation of the Works of Nature" initially published by Song Yingxing in 1637 during the Ming Dynasty, once again documented the history of using bamboo pipelines to transport natural gas [6].

The origin of China's modern petroleum and natural gas industry dates back to the mid-19th century. In 1850, the Muzi Well in Zigong, Sichuan, completed drilling to a depth of 1200 m, making it the world's deepest gas well at the time. The gas field it belonged to was one of the earliest discovered and exploited by humans. In 1878, the first oil well in China was drilled in Miaoli, Taiwan, establishing the first petroleum management agency in China—the Miaoli Mineral Oil Bureau. In 1907, the first onshore oil well, Yan Yi Well, was drilled in Yanchang, Shaanxi, with a depth of 120 m, marking the end of China's history without industrial oil production on land. Subsequently, the first oilfield on land in China, the Yanchang Oilfield, was established. In 1909, an industrial oil well was drilled in Dushanzi, first in Xinjiang, becoming one of the birthplaces of China's petroleum industry. In 1936, Dushanzi Refinery was established, initiating large-scale exploration and development of the Dushanzi Oilfield, and vigorously promoting the petroleum refining industry. Yumen Oilfield was the largest and highest-yielding oilfield developed in old China. From the discovery of the first oil well in Laojun Temple in 1939–1948, Yumen Oilfield produced a total of 455,000 tons of crude oil, accounting for approximately 72.3% of the national crude oil production from 1904 to 1948 [7]. At that time, petroleum and natural gas were mainly transported by mules and horses, as well as horse-drawn and donkey-drawn carts, with very limited transport capacity.

China's onshore oil and gas resources are mainly distributed in the Songliao Basin, Tarim Basin, Bohai Bay Basin, Ordos Basin, Junggar Basin, Qaidam Basin, Pearl River Mouth Basin, Sichuan Basin, and East China Sea Shelf Basin. Offshore oil and gas resources are primarily found in the Bohai Sea and South China Sea. Oil and gas fields are typically located in sparsely populated deserts, vast grasslands, or expansive oceans, while the consumption market is concentrated in densely populated, economically developed large and medium-sized cities and their surrounding areas. On the demographic map of China, there is a boundary line known as the "Hu Huanyong Line," proposed by geographer Hu Huanyong in 1935, also known as the Aihui-Tengchong Line (now Heihe-Tengchong Line). The 2020 data from the seventh national census indicate that 43.8% of the land in the southeastern part of the Hu Line is inhabited by 93.5% of the population, while the northwestern part, comprising 56.2% of the land, is home to only 6.5% of the population (Figs. 17.1 and 17.2). This results in the inevitable demand to transport oil and gas resources from

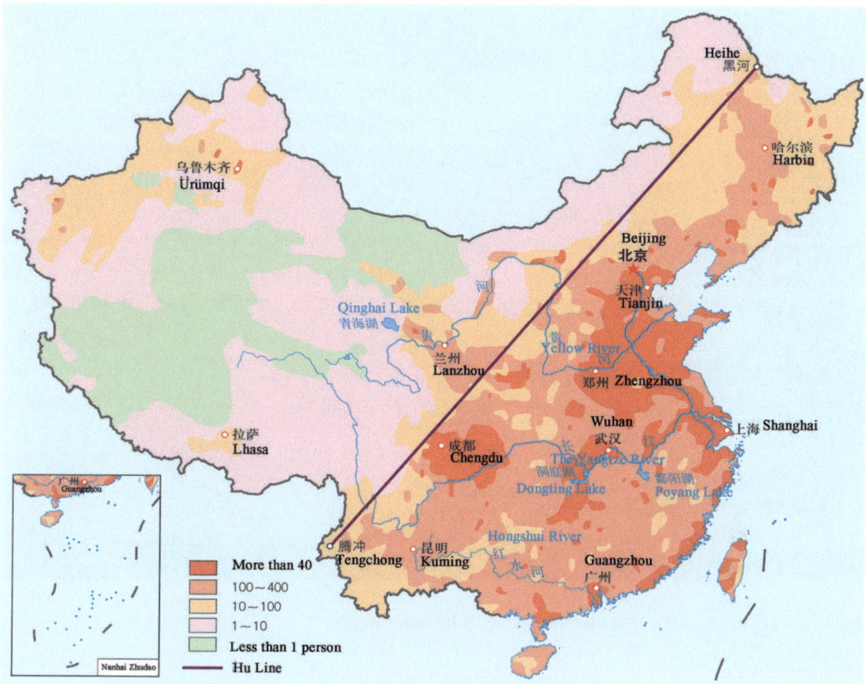

Fig. 17.1 Hu Huanyong line and China's population density map

the northwest to the southeast. The oil and gas pipeline industry has flourished in response to the increasing scale of oil and gas transportation, especially since China became a net oil-importing country in 1993.

In the present day, the transportation industry, accompanying human civilization, has evolved into a comprehensive transportation system that integrates five major sectors: road, rail, water, air, and pipeline. This system features organic integration, appropriate division of labor, and rational distribution, effectively reducing the distances between people, sky and earth, sea and land, as well as resources and markets. Hidden underground, oil and gas pipelines, as the fifth transportation sector, have achieved a scale of mileage in China that is on par with railways and highways. They undertake around 70% of onshore crude oil and approximately 95% of natural gas transportation, serving as the energy arteries that ensure the nation's economic development and the well-being of its people.

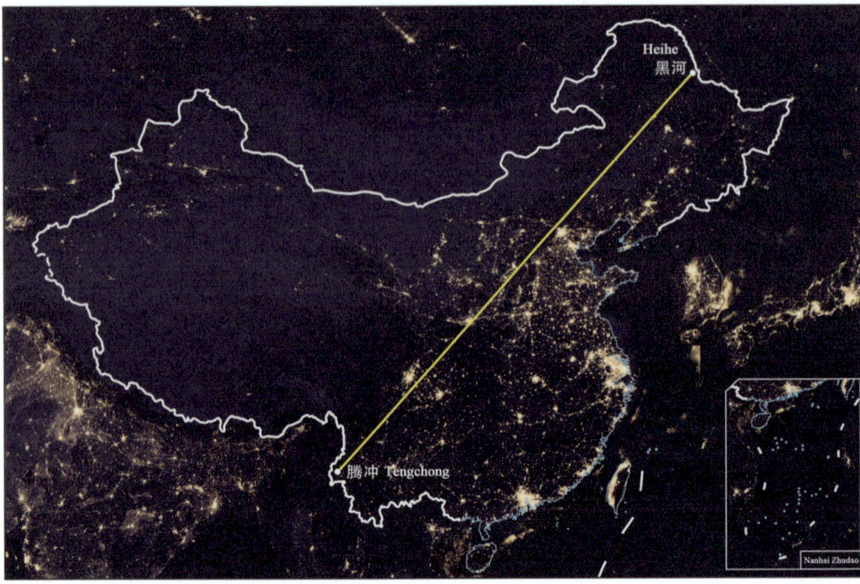

Fig. 17.2 Hu Huanyong line and China night lighting map

Thirty Years of Arduous Efforts Have Laid the Foundation for China's Pipeline Industry

The modern transportation of oil and natural gas has experienced the intersecting development of various modes, including roads, railways, water transport, and pipelines. Compared to the other four modes of transportation, pipeline transportation has significant advantages, such as occupying less space, requiring lower investment, quick construction, large transportation capacity, low cost, convenient operation, ease of automation, minimal environmental pollution, and resilience to natural conditions like surface traffic and climate. Therefore, in today's world, onshore and shallow-sea oil and natural gas resources mainly rely on pipeline transportation.

In the 1860s, with the discovery and industrial exploitation of the oil-producing region in Pennsylvania, barges became the first "long-distance" transportation tool for oil, later evolving into large oil tankers. Simultaneously, between the oil-producing areas and export terminals, people attempted to lay metal pipelines, constructed by connecting iron pipes with bolts, as a substitute for horse-drawn carriages to transport crude oil, resulting in significant cost savings. The advent of welding technology in the late 1920s laid the technical foundation for the development of modern pipelines, and World War II served as a powerful driving force for the large-scale construction of oil pipelines. To ensure energy supply during the war and avoid enemy attacks, the United States laid two large-caliber long-distance oil pipelines between 1941 and 1943, considered the beginning of modern oil pipeline transportation [8].

In 1944, China's first crude oil pipeline was established in the Yumen Oilfield, designed to transport crude oil from the field to the refinery. The pipeline was 4.5 km long with a diameter of 120 mm. The design and construction were overseen by Mr. Weng Xinyuan (pictured in Fig. 17.3), a key figure in the foundation of China's oil and gas storage and transportation industry. While this pipeline may seem inconsequential now, it was crucially built during a period of extreme material scarcity and the wartime conditions of the resistance against Japan, characterized by the slogan "One drop of gasoline, one drop of blood." The first section of the Colonial Pipeline built in the United States in 1941 had a diameter of only 210 mm. More importantly, the pipeline transported easily solidifying crude oil with a pour point as high as 10 °C, which was not being extracted in the United States at that time. To insulate the pipeline, builders not only covered the outside of the pipes with cotton felt but also mobilized local communities to collect dry horse dung, which was spread above and below the pipeline in the trench. While these practices might have become anecdotes, they were scientifically sound approaches based on practical considerations [9].

Fig. 17.3 Mr. Weng Xinyuan, Founder of China's oil and gas storage and transportation industry

China's first finished oil pipeline was an international pipeline built during the Sino-Japanese War. It was jointly planned by the United States, the United Kingdom, and China, considered one of the greatest military engineering projects of World War II. Stretching 3,218 km, the pipeline originated from Calcutta, India, passed through Myitkyina in Burma, and entered China, ultimately reaching Kunming. Despite being operational for only seven months from May to November 1945, this pipeline, which transported approximately 100,000 tons of gasoline, diesel, and lubricating oil, equivalent to a year and a half of road transportation at that time [10].

In the early days of the People's Republic of China, the petroleum industry was significantly underdeveloped. After a recovery period from 1950 to 1953, during which crude oil production exceeded the highest annual output before the founding of the PRC, it could only meet 37% of the country's demand. The first Five-Year Plan was implemented in 1953, emphasizing the importance of petroleum exploration and development. In 1955, the Karamay Oilfield, China's first large oilfield, was discovered, marking progress in oilfield exploration and development. In 1958, the first batch of finished oil from the Dushanzi Refinery was shipped to the mainland, showcasing its oil refining capabilities [11]. The Karamay-to-Dushanzi crude oil pipeline was subsequently established as the first long-distance crude oil pipeline in New China, officially commissioned on January 10, 1959, with a total length of 147.2 km, forming the petroleum production–transportation-refining-marketing chain for the Karamay Oilfield. In 1970, with the development of the Daqing Oilfield, the construction of the Northeast Basan crude oil pipeline commenced, becoming a historical monument in China's pipeline industry. By 1980, China had accumulated more than 6,500 km of constructed crude oil pipelines.

The development of finished oil pipelines began with the construction of short-distance airport pipelines. Since 1958, internal finished oil pipelines were successively built at Capital Airport, Shanghai Airport, and Guangzhou Airport. In 1976, the first long-distance finished oil pipeline in China was completed in the Qinghai-Tibet Plateau, specifically the Golmud to Lhasa finished oil pipeline, spanning a total length of 1,080 km. It remains the world's highest-altitude finished oil pipeline to date. By 1980, China had cumulatively constructed over 1,100 km of finished oil pipelines.

The development of natural gas pipelines started with the construction of the Bayu Gas Transmission Pipeline in 1963. Around 1980, the Sichuan Gas Field's main gas transmission line—the Southern Semi-Circular Natural Gas Pipeline System—was essentially completed. By 1980, China had accumulated over 2,700 km of constructed natural gas pipelines.

In 1975, China's first 50,000 cubic meters floating roof oil tank was built at the Chenshan Terminal in Shanghai. Subsequently, dozens of similar-sized floating roof oil tanks were constructed by petrochemical enterprises, oil fields, and ports. The large-scale construction of petroleum storage facilities enhanced the flexibility of petroleum pipeline transportation [12].

Over 30 years of arduous exploration and practice have written a strong prologue for China's oil and gas pipeline industry. The underground oil and gas pipelines, resembling a series of steel dragons, with the bold spirit of "I contribute oil for the motherland" and unwavering determination, continuously transport oil and natural gas to various parts of the motherland, contributing the "pipeline power" to the country's development.

The Basan Pipeline Opens a New Chapter in China's Pipeline Industry

In the first half of the twentieth century, a major debate unfolded among Chinese and foreign geologists regarding the "Theory of China's Poor Oil Resources" and the "Continent-Source Oil Generation Theory." In 1913, Standard Oil Company (now ExxonMobil) conducted petroleum geological surveys in several regions of Shandong, Henan, Shaanxi, Gansu, Hebei, Northeast China, and Inner Mongolia but failed to find economically viable oil reserves. Influenced by this, American geologists published articles asserting the "Theory of China's Poor Oil Resources." During Japan's invasion of Northeast China, extensive exploration efforts also ended in failure, seemingly confirming the correctness of the "Theory of China's Poor Oil Resources." However, contemporary China produced a group of renowned geologists, including Li Siguang, Weng Wenhao, Pan Zhongxiang, and Huang Jiqing, who, grounded in the actual geological conditions of China, put forth scientific predictions about the country's oil and gas prospects. Their "Continent-Source Oil Generation Theory" refuted the foreign experts' assertion that "oil can only be generated in marine sedimentary strata" and comprehensively rejected the notion of "China's poor oil."

On September 26, 1959, the first gusher erupted in Songji 3rd Well, marking the birth of the Daqing Oilfield. China acquired a world-class supergiant oilfield, shedding the label of "poor oil" and transforming the landscape of the petroleum industry. The Daqing Petroleum Battle left a glorious chapter in the long history of the Chinese nation, creating the "Daqing Spirit" and the "Ironman Spirit," both centered around patriotism. As then Minister of Petroleum Industry and Commander-in-Chief of the Daqing Petroleum Battle, later Vice Premier of the State Council, Yu Qiuli, remarked: "Except for Daqing, no other enterprise's birth and development are so closely connected to the fate and spirit of the Chinese nation." [13].

By 1966, the production of Daqing Oilfield had reached 10 million tons, relying primarily on railway transportation. Despite dispatching a train of crude oil tank cars approximately every 17 min on average, there still existed a considerable gap between production and transportation capacity. Shanhaiguan, as a vital transportation hub for southbound traffic to North China, posed a bottleneck for railway transportation and became a critical point of congestion in the Daqing oil transportation system. Consequently, despite a nationwide energy shortage, Daqing Oilfield had to resort to closing wells and restricting production. In 1970, Premier Zhou Enlai and other leaders, including Vice Premier Li Xiannian, in charge of national economic construction at the time, made several decisions after extensive research.

They concentrated resources and funds to urgently construct the Daqing crude oil export pipeline. A joint notice was issued by the State Council and the Central Military Commission, instructing the original Shenyang Military Region and the Fuel and Chemical Industry Department (abbreviated as "Fuel Chem Department") to jointly oversee the pipeline construction. Subsequently, the Northeast Pipeline Construction Leadership Team was formed, with Lieutenant General Xiao Quanfu, former Deputy Commander of the Shenyang Military Region, serving as the team leader. Zhang Wenbin, former Deputy Minister of the Ministry of Petroleum Industry, and Lieutenant General Li Shaoyuan, former Deputy Political Commissar of the Shenyang Military Region, and Lieutenant General Luo Kunshan, former Deputy Chief of Staff of the Shenyang Military Region, served as deputy team leaders. The team set up an Engineering Command, with Wang Yunwu as the commander and Deng Cong as the political commissar. Wang Peide, Zhang Zhenyong, Li Zhanbiao, Tang Zhenhua, Zhang Zhicheng, and Zhang Fulu served as deputy commanders. Soon after, Wang Yunwu resigned, and Wang Peide took over as the commander.

On August 3, 1970, the Northeast Pipeline Construction Leadership Team held its first meeting in Shenyang, with 28 participants, and officially named the Northeast Oil Pipeline Construction Project as the "Basan Project." On August 15 of the same year, after consultations on the construction plan, funds, materials, equipment, and other issues with the State Planning Commission, the State Construction Commission, the Ministry of Finance, the Ministry of Metallurgy, the First Ministry of Machine Building, the Ministry of Foreign Trade, the Fuel and Chemical Industry Department, and the Bureau of Materials, the summary of the meeting on the urgent construction of the Northeast Oil Pipeline was issued. The Daqing to Fushun pipeline (Qingfu Line) was formally included in the national plan as the first phase of the "Basan Project," aiming for completion and operation in 1971.

Constructing a long-distance, large-diameter oil pipeline for the first time in China presented numerous challenges. There were no established processes, technical data, or practical experience. There were no professional design institutions or personnel, no steel and pipe fabrication lines, no specialized equipment for oil transportation, a lack of expertise in pipeline research, and insufficient support in terms of electricity and communication. The construction time was also restricted by the seasons. Despite these obstacles, the "Basan Project" was crucial for the normal operation of the national industrial production system. It was not just a pipeline but a new path for national development that had to be forged through practical experience. On August 20, 1970, Xiao Quanfu issued the command: "Use conditions if available, create conditions if not," initiating a "Liaoshen Campaign" in the construction of the oil pipeline.

In response to Xiao Quanfu's call for efforts, pipeline construction experts Tang Zhenhua, chemical engineering expert Zhang Zhaoming, mechanics expert Pan Jiahua, and others left locations like "cowsheds" and "reeducation-through-labor farm" to engage in research and analysis for the "Basan Project." The local governments of the three northeastern provinces provided strong support, and the local population along the pipeline considered participation in the construction of the "Basan Project" as an honorable duty. A massive workforce of 200,000 laborers quickly

Fig. 17.4 Construction site of the "83 Project"

organized and dispersed to various construction sites along the "Basan Project" (see Fig. 17.4). The original Shenyang Military Region rapidly mobilized ten infantry divisions, three engineer corps, two pontoon bridge battalions, and some special forces units in communication, surveying, and diving, among others, to participate in the "Basan Project," undertaking the most crucial, challenging, and dangerous construction tasks. The Fuel and Chemical Industry Department transferred over 500 core pipeline construction personnel and more than 100 technical experts from the Sichuan Oil Pipeline Bureau to provide support for the "Basan Project."

The spirited "Basan Project" campaign was launched on the vast black soil. The harsh seven-month winter in the Northeast was approaching, with temperatures dropping to over minus 40 degrees Celsius, but the severe cold couldn't dampen the enthusiasm of the builders. The 1.7-ton steel pipes were transported by shoulders and carts (Fig. 17.5). After working continuously for more than ten hours, the warm sweat-soaked back was chilled by the biting cold wind. However, no hardship could shake their belief and determination to "not see oil flow, swear not to turn back." The most challenging part was crossing the Nen River. Due to the buoyancy of the river water, the pipeline could not sink to the riverbed, posing a serious challenge to the construction. At this time, it was already late autumn, and once the river froze in winter, it would lead to a significant delay in the entire project. The builders came up with a solution: they added another layer of pipes outside the original pipeline, injecting cement into the gap between the two layers to increase the weight, allowing the pipeline to sink to the riverbed. Finally, on November 5, 1970, just before the severe winter arrived, the pipeline successfully crossed the Nen River.

Over a span of 13 months, on November 7, 1971, crude oil from Daqing Oilfield gushed out in Fushun, nearly 600 km away, marking the completion and oil transmission of China's first large-diameter, long-distance oil pipeline. In that era of underdeveloped productivity, every meter advanced by the "Basan Project" was exceptionally challenging, and the entire pipeline embodied the countless efforts and sweat of the

Fig. 17.5 Scene of oxcart carrying pipes during the "Basan Project"

builders. Subsequently, from 1972 to 1975, the second phase of the "Basan Project" saw the completion of pipelines connecting Daqing to Tieling, Fushun to Anshan, Tieling to Qinhuangdao, Tieling to Dalian, and Dandong to North Korea (Fig. 17.6). During this period, a pipeline was also laid from Panjin to Jinxi, transporting crude oil from the Liaohe Oilfield. Additionally, a pipeline was laid from Fushun Petroleum Second Factory to Liaoning Power Plant, supplying fuel oil for the power plant.

From August 1970 to September 1975, the "Basan Project" laid a total of 2,471 km of crude oil pipelines, forming a Northeast crude oil pipeline network with Daqing as the starting point and Tieling as the hub. This network included three major oil arteries: Daqing to Fushun, Daqing to Qinhuangdao, and Daqing to Dalian. Among them, the transported crude oil could be transferred via Qinhuangdao Port and Dalian Xingang for maritime shipping. Along the route, there were 11 crude oil transfer points, three railway transfer and loading bridges, and a 7.1-km railway dedicated pipeline. The combination of pipeline transport, maritime shipping, and railway transport could supply oil to 28 refineries in 13 provinces and cities nationwide.

The completion and operation of the "Basan Project" made pioneering contributions to China's pipeline industry, filling the gap in long-distance, large-diameter pipeline construction. It cultivated a specialized pipeline workforce that integrated design, construction, operation, research, and management, creating rich theoretical and practical experiences in pipeline construction. The project established a systematic set of pipeline technical standards with distinctive Chinese characteristics. Concurrently, it added a new force to China's transportation industry, marking a historic shift from primarily railway transportation for Chinese crude oil to predominantly pipeline transportation. The freed-up railway capacity was redirected to transport materials such as steel, timber, and grain crops, driving the development of industries like metallurgy, pipe manufacturing, corrosion prevention, and mechanical equipment, thereby stimulating economic vitality.

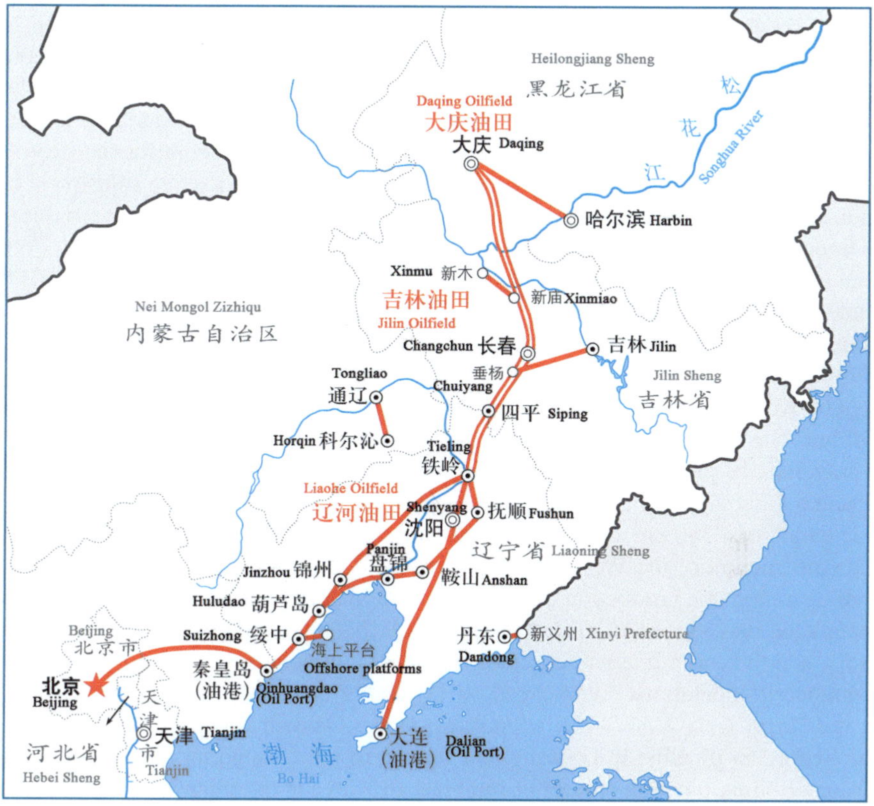

Fig. 17.6 Schematic diagram of the pipeline route of the "Basan Project"

The "Basan Project" gave rise to the "Basan Spirit" with the motto "Dedicated to the pipeline industry, home is wherever I am, pride in hardship, and delight in field operations." This spirit continues to inspire generations of pipeline professionals to forge ahead.

The West–East Gas Pipeline Sets a New Milestone for China's Pipeline Industry

At the turn of the century, the Central Committee of the Communist Party made a significant decision to implement the "Western Development" initiative. Over 20 years have passed, and the economic and social development in the western regions has achieved historic results, expanding the strategic development space for the country and laying the foundation for a new pattern of Western Development in the new era, guided by the Belt and Road Initiative.

The West–East Gas Pipeline, conceived since 1998, stands as a landmark project in China's implementation of the Western Development strategy. On February 14, 2000, Comrade Zhu Rongji presided over the Prime Minister's Office meeting, explicitly stating that the West–East Gas Pipeline Project is a significant endeavor to turn the natural gas resources in Xinjiang into an economic advantage for the diverse ethnic groups in Xinjiang, as well as a crucial measure to promote the adjustment of industrial and energy structures, and enhance economic benefits along the provinces and regions along the pipeline. On August 23, 2000, the State Council held its 76th executive meeting, approving the West–East Gas Pipeline Project. By August 2014, three pipelines—Line 1, Line 2, and Line 3—had been completed for the West–East Gas Pipeline project (see Fig. 17.7), making it the natural gas pipeline project, benefiting the largest population [14].

The West–East Gas Pipeline Project's Line 1 starts from the Tarim Basin Lunnan Oil and Gas Field in Xinjiang and extends eastward to Baihe Town in Shanghai. It traverses 10 provinces and municipalities, including Xinjiang, Gansu, Ningxia, Shaanxi, Shanxi, Henan, Anhui, Jiangsu, Shanghai, and Zhejiang, with a total main-line length of 4,380 km and a total investment of 46.3 billion yuan. On July 4, 2002, the construction of the West–East Gas Pipeline Line 1 officially commenced. The trial operation and production of the Jingbian to Shanghai section were successfully completed on October 1, 2003. Gas supply to Shanghai commenced on January 1, 2004, and the entire line was completed and put into operation on October 1, 2004. Commercial operations commenced on December 30, 2004. The main gas source is the Tarim Kela-2 gas field, connecting to the Qaidam Basin gas field through the Seninglan pipeline in Gantang and linking to the Changqing gas field and the Shanxi-Beijing pipeline system in Jingbian. In Xinyang, it connects to the Huaiwu and Zhongwu lines. Jinjing line was constructed in parallel to facilitate connectivity between the North China and East China markets (Fig. 17.8).

With the increasing gap between the supply and demand of natural gas resources in China, the introduction of foreign natural gas resources became an inevitable choice. In 2006, China successively signed natural gas purchase and sales memoranda or agreements with Kazakhstan and Turkmenistan, initiating the West–East Gas Pipeline Project's Line 2. On February 22, 2008, the West–East Gas Pipeline Project's Line 2 officially started construction (Fig. 17.9). By the end of 2009, the single line of the Central Asia Natural Gas Pipeline and the western section of the West–East Gas Pipeline Project's Line 2 were completed and put into operation as scheduled, bringing Central Asian natural gas into western provinces of China. In 2011, the West–East Gas Pipeline Project's Line 2 was fully connected, primarily sourcing natural gas from Central Asian countries such as Turkmenistan and Kazakhstan, with domestic gas serving as backup and supplementary supply. The West–East Gas Pipeline Project's Line 2 is China's first cross-border natural gas pipeline project, connecting to the Central Asia Natural Gas Pipeline. The mainline enters China at Horgos in Xinjiang, passing through Dushanzi, Urumqi, Shanshan, Hami, and others. In Gansu, it goes through Hongliu, Jiuquan, Shandan, Wuwei, and Zhongwei in Ningxia, crossing the Yellow River and running parallel to and connecting with the West–East Gas Pipeline Project's Line 1 in the process. It continues southeast through

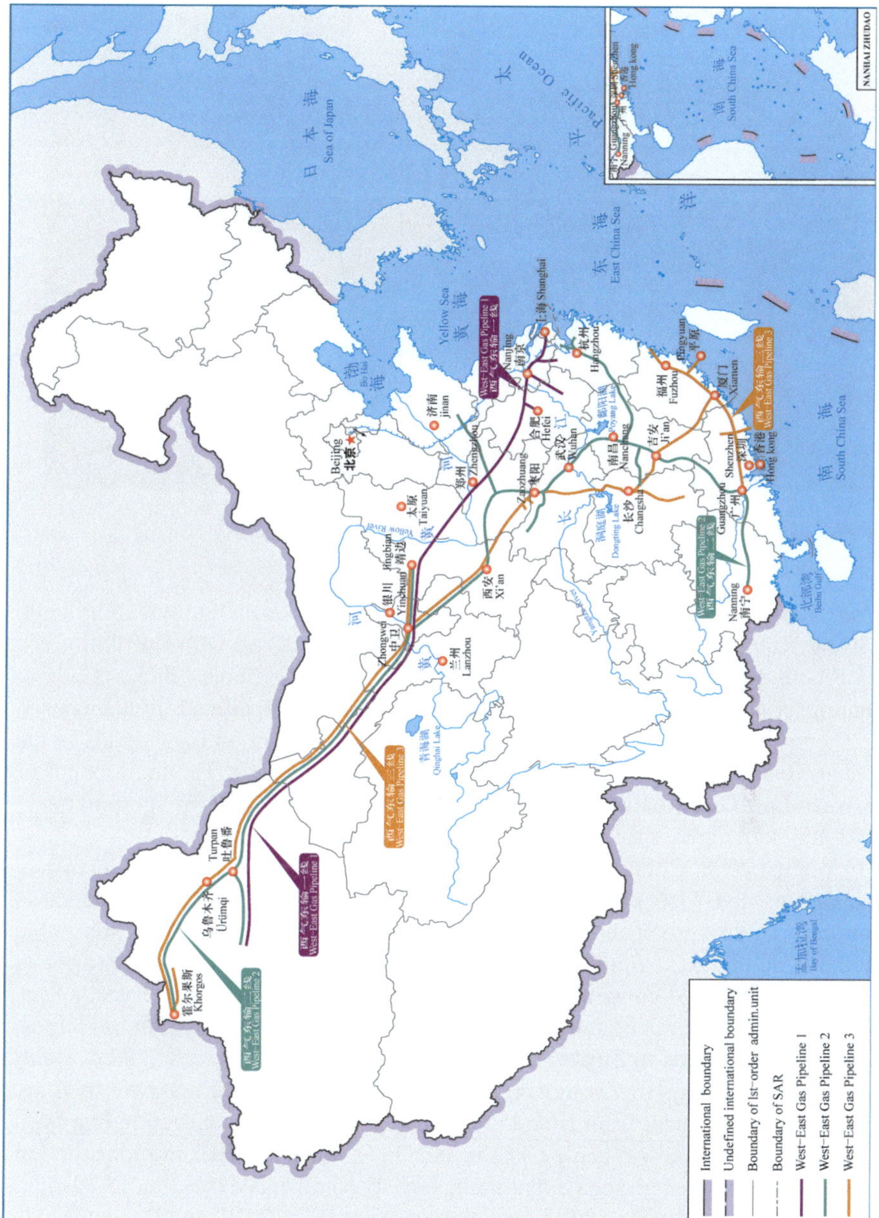

Fig. 17.7 West–East gas pipeline route

Fig. 17.8 The construction site of the West–East gas pipeline project-line 1

Xi'an, Nanchang, Ganzhou, reaching Guangzhou, with a branch line extending south to Hong Kong. The Nanchang branch line is interconnected with the West–East Gas Pipeline Project's Line 1 in Shanghai. The Nanning branch line extends to Nanning, connecting to the China-Myanmar Natural Gas Pipeline. Simultaneously, the Zhongwei-Guiyang Line is being constructed to connect with gas fields in the Sichuan Basin and the China-Myanmar Natural Gas Pipeline. The total length of the West–East Gas Pipeline Project's Line 2 and its branch lines is approximately 9,102 km, with a total investment of 142.2 billion yuan.

In August 2008, China signed additional supply agreements with Turkmenistan, Kazakhstan, and Uzbekistan, expanding on the existing natural gas purchase contracts. These agreements entailed an annual increase of 25–30 billion cubic meters of natural gas supplied to China. Concurrently, the coal-to-gas project in Xinjiang commenced construction, and the West–East Gas Pipeline Project's Line 3 was initiated. The mainline of this project starts from the first station in Horgos, Xinjiang, and extends to Fujian. It was constructed in three segments: the western segment from Horgos to Zhongwei, the central segment from Zhongwei to Ji'an, and the eastern segment from Ji'an to Fuzhou. Additionally, a connection line from Guangzhou to Fuzhou was being constructed. On October 16, 2012, the construction of the western and eastern segments simultaneously commenced (see Fig. 17.10), and they were completed and put into operation on August 25, 2014, and December 12, 2016, respectively. The central segment, from Zhongwei to Zaoyang, commenced construction on September 23, 2021. The section from Zhongwei to Zaoyang is roughly parallel to the West–East Gas Pipeline Project's Line 2, later passing

Fig. 17.9 Construction site of the West–East gas pipeline project-line 2

through Changsha to Ji'an. Upon completion, the project will significantly enhance the resource regulation capability and operational reliability of the West–East Gas Pipeline system.

The West–East Gas Pipeline Project is a major energy artery spanning across China from east to west. It traverses vast deserts, crosses the Tianshan Mountains, spans the Yellow River, and passes through the Yangtze River. Confronting challenges posed by various topographical features such as deserts, plateaus, mountains, plains, and

Fig. 17.10 Construction site of the West–East gas pipeline project-line 3

water networks, as well as diverse climates including high-altitude and low-oxygen conditions, the project's scale and complexity are unparalleled in the world. The construction of this project has transformed the resource advantages of the western region into economic strengths. It is instrumental in advancing energy cooperation under the "Belt and Road" initiative, adjusting China's energy structure, driving the development of related industries such as machinery, electricity, chemicals, metallurgy, and construction materials, promoting the transformation and upgrading of industrial structure. Simultaneously, this project declares to the world that China not only has the capability to construct a world-class long-distance gas pipeline but also, through technological progress and modern management, has propelled the development of the Chinese pipeline transportation industry, enabling China to possess the ability to construct and manage pipelines at the world's advanced levels.

China's Criss-Crossing Oil and Gas Pipeline Network

Since the founding of the People's Republic of China, the oil and gas pipeline industry has undergone a glorious journey from weakness to strength, from introduction and absorption to independent innovation, from "local lines" to a "national network," and from focusing on domestic connections to linking overseas. According to data released by the China National Petroleum and Natural Gas Pipeline Network Group as of the end of 2020, the total length of onshore long-distance oil and gas pipelines in China has reached 144,000 km, making China the third-largest pipeline country globally after the United States and Russia.

In the early days of the People's Republic of China, various industries were in need of development. The Karamay Oilfield, China's first major oilfield, was discovered in 1955, and in 1958, the country's first crude oil pipeline, the Karamay to Dushanzi Crude Oil Pipeline, was completed. In 1963, China's first gas pipeline, the Sichuan-Ba Gas Pipeline (now from Chongqing Ba'nan District to Chongqing), was completed. In 1976, China's first refined oil pipeline, the Golmud to Lhasa Refined Oil Pipeline, was completed. The "83" pipeline project initiated in 1970 laid the foundation for the pipeline transportation system, nurtured talents in pipeline design, construction, and operation management, established Chinese characteristics in pipeline technology and management standards, and developed supporting production capabilities for materials and equipment.

In the 1980s, under the guidance of the general principles of reform and opening-up in China, the focus shifted towards introducing and absorbing advanced foreign technology and management experience, becoming a driving force for the development of China's oil and gas pipeline industry. In 1986, relying on the technical expertise of foreign companies, China completed the technological transformation of the airtight transport system for the TieLing to Dalian crude oil pipeline. The first airtight and automated crude oil pipeline, the Dongying to Huangdao double-line crude oil pipeline, was built, achieving automated oil transportation and airtight delivery. Additionally, two large floating roof oil tanks with a capacity of 100,000 cubic meters were constructed in Qinhuangdao. In 1996, through collaborative design with foreign companies, China completed the Kuerle to Shanshan crude oil pipeline, featuring X65 steel, high pressure, long-distance transport, and roomy station spacing. In 1997, the high-pressure and long-distance Shanxi-Beijing gas pipeline was completed, marking the starting point for China's oil and gas pipeline technology catching up with the world's advanced levels. Concurrently, the pipeline industry's technological and management standards began to align with international practices.

In the 21st century, China's oil and gas pipeline industry entered an era of independent innovation, witnessing a flurry of large-scale pipeline construction with rapidly evolving technological upgrades. Despite setbacks in international collaboration, China achieved the construction and operation of its first world-class natural gas pipeline project designed and built independently in 2004—the West–East Gas Pipeline (Line 1). It utilized X70 steel with a diameter of 1016 mm and a design pressure of 10 megapascals. In 2006, the Shanxi-Beijing Gas Pipeline (Line 2) was completed and put into operation, also using X70 steel with a diameter of 1016 mm and a design pressure of 10 megapascals. In 2009, the Sichuan-East Gas Pipeline mainline was fully connected, spanning challenging terrains such as the Bashan Mountains, the Wuling Mountains in western Hubei, the Jianghan Plain, and the Yangtze River Delta region—making its construction difficulties rarely seen in the world. In 2011, the West–East Gas Pipeline (Line 2) was completed and put into operation, featuring an increased diameter of 1219 mm and an upgraded design pressure of 12 megapascals. The entire line adopted X80 steel, marking a pioneering achievement worldwide. In 2015, the Shanxi-Beijing Gas Pipeline (Line 3) was completed and put into operation, continuing to use X70 steel with a diameter of 1016 mm and a design pressure of 10 megapascals. In 2018, the Shanxi-Beijing Gas Pipeline (Line 4) was completed and put into operation, utilizing X80 steel with a diameter of 1219 mm and a design pressure of 12 megapascals. The total mileage of Lines 1–4, along with their supporting connecting lines, exceeded 5000 km, establishing a multi-channel, multi-source transmission and storage system. It formed China's most reliable gas supply system around the capital, often referred to as the "lifeline of gas supply to the capital." In 2019 and 2022, the Northern (Heihe-Changling) and Central (Changling-Yongqing) sections of the China-Russia Eastern Gas Pipeline project were successively put into operation (Figs. 17.11 and 17.12), expanding the in-service X80 steel pipeline diameter to 1422 mm. Simultaneously, in accordance with international standards, China comprehensively promoted the localization of pipeline materials and equipment, achieving the production of gas turbines, variable-frequency adjustable-speed motors, centrifugal compressors, various valves, automation control systems, and more—all made in China.

With the deepening of international cooperation in the oil and gas sector, China has comprehensively completed the construction of four major oil and gas channels. The Northwest Channel: In 2005, China's first cross-border energy import channel, the China-Kazakhstan crude oil pipeline, was completed and put into operation. In 2009, the China-Central Asia natural gas pipeline went through trial operations. The Northeast Channel: In 2010, the China-Russia crude oil pipeline was completed (see Fig. 17.13), and in 2017, the China-Russia crude oil second-line pipeline was

Fig. 17.11 The construction site of the China-Russia east route natural gas pipeline project

Fig. 17.12 Interior view of the China-Russia East line Heilongjiang shield crossing project tunnel

completed. In 2019 and 2022, the China-Russia East Line natural gas pipeline's northern and middle segments were put into operation. The Southern Channel: In 2013, the China-Myanmar oil and gas pipeline was fully connected (see Fig. 17.14). The Eastern Channel: Oil and gas from Africa, the Middle East, and Australia are transported by sea and land on the eastern coast. In just over a decade, China has comprehensively built the four major oil and gas strategic channels in the Northwest, Northeast, Southwest, and offshore areas "from scratch," achieving a comprehensive development of both oil and gas in these regions.

Fig. 17.13 The construction site of the second line of the China-Russia crude oil pipeline project

Fig. 17.14 China-Myanmar oil and gas pipeline Lancang river project

Today, China's crude oil pipeline network has essentially connected the Northwest with the Southwest, the Northeast with North China, and established channels for supplying inland areas from coastal regions after offshore oil landings. This has formed a pattern of "Northwest heading southeast, Northeast heading south, and coastal delivery to the inland." The refined oil pipeline network has established the main framework of trunk lines, including "Northwest to Southwest, Northwest to Central China, Northeast to North China to Central China." With the China-Myanmar crude oil pipeline as a backbone, it has created two major refined oil pipeline networks in Yunnan and Guangxi, as well as coastal refined oil delivery channels. Main trunk pipelines such as West–East Gas Pipeline (West–East Gas Transmission Project), China-Russia East Line, Shaanxi-Beijing Gas Pipeline, Sichuan Gas Transmission to the East, and China-Myanmar Natural Gas Pipeline links the West–East Gas Pipeline and Shaanxi-Beijing Gas Pipeline via Jining Connection Line; via Lanzhou-Yinchuan natural gas pipeline connects the Se–Ning–Lan natural gas pipeline, West–East Gas Pipeline, and Shaanxi-Gansu-Ningxia Gas Field to the Yinchuan natural gas pipeline; via Zhongwu natural gas pipeline connects the West–East Gas Pipeline and the Sichuan–Chongqing pipeline network; via Zhonggui natural gas pipeline connects the West–East Gas Pipeline, the Sichuan–Chongqing pipeline network, and the China–Myanmar Natural Gas Pipeline; via Yongqing–Tangshan–Qinhuangdao gas transmission pipeline connects the Shaanxi–Beijing Gas Pipeline and the China–Russia East Line natural gas pipeline, achieving the connection of the gas-producing regions of Chuanqing, Changqing, and Northwest with the gas-consuming markets in the Northeast, Northwest, Southwest, and the eastern coastal areas. This has formed a network pattern of "West–East Gas Transmission, North–South Gas Transmission, and Coastal Gas Landing." From then on, the oil and gas pipeline network in China has gradually taken shape, connecting overseas, covering the entire country, spanning east and west, running through north and south, and closely following the regional network development [8].

With the continuous advancement of reform, openness, and market-oriented reforms in China, the operational structure of the oil and gas pipeline network has gradually evolved. In 1991, China National Petroleum Corporation (CNPC) and the Beijing Municipal Government jointly invested to establish Beijing Natural Gas Pipeline Transmission Company, responsible for constructing the Shaanxi-Beijing Line 1 gas pipeline. This marked the first collaboration between a central state-owned petroleum company and a local government in the construction of a large-scale natural gas infrastructure. As gas demand increased along the pipeline and in Beijing, and in response to the national natural gas pipeline network layout requirements, subsequent pipelines including the Shaanxi-Beijing line 2, 3 and 4 were successively built, becoming a crucial component of China's natural gas pipeline network. In 2000, when the West–East Gas Pipeline project commenced, China initially sought foreign investment. Through international open bidding, negotiations were held with a consortium composed of Shell, Exxon, and Gazprom, but the foreign consortium's demands in terms of fixed returns on investment, gas sources, and market cooperation were deemed excessive, leading to the failure of the collaboration. In May 2012, CNPC, the National Social Security Fund Council, Beijing Guolian Energy Industry Investment Fund (Limited Partnership), and Baosteel Group signed the "West–East Gas Pipeline Line 3 Project Joint Venture Cooperation Framework Agreement." This marked the first time that CNPC introduced social and private capital to participate in the construction of the West–East Gas Pipeline Line 3 project, guiding and promoting private capital to focus on sectors and areas related to national welfare and development. In December 2019, the Central Committee of the Communist Party of China decided to establish the National Oil and Gas Pipeline Group Corporation, unifying the backbone pipelines of CNPC, China Petrochemical Corporation, China National Offshore Oil Corporation, and provincial-owned enterprises under the National Oil and Gas Pipeline Group Corporation for unified allocation, operation, and management. This marked the beginning of a new journey towards a "national integrated network." The National Oil and Gas Pipeline Group Corporation is actively working to build an "*X*+1+*X*" natural gas market system, where the two "*X*"s represent the diversification of upstream market entities and gas sources, the diversification of downstream market distribution entities and end customers, and the "1" represents the midstream oil and gas pipeline transportation and distribution system.

The Future of China's Oil and Gas Pipeline Transportation

Starting from the "Tenth Five-Year Plan," China began formulating national five-year development plans for oil and gas pipelines. In 2017, the National Development and Reform Commission (NDRC) and the National Energy Administration (NEA) released the "Medium and Long-Term Plan for Oil and Gas Pipeline Networks." This marked the first time that China promulgated a national-level development plan for oil and gas pipeline networks, highlighting the crucial role of such networks in the development of the national economy.

Throughout human history, each industrial revolution has signaled a reconstruction of the industrial landscape. Today, the Fourth Industrial Revolution, characterized by the combination of information and communication technology with virtual systems in cyberspace, is imminent. Under the drive towards "carbon peak" and "carbon neutrality" goals, energy structural adjustments are poised for development.

The regional imbalance in the energy landscape, the competitiveness of new energy sources, and the revolutionary nature of new technologies will drive significant changes in the energy structure. Petroleum will gradually transform into a production raw material, and natural gas will develop synergistically with new energy sources. This will gradually form an energy supply system dominated by non-fossil energy sources, integrating and developing with traditional fossil energy sources such as coal, oil, and gas. In the future energy system, the importance of natural gas pipelines remains unquestionable. Renewable resources such as wind and solar energy have intermittent, uncertain, and unstable utilization characteristics over time and quality. Natural gas pipelines will play a crucial role in addressing these issues, serving as a safety net in the process and shouldering a more significant historical mission in energy supply and consumption. Therefore, whether in the present or the future, oil and gas pipelines are the primary infrastructure ensuring the supply of oil and gas, and expediting the construction of a modern oil and gas pipeline system is imperative.

Since the implementation of the oil and gas "going global" strategy in 1993, Chinese petroleum enterprises have established five major oil and gas cooperation zones in more than 40 countries across five continents [15]. This has laid the foundation for the preliminary construction of four major strategic oil and gas import channels, namely, Northeast, Northwest, Southwest, and Offshore. From the perspective of global geopolitics and the oil and gas supply-demand landscape, countries and regions along the "Belt and Road" are the main sources of China's oil and gas imports. Therefore, considering the "Belt and Road" import channels as a crucial component of the oil and gas pipeline network layout, strengthening both domestic and international connectivity, promoting bidirectional openness between the east and west, facilitating energy interconnectivity among countries and regions along the route, and collectively building the "Energy Silk Road" are essential. This approach will enhance the level of mutual complementarity in oil and gas supply and demand, playing a significant role in ensuring China's energy security under open conditions.

Current pipeline technology cannot yet support the large-scale storage and long-distance transportation of hydrogen and carbon dioxide. This is a crucial focus area in the field of oil and gas pipelines to achieve the goals of "carbon peak" and "carbon neutrality." Hydrogen is a clean, zero-carbon secondary energy source with high energy density, presenting significant development potential in future new energy systems. It has become a global research hotspot. In 2014, the Fifth Assessment Report of the Intergovernmental Panel on Climate Change (IPCC) pointed out that Carbon Capture, Utilization, and Storage (CCUS) is one of the key technologies essential for achieving climate change mitigation goals. The entire process involves capturing, transporting, utilizing, and storing carbon dioxide (CO2) [16]. This includes extracting and purifying CO2 from industrial processes, energy utilization, or the atmosphere, followed by recycling for reuse or permanent storage, aiming to achieve permanent carbon sequestration.

The Energy Internet is an inevitable trend in the future development of energy. It serves as a backbone network that integrates the Internet of Things (IoT), power grids, and oil and gas pipelines. It deeply integrates the production, transportation, storage, and trade of fossil and renewable energy sources, constructing a new model for the development of the energy industry with horizontal multi-energy complementarity and vertical optimized configuration [17, 18]. This reshapes the architecture of energy supply and demand. The Energy Internet is a vast intelligent energy system, and the intelligent development of oil and gas pipelines needs to be integrated into it to comprehensively enhance the safety and efficiency of the pipeline network.

There is no doubt that the oil and gas pipeline network is a crucial infrastructure for China's initiatives such as the "Belt and Road" and the energy revolution. It serves as a key link in coordinating the development of the upstream and downstream of the oil and gas industry and is an essential component of China's modern energy system and comprehensive transportation network. In recent years, China's oil and gas pipeline construction has made significant progress and achieved positive results, with the network scale ranking among the world's best and the technological equipment reaching international advanced levels. However, considering the higher requirements for ensuring national energy security and supporting the goals of the "Two Centenary Goals," the development of the oil and gas pipeline network remains a significant and challenging task.

References

1. Zhou Zhenhe and Zhang Li. Annotated Explanations of the Geographical Records in the History of the Han Dynasty. Nanjing: Phoenix Publishing House, 2021.
2. Li Fang. The Extensive Records of the Taiping Era. Compiled and translated by Gao Guang and Wang Xiaoke. Beijing: Zhonghua Book Company, 2021.
3. Shen Kuo. Dream Pool Essays. Translated by Zhu Yuchen. Beijing: Zhonghua Book Company, 2016.
4. Bo Lanxi. Comprehensive Records of the Unification of the Yuan Dynasty. Beijing: Zhonghua Book Company, 1966.
5. Chang Qu. Records of Huayang Prefecture. Jinan: Qilu Book Press, 2010.
6. Song Yingxing. Heavenly Creations. Translated by Yang Weizeng. Beijing: Zhonghua Book Company, 2021.
7. Writing Group of A Century of Development History of China's Petroleum Industry. A Century of Development History of China's Petroleum Industry. Beijing: China Petrochemical Press, 2021.
8. Huang Weihe, Zheng Honglong, and Li Mingfei. Development History and Prospects of China's Oil and Gas Storage and Transportation Industry [J]. Oil & Gas Storage and Transportation, 2019, 38(1): 1–11.
9. Yan Dafan. Editorial: Reflection on the 70th Anniversary of Higher Education in Oil and Gas Storage and Transportation in China [J]. Oil & Gas Storage and Transportation, 2022, 41(6): 1–10.
10. Huang Weihe. Development of Oil and Gas Storage and Transportation Technology in China [J]. Oil & Gas Storage and Transportation, 2012, 31(6): 411–415.
11. General Compilation Committee of China Oil and Gas Field Development Chronicles. China Oil and Gas Field Development Chronicles: Xinjiang Oil and Gas Region. Beijing: Petroleum Industry Press, 2011.
12. Writing Group of China Oil and Gas Pipelines. China Oil and Gas Pipelines. Beijing: Petroleum Industry Press, 2004.
13. Guo Gangyan and Dang Suimei. An Introduction to Daqing Spirit and Iron Man Spirit. Beijing: Petroleum Industry Press, 2020.
14. Editorial Board of China Petroleum West-East Gas Transmission Pipeline Chronicles. Chronicles of China Petroleum West-East Gas Transmission Pipeline Company (2000–2016) [G]. Shanghai: Shanghai Academy of Social Sciences Press, 2018.
15. National Energy Administration. China Has Preliminarily Established Five Major Overseas Oil and Gas Cooperation Zones [EB/OL]. (2012-09-29) [2022-05-04]. http://www.nea.gov.cn/2012-09/29/c_131881502.htm.
16. Huang Weihe, Gong Jing, and Wang Jun. The Tasks of Oil and Gas Storage and Transportation Discipline under the Vision of Carbon Neutrality [J]. Oil & Gas Storage and Transportation, 2022, 41(6): 1–10.
17. Huang Weihe, Shen Xin, and Hao Yingpeng. Prospects for China's Oil and Gas Pipeline Network and Energy Internet Development [J]. Journal of Beijing Institute of Technology (Social Sciences Edition), 2019, 21(1): 1–6.
18. Sun Yuanzhang, Zhang Pengcheng, Ke Deping, et al. Planning, Design, and Optimal Operation of Energy Internet Systems under Carbon Peak Conditions [J]. Southern Power System Technology, 2022, 16(1): 1–13.

Chapter 18
Diverse: Wonders of Transportation Creation Throughout the Ages

Throughout history, the Chinese nation has thrived, persistently overcoming challenges and conquering nature for a better life. From the difficulties described in the ancient saying "the road to Shu is as hard as ascending to the sky" to the transformation of impassable chasms into accessible routes, diligent and resilient workers, along with pioneers in construction, have overcome unimaginable difficulties, creating one transportation miracle after another in history.

In the Sichuan Waterway: Boatmen Sailing Against the Current

Historic Waterways

During the pre-Qin period, the Shu region lacked iron industry and was unable to establish mountain passes, relying solely on waterways for transportation. According to the "Classic of Documents—Tribute of Yu," it is recorded: "Hua Yang's Black Water belongs to Liangzhou. The Min and Bao have been organized, the Tuo and Qian have been navigated… Westward they flow through Huan, floating on the Qian, crossing the Mian, entering the Wei, and flowing into the Yellow River." [1]. This indicates that during that time, the Minshan and Baozhong mountains were already cultivated, the Tuojiang and Qianshui (present-day Jialing River) were navigable. When people made tributes, they could follow the Huan River (present-day Bailong River), pass through the Qianshui (present-day Jialing River) and Mianshui, travel overland to the Weishui, and then cross the Yellow River.

In relation to the rivers of Ba Shu, Du Fu depicted the treacherous and vast waters of the Three Gorges in his poem "Waiting for Yan Liu, the Imperial Envoy of Yuzhou, Who Has Not Yet Arrived to Descend the Gorge." The lines go: "News

© China Communications Press Co., Ltd 2025

H. Lu (ed.), *A Brief History of Transportation in China*,
https://doi.org/10.1007/978-981-97-9851-3_18

arrives, mounted on a dappled steed; by the riverside, I wait to this day. Unaware if the clouds and rain have cleared, I waste away in short and long verses. The mountains, with their black water, stretch far and wide; the river connects to the depth of White Emperor. Passing through a single pillar, I linger with eyes gazing upon the ascent." [2]. Liu Yuxi expressed his sighs about the riverside in his poem "By the West Gate of the City": "Before the west gate of the city, the waves pile up year after year, unstoppable by the ages." [3]. Throughout various dynasties, efforts were made to modify the rivers of Ba Shu. By the time of the War of Resistance Against Japan, there was a slight improvement in water transportation conditions. However, the lives of the boatmen remained harsh. In 1940, Tian Yiwen wrote the prose "Song of the River" in Shapingba, depicting the boatman's living environment: towering waves, dangerous whirlpools everywhere, swift currents in some places, and hidden reefs; the river behaved capriciously, seemingly angry, laughing wildly, and singing as if in a trance. The boatman had to navigate through these turbulent waves. If not careful, cunning whirlpools and hidden reefs would claim the boatman's life [4].

The Amazing Life of a Boat Tracker

The boat tracker must use a towline, which is mostly specially made, thick, tough, and sturdy. Some are made of nylon, while others are made of bamboo strips (Fig. 18.1). Poor trackers, who make a living by towing, sell their strength, and their hands and shoulders become as hard as machines, so much so that over the years, the towline can leave deep grooves on hard rocks (Fig. 18.2). Whenever they carry the towline on their shoulders, they bend their backs and crawl, with their bodies lowered close to the sand and stones, walking on both feet and hands.

Waterway During War of Resistance Against Japan

During the War of Resistance Against Japan, numerous shipping enterprises, primarily led by Lu Zuofu's Minsheng Company, utilized the Yangtze River to transport a significant number of personnel and supplies from the lower reaches of the Yangtze to the vast rear areas of Sichuan. Simultaneously, they transported large contingents of troops from the rear to the frontline of the Anti-Japanese War, transforming the Yangtze River into a vital lifeline for the war effort and an invaluable golden waterway.

After the full outbreak of the War of Resistance Against Japan, the entire country underwent a mass westward migration to the rear areas, and Sichuan became the center for various aspects of wartime Chinese politics, economy, and military. At that time, Sichuan lagged behind the coastal regions in various developments and found it difficult to adapt to this change, particularly facing the challenging transportation conditions famously known as the "Difficulties of the Shu Road." In order to construct

Fig. 18.1 Bamboo splint towline

Fig. 18.2 Deep grooves left by towline

and consolidate the rear areas, which were advantageous for the war effort, various sectors inevitably sought to utilize cost-effective, efficient, and low-priced water transport. This led to significant development in the navigation of the upper reaches of the Yangtze River, specifically the rivers in the Chuanjiang system (generally referring to the tributaries of the Yangtze River above Yichang). This development played a positive role in supporting the war effort.

To develop water transportation, the first requirement is to have navigable river waterways. However, most of the major tributaries of the Chuanjiang system have rapid currents, numerous reefs, and numerous shoals, making the riverbed conditions extremely complex. In the past, only certain river sections were developed for navigation by the Minsheng Company. Many rivers have not undergone surveying, improvement, or trial navigation, and hydrological data was also relatively scarce. Developing navigation on the Chuanjiang system faced numerous difficulties. In order to develop water transport on the Chuanjiang system, in 1939, the National Government entrusted the Ministry of Communications to conduct surveys, improvements, and trial navigation on the major tributaries of the Chuanjiang system.

The original route for leaving Sichuan by water could go from the Chuanjiang system through the Three Gorges to Yichang, then downstream to Wuhan, or from Yueyang into Dongting Lake, connecting to various waterways leading to Hunan. However, due to the loss of cities like Yueyang and Yichang during the war, the main transportation routes from Sichuan to the outside world were blocked, necessitating the exploration of new water routes. Chongqing, as the wartime capital, needed support for both the front lines and external assistance. Therefore, transportation from the southeast and northwest became crucial. In addition, it was necessary to establish water-land transportation to remedy the situation. Hence, the Chuan-Xiang and Chuan-Shan water-land transportation routes were initiated. During the war, there was an extreme shortage of automotive fuel, parts, etc. As a result, water-land transportation aimed to fully utilize waterways and manpower to achieve economic objectives [5].

To this end, the Ministry of Communications sent representatives for multiple inspections, held several special meetings for discussion, and ultimately selected the water route as follows: from the Yangtze River to Fuling, turning into the Wujiang, sailing upstream to Gongtan or Pengshui against the current, undergoing a section of overland transport to Longtan, then entering the waterway, traveling along the Chenjiang and its tributaries to places like Wanling and Changde. To open this waterway as soon as possible, it was necessary to focus on the Wujiang, known for its many reefs. The project involved clearing numerous visible and hidden reefs along the 200-km water route from Fuling to Gongtan. The improvement project began in 1939 and was completed in 1941. It involved blasting 92,000 cubic meters of rocks on the water surface, underwater blasting of 17,000 cubic meters of hidden reefs, completing more than 37,000 cubic meters of earthwork for the construction of the passage, and setting up multiple weir stations as auxiliary measures. This made the 200-km waterway from the mouth of the Yangtze River to the Wujiang, upstream, a major water route from Sichuan to Shaanxi. Ships could navigate, and wooden boats could smoothly travel on the important tributary of the Yangtze River, the Jialing River. Chongqing, as the wartime capital, was located at the confluence of the Jialing River and the Yangtze River. Therefore, the development of Jialing River transportation was of great significance for wartime Chongqing's transportation and connection to the southwest and northwest regions. Starting in 1939, efforts were made to dredge the section from Hechuan to Guangyuan. New towing paths were opened in places with steep cliffs on both banks, totaling 14 km. Various dangerous shoals were treated to varying degrees in over 50 locations, and 110 navigational aids were set up. This allowed boats with a payload of 14 tons to pass through even during shallow water periods. Subsequently, the upper reaches of the Jialing River were gradually improved. In addition to the Jialing River, there was also the Minjiang River in Sichuan. It could travel downstream from Chengdu to Leshan and Yibin, directly reaching Chongqing, thus playing a significant role in shipping. Therefore, the river section below Leshan became a key area for improvement. Rocks were blasted, and winches were installed, allowing ships to sail from Chongqing to Leshan and sometimes even to Chengdu.

The aforementioned projects ensured the smoothness of the wartime water route into Sichuan, turning the waterway into an indestructible transportation lifeline.

With the outbreak of the war, the tranquility and order that once prevailed were shattered. People's national spirit was ignited, and they willingly contributed their modest efforts to the war. Even when facing significant dangers while towing barges, the boat trackers fearlessly embraced the challenges. While boat trackers lacked the ability to transform their working environment, they voluntarily undertook the most arduous manual labor in transporting war materials. Boat trackers who wholeheartedly engaged in towing not only aimed to earn a modest income but, more importantly, sought to rescue the national industry that urgently needed relocation in times of crisis. On the Yangtze River, boat trackers faced the relentless force of the river, wrestling and becoming one with its currents.

It is evident that wartime Chongqing and the water transportation in Sichuan served as the lifeline and major artery for military transportation, making irreplaceable contributions during this crucial period.

"Openly Repair the Plank Road, But Secretly Cross the Wei River." The Ingenious Achievements of Ancient Transportation

The primary function of the plank road is to facilitate communication, as stated by the Prime Minister of Qin, Fan Ju, in the "Strategies of the Warring States": "The plank road spans a thousand miles to Shu Han, making all under heaven fear Qin." [6]. In China, ancient plank roads, along with the Grand Canal and the Great Wall, are recognized as the three outstanding ancient constructions, playing crucial roles in military defense, transportation of goods, and civilian life. The saying "Openly Repair the Plank Road, but Secretly Cross the Wei River" is widely known and represents one of the famous military strategies in ancient China.

A plank road originally referred to a road built along cliffs and precipices, also known as a gallery road or a reversible path. As early as the Warring States period, plank roads were constructed in China. For instance, during the reign of King Hui of Qin, the Baoxie Plank Road was built from Baocheng in Shaanxi to Meixian (present-day Meixian County) in a diagonal valley. During the Qin invasion of Shu, the Jinniu

Road was constructed, later known as the Southern Plank Road (see Chap. 3 for details). While modern roads have formed a network, in mountainous areas with restricted transportation, similar plank roads still exist for pedestrian and livestock passage.

In a broad sense, the elevated passages between ancient high-rise buildings in China are also called "Zhandao" (plank road or boardwalk). The term "Zhandao" has a broader and more common meaning today, especially in the real estate industry. For example, a charming staircase-like wooden path in a garden is often referred to as a "wooden boardwalk."

To navigate through impassable mountain valleys, people carved diamond-shaped holes into cliffs separated by rivers. Stone or wooden stakes were inserted into these holes, and wooden or stone planks were placed horizontally above, allowing for walking and carriage passage. This is what is known as "Zhandao" or boardwalk. To prevent these stakes and planks from decay due to rain, people built pavilions (also called "Langting") above the boardwalks. The connecting paths between the boardwalk and the pavilion are known as "Zhan Ge Zhi Dao", shortened to "Zhandao."

It is said that the Jinniu Road, built by the Qin Dynasty, is considered the earliest Zhandao in China. The Jinniu Road, also known as "Shu Zhan", ran from the north in Shaanxi's Mianning County to the south at the Dajiang Pass in Bazhong, Sichuan, serving as an ancient transportation artery between Sichuan and Shaanxi. The segment from Guangyuan in northern Sichuan to Ninqiang in southern Shaanxi was particularly treacherous, prompting the poet Li Bai to exclaim, "The road to Shu is as difficult as ascending to the sky!".

"Openly Repair the Plank Road, but Secretly Cross the Wei River." In this allusion, "Chen Cang" refers to the ancient name for Baoji, specifically the Chen Cang ancient ferry on the north bank of the Wei River. "Plank road" refers to the ancient transportation route that crosses the Qinling Mountains, connecting the Guanzhong region to Hanzhong, Bashu. It consisted of the ancient Qinling road, Baoxie road, and Lianyun plank road, constructed over cliffs, swamps, and challenging terrains. The plank road exits in Guanzhong at Xieyuguan, approximately 70 km from the Chen Cang ancient ferry. Figure 18.3 depicts the marvelous achievement of ancient transportation—the ancient plank road.

Hanging Wall Highway: Engineering Marvel on the Lofty Cliffs of the Taihang Mountains

The Hanging Wall Highway is a miraculous project carved out on cliffs and precipices. Such projects are extremely perilous and challenging, and are rarely seen in the history of road construction worldwide.

In the vast expanse of the 800 miles of the majestic Taihang Mountains, there are a total of seven Hanging Wall Highways carved into the cliffs. Among them, the

Fig. 18.3 Ancient plank road [7]

most extensive and spectacular is undoubtedly the Xiyagou Hanging Wall Highway. Xiyagou is located deep within the Taihang Mountains, at the easternmost end of Lingchuan County, where the borders of Shanxi and Henan provinces meet. The people of Xiyagou, relying on their own hands, self-reliantly and tirelessly worked for 30 years. Using hammers and chisels, they carved out a "Hanging Wall Highway" stretching 7.5 km along the cliff, creating the renowned "Xiyagou Spirit" in the heart of China. For the convenience of construction and to allow natural light after opening, the Hanging Wall Highway was closely carved into the cliff, with windows on one side spaced more than ten meters apart. From a distance, the line of side windows outlines the direction of the tunnel (see Fig. 18.4).

In July 2009, the Xiyagou Hanging Wall Highway proudly topped the list of "Typical Rural Roads" in China Road Atlas. In September 2011, it was selected as one of the sixty major landmarks of New China, and in June 2010, it was chosen as one of the top ten new discoveries in Shanxi's Three Jin regions. The Xiyagou Hanging Wall Highway carries rich historical information and spiritual connotations. With distinctive characteristics of its time and irreplaceable qualities, it vividly reflects the indomitable spirit of the Chinese people in overcoming challenges and difficulties.

Fig. 18.4 Xiyagou hanging wall highway [8]

Xiyagou Village is composed of 17 natural villages, with over 200 households and more than 800 people. The ancestors of the villagers sought refuge in the mountains during ancient wars and disasters. Surrounded by mountains, the village is characterized by steep cliffs both uphill and downhill. For centuries, the villagers led a primitive life, engaging in activities such as clearing land, farming, planting trees, herding, and cutting firewood, creating a secluded paradise. After the founding of the People's Republic of China, they aspired to a new life. However, the isolated living environment made it difficult for them to escape hardship: large quantities of fruit rotted, medicinal herbs turned into manure, herds of pigs and sheep couldn't be driven down the mountain, and people succumbed to illness on the way down. In

1962, a deputy secretary of the county committee visited the village, riding a horse to the cliff's edge and being unable to find a way down. This incident frightened the horse as well. Upon returning, he allocated 3,000 yuan to Xiyagou for road construction. From then on, the villagers embarked on a difficult and lengthy journey of road construction.

The first time, they dug a small path on the cliff, which only the brave dared to walk. They attempted to drive pigs down the mountain, but 27 pigs fell and died after covering a little over a mile, with 13 casualties. The second time, they changed the direction of the road, but halfway through, they couldn't proceed and inadvertently attracted wolves from the mountains, earning it the name "Wolf Road." During the third attempt at road construction, they tried to drill a hole to get through. Unfortunately, after digging about 100 m, they faced the challenge of disposing of the excavated soil and couldn't ventilate the smoke. As a result, this hole turned into a "sheep cave."

Three Falls and Three Rises, the struggle and exploration of the people of Xiyagou in the past 20 years have been a spiritual torture of hope and disappointment, a pouring and hard work of sweat and effort. Despite three failed attempts to build roads, the people of Xiyagou were not willing to accept defeat and never gave up. The spring breeze of reform and opening up once again ignited the hope of the people of Xiyagou to step out of the mountains. On January 1, 1982, the village party branch once again raised the flag of road construction. They invited engineers and formulated a bold and innovative road construction plan: "Follow the mountain, drill holes along the cliff, clear debris through skylights, and spiral ascent." In order to raise funds, they sold their meager possessions. The village committee sold collective trees, cattle, sheep, and breeding rooms, even mortgaging a few offices for loans. Determined to make a breakthrough, they formed a party member vanguard team, a youth assault team, a bachelor death squad, and a women's support team. The entire village, men, women, young, and old, threw themselves into road construction. With the spirit of moving mountains, they spent more than 10 years and finally built a road that allows motor vehicles to travel smoothly on the dangerous cliffs of Taihang Mountains—the Xiyagou Hanging Road. The Hanging Road is 7.5 km long, with a tunnel length of 1380 m, costing 596,000 yuan, mobilizing 108,000 voluntary workers, generating 230,000 cubic meters of earthwork, and building five stone bridges.

Xiyagou has opened up a road to the outside world, expanded the living space for villagers, and propelled Xiyagou Village into the new era of reform and opening up. Subsequently, Xiyagou Village also constructed roads near the border with Huixian County, Henan Province. This facilitates the transportation of agricultural and sideline products and various logistics, further benefiting the economic development of the village. The connection with Henan promotes convenience for residents of both provinces. With the completion of roads in 17 natural villages, various vehicles and goods can reach the doorstep of residents, significantly influencing the thoughts, concepts, and lifestyles of the people in the valley. Additionally, this road has led to a noticeable increase in tourists, rapidly boosting the development of the tourism industry along the route, increasing the income of the residents, and truly becoming a path to prosperity.

Walking in the Clouds: An Upper-Span Bridge in a Deep Valley with a High Drop

Guizhou, with tens of thousands of bridges spanning from north to south, forms the unique lifelines of plateau transportation. It is the Chinese province with the highest number of bridges, boasting 21,000 highway bridges with a total length of 3,176 km. Guizhou has over 40 bridges among the world's top hundred highest bridges, and it dominates half of the world's top ten highest bridges. Virtually encompassing all types of bridges found in the world today, Guizhou can be considered a world-class "bridge museum." Located in the southeastern part of the southwestern region of China, Guizhou is often described as "eight mountains, one river, one patch of farmland." With 92.5% of its area being mountainous and hilly, Guizhou is the only Chinese province without plains, directly leading to the need for the construction of numerous bridges.

Guizhou's terrain is characterized by undulating mountains and numerous rivers, and the construction of transportation infrastructure in the past was extremely challenging, resulting in a difficult life for the majority of the population. In order to promote economic development, various stakeholders, including the government, enterprises, society, research institutes, and more, collaborated to construct numerous bridges that varied in type, were technologically complex, and presented great difficulties. These bridges served as connectors, linking the mountains together, bringing about significant changes in the lives, travel, and economic development of the people in the mountainous regions of Guizhou. The bridges built in Guizhou establish connections and pave the way for roads. Spanning across mountains, traversing perilous streams, these bridges provide crucial support for the "every county connected by expressways" initiative in Guizhou. Additionally, they offer convenient transportation for tourists visiting Guizhou, swiftly propelling economic development and significantly improving the efficiency of transportation between major cities in the province.

The "Industry Bridges," "People's Bridges," and "Prosperity Bridges" connect urban and rural areas, facilitating the flow of people, goods, and information between Guizhou and the outside world. With the support of these bridges, Guizhou's expressway network covers the entire province, connecting it to the rest of the country. Starting in 2014, with the completion and opening of the Guizhou-Guangzhou High-Speed Railway, the Changsha-Kunming section of the Shanghai-Kunming High-Speed Railway, and the Chongqing-Guiyang Railway, Guiyang rapidly became a key railway hub in the southwestern region. The three-dimensional transportation network continues to expand, propelling Guizhou's poverty alleviation efforts onto the "fast track." In 2018, the total length of operational expressways in Guizhou reached 6,453 km, ranking seventh in the country, with 18 highways providing access to other provinces.

After successfully achieving "county-wide access to expressways, village-wide access to highways, and township-wide access to asphalt roads," Guizhou's GDP has maintained double-digit growth, propelling economic development to new heights.

In 2017, the province led the nation in economic growth, and its overall economic size climbed to the 25th position nationwide. By 2018, Guizhou's total economic output reached 1.48 trillion yuan, with a per capita income exceeding 40,000 yuan.

With the completion of the bridges, the industrial structure in Guizhou has undergone adjustments. Adhering to the dual principles of development and ecology, Guizhou has accelerated the layout and expansion of the tourism industry, gradually transitioning from a "summer resort" to a "tourism highland." The annual growth rate of tourist numbers is 30%, and in 2017, the total number of tourists received in the province ranked second nationwide. Supported by expressways and high-speed railways, the mountainous and distinctive tourism industry belt connects and drives 100 scenic spots, 100 mountainous outdoor sports tourism bases, and 1000 unique tourism villages and hamlets throughout the province, forming 20 key and exquisite routes within the province and connecting with other provinces. Guizhou has improved its transportation infrastructure and, seizing historical opportunities, has taken the lead in developing its transportation big data and its applications, positioning itself at the forefront nationally. Numerous bridges spanning precipitous terrain have facilitated the integration of transportation and land use in the challenging topography of Guizhou, playing a crucial role in radiating and propelling urban civilization. They have become a significant symbol showcasing China's comprehensive national strength.

Between mountains and rivers, within cities, Guizhou's bridges have turned the dream of transforming the plateau into plains into reality. Guizhou's bridges continuously set records and have achieved several world firsts: the Yachi River Bridge on the Gui-Qian Expressway—the world's largest-spanning steel truss cable-stayed bridge (Fig. 18.5); the Beipanjiang Bridge on the Bidu Expressway—the world's highest bridge (565 m) (Fig. 18.6); the Balinghe Bridge on the Zhensheng Expressway—the world's first kilometer-scale steel truss cable-stayed bridge in mountainous areas (Fig. 18.7); the Beipan River Bridge on the Shuipan Expressway—the world's largest-spanning reinforced concrete beam bridge (Fig. 18.8, see Chap. 22 for details); and the Hezhang Bridge on the Biwei Expressway—the world's tallest pier for a girder bridge (195 m) (Fig. 18.9). These bridges, in various forms, embody the hard work and intelligence of Chinese bridge builders, showcasing the human miracles created by institutional advantages.

Desert Wonders: Taklimakan Desert Highway

The Taklamakan Desert is the largest desert in China, often referred to as the "Sea of Death" due to its extremely harsh natural conditions. However, within this "Sea of Death," there are four roads that traverse the entire desert. The completed roads crossing the Taklamakan Desert include the Hotan to Alar City road, the Alar to Tazhong Town road, the Lunnan Town to Minfeng County road, and the Tazhong Town to Qiemo County road (Fig. 18.10). These routes traverse the Taklamakan Desert from south to north, not only bringing significant convenience to the cities on

Fig. 18.5 Yachi River Bridge on Guizhou-Guizhou expressway [9]

Fig. 18.6 Beipan River Bridge on Bidu expressway [10]

the desert's edge but, more importantly, opening up transportation channels for the vast oil and gas resources within the desert. This infrastructure has laid the foundation for the extraction of these resources, transforming this once lifeless "Sea of Death" into a "Sea of Gold" rich with wealth.

The Taklamakan Desert Highway in Xinjiang, China (also known as the Tarim Desert Highway), is the world's longest graded highway traversing a shifting sand desert. It is also the earliest desert highway in China. This region has been the

Fig. 18.7 Baling River Bridge on Zhensheng expressway [11]

Fig. 18.8 Beipan River Bridge on Shuipan expressway [12]

center of the ancient Silk Road since ancient times and is now a major battleground for oil exploration and development. It serves as a key oil and gas field base for China National Petroleum Corporation (CNPC) and China Petrochemical Corporation (SINOPEC), passing through Lunnan Oilfield, Tahe Oilfield, and Tazhong Oilfield. The highway has facilitated the development of the local economy and has become one of the destinations for numerous tourists seeking to explore the area.

However, building a highway in such an extreme environment is not an easy task and, to some extent, could be considered an impossible mission. Constructing a road in a desert covering hundreds of thousands of square kilometers with shifting sand

Fig. 18.9 Hezhang Bridge on Biwei expressway [13]

dunes poses significant challenges. The primary obstacle is ensuring that the road built one day will not be buried by the desert the next. Additionally, the geological conditions in the desert are extremely soft, causing the road foundation to deform quickly on the loose soil. Coupled with the vast temperature differences between day and night and the extremely arid climate, each of these factors presents a formidable challenge that must be overcome in the construction of a desert highway. Nevertheless, China's builders had demonstrated the diligence and ingenuity of the working people in the face of such enormous challenges.

China's unique combination of "reed fences" and "reed grids" for sand fixation, along with the construction technology of "strong foundation and thin surface" for the road, has reached a world-leading level, significantly addressing the challenges in desert highway construction. Within dozens of meters on both sides of each road segment, reed grids are used to fix the shifting sand dunes, reducing the mobility of the yellow sand. Additionally, salt-resistant desert plants are planted to further stabilize the sand. To ensure the survival of plants in the desert, advanced drip irrigation technology is employed, and maintenance stations are established every 2 km to maintain the vegetation and road surface. The caretakers at these maintenance stations, mostly married couples, stand guard day and night, contributing significantly to the lifeline of these desert highways for decades.

Located in the heart of the desert, Ta Zhong Town serves as a crucial supply station for the desert highways in the desert hinterland. The 216 National Road, 233 Provincial Road, and Desert Highway No. 1, all crossing the desert, converge at this town. Ta Zhong Town is not only an essential supply station in the desert but also the location of the Ta Zhong Oilfield. The construction and opening of several desert highways have significantly increased the transportation efficiency of the Ta Zhong

Fig. 18.10 Hotan-Alar City, Alar-Tazhong Town, Lunnan Town-Minfeng County, Tazhong Town-Qiemo County Highway [14, 15]

Oilfield, which exploits abundant oil and gas resources. With the support of these highways, the value of this desert region has been greatly enhanced.

The Taklamakan Desert Highway is a grand project that conquers the "Sea of Death," turning a millennium-long dream into reality and setting the Guinness World Record for the longest graded highway across a shifting sand desert.

Heavenly Road to Xizang: New Passage to Xizang from Qinghai, Sichuan and Yunnan

The core route of the "New Yunnan-Xizang Passage" (Fig. 18.11) is the B-C-R Highway, which runs from Dali in Yunnan through Gongshan County to Rannu Town in Bomi County, Xizang, where it connects with the southern route of the Sichuan-Xizang Highway, namely National Highway 318.

The passage, starting from Dali City, Dali Bai Autonomous Prefecture, Yunnan Province, follows the Dali-Baoshan Expressway to Jinchangling, passing through Yunlong, Lushui, Fugong, and Gongshan, and concludes in Zayü County, Xizang (located near the China-India border). The total length is 786 km, with the Xizang section being 263 km. The entire route is open to traffic (see Fig. 18.9). Within Yunnan Province, the road spans 551 km, while the Xizang Autonomous Region section covers 235 km. The segment from Dali City to Jinchangling (127 km) is part of the Hangrui Expressway, the stretch from Jinchangling to Liuku (97 km) is

Fig. 18.11 New Yunnan-Xizang passage [16]

a secondary road, the Liuku to Buzhongluo segment (292 km) is a tertiary road, and the Buzhongluo to Zayü County Sect. (270 km) is a simple road.

The section from Jinchangling to Liuku is being reconstructed into an expressway, the Liubing Highway, specifically the Liuku to Bingzhongluo segment, is being upgraded to a secondary road, and the B-C–C Highway, which includes the Bingzhongluo to Zayü County stretch, is being upgraded to a tertiary road. Additionally, most segments of this route through the Nujiang low-altitude area have an average elevation of 2000 to 3000 m, ensuring year-round vehicle access.

The saying "To get rich, build roads first" resonates with the hardworking and courageous people of Nujiang who, for centuries, relied on cableways to cross the river. The Dian-Zang New Passage has not only brought prosperity to the people and strength to the province but has also complemented the development of hydroelectric power in the Nujiang region, playing a significant role in the economic and social advancement of the Nujiang area.

This passage has significantly shortened the driving distance. The Dian-Zang Highway, completed and opened in 1976, stretched from Kunming through Dali, Lijiang, and Diqing to Zayü County in Xizang, covering a total length of 1729 km. With the construction of the Dian-Zang New Passage, the route from Kunming through Dali, following the Nujiang River north to Zayü County in Xizang, spans 1093 km, shortening the distance by 636 km. Traveling along the Dian-Zang New Passage from Kunming through Dali, Nujiang, Zayü, Linzhi, to Lhasa covers a total distance of 2055 km. Compared to the existing Dian-Zang Highway from Kunming through Dali, Lijiang, Diqing, entering Xizang and reaching Lhasa, which spans 2314 km, the new route shortens the distance by 259 km.

The passage ensures year-round accessibility. Currently, the Yunnan-Xizang Highway from Kunming to Zayu County in Xizang involves traversing high-altitude regions, spanning three major river systems—Jinsha River, Lancang River, and Nu River—crossing over ten mountain passes above 4,000 m. The road conditions are harsh, and there are issues with seasonal and time-dependent closures. With the construction of the new Yunnan-Xizang passage, the section from Lushui County in Nujiang Prefecture, Yunnan, to Zayu County in Xizang only has two mountain passes, namely Norla Mountain and Qimala Mountain, both exceeding 4,000 m in altitude. Once tunnels of over 1,600 m in length are established for each, year-round accessibility can be achieved.

The passage significantly enhances national defense mobilization capabilities. The Yunnan-Xizang New Passage runs through Yunnan's Nujiang Prefecture and Xizang's Nyingchi City, located at the junction of China with Myanmar, India, and the Yunnan-Xizang region. In the face of changing domestic and international situations, Nujiang Prefecture is on the front line of "infiltration and anti-infiltration, subversion and anti-subversion, competition and anti-competition." Zayu County in Nyingchi City is the site of the famous Walong Battle in the Sino-Indian Self-Defense Counterattack in 1962. Both of these regions hold crucial positions in national defense and security strategy. After the completion of the Yunnan-Xizang New Passage, the travel time from Kunming to Zayu County will be reduced from the current 8–9 days on the existing Yunnan-Xizang Highway to 2–3 days, significantly shortening transportation time, enhancing transportation guarantee capabilities, and significantly strengthening national defense mobilization capabilities.

The passage significantly enhances development driving force. On the one hand, the section of the Yunnan-Xizang New Passage from Nujiang's Lushui County to Zayu County in Xizang is located on the border, holding crucial significance for Yunnan Province's "bridgehead" construction. It can radiate the three major regions of China with Myanmar, India, and the Yunnan-Xizang region, accelerating the formation of the Northwest Yunnan tourism grand loop and economic circle. On the other hand, Nujiang Prefecture and Nyingchi City are both economically underdeveloped border regions. These two regions harbor rich water, mineral, biological, and tourism resources. However, due to factors like lagging transportation, these abundant resources have not been effectively developed and utilized. The Yunnan-Xizang New Passage can connect the world-renowned "Nujiang Grand Canyon" and the "Snowy Small Jiangnan," significantly improving the infrastructure of both regions and forcefully promoting leapfrog development in their economic and social aspects.

The Magnificent Jade Ribbon Soaring in the Sky—Large-Scale Three-Dimensional Intersection Facility Connecting City and Highways

Elevated bridges have become symbols of urban landscapes at different stages of urbanization, serving various functions in the ebbs and flows of urban development. The crisscrossing and well-organized roads represent the progress of human urbanization.

The world's earliest interchange bridge appeared in the United States. As early as 1928, American architects built the world's first cloverleaf-shaped highway interchange at the intersection of two roads in New Jersey. In 1930, Chicago erected an interchange bridge. From 1931 to 1935, Sweden also built interchange bridges in several cities. Today, the builders of China's transportation system are constructing interchange bridges across the vast land, providing safe and efficient transportation arteries for modern cities and numerous highways.

Beijing's Dazzling Array of Overpasses

The first true modern overpass in China is the Fuxingmen Flyover in Beijing. Completed in October 1974, it is situated at the intersection of West Second Ring Road, Fuxingmen Inner Street, and Fuxingmen Outer Street in Beijing. It is a cloverleaf-style interchange. Its completion marked a new milestone in China's urban road bridge design, construction, and traffic management.

Fig. 18.12 Overpass in Beijing

Beijing has nearly 300 bridges of various sizes, with over 200 of them being flyovers, accounting for more than 70% of the total number of urban flyovers in the country. Many medium and large intersections in Beijing have flyovers, especially along the Second Ring Road and Third Ring Road, where the goal of achieving traffic signal-free flow has been realized. These flyovers, like majestic dragons and tigers, soar over the urban roads, showcasing the grandeur and charm of the capital as a modern metropolis (Fig. 18.12) [17].

World's Record: Chongqing Huangjuewan Overpass

Huangjuewan Overpass (Fig. 18.13) connects eight directions, including Guangyang Island, Jiangbei Airport, Nanan, Dafosi Bridge, Chaotianmen Bridge, Danzishi, Sigongli, and Chayuan. It comprises a total of 15 ramps, including 13 elevated ramps and 2 roadbed ramps. The total length of the overpass ramps is 16,414 m, with the elevated bridge measuring 6,770 m and the road and connection sections measuring 9,644 m. It is considered the "most complex overpass in Chinese history" and is currently the most powerful "hub-type" overpass in the central urban area of Chongqing. With an elevation difference of 37 m from the ground to the highest point, equivalent to a 12-story building, it covers an area of 616 mu (approximately 0.41 square kilometers). With continuous traffic and grandeur, it has become a "popular" architectural landmark in Chongqing, especially enchanting when illuminated at night.

Fig. 18.13 Huangjuewan urban overpass

The Huangjuewan Overpass provides significant travel convenience for the residents, playing a crucial role in improving the southern transportation network of the main urban area in Chongqing and promoting economic development in the surrounding regions.

References

1. The Book of Documents. Translated and annotated by Wang Shishun and Wang Cuiye. Beijing: Zhonghua Book Company, 2012.
2. Du Fu. Collected Works of Du Gongbu. Shanghai: Shanghai Ancient Books Publishing House, 2003.
3. Liu Yuxi. Collected Works of Liu Mengde. Shanghai: Shanghai Ancient Books Publishing House, 2013.
4. Tian Yiwen. The Collection of Firefly-Carrying Sacks. Guangzhou: Huacheng Publishing House, 1984.
5. Chi Xianghua. Wooden Boat Transportation on the Chuanjiang River During the War of Resistance Against Japan [D]. Chongqing: Southwest University, 2008.
6. Strategies of the Warring States. Shanghai: Shanghai Ancient Books Publishing House, 2008.
7. Wang Xiaohong. Magnificent Shu Path: A Living Fossil of Ancient Land Transportation [N]. Huaxi Metropolis Daily, 2016-10-29(5).
8. Baidu Encyclopedia. Taihang Mountain Xiyagou Cliff-Side Highway [EB/OL]. [2022-05-08]. https://baike.baidu.com/item/%E5%A4%AA%E8%A1%8C%E5%B1%B1%E9%94%A1%E5%B4%96%E6%B2%9F%E6%8C%82%E5%A3%81%E5%85%AC%E8%B7%AF/9315337?fr=aladdin
9. People's Network. Guizhou-Qianxinan Expressway's Yachihe Bridge Welcomes Returning Peak Traffic [EB/OL]. (2020-10-08) [2022-05-08]. https://m.sohu.com/a/423191738_114731
10. Laobaixiao Bridge Collection. Beipanjiang Bridge on Guanxing Highway in Guizhou Province [EB/OL]. [2022-05-08]. http://www.china-qiao.com/ql26/26ql/gzql010.htm
11. Colorful Guizhou Network. Balihe Bridge, Where the Most Beautiful Wisdom Meets the Most Beautiful Scenery [EB/OL]. (2022-01-24) [2022-05-08]. https://baijiahao.baidu.com/s?id=1722790427525090562&wfr=spider&for=pc
12. Colorful Guizhou Network. Hua Chunying Praises Guizhou Again, This Time for Its Bridges! Let's See How Amazing Guizhou's Bridges Are [EB/OL]. (2021-10-18) [2022-05-15]. http://whhly.guizhou.gov.cn/xwzx/tt/202110/t20211018_70924559.html
13. Colorful Guizhou Network. This is Guizhou, With the Most Beautiful Scenery You Desire [EB/OL]. (2020-07-20) [2022-05-15]. https://www.sohu.com/a/408698066_99994054?_trans_=000014_bdss_dk315wh
14. Baidu. A Highway That Cost 37 Billion Yuan and Employed Over 30,000 People to Build, Yet Now Hardly Sees a Single Car [EB/OL]. (2020-08-24) [2022-05-15]. https://baijiahao.baidu.com/s?id=1675893625544262893
15. China Forestry Association. Pursuit of Excellence: High-Quality Development of Forestry and Grassland - Korla in Action [EB/OL]. (2019-11-25) [2022-05-15]. https://www.sohu.com/a/355885506_674339
16. Sina.com. The Bing-Cha-Ran Route: The Little Seventh of Tibet Routes, Renowned as the Off-road Heaven Among the Seven Routes to Tibet [EB/OL]. (2021-04-20) [2022-05-28]. http://k.sina.com.cn/article_7514230677_1bfe20f9500100uzk2.html
17. Guiyang Network. Guizhou is Going Viral, This Miracle Stuns the World [EB/OL]. (2020-06-11) [2022-05-18]. https://baijiahao.baidu.com/s?id=1669170634601461328&wfr=spider&for=pc

Chapter 19
Transport in the Treasured Island: The Roller-Coaster Development

Treasured Island Taiwan has been a part of China's territory since ancient times. Originally connected to the mainland without any strait, it later became an island due to changes in the Earth's crust, forming a strait. Taiwan is part of the many islands in East Asia, surrounded by water on all sides. It has a unique historical development and island transportation system, holding a strategically important position.

Geographical Location, Historical Evolution, and Current Situation of Taiwan

Taiwan is located on the southeast edge of China, with the Taiwan Strait to the west facing Fujian Province, at its closest point only 130 km away. The entire province strategically controls the gateway to the western Pacific Ocean and is known as the 'Seven Provinces Barrier' of our country. The West Pacific shipping route serves as both a maritime corridor connecting Northeast Asia and Southeast Asia and a vital passage linking the Pacific and communicating with the Indian Ocean, making it one of the busiest maritime routes in the world.

Taiwan Province includes Taiwan Island, surrounding islands, and the Penghu Archipelago, consisting of more than 80 islands, with a total area of nearly 36,000 square kilometers. Among them, the main island of Taiwan is 394 km long from north to south, 144 km wide at its widest point from east to west, and has a coastline of 1,139 km, making it the largest island in our country.

The main island of Taiwan is a mountainous island, with mountains and hills accounting for two-thirds of its area and plains occupying less than one-third. The island's topography is characterized by higher elevations in the central region, forming the features of mountainous terrain in the east, hilly terrain in the central part, and plains in the west, with the Central Mountain Range running north to south serving as the watershed.

© China Communications Press Co., Ltd 2025

H. Lu (ed.), *A Brief History of Transportation in China*,
https://doi.org/10.1007/978-981-97-9851-3_19

In ancient times, Taiwan was connected to the mainland. Later, due to tectonic movements, the connected part sank into the sea, forming the strait, and Taiwan Island emerged. According to historical records, mainland residents migrated eastward to Taiwan, cultivating and developing the island, dating back over 1,700 years to the Three Kingdoms period. In the Sui Dynasty, Emperor Yang of Sui sent envoys to "observe foreign customs" in Taiwan three times. During the Tang and Song dynasties, coastal residents, especially those from Fujian, including Xiamen, Zhangzhou, and Quanzhou, fled to Penghu or migrated to Taiwan to escape conflicts. In the first year of the Yuan dynasty (1335), the Yuan government officially established an "inspection office" in Penghu, overseeing civil affairs in Penghu and Taiwan, under the jurisdiction of Tong'an County, Fujian (present-day Xiamen). During the Ming Yongle period (1403–1424), the navigator "Eunuch Sanbao" Zheng He led a large fleet on voyages to various countries in Southeast Asia, making a stop in Taiwan. After the 15th century, Wokou pirates continuously harassed the southeastern coastal areas of China, leading the Ming dynasty to station military forces in Keelung and Tamsui. In the early 17th century, Dutch colonizers invaded Taiwan, making it a Dutch colony. In the first year of the Chongzhen era (1628), Zheng Zhilong led residents from Fujian and Guangdong to relocate to Taiwan, engaging in farming, trade, and organizing armed forces to resist Wokou and Dutch incursions. In the 18th year of Shunzhi (1661), Zheng Chenggong, also known as Koxinga, led 25,000 soldiers and hundreds of warships, marching from Jinmen to Taiwan. In the first year of Kangxi (1662), Koxinga forced the Dutch governor to sign a surrender, reclaiming Taiwan. In the 22nd year of Kangxi (1683), the Qing government dispatched Admiral Shi Lang, leading over 20,000 soldiers and 200 warships, to advance from Tongsan to Penghu and Taiwan, defeating the Zheng forces. Zheng Chenggong's grandson Zheng Keshuang surrendered to the Qing government. In the 23rd year of Kangxi (1684), the Qing government established the Taiwan-Fujian Defense Circuit and Taiwan Prefecture, which were under the jurisdiction of Fujian Province. During the Sino-French War from the 10th to the 11th year of Guangxu (1884–1885), French forces attacked Taiwan but suffered heavy losses under Liu Mingchuan's command. In the 11th year of Guangxu (1885), the Qing government designated Taiwan as a separate province, making it the 20th province of China, with Liu Mingchuan appointed as the first governor of Taiwan. In the 20th year of Guangxu (1894), Japan initiated the First Sino-Japanese War, and the following year, after China's defeat, was forced to sign the humiliating Treaty of Shimonoseki on April 17, ceding Taiwan to Japan. In August 1945, Japan was defeated in World War II, and Taiwan was returned to China.

Today, the population of Taiwan is approximately 23 million people, making it one of the regions with relatively high population density in the world. Aside from the original 320,000 indigenous residents, the rest of the population mainly originated from the mainland, with a significant portion coming from the regions of Fujian, including Xiamen, Zhangzhou, and Quanzhou. As a result, the primary language of the Taiwanese people is Minnan (Southern Min) dialect. Taiwan has preserved ancient Chinese traditional culture, evident in the abundance of historical sites, temples, and the rich collection of artifacts in the National Palace Museum. The temples in Taiwan reflect the residents' devout religious beliefs, with Buddhism and Taoism being the main religions. Additionally, there are followers of Christianity, Catholicism, Islam, and other religions in Taiwan.

Taiwan's Highway Development Overview

During the period of Japanese invasion and colonial rule, Taiwan's highways experienced significant development. According to statistics, in 1946, the total length of highways in Taiwan reached 17,272 km, including provincial roads (1,138 km), county roads (2,601 km), and township roads (13,533 km). However, only about 40% of these routes were actually open to traffic in the early post-war period. From the liberation of Taiwan to 1950, the focus was mainly on restoring existing highways. In the 1950s to the mid-1960s, Taiwan initiated new highway construction projects, including the East–West Cross-Island Highway, the MacArthur Highway (from Taipei to Keelung), the Taitung Highway, and the Hualien Highway. By 1964, the volume of highway freight transportation exceeded that of railways, marking the gradual replacement of railways by highways as the most important land transportation mode in Taiwan. In July 1968, Taiwan formulated the "Taiwan Highway Ten-Year Construction Plan." Starting from the 1970s, Taiwan prioritized transportation in various significant economic development plans and began the construction of expressways. The highway system gradually modernized, and there were considerable advancements in traffic automation and road safety.

Taiwan's highways are classified based on transportation functions into expressways, rapid roads, major arteries, minor arteries, and urban streets. In 1997, the total length of highways in Taiwan reached 20,185 km.

Highway transportation includes passenger and freight transportation. Passenger transportation is jointly operated by Taiwan Motor Transport Co., various private passenger transport companies, and county/city public bus transport entities. Freight transportation is undertaken by private companies. Passenger volume has gradually declined since the mid-1980s, while freight volume has fluctuated significantly and has also shown a decreasing trend in recent years.

Expressways: The first expressway built in Taiwan is the Zhongshan (North–South) Expressway. This is a major artery for north–south land transportation in Taiwan, stretching from Keelung in the north to Kaohsiung in the south, with a total length of 381.7 km. The road has 4 to 8 lanes, and the average driving speed is 90–110 km per hour. It takes only 4 h to travel from the south to the north, which is twice as fast as ordinary highways. Its branches connect to the international ports of Keelung, Taichung, and Kaohsiung, as well as the international airports of Taoyuan and Kaohsiung. Since the 1990s, Taiwan has successively constructed the Northern Second Expressway, the Northern Yilan Expressway, and several other transverse expressways. Photos of Taiwan's expressways are shown in Fig. 19.1.

The Island Loop Highway: A densely woven network of roads covering the entire island. Starting from Keelung in the north, passing through Taipei, Hsinchu, Miaoli,

Fig. 19.1 Photos of Taiwan's expressways

Changhua, Tainan, Kaohsiung, and other counties and cities, it turns towards Feng-gang in Pingtung, then bends northeast, crossing the southern end of the Central Mountain Range, passing through the eastern counties of Taitung, Hualien, and Su'ao, and reaching Yilan; from Yilan, it splits into two lines, one heading to Taipei and the other to Keelung, completing a loop around the island, with a total length of 1103 km, interconnecting with other roads across the land.

Trans-Central Highway: Crossing the north–south Central Mountain Range and connecting with the eastern and western branches of the Island Loop Highway, there are a total of 6 roads, known as the "Old Three Roads" and "New Three Roads." The "Old Three Roads" were expanded from roads opened during the Qing Dynasty and its Japanese occupation and colonial rule, comprising three roads in the north, central, and south. The Northern Trans-Central Highway starts from Daxi in Taoyuan County, passing through Sanmin, Fuxing, Gaopo, Baliang, Chiduan, and connecting with the Yilan branch, with a total length of 120 km. The Central Trans-Central Highway, also known as the East–West Trans-Central Highway, starts from Dongshi Township in Taichung County, ascends along the Dajia River, passing through Guguan, Darong, Lishan, Hehuan Pass, Guanyuan, Tianxiang, and Xi Pan to Taroko, crossing the central part of the Central Mountain Range in Taiwan, connecting with the Suhua Highway, with a total length of 194 km (see Fig. 19.2). The Southern Trans-Central Highway starts from Yujing in Tainan County, passing through Jiaxian, Taoyuan, Meishan, Yako, Xiangyang, Xinwu to Haiduan, connecting with the Hualien Highway, with a total length of 182.6 km. The "New Three Roads" of the Trans-Central Highway refer to the three radiating segments completed in 1986 with Yushan Park as the central point, namely from Chiayi to Yushan, from Shuili to Yushan, and from Yushan to Yuli, with a total length of 285 km. The opening of these

Fig. 19.2 Trans-central highway

three trans-central highways is beneficial for the development of central mountain resources and the tourism industry.

The North–South Trunk Highways: Multiple mainline highways running from north to south. The first trunk highway originates from Taipei, passing through Banqiao, Fengyuan, Taichung, Nantou, Dapu, Yujing, and Qishan, extending to Pingdong, with a total length of 470 km. The second trunk highway starts from Hsinchu, traversing Miaoli, Fengyuan, intersecting with the first line in Taichung, then turning south to Yutian and merging with the Western Island Loop at Changhua, proceeding directly to Tainan, covering a total distance of 234 km. The third trunk highway begins in Taichung County, passing through Puli, Daguang, Shuili, and Heshe, crossing the Yushan watershed along the Nanzih Xianxi, descending through Jiaxian, and converging with the first line in Qishan, with a total length of 258 km. These three mainlines are situated in western Taiwan, roughly parallel to the expressways. To strengthen the transportation network in the western corridor, Taiwan planned the construction of 12 new east–west four-lane or higher-grade highways in the early 1990s. The total length is 360 km, including the Wanli Ruibin Line, Bali Xindian Line, Guanyin Dawan Line, Nanliao Zhudong Line, Houlung Wenshui Line, Changbin Taichung Line, Hanbao Caotun Line, Taixi Gukeng Line, Dongshi Jiayi Line, Beimen Yujing Line, Tainan Guanmiao Line, and Kaohsiung Chaozhou Line.

Coastal Highway: A roadway running along the coastline, with a total length of 887.4 km, divided into four segments—East, West, South, and North. The Northern Coastal Highway spans a total length of 262 km, divided into two parts: one part is the Guandu-Su'ao Line, passing through Tamsui, Keelung, Ruibin, and Donggang;

the other part is the Guandu-Xiangshan Line, passing through Bali, Xiafu, Zhuwei, Jiugang, and Niupu. The Western Coastal Highway, starting from Wuci, passes through Zhonggang, Xingang, Xigang, Taixi, Jinhu, Budai, Tainan, Zuoying, Kaohsiung, Linyuan, Donggang, Linbian to Fangliao, with a total length of 281 km. The Eastern Coastal Highway, with a total length of 194 km, begins in Hualien, passing through Fengbin, Changbin, Chenggong, Fuyuan, Beinan, Taitung, and Zhiben. The Southern Coastal Highway, spanning 89 km, begins in Fenggang and passes through Dongcheng, E-Luan-Bi, Gangkou, Gangzi, to Anshuo. The Coastal Highway can be considered the peripheral route of the Ring Around Taiwan.

In the early 1990s, Taiwan formulated the "West Coastal Highway Upgrade to Expressway Plan" and completed it in 1996, inaugurating the first expressway in the province running from north to south. This highway begins at Guandu Bridge in the north and extends to Beipi Bridge at the junction of Kaohsiung County and City, covering a total length of 375 km. Among them, the Taipei Guandu Bridge to Xiangshan section of Route 15 is 82 km, the Xiangshan to Jianan section is 79 km, and the Jianan to Beipi Bridge section of Route 17 is 209 km.

The Liaison Highways serve as connecting branch roads linking the Island Ring Highway, Transversal Highways, Longitudinal Highways, and Coastal Highways, forming a comprehensive and densely distributed transportation network across the entire island. There are a total of 108 Liaison Highways with a combined length of 3117 km.

The Developmental History of Traditional Railways in Taiwan

The earliest railway constructed in Taiwan was the Keelung to Badouzi line, which was built in the second year of the Guangxu era (1876). In the thirtieth year of the Daoguang era (1850), the British envoy had proposed mining coal in Keelung to the Qing government, but the request was rejected. Later, to meet the demands of the Fuzhou Shipyard, Shen Baozhen, in the thirteenth year of the Tongzhi era (1874), accepted suggestions from Japanese and Italian advisors and submitted to the Imperial Court to employ modern methods for coal mining. The construction of the Badouzi mine officially began in the second year of the Guangxu era (1876). To transport the mined coal, plans were made to build a light railway, designed by the British engineer Moulton (Wusong Railway Limited Company), who had previously laid the Woosung Railway in Shanghai. Moulton arrived in Keelung in the second year of the Guangxu era (1876) and planned the route from the mine to the port. After the Mudan Incident in the thirteenth year of the Tongzhi era (1874), the Qing government implemented a policy of having the Fujian Governor reside in Taiwan for half a year and in Fujian for the other half to strengthen the connection between Taiwan and Fujian. Ding Richang succeeded Wang Kaitai as the governor after his passing, and arrived in Taiway in the eleventh month of the second year of the

Guangxu era (December 1876). On the eighteenth day of the eleventh month of the second year of Guangxu Era (January 2, 1877), he visited Keelung to inspect the coal mine, including the proposed light railway. A report on the twenty-sixth day of the twelfth month of the second year of the Guangxu era (February 8, 1877) in the "Shen Bao" stated, "News from Taiwan. Hong Kong reports that on the second day of the first month of the Gregorian calendar, Deputy Governor Ding Yusheng has moved his office to Jilong to inspect the coal mine and assess the coastal defense situation along the eastern shore. After the inspection, he returned and arrived at the Taiwan city by land. Before his arrival, the people had already purchased foreign cloth to make flags and cleaned the streets in preparation. It is also heard that foreign machinery has been used to mine coal, and a railway from the mine to the beach has been laid, and trains can now run." However, initially, this light railway used animal power instead of steam locomotives. In the first month of the third year of the Guangxu era (March 15, 1877), a report from "Shen Bao Fuzhou Miscellaneous" stated that the railway "carrying coal cars is very fast, seemingly without the need for human labor, and there are hundreds of spectators every day." After Liu Mingchuan built the Taiwan Railway, in 1888, steam locomotives (manufactured by the German Hohenzollern locomotive factory, serial number 447) were introduced to this railway.

Apart from this railway, Fujian Governor Ding Richang also submitted a memorial titled "Proposal for the Preparation of the Taiwan Wheel Road" in the third year of Guangxu (1877). "Wheel road" was one of the terms used during the Qing Dynasty to refer to what is now called "railway." At that time, trains were also referred to as "firewheel carriages," later omitting the character "wheel." However, there was also the term "wheel car" during that period (similarly, the term "steamship" from the same era became "ship"). In this memorial, Ding Richang used the term "Taiwan wheel road," and the only occurrence of "Taiwan railway" is likely a typographical error. He planned to lay a railway from Anping in Taiwan City (present-day Anping District, Tainan City) to the area around Dagouqi (present-day Gushan District, Kaohsiung City) to facilitate the transport of military supplies. To achieve this, he had the dismantled track materials of the Wusong Railway transported from Shanghai to Taiwan and persuaded Lin Weiyuan and Lin Weirang of the Lin family in Banqiao to donate 500,000 silver yuan [1]. However, this plan eventually came to a halt due to insufficient funds, the inadequate length of Wusong Railway tracks, and the inability to supplement them. The donation from the Lin family was later redirected to Henan for disaster relief, and the tracks of the Wusong Railway were left exposed in Anping Harbor, causing surface rust. In September of the ninth year of Guangxu (October, 1883), they were transported back to Shanghai by the China Merchants' Company.

The locomotive with the distinctive "Magnetic Elec." (掣电号) signage (Fig. 19.3), manufactured in the United Kingdom, is seen operating on the Taipei-Hsinchu route in the nineteenth year of Guangxu (1893). The photograph's capture is estimated to be between 1902 and 1905. In recently discovered original photos, the copper numeral "10" is clearly visible on the carriage body, while the characters "掣电" are absent (Fig. 19.3).

After Ding Richang, the first governor of Taiwan, Liu Mingchuan, submitted the "Draft for the Establishment of a Railway for Business" on the twentieth day of

Fig. 19.3 The locomotive with the signage "Magnetic Elec"

third month of the thirteenth year of the Guangxu era (April 13, 1887), seeking permission to build railways in Taiwan [2]. After the Qing government approved it on twenty-eighth day of the fourth month of the same year (May 20, 1887), Liu Mingchuan established the "Taiwan Railway Business General Bureau" to construct railways on the island. To address the funding issues from Ding Richang's era, Liu Mingchuan considered adopting the "official supervision and commercial operation" model used by the Kaiping Railway. Before his submission, the Shanghai "Shen Bao" had reported Taiwan's intention to build railways on sixteenth day of the fourth month of the eleventh year of the Guangxu era (May 19, 1886), and former British Consul in Taiwan, Robert Hart, expressed a willingness to handle the matter. The American company Taggart & Company also sought to undertake the project. Eventually, the section from Keelung to Taipei was handled by the Shanghai firms Threlfall & Company and Jardine, Matheson & Co. [3]. Due to fundraising issues, apart from ordering passenger cars from the United Kingdom and Germany and constructing iron bridges by the Bureau of Commerce, the land was acquired by the government. The project was executed by the Military Arsenal, and facilities such as stations and depots were also constructed by the government. Later, as the "commercial operation" approach proved unfeasible, the project ultimately reverted to the "official supervision and official operation" route.

On eighteenth day of the seventh month of the fourteenth year of the Guangxu era (August 25, 1888), the locomotives "Tengyun" and "Yufeng" (manufactured by the Duesseldorf-Hoerder-Huette Locomotive Works in Germany, with factory serial numbers 444 and 445 respectively) began trial runs on the completed section from Dadaocheng to Xikou (now Songshan District, Taipei). On the sixteenth day of the eleventh month (December 18), the section from Taipei to Xikou officially commenced operations. The route later extended through Nangang, Shui-fan-jiao (now Xizhi District, New Taipei City), Wudu, and Liudu. By the 16th year of Guangxu (1890), the challenging Shihkuliang Tunnel was finally pierced through. According to an article titled "Mountain Tunnel Completed" in the Shanghai "Shen Bao" on

May 20th, 1890, the tunnel was "over a hundred Zhang long, three Zhang high, and two Zhang wide, which is quite remarkable." The northern section in Keelung was completed in tenth month of the 17th year of Guangxu (November 1891). When the second Governor of Taiwan, Shao Youlian, took office, he disembarked in Keelung. On twenty-first day of October (November 22, 1891), at dawn, he took the first-class train to Taipei, arriving at 9:45 am.

On the other hand, the railway section from Taipei to the south was also simultaneously planned and constructed. In sixth month of the 18th year of the Guangxu reign (July 1892), the railway had extended from Taoziyuan to present-day Zhongli and Yangmei. In June 1893, the railway further extended to the outskirts of Hsinchu County, and the operational section of the Taiwan Railway extended to Hsinchu in the eleventh month of 19th year of Guangxu (October 1893). However, due to financial issues, Shao Youlian reported in first month of the 20th year of Guangxu (March 1894) that the railway would only be laid as far as Hsinchu [22]. Even before Shao Youlian took office, on second day of the eighth month in the 17th year of Guangxu (September 4, 1891), the Shanghai "Shen Bao" had reported that due to financial problems, the construction of the central and southern sections of the Taiwan Railway was temporarily suspended. During the suspension of railway construction, the railway embankment from Hsinchu to Hsiangshan had already been built, and it wasn't until the Japanese colonial period that rails were laid and the railway was put into use.

After the signing of the "Treaty of Shimonoseki" on the twenty-third day of the third month of the 21th year of Guangxu (April 17, 1895), Taiwan was ceded to Japan by the Qing government. Taiwan resisted the Japanese takeover, leading to the outbreak of the First Sino-Japanese War. In the early stages of the war, the Taiwan Railway was used to transport troops to engage with the Japanese forces in Keelung. According to reports from diplomatic sources, in early June, the Keelung railway terminal (referred to as "railway pier" during the Qing Dynasty, as shown in Fig. 19.4) was crowded with evacuating civilians. As soon as soldiers disembarked, people would crowd onto the trains to escape to Taipei. At that time, there were about 12,000 soldiers in and around Keelung. After the loss of the Shiqiuling Fort on June 4, the railway facilities in Keelung were taken over by the Japanese military. At that time, only three passenger cars and a few freight cars remained in Keelung, and there were no locomotives available [4].

After the first "Taiwan Governor-General" Kabayama Sukenori landed in Keelung during the period of Japanese invasion and colonial rule, he ordered engineer Ozawa Yasumasa to investigate the condition of the Taiwan Railway. On May 15, 1895, during the 21st year of the Guangxu reign (May 15, 1895), after the Japanese forces entered Taipei, Ozawa Yasumasa found three locomotives at the Taipei railway terminal, but only one could be repaired. After a trial run on June 9, railway traffic between Taipei and Keelung officially resumed the next day. In addition, the Taiwan Railway Line District Command (under the "Governor-General's Office" Army Bureau of Engineers) and the Railway Department (under the "Governor-General's Office" Department of Transportation and Telegraphs) responsible for railway affairs were quickly established. Although the railway between Taipei and Keelung had

Fig. 19.4 Taipei train station, constructed during the Qing Dynasty

resumed operation, due to the poor condition of the tracks and trains, initially, only one round-trip train could operate each day, which could not meet the transportation needs of the Japanese military. In an incident of overloaded freight, the locomotive couldn't bear the load, and workers had to assist in pushing locomotive to continue forward. The Japanese then referred to the Taiwan Railway as the "rear-pushed train" and the "pulmonary disease train."

The railway section from Taipei to Hsinchu gradually came under the control of the Japanese military. After occupying Hsinchu on June 22, 1895, the Japanese forces first repaired the section from Hsinchu to Dahuoku. This segment resumed operations on June 28 and was utilized for transporting food supplies. Additionally, on June 29, Shoyama Yasumasa took a train from Taipei to Hsinchu. This trip served not only to repair the railway but also to test its operational status. The entire Taiwan Railway resumed operations on July 10.

Subsequently, until March 1897 when it was transferred to the "Taiwan Governor-General's Office, Civil Affairs Bureau, Telegraph Department Temporary Railway Section," the Taiwan Railway was managed by the "Railway Corps," essentially functioning as a military railway [23]. Apart from overseeing the existing railways left by the Qing Dynasty, the "Railway Corps" also started laying tracks for the Tai-car system (tracks obtained from military railways no longer needed by the Japanese army after the First Sino-Japanese War). The first completed section was from Hsinchu to Changhua, gradually extending southward afterward [5]. By 1895, the section from Tainan to Dagou (Kaohsiung) was already completed, and by February 1898, the Tai-car tracks from Hsinchu to Dagou were finished. This Tai-car system, also known as the "Army Road," roughly followed the Qing Dynasty official road, which had been renovated by the Japanese army. As military conflicts diminished and civilian demand increased, the system was opened to the public for passenger transport and freight services, until the decision by the Taiwan Governor-General's Office to construct a new trunk line railway. Later, the "Army Road" was dismantled and converted into a highway, now known as the National Highway 1 in Taiwan.

Before the Taiwan Railway was separated from Japanese military administration during the Qing Dynasty, the first Japanese "Governor-General of Taiwan," Kabayama Sukenori, expressed the urgent need to build a formal trunk line railway connecting the north and south of Taiwan to suppress anti-Japanese forces and establish control over the island. After several failed attempts at private fundraising, the "Governor-General's Office" decided in 1898 to construct the trunk line officially. During this period, the Japanese authorities also improved the northern section of the railway. Due to concerns about the steep slope near the Qing Dynasty's Shihkuliang Tunnel, the Japanese authorities changed the route and started digging the Chutzupi Tunnel in May 1896. It was completed on February 2, 1898, with an opening ceremony held on February 13 at the north entrance of the tunnel. Although only a single-track railway was laid at the time, the tunnel reserved space for another track. In addition to the Chutzupi Tunnel project, improvements in the northern section included straightening the route, the construction of the Tsushunwan Tunnel (now Nangang Tunnel), and realignment of the Wudu–Baifu section. After the railway improvements, four trains per day were running by 1899. The second-generation Taipei Station, inaugurated on August 25, 1901, is shown in Fig. 19.5.

During the period of Japanese invasion and colonial rule in Taiwan, the construction of the Governor-General's Office's mainline route roughly followed the early surveyed "Mountain Line." However, due to various factors related to time and space, certain sections of the route experienced changes. For example, the original plan for the route from Hsinchu to Miaoli's Zaoqiao intended to pass through Toufen, but this would have resulted in excessive curvature, prompting a change to the coastal route through Chunggang (modern-day Zhunan). Similarly, the route from Taichung to Yunlin's Linne originally planned to go through Nantou City. Still, due to minimal goods requiring transportation in this area and the loss of economic benefits of not passing through the economic hub of Changhua in central Taiwan, the route was altered to pass through Changhua.

Fig. 19.5 The second-generation Taipei Station, inaugurated on August 25, 1901

The construction of the railway proceeded simultaneously in the northern and southern parts. The Tamsui Line (Taipei-Tamsui Line) was initially laid as a auxiliary line to transport railway construction materials arriving by sea. The mainline's Taipei to Hsinchu section was completed up to Hsiangshan and Chunggang (modern Zhunan) stations by August 10, 1902. The Chunggang to Miaoli section was finished on May 25, 1903, and progressed to Sanxia River (modern Sanyi Township, Miaoli County) in 1904. The southern section of the mainline started laying tracks to the north from Tako in November 1899, with the Tako-Tainan section becoming operational on November 29, 1900. By May 15, 1901, the operating section extended to Shanhua Station [6]. By 1904, the southern section extended to Changhua, and the northern section reached Sanxia River, leaving only the most challenging central section. The construction of this central section was difficult due to the need to cross the Da'an River and Dajia River between Sanxia River and Huludun. Many tunnels had to be excavated through mountainous terrain. Although "A Century of Taiwan Railways" and "Cross-Island Railway: Taiwan Railways" both mention the extension of the southern section to Changhua in 1904, the actual completion of the railway from Erhpaishui north to Changhua did not occur until March 15, 1905, with operations commencing on March 26. During the construction of the final central section of the mainline, the Russo-Japanese War broke out on February 8, 1904. Due to the severe damage suffered by the Russian Pacific Fleet, Tsar Nicholas II ordered the Baltic Fleet to move to East Asia for support. As their route would pass through Taiwan, Japan utilized temporary military funds to construct a light railway from Pekungkeng to Huludun (previously a light railway already existed from Pekungkeng to Sanchatian). This line, completed in 1906, served both military transport and the carriage of construction materials and passengers. To climb onto the Huali Plateau south of the Da'an River, a specially designed curved ramp was incorporated. The completion of Tunnel No. 9 in March 1908 allowed the entire mainline to be fully completed and opened for service on April 20 of the same year.

In the year following the completion of the North–South Railway (1909), the Taiwan Governor-General's Office established the "Hualien Construction Office," a railway management department for the Taiwan region, initiating the construction of the Taitung Line. In 1916, plans for the Yilan Line and Fangliao Line were proposed by the Taiwan Railway Administration. For the Yilan Line, after approval by the legislature, construction offices were established in Yilan and Rueifang in 1917, commencing the construction of the southern segment (Su'ao to Yilan) and northern segment (Badu to Houtong) on October 1. However, due to factors such as the inflation of prices and wages caused by World War I (1914–1918), the Yilan Line project faced delays for several years. It was finally declared completed in October 1924 after the opening of the Tsao-ling Tunnel, allowing goods from the Lanyang region to be swiftly transported to Keelung Harbor and Taipei by railway.

Regarding the Fangliao Line, as its predecessor, the Fengshan Branch Line, had extended to Aqian (modern-day Pingtung) in October 1912, the construction of the Fangliao Line began from Aqian in 1917. The project started progressing in December of the same year but faced budgetary issues similar to the Yilan Line, leading to a temporary suspension of the project. It resumed in 1922 and was completed in October 1923. The Donggang Line began operations on July 19, 1940.

As for the Taitung Line, the project, which had experienced a hiatus along with the Yilan and Fangliao Lines, commenced its second phase in 1922. The section from Hualien Port to Pushekeh had been in operation for several years but operated at a deficit annually. However, to complete the island-wide railway and promote regional development, the project persisted. In 1926, the Taitung Line was completed in its entirety.

The Coastal Line construction project began in August 1919, with construction offices established at both ends in Zhunan and Taichung. By December 15, 1920, the section from Wangtian (modern-day Chenggong Station) to Qingshui had commenced operations, initially referred to as the "Wangtian Branch Line." The subsequent phases of the project were completed by October 1922. After the completion of the Coastal Line, long-distance trains were rerouted to the more time-efficient Coastal Line.

In addition to the planned routes of the Island Line, there were other branch lines operated by the authorities, some of which were acquired from privately laid railways, while others were taken over or newly constructed by the Taiwan Railway Administration. Among the early completed passenger branch lines under the jurisdiction of the Taiwan Governor's Office was the Tamsui Line, which commenced operations on August 25, 1901. In the 1920s, the Taiwan Railway Administration acquired the Jiji Line and the Pingxi Line, both of which were originally private railways, adding two more branch lines to those managed by the Taiwan Governor's Office. Apart from the main railways handling general passenger and freight transport, the three major forest stations directly operated by the Taiwan Governor's Office also had forestry railways (later operated by the Taiwan Development Corporation). Among them, the most notable is the Alishan Forest Railway.

With the increasing transportation demand and facing competition from road transport, the dualization of railways became a priority. The first dualization project was the Keelung-Taipei section. The dualization from Keelung to Badu was completed from April 1912 to April 1914, and in December 1917, the railway from Badu to Taipei was also dualized. Later, in December 1926, the Taiwan Region Transportation Authority proposed a comprehensive improvement plan. This plan included not only the dualization of the entire line but also improvements to existing routes and bridges. Additionally, plans for the improvement of the Taipei, Keelung, and Kaohsiung stations were proposed, but only the dualization project budget received approval, and the plan for elevating Taipei Station was canceled. The dualization of the entire Island Line was originally expected to take 10 years, with the first 5 years focused on dualizing the Taipei-Hsinchu and Tainan-Kaohsiung sections, followed by the dualization of the Chiayi-Tainan and Changhua-Chiayi sections in the next 5 years. However, due to financial difficulties and other factors, the Tainan-Kaohsiung section was extended from 5 to 10 years and was completed in October 1935.

To cope with the competition from road transport, the Taiwan Regional Railway Authority attempted to improve train operations and equipment, enhance service quality, simplify freight transport methods, and encourage people to use trains for tourism. Later, the Taiwan Regional Railway Authority considered diversifying into road transport. However, instead of nationalizing the entire road transport industry, it was seen as a supplementary or alternative transportation method to avoid destructive competition between road and rail transport. Outside of this scope, private entities could continue to operate. In order to prevent internal competition, road transport operators began acquiring and merging with each other or signing private agreements. Private railway and track operators also started diversifying into road transport. However, the Taiwan Regional Railway Authority's acquisition plans faced obstacles. Eight companies in the northern region believed that the Railway Authority was arbitrarily setting condolence payments to seize private businesses. They formed a united company to resist, and with the mediation of the governors of Taipei and Hsinchu, the condolence payment issue was resolved. Operations in road transport began on July 26, 1933. In the central region, seven companies also expressed dissatisfaction with the Taiwan Regional Railway Authority. They sought mediation from the governor of Taichung, but companies like Pinghe, Showa, Asahi, and Yuanchang remained dissatisfied. Later, upon learning that regardless of the negotiation results, the Taiwan Regional Railway Authority would proceed with operating state-owned buses, the Central Region Motor Vehicles Association only began operating road transport services on September 3, 1934, after concerted efforts by Watanabe Keiji, head of operations at the Taiwan Regional Railway Authority, and Maruoka Michio, head of the motor vehicle department. Similar issues arose in the southern region, triggering opposition from industry stakeholders. However, the Taiwan Regional Railway Authority remained adamant, officially commencing the operation of state-owned buses on September 5, 1936. Companies like Gong'an and Kaohsiung knew there was no room for negotiation and had to accept the decision, while companies like Tainan and Datong realized, after pleading, that they had no other choice. They expressed on September 22 of the same year that they "intended to sacrifice for the national policy and agreed to the acquisition." The decrease in revenue from railway passenger transport was indeed compensated by diversifying into buses.

In addition, the Taiwan Regional Railway Authority initiated intermodal transportation services, allowing passengers and cargo owners to utilize a single document for seamless transportation across various modes of transport. The origin of intermodal services in Taiwan can be traced back to the 1910s. In response to the needs of pilgrimages to the Beigang Chaotian Temple, the Taiwan Regional Railway Authority facilitated cooperation between the Taiwan Railway and sugar companies like Dajapan, Shinko, and Toyo. They collaborated to sell round-trip tickets from stations north of the Taichung Station on the Main Line to Beigang. After the completion of the Taitung Line, starting from April 1926, the Taiwan Regional Railway Authority signed an agreement with Osaka Shosen Kaisha to implement intermodal passenger and baggage transport between key stations on this line and the Taitung Line, using the regular ship "Miyazaki Maru" (Suoao to Hualien Harbor).

However, apart from the intermodal service between the Taitung Line and Osaka Shosen, intermodal transportation between different modes of transport in Taiwan was not widespread until the early 1930s. With the development of road transportation, private railways actively engaged in intermodal transportation with government-owned railways to compete with road transport. In 1936, the scope of intermodal transportation expanded to include the Taiwan island-wide transport of Nippon Koku Yuso Kabushiki Kaisha, forming an intermodal network encompassing Taiwan by land, sea, and air. Taiwan's international intermodal operations began in 1910 with a contract with Osaka Shosen to handle Taiwan-Japan intermodal baggage transport. From April 1, 1912, intermodal passenger and baggage transport was implemented at five stations in Taiwan (Taipei, Taichung, Chiayi, Tainan, and Takao) and 17 locations in Japan (Nagasaki, Moji, Kobe, etc.) through the cooperation of Nippon Yusen and Osaka Shosen. On May 1 of the same year, the scope expanded to include 15 stations in Japan (Shinbashi, Hirano, Nagoya, Kyoto, Osaka, etc.). The intermodal scope continued to grow, and by 1931, intermodal passenger transport was available between 27 stations in Taiwan and 155 stations in mainland Japan. Cargo intermodal transportation to Japan began in the same year, and intermodal passenger transport with Korea and Manchukuo started in 1934. After 1936, air transport became part of the intermodal range between Taiwan and Japan. On September 20 of the same year, the Taiwan region, Japan, Korea, and Manchukuo formed an intermodal network by land, sea, and air. However, compared to the development of intermodal transportation between Taiwan and the mentioned regions, intermodal services between the Taiwan region and Penghu Magong only appeared on June 15, 1936. The formation of a transportation intermodal network expanded Taiwan's trade market, fostering the development of various industries and promoting social and cultural exchanges.

In 1958, Taiwan began researching the modernization of its mainline railways. Due to the high cost of electrification and the insufficient power supply from the Taiwan Power Company, the decision was made to opt for diesel-electric power. On August 7, 1959, the "87 Flood" caused severe damage to the railways and roads between Hsinchu and Tainan. By June 1, 1949, the Hsintien Line had been acquired by Taiwan Railways, becoming one of its branch lines. In 1962, Taiwan Railways implemented a dual-track automatic block signaling system between Taipei and Zhunan and gradually introduced the Centralized Traffic Control system to the entire North–South mainline in the 1960s to improve operational efficiency.

During the period of Japanese invasion and colonial rule, the existing plan for the island-wide railway, which had been initiated, saw further progress with the commencement of the North Return Line on December 25, 1973. Concurrently, another significant railway project initiated was the electrification of the Main Line. Starting from 1992, the Northern Coastal Line underwent various improvement, including double-tracking and electrification, resulting in the excavation of numerous new tunnels due to realignment. The improvement projects were completed by the end of June 2003, and the inauguration ceremony was held on July 4 of the same year in Hualien.

The North Return Line connected the previously independent Yilan Line and Taitung Line. However, due to the 762-mm gauge of the Taitung Line during the Japanese invasion and colonial rule, which rendered the locomotives incompatible, the East Extension Project began in 1978. Apart from changing the gauge, the East

Extension Project also involved realigning certain sections with suboptimal conditions. For example, the original line from Lu Ye, passing through Zhongxing, Chulu, and Binlang stations, entering the old Taitung Station, was later changed to the route from Lu Ye, passing through Shanli, to the new Taitung Station. Throughout the construction process of the entire project, the Taitung Line continued its operations. On June 25, 1982, a seamless transition took place overnight, and the gauge of the Taitung Line was changed from 762 to 1067 mm. However, due to issues encountered during the construction of the Taroko Express Tunnel, where a layer of silt was encountered, the old tunnel was still used in 1982. After the new tunnel replaced the old one in 1984, the East Extension Project was considered truly completed.

Less than a year after the completion of the North Return Line, the final section of the Round-the-Island Railway—the South Return Line—also began construction on July 1, 1980, and officially commenced operations in October 1992.

At the same time as the formation of the Round-the-Island Railway, the Yilan Line started its double-tracking in July 1980, completing it in 1986, and later electrified in July 2003. Other improvement projects, such as the Kaohsiung to Pingtung railway double-tracking (1983–1987), electrification, and the realignment and double-tracking of the Mountain Line between Zhunan and Fengyuan (1987–1998), as well as the Taipei Railway undergrounding project (1983–1989), were also carried out successively.

The Mass Rapid Transit (MRT)[1] System and the Taiwan High-Speed Rail (THSR)

Taipei Metro

In the past, Taiwan did not have a standard gauge (1435 mm) railway until the construction of the Taipei Metro (Tamsui Line) began, adopting the standard gauge. The Muzha Line of the Taipei Metro was the first metro line to operate in Taiwan, and it commenced trial operations on March 28, 1996 [7]. Notably, it is a fully automated (driverless) medium-capacity line, primarily constructed as an elevated bridge structure. Subsequently, Taipei developed underground high-capacity metro lines with trains consisting of six carriages with distributed power, operating with drivers (Fig. 19.6).

By the end of 2000, the Taipei Metro rail network had a total length of 65.1 km, consisting of initial cross-shaped routes such as Muzha (**Fig. 19.7**), Tamsui, Xindian, Nangang, and Banqiao [8]. After the opening of the Neihu Line (later combined with the Muzha Line and renamed the Wenhu Line [9]), all 12 administrative districts of Taipei City had metro lines passing through. With the opening of the Luzhou Line in 2010, the operational mileage of the Taipei Metro exceeded 100 km. By 2020, the

---

[1] Rapid transit generally refers to the subway or metro system.

Fig. 19.6 The Taipei Metro adopts 6-car heavy-capacity electric multiple unit train

total mileage had reached 152 km, with 131 stations. The daily ridership exceeded 1.9 million, and the annual ridership reached 696 million.

Due to the main international airport serving Taipei being located in Taoyuan, its metro lines are integrated into a comprehensive transportation system. Traveling from Taipei Main Station (a station shared by the metro, high-speed rail, and traditional Taiwan Railways) to Taoyuan Airport takes only 35 min via the MRT.

The successful operation of the MRT in Taipei has led various county and city governments in Taiwan to undertake the construction of MRT lines. In the early 1990s, there was optimistic anticipation regarding rail transit, with plans to build MRT systems in cities other than Taipei and Kaohsiung (such as Taoyuan, Hsinchu, Taichung, Chiayi and Tainan). The goal was to alleviate commuting pressure within cities and support the development of high-speed rail stations on the outskirts, connecting them to the city centers. However, as Taiwan's government faced rapid financial deterioration in the early 2000s, it became financially incapable of large-scale MRT construction. This led the transportation authorities in Taiwan to adopt a more conservative stance towards new MRT construction projects. They rigorously scrutinized proposals submitted by local governments, resulting in the majority of plans remaining on paper.

Faced with the demand for MRT construction from local governments and considering the sustainable operation of MRT systems, the transportation authorities in Taiwan implemented the "Mass Rapid Transit (MRT) System Construction and Surrounding Land Development Plan Application and Review Operation Guidelines"

Fig. 19.7 The earliest operating metro line in Taipei, the Muzha Line (now part of the Wenhu Line)

on April 11, 2011. They modified the original two-stage review process to a three-stage review, raising the requirements for each stage. Plans that pass through all three stages of review are then eligible for budget allocation and subsequent construction in subsequent years. The new review mechanism specifically establishes a threshold for the financial self-sufficiency rate and encourages compensating for insufficient construction funds through the benefits of land development along the MRT lines.

Taiwan High Speed Rail

Taiwan High Speed Rail (abbreviated as "THSR") spans the densely populated western corridor of Taiwan, with a total length of 349.5 km. It operates at a maximum speed of 300 km per hour, connecting the northern and southern regions of Taiwan and forming a daily living circle. The shortest travel time between Taipei and Kaohsiung, the two major cities in the north and south, is only 105 min [10]. Since its opening on January 5, 2007, Taiwan High Speed Rail has become a crucial long-distance transportation tool in western Taiwan and a significant indicator in Taiwan's railway transport industry. As of 2019, there are 130 to 162 round-trip services daily between the north and south, with an average daily passenger volume of 184,000 people [11]. During weekends and long holidays, the daily average can reach up to 300,000 people [12]. The highest single-day ridership recorded was approximately 318,000 people on September 15, 2019 [13].

The Taiwan High Speed Rail is Taiwan's first public project developed under the Build-Operate-Transfer (BOT) model, with a construction cost of approximately NT$450 billion (around $14.5 billion USD). The construction and operation responsibilities are undertaken by the Taiwan High-Speed Rail Corporation, with a franchise period spanning from 1998 to 2067, totaling 70 years. After the franchise period expires, the operational rights will be returned to the government without compensation. The Taiwan transportation authority is required to identify the organization to continue operations by the year 2065 and inform the Taiwan High-Speed Rail Corporation accordingly. The department managing the affairs of the Taiwan High-Speed Rail is the Railway Bureau, a subsidiary under the Taiwan Ministry of Transportation, originally known as the High-Speed Rail Engineering Bureau. This bureau was involved in the early stages of preliminary planning and later took on responsibilities for the construction, operational supervision, and the development of MRT systems connecting to high-speed rail stations.

The idea of constructing a high-speed rail system in Taiwan began with the "Special Study on Developing Super Railways" in 1974. However, at that time, the completion of the National Freeway No. 1 (Zhongshan Highway) led the government to be less proactive about building high-speed rail. Subsequently, considering the increasing intercity transportation demand in western Taiwan, the Taiwan Ministry of Transportation initiated the "Feasibility Study of the Taiwan Western Corridor High-Speed Railway" in 1987. In 1990, following approval from the Taiwan administrative authorities, the "Taiwan North–South High-Speed Railway Construction Plan" was launched, marking the commencement of the Taiwan High-Speed Rail project. Originally planned to be completed within six years, the project faced delays due to prolonged preliminary work, such as determining funding sources and system specifications. The construction officially started in 1999, and the project's financing model shifted from annual government budget allocation to private sector investment participation.

On July 1, 2016, the South Link (Nangang to Taipei) officially commenced operations, marking the full opening of the Taiwan High-Speed Rail system. Starting from that day, the northern terminal station was changed to Nangang Station, which helped alleviate passenger flow at Taipei Station. In addition, with the change in the departure station, the Taipei Underground Section train dispatch project, planned from the beginning, was also officially completed simultaneously. Taipei Station underwent platform door installation to replace the platform control operation mode. The minimum departure interval from the originating station was reduced from 6 to 5 min.

On November 14, 2019, Taiwan High-Speed Rail introduced the "Kanahei Painted Train," which was exceptionally beautiful.

Taiwan High-Speed Rail has established a competitive relationship with other transportation modes in western Taiwan, with the aviation industry being the most affected. There used to be a considerable number of scheduled flights between Taipei and Kaohsiung, but after 2013, all domestic flights on the west side of Taiwan were suspended [14]. Following the addition of Yunlin Station to the high-speed rail network, there were reports in 2018 that the Taiwan Daily Transportation

Company, specializing in Yunlin County's inter-county services, was considering ceasing operations [15].

In August 2010, Taiwan High-Speed Rail achieved a milestone by surpassing 100 million cumulative passengers. By December 2012, it had exceeded 200 million passengers, and in January 2015, the ridership surpassed 300 million [16]. On February 22, 2015, Taiwan High-Speed Rail set a new record by transporting over 239,000 passengers in a single day. As of 2015, the cumulative ridership of Taiwan High-Speed Rail was approximately 348 million passengers [17].

Outbound Transportation of the Beautiful Island: Sea and Air Transport

Commercial Ports

Taiwan is surrounded by the sea, and port development began in the 17th century, making it the only external transportation port at that time. By the tenth year of the Xianfeng era (1860), the Qing government opened four Taiwan ports, namely Tainan Port (Anping), Hobe Port (Danshui), Keelung Port (Keelung), and Qihou Port (Gaoxiong, now known as Kaohsiung), as commercial ports, based on the Tianjin Treaties with England and France. This historical event, known as the "Opening of Taiwan's Ports," marked the beginning of Taiwan's construction of a modern international trade system [18].

During the period of Japanese colonial rule in Taiwan, the Japanese authorities designated Keelung and Kaohsiung as external ports for the north and south of Taiwan. They implemented modernized port construction plans and established a comprehensive port management system. The management of the ports was overseen by the "Transportation Bureau Maritime Department" under the Taiwan Governor-General's Office at that time. "Port Bureaus" were established in Taipei Prefecture and Kaohsiung Prefecture to serve as operational agencies for the respective ports. In 1943, during the wartime regime, to unify the management of the ports, the Japanese authorities established the "Port Bureau," directly under the Taiwan Governor-General's Office, for Keelung and Kaohsiung ports, consolidating the authority for port management [19]. Prior to the development of the civil aviation system, Keelung Port (Fig. 19.8) and Kaohsiung Port (Fig. 19.9) served as Taiwan's major gateways for external connections.

Taiwan currently has seven international commercial ports, including four major ports and three auxiliary ports. Among them, Kaohsiung Port ranks as the world's sixth-largest container port (see Fig. 19.10). Additionally, CPC Corporation, Taiwan, established Yongan Port in Yong'an District, Kaohsiung City, for the import of liquefied natural gas (LNG) and is considered an important auxiliary port in the international commercial port system. Yongan Port is specifically an energy port managed by

Fig. 19.8 Keelung port

Fig. 19.9 Kaohsiung port

CPC Corporation, Taiwan, with the transportation authorities in Taiwan responsible only for maritime administration affairs.

Industrial Ports

Today, the Taiwan region has three industrial zone-specific ports: Maoliao Industrial Port in Mailiao Township, Yunlin County, established by Formosa Plastics Corporation, mainly responsible for the import and export of raw materials and finished products for the Formosa Petrochemical Complex; Heping Industrial Port in Heping Village, Xiulin Township, Hualien County, built and operated by Taiwan Cement Corporation, tasked with the import and export of sand and gravel from eastern Taiwan and raw materials and finished products for Taiwan Cement Corporation; and the Guantang Industrial Zone Port, built and operated by CPC Corporation, Taiwan.

Fisher Ports

Along the coast of Taiwan, each county and city has fishing ports, categorized into two types: Class 1 Fishing Ports and Class 2 Fishing Ports, based on their size [20]. Among them, Keelung City's Badouzi Fishing Port, Yilan County's Nanfang'ao Fishing Port, Kaohsiung City's Xingda Fishing Port, and Taichung City's Wuci Fishing Port are the largest in scale. Xingda Fishing Port is a large-scale deep-sea fishing port in Southeast Asia, but its operational results have not met expectations.

The fishing port management unit is the Fisheries Agency of the Taiwan region, and the local fishermen's associations in the port area are the actual operating entities. Currently, there are a total of 9 Class 1 fishing ports and 215 Class 2 fishing ports in Taiwan. The following fishing ports are managed by the respective local authorities: Qianzhen Fishing Port (Kaohsiung City Government), Nanfang'ao Fishing Port (Yilan County Government), Badouzi Fishing Port (illustrated in Fig. 19.10) and Zhengbin Fishing Port (Keelung City Government), Hsinchu Fishing Port (Hsinchu City Government), Wuci Fishing Port (Taichung City Government), and Donggang Salt Pan Fishing Port (Pingtung County Government).

Taiwan has multiple fishing ports with marinas for private yachts. Private yacht owners need to apply for berthing at 11 fishing ports, including Xingda Fishing Port and Gushan Fishing Port (illustrated in Fig. 19.11), and Qijin Fishing Port in Kaohsiung City, Badouzi Fishing Port (Bisha Port Area) in Keelung City, Hsinchu Fishing Port in Hsinchu City, Anping Fishing Port and Jiangjun Fishing Port in Tainan City, Xingang Fishing Port and Jinzun Fishing Port in Taitung County, and Qimei Fishing Port and Jibei Fishing Port in Penghu County [21]. In addition to these piers, there are also two types of piers: transportation boat piers and yacht piers in Taiwan.

Fig. 19.10 Badouzi Fishing Port in Keelung

Fig. 19.11 Gushan Fishing Port in Kaohsiung

References

1. Zhao Chunchen. Ding Richang's Collected Works. Shanghai: Shanghai Ancient Books Publishing House, 2010.
2. Liu Mingchuan. Liu Zhuangsu's Memorials to the Throne. Beijing: Zhaohua Publishing House, 2018.
3. Zhang Jianfeng. Memories of the Cession of Taiwan in 1895. Taipei: Strait Academic Press, 2010.
4. Zhong Zhizheng, Zhang Jianfeng. Research on the 1895 War: The Battle for Taiwan You Didn't Know [J]. Strait Academic Journal. 2009(7): 192–197.

5. Huang Zhiwei. The Story of Provincial Highway No. 1 in Taiwan. Taipei: Rugu Press, 2011.
6. Lin Miaoxian. The Evolution of Taiwan Railway Administration's Stations. Nantou: Taiwan Historical Research Institute, 2006.
7. Su Zhaoxu. Encyclopedia of Taiwan Railways - TRA, HSR, MRT (Complete Edition). New Taipei City: Renren Publishing House, 2010.
8. Wang Shaowei. Evolution of the MRT Systems in Taiwan's Metropolitan Areas [J]. Railway Information, 2011(1): 66–73.
9. Taipei Rapid Transit Corporation. Considering the Administrative District Names Familiar to the Public for the Starting and Ending Points of MRT Routes, the Muzha Line and Neihu Line are Officially Merged and Renamed as Wenshan Neihu Line (Abbreviated as Wenhu Line) [EB/OL]. (1998-10-08) [2022-05-08]. https://www.metro.taipei/cp.aspx?n=B27E7BC8CBBCDA61
10. Wang Shufen. Nangang Becomes the Terminal Station of High-Speed Rail, Direct Access to Kaohsiung in 105 Minutes [EB/OL]. (2016-05-27) [2020-01-04]. https://tw.news.yahoo.com/%E5%8D%97%E6%B8%AF%E8%AE%8A%E9%AB%98%E9%90%B5%E7%AB%AF%E9%BB%9E%E7%AB%99-%E7%9B%B4%E9%81%94%E9%AB%98%E9%9B%84105%E5%88%86%E9%90%98-031113343.html
11. Xie Jiazhen. High-Speed Rail Passenger Milestone! The 600 Millionth Rider Can Ride for Free for a Year [EB/OL]. Newtalk News. (2019-12-23) [2020-01-04]. https://newtalk.tw/news/view/2019-12-23/344399
12. Wang Yihong. High-Speed Rail Hits Over 300,000 Passengers on April 7th, April Revenue May Reach a New High [EB/OL]. (2019-04-08) [2022-05-09]. https://ec.ltn.com.tw/article/breakingnews/2752099
13. Xiao Yuxin. High-Speed Rail Hits Over 318,000 Passengers During Mid-Autumn Festival Holiday, Setting a Record Since Its Opening in 2007 [EB/OL]. (2019-09-17) [2022-05-09]. https://news.ltn.com.tw/news/life/breakingnews/2918072
14. Su Zhaoxu. The New Era of High-Speed Rail. New Taipei City: Renren Publishing House, 2005.
15. Zheng Weiqi. Japanese-Era Passenger Transport to Cease, Public Transportation Office Suggests Consolidation and Route Adjustments for Improvement [EB/OL]. (2018-08-04) [2022-05-04]. https://news.ltn.com.tw/news/life/breakingnews/2509347
16. Apple Daily Instant News. High-Speed Rail Delivers Good News, Passenger Trips Surpass 300 Million [EB/OL]. [2022-06-17]. https://web.archive.org/web/20160304200412/http://www.appledaily.com.tw/realtimenews/article/new/20150806/663947
17. Taiwan High Speed Rail Corporation. Overview of Operating Year. (2015-12-14) [2022-06-17]. https://web.archive.org/web/20110216210229/https://www.thsrc.com.tw/tc/about/ab_operate_annual.asp
18. Lin Zhaowu, Zhou Huijing. Taiwan's Opening to Trade: The First Glory of the Commodity Economy [J]. CommonWealth Magazine, 1991(11): 1.
19. Xie Junze. From Takau to Kaohsiung: The Construction and Management of Kaohsiung Harbor During the Japanese Colonial Period (1895-1945) [J]. Taiwan Documentary Quarterly, 2011(6): 211-244.
20. Wikipedia. Taiwan Ports [EB/OL]. [2022-05-15]. https://zh.wikipedia.org/wiki/%E8%87%BA%E7%81%A3%E6%B8%AF%E5%8F%A3
21. Council of Agriculture, Executive Yuan. Yachts May Apply to Dock at Eleven Fishing Ports Including Kaohsiung's Gushan Fishing Port (Amended by Council of Agriculture, Executive Yuan Order No. 0931341760, May 14, 2004) [EB/OL]. (2014-07-14) [2022-06-17]. https://www.coa.gov.tw/theme_data.php?theme=communique&id=1412
22. First Historical Archives of China. Imperial Verdicts and Memorials of the Guangxu Reign. Beijing: Zhonghua Book Company, 1996.
23. Wataru Tobibe. Taiwan Railway Reader. Translated by Huang Defeng. Nantou: Taiwan Historical Research Institute, 2006.

Chapter 20
Leap of Civilization: The Road to Sustainable Urban Transportation

Humanity has traversed various stages of civilization, from primitive civilization to agricultural civilization, and then to industrial civilization. Currently, we are entering a new phase of development characterized by ecological civilization. Prioritizing ecology and pursuing green development has become a national strategy and an inevitable choice. At this moment, reflecting on the journey of green development in Chinese urban transportation, consolidating and summarizing the consensus on green development is of great significance for achieving the lofty goals of high-quality urban development and building a transportation powerhouse.

As people transition from rural areas to cities, and as urbanization and motorization become an unstoppable historical trend, the rapid modernization of cities becomes a distinctive feature of our era. However, facing the explosive expansion of cities and the environmental pollution and excessive energy consumption caused by traffic congestion, the harmony between humans and nature has become a new aspiration in this era. Following the historical development trajectory, traversing through time, and experiencing the changes in urban transportation through each leap of civilization, let's look back at the green development path of urban transportation. The transportation system should serve everyone, not only ensuring smooth mobility for people and goods but also leaving a mark of civilization in the long history of human social development.

Walking: The Best Traffic is no Traffic

Era Dominated by Walking

For thousands of years, whether in local living or long-distance travel, walking has been the most important mode of transportation.

© China Communications Press Co., Ltd 2025

H. Lu (ed.), *A Brief History of Transportation in China*,
https://doi.org/10.1007/978-981-97-9851-3_20

Before the era of reform and opening up, it was challenging for rural populations to migrate to urban areas. The cities were small, and the level of urbanization was low, with both in a relatively stable state. By the end of 1978, there were only 193 cities in the country, and the urbanization level was 17.92%. In 1981, the total urban built-up area in the country was only 7,438 square kilometers, less than the sum of the built-up areas of six cities today: Chongqing, Beijing, Guangzhou, Shanghai, Dongguan, and Tianjin. Beijing focused on development within the Second Ring Road, and Shenzhen was just a small coastal fishing town. Even with the gradual deepening of reform and opening up, the urbanization level only reached 29.04% by the end of 1995 [1].

Since the establishment of the People's Republic of China, urbanization policies in the country have long prioritized the development of medium and small cities. In 1955, the National Planning Commission first proposed, "In the future, newly built cities should primarily focus on medium and small towns and workers' towns. A few medium-sized cities should be constructed under possible conditions. Large cities should not be built unless there are special reasons." The reform and opening-up that began in 1978 brought about a new change in the concept of urban development. Before the Third Plenary Session of the Eleventh Central Committee of the Communist Party of China (December 1978), the State Council held the Third National Urban Work Conference in March 1978. The important outcome document of the conference, "Opinions of the Central Committee of the Communist Party of China on Strengthening Urban Construction" (Document No. 13 of the Central Committee [1978]), clearly stated that "cities of all sizes throughout the country are the base for the development of modern industry and the center of regional politics, economy, and culture." It emphasized that "urban work must adapt to the needs of high-speed development of the national economy and contribute to achieving the overall tasks of the new era." In 1980, the National Urban Planning Work Conference proposed the idea of "controlling the scale of large cities, developing medium-sized cities reasonably, and actively developing small cities" [2], which was approved by the State Council [3]. After entering the 1990s, the speed of urbanization and the expansion of city scale accelerated, with urban population and industries concentrating in large cities, and urban clusters gradually emerging.

For a considerable period after the founding of the People's Republic of China, due to the generally small spatial scale of urban areas and compact land use, walking and cycling were the primary modes of transportation for daily travel. In the 1980s, the mode share of walking and cycling in Shanghai for daily travel was as high as 73%; in 1986, Beijing had a mode share of 67.8% for walking and cycling. In the early years of the People's Republic of China, only a few large cities had public buses operating on their streets. In some cities, public bus companies were established only after the 1970s. Before the early 1980s, public buses and trolleys were almost the only means of public transportation in cities nationwide. By the end of 1988, Shenzhen had only 24 bus routes, and in some medium and small cities, the number of bus routes was even in the single digits, making non-motorized transportation the absolute majority.

The "Neighborhood-Style" Residential Community Conducive to Walking

In the early days of the People's Republic of China, the country planned to vigorously restore the economy, focusing on industrialization construction, and there was an urgent need for housing. The key decisions during this period were based on the strategic shift from rural to urban areas proposed at the Second Plenary Session of the Seventh Central Committee in March 1949, which emphasized transforming cities from consumer-oriented to production-oriented. Building productive cities not only required the influx of a large number of multidisciplinary researchers but also an increase in the number of industrial workers. Coupled with the pressing reality of inadequate housing for urban residents, addressing housing needs and improving housing conditions became an imminent and significant issue. In February 1951, Chairman Mao Zedong issued instructions regarding the "housing shortage" issue: "The current shortage of housing in major cities has caused great dissatisfaction among the people. We must systematically build new houses, repair old ones, and meet the needs of the people." [4].

Major industrial cities across the country embarked on large-scale construction of workers' dormitories, with the workers' new villages as typical representations. In 1951, the Shanghai Municipal Government established the "Shanghai Workers' Housing Construction Committee," deciding to construct workers' residences that year as "the beginning of larger-scale construction of workers' housing in the future." This initiative aimed to address the housing difficulties of Shanghai's three million industrial workers. Cao Yang New Village (Fig. 20.1), located in the northwest corner of Shanghai, began construction in September 1951 and was completed in April 1952. It stands as the earliest prototype of workers' new villages in Shanghai and is considered the eldest son of the new workers' villages in the People's Republic of China. The first phase of the project covered an area of 13.3 hectares (0.133 square kilometers) and included 167 units with a total construction area of approximately 30,000 square meters. The village comprised 1,002 households, with an average living space of 4 square meters per person. Cao Yang New Village was centered around an elementary school, and neighborhoods were arranged within a 600-m service radius. Residents could reach essential facilities such as the post office, bathhouse, cooperative, cultural center, and more within a 7–8-min walk [5].

This residential area form was influenced by the "Neighborhood Unit" theory popular in the 1920s in various Western countries, with the construction of residential areas based on the unit surrounded by main urban roads. In the residential area, primary schools and public buildings are set up, featuring a community center, internal community streets, and at least 10% of community land designated as public open space or parks. In 1952, the Ministry of Railways also constructed a "Neighborhood Unit" residential area near Zhenwumiao in the Xicheng District of Beijing. The "Neighborhood Unit" residential areas in Beijing and Shanghai complemented each other, becoming exemplary and representative residential areas nationwide.

Fig. 20.1 Shanghai Caoyang new village master plan

In April 1955, in response to an invitation from the Chinese government, a Soviet expert group led by Bodleyev was formed to guide various aspects of urban construction, including urban planning and public transportation in Beijing. Amid the prevailing trend of a unilateral focus on learning from the Soviet Union, the "Neighborhood Unit" was replaced by the "Unitary" and "Community" residential models. In housing construction, the "Community" system was adopted in new areas based on the 9–15 hectares (0.09–0.15 square kilometers) neighborhood construction principles specified in the 1933–1935 Moscow Reconstruction Master Plan. During the first five-year plan of the People's Republic of China from 1953 to 1957, the Baiwanzhuang residential area in Beijing (Fig. 20.2), modeled on the "Neighborhood Unit" structure, borrowed from the Soviet "Community" practice and employed the Soviet-style "Unitary" residential building model. Many institutional compounds, such as the construction ministry's compound and the central party school's compound, also adopted the "Community" planning paradigm.

In 1957, under the guidance of the Soviet expert group, Beijing's residential areas were constructed based on the "residential community" as the basic unit, replacing the previous "neighborhood" system. The residential community is a fundamental structural unit in urban residential area planning, and each community can be divided into several residential clusters. Its basic characteristics include a clear delineation

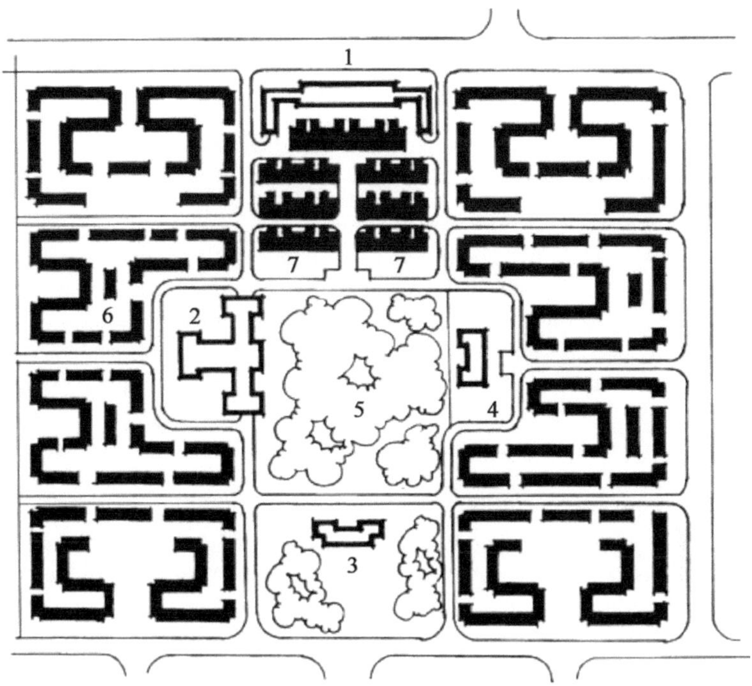

Fig. 20.2 Beijing Baiwanzhuang residential area floor plan [6]. 1—Office building; 2—Shopping mall; 3—Elementary school; 4—Kindergarten; 5—Central green space; 6—Boiler room; 7—Row houses

of boundaries, not being divided by major urban roads, and the provision of public service facilities within the community. Residents can meet their daily needs without leaving the community. The residential community emerged as a new form of living after the "neighborhood" system and remained the primary form of residential area construction until the late 1970s [7].

Although diverse residential areas have flourished after the reform and opening-up, the layout theory of the "neighborhood" residential community, based on Western and Soviet residential area planning theories, with Chinese characteristics, has continued. According to the principle of meeting residents' basic life needs within a reasonable walking distance, the "Urban Residential Area Planning and Design Standards" (GB 50,180—2018) categorizes residential areas into four levels: 15-min living circle residential area, 10-min living circle residential area, 5-min living circle residential area, and residential neighborhood. A reasonable walking distance has become an important criterion for assessing the livability of residents' living environments and the quality of urban space utilization.

The Development of Bicycles and the Evolution of Urban Transportation

The Development of Bicycles and the Evolution of Transportation Structure

In the late 1970s, endless streams of bicycles filled the city roads. Bicycles became a symbol of modernization internationally and marked China's first mass-produced industrial product. During peak commuting hours, the impressive flow of bicycles earned China the title of the "Bicycle Kingdom." Today, China remains the world's largest exporter, producer, and owner of bicycles, truly deserving the moniker of the "Bicycle Kingdom."

Impacted by modern transportation, including bicycles, the urban traffic structure in Chinese cities underwent a profound transformation over the decades around the twentieth century. Taking Shanghai as an example, on June 14–16, 1889, Meign, the director of the Public Concession Bureau Works Department, conducted a survey of the number of vehicles and pedestrians passing through Waibaidu Bridge from 8 to 20 h. The daily averages were 11,770 pedestrians, 6,984 rickshaws, 544 carriages, 9 sedan chairs, and 11 horse riders. In contrast, on May 17–18, 1926, at the same location, the Works Department conducted another survey of the number of various vehicles and pedestrians from 7 to 19 h. The daily averages, excluding pedestrians (50,823) and rickshaws (14,600), showed a notable absence of carriages, sedan chairs, and horse riders on the streets. Instead, bicycles (3,459), motorcycles (194), cars (3,764), buses (172), and trams (922) had taken their place. This revolutionary change on the city roads in Shanghai over the 37-year period from 1889 to 1926 was also observed to varying degrees in other Chinese cities [8].

After the outbreak of the Pacific War, due to the shortage of energy, urban public transportation nearly came to a standstill. Residents' rigid travel needs could only be met through human-powered transportation. On January 10, 1942, public buses in Shanghai ceased operation completely, and the tram industry "was also affected by the power-saving plan, leading to a reduction in vehicles and shortened operating hours" [8]. For the majority of people, the only reliable mode of transportation was the bicycle. As a lightweight vehicle that could navigate without any external energy, the bicycle's advantages were evident, and its demand skyrocketed, making it the most popular means of transportation at that time. In the 1960s and 1970s, bicycles played a crucial role in people's lives, becoming not only the most prevalent mode of transportation on urban roads but also a symbol of a family's wealth. It even became a primary material condition for a woman's marriage.

In the 1980s, owing to the high-density development of Chinese cities, a significant decrease in bicycle prices due to domestic production, lagging development of public transportation, and the generally flat terrain in most Chinese cities, China had the highest number of bicycles in the world [9]. During this period, there were over 60 bicycle manufacturing factories and over a thousand bicycle components factories,

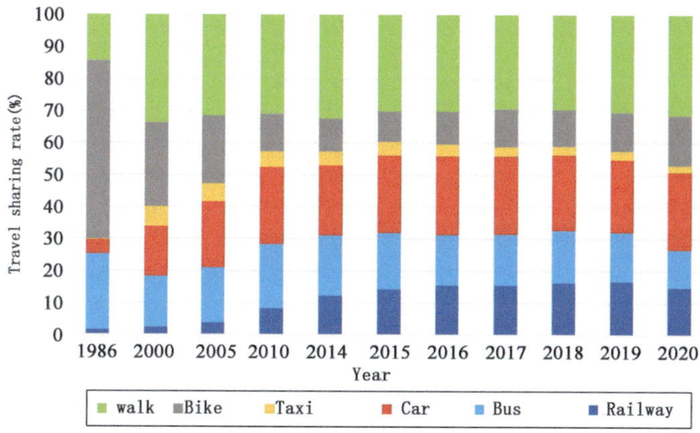

Fig. 20.3 Changes in Beijing's transportation structure over the years [10]. *Note* Data before 2000 is based on the eight urban districts, from 2000 to 2014 within the Sixth Ring Road, and after 2014 in the six urban districts

essentially forming a complete production system. The bicycle mode share in urban transportation also reached its peak, with the bicycle mode share in Beijing around 54% in 1986 [10] (Fig. 20.3), indicating that over half of the residents used bicycles for commuting.

However, the era of motorization in China is gradually unfolding. With the rapid economic growth, the social trend of private car ownership, and urban expansion, bicycles are gradually marginalized, making way for motorized vehicles in various aspects of urban transportation planning, construction, and management. In order to increase the capacity of motorized lanes, non-motorized lanes in some cities are continuously narrowed down or even eliminated. The remaining non-motorized lanes on certain roads are often occupied by arbitrarily parked motorized vehicles. Due to the continuous deterioration of the cycling environment, the mode share of bicycles had steadily declined from 1990 to 2010. In Shenzhen, for example, the mode share of bicycles decreased from approximately 30% in 1995 to about 4% in 2005, and in 2015, Beijing's mode share for bicycle trips was only 9.5%.

With the establishment of fundamental national policies such as environmental protection, resource conservation, rational land use, and the protection of arable land, bicycles have gained attention for their health, environmental, economic, and convenient advantages. Traffic authorities have promoted bicycle commuting through a series of green travel initiatives, demonstration projects, and policies. Additionally, the rapid rise of public bicycles in 2007 and shared bicycles in 2015 has made China the fastest-growing and largest region for public bicycle development worldwide, contributing to the resurgence and gradual revitalization of bicycles.

Road Features in Bicycle Traffic and Bicycle Renewal

In the cross-section of urban roads in Chinese cities, the design of "Three-Lane Roads" and "Four-Lane Roads" (commonly known as "Three Boards" and "Four Boards") includes side separators, enhancing safety and efficiency for cycling. This design is particularly friendly to bicycle traffic and has unique characteristics in China. On the eve of the establishment of the People's Republic of China, the Urban Planning Committee was established to oversee the road system, and a specialized committee was responsible for designing road cross-sections. By 1957, the basic forms of road cross-sections in Beijing were unified into three types: "One Board, Two Boards, and Three Boards." This imaginative naming convention is not found in modern Chinese literature, Western, Japanese, or Soviet road design standards, making it a pioneering innovation in China's urban planning and development history on a global scale [11]. The "Four Boards" design involves adding a median strip to the "Three Boards." The first "Three Boards" road in Beijing was Sanlihe Road, completed in 1957, with 4-m-wide non-motorized vehicle lanes (commonly known as "slow lanes") on both sides and poplar trees planted in the median strip, achieving excellent landscape effects, safety standards, and service efficiency. Subsequently, "Three Boards" became the main cross-sectional form for Beijing's main roads. The urban planning principles established by Beijing, including road hierarchy and cross-section design methods, have become exemplary and have been widely promoted nationwide. Their influence persists to this day, laying a solid infrastructure foundation for China to develop as the "Kingdom of Bicycles."

The integration of public bicycles and shared bicycles combines the three elements of cleanliness, intelligence, and sharing, driving the revival of bicycle transport in China and holding profound significance for rebuilding bicycle culture. With the rise of public bicycles worldwide since the late 20th century, the operation of a public bicycle system, initiated by local governments, began in Hangzhou in 2007. In less than a decade, China rapidly became the country with the most extensive distribution of public bicycles, with over 237 cities establishing public bicycle systems and operating around 750,000 bicycles [12]. In 2016, the promotion of shared bicycles such as Mobike and Ofo sparked a new wave of enthusiasm for bicycle commuting. As of July 2017, there were nearly 70 shared bicycle operators nationwide, deploying over 16 million bicycles, registering more than 130 million users, and providing services to over 1.5 billion people [13]. These two waves of bicycle enthusiasm not only cultivated residents' habits of using bicycles for transport but also created prerequisites for local governments to add non-motorized lanes and dedicated bicycle lanes.

At the same time, electric bicycles are increasingly being used for residents' daily commuting, and the demand is booming. Data shows that as of 2021, the social ownership of electric bicycles in China has approached 300 million, growing threefold within a decade. The overall production trend is on the rise, with an annual output exceeding 30 million units. In the period from January to August 2020, the business revenue saw a year-on-year growth of 23.4%. Users aged 30–39 and 40–49 are the main demographic groups, accounting for 47 and 31%, respectively. This indicates that electric bicycles remain essential as a means of transport for daily life in China, especially among the middle-aged population who use them for commuting, transporting children, leisure, and entertainment [14]. Non-motorized traffic lanes need to adapt to this new trend.

Development of Public Transportation Takes Priority as a Consensus

The Policy Context of Prioritizing Public Transportation Development

The strategic concept of prioritizing the development of public transportation is not a product of the 21st century. As early as October 1982, when urban transportation was still focusing on construction issues and exploring ways to improve the utilization of existing road systems and enhance capacity, the idea of "prioritizing the development of public transportation, appropriately controlling the growth of bicycles" was first proposed at the 4th National Symposium on Urban Transportation Planning. In 1985, the importance of urban public transportation was officially emphasized in the "Report on Reforming Urban Public Transportation," issued as a State Council document, stating that "solving urban traffic congestion problems requires comprehensive management, strict control of various vehicles, vigorous development of public transportation, and an increase in passenger vehicles." By the end of 1983, the total number of urban public transportation passenger vehicles nationwide was 44,749 (including 34,103 buses, 4,192 trackless trolleys, 321 track trolleys, 5,965 taxis, and 168 subway cars), with an annual passenger volume of 23.1 billion.

In the 1990s, issues related to public transportation policies garnered increasing attention. In December 1994, during the 14th Annual Conference on Urban Transportation Planning, participants raised the issue that solving urban passenger transportation problems must primarily rely on public transportation. The development of public transportation should not only receive policy support but also undergo reforms in operational mechanisms and composition. Public transportation should evolve to meet diverse transportation needs. It was also suggested that in large cities with populations exceeding one million, various approaches should be employed to gradually develop rail transportation and make it the backbone of urban public transportation.

In the early 21st century, the "Priority Development of Urban Public Transportation" (abbreviated as "Public Transit Priority") strategy was initially proposed by industry regulatory department. It was later explicitly endorsed by the State Council office. In March 2004, the "Opinions on Prioritizing the Development of Urban Public Transportation" (Jian Cheng [2004] No. 38) was released by the Ministry of Construction, stating that "urban public transportation is an important urban infrastructure, a social welfare project related to the national economy and people's livelihood," and emphasizing that "public transit priority means 'people's priority.'" In

September 2005, the "Notice of the State Council Office on Forwarding Opinions on Prioritizing the Development of Urban Public Transportation by the Ministry of Construction and Other Departments" (Guo Ban Fa [2005] No. 46) was issued, further highlighting that urban public transportation is a crucial infrastructure closely related to the production and life of the people. Prioritizing the development of urban public transportation is considered a crucial means to improve the efficiency of traffic resource utilization and alleviate traffic congestion. Subsequently, both central and local governments introduced a series of economic and industrial support policies to expedite the construction of public transportation systems, thereby establishing public transit priority as a development strategy.

Under the vigorous promotion of the public transport priority development strategy and the active participation of local governments, the scale of investment in and construction of urban public transport infrastructure, including routes, vehicles, and depots, has expanded significantly. By the end of 2011, China had 34,000 operational bus and trolleybus routes nationwide, with a total operational route length of 673,000 kilometers (six times that of 2002), 453,000 operational vehicles (1.8 times that of 2002), and depot areas totaling 52.31 million square meters. Additionally, the country operated 58 urban rail transit lines with a total operational length of 1,700 kilometers (7.7 times that of 2002) and 9,945 rail vehicles (10.1 times that of 2002) [20]. In December 2012, in response to new challenges arising from urbanization and to accelerate the transformation of urban transport development patterns, the State Council issued the "Guiding Opinions on Prioritizing the Development of Public Transport in Cities" (Guofa [2012] No. 64), which further emphasized that "public transport development should be placed at the forefront of urban transport development."

National Public Transit City Development Project

In November 2011, the Ministry of Transport first proposed to organize and implement the National Public Transit City Construction Demonstration Project. A total of 87 cities, including Beijing, were subsequently identified as cities for the construction of public transit cities. As of May 2022, five batches comprising 46 cities have been awarded the title of "National Public Transit City Construction Demonstration City." Since the initiation of the National Public Transit City Construction Demonstration Project, transport authorities at all levels and cities participating in the construction initiative have actively contributed. The capacity to provide public transportation services for people's livelihoods and promote urban economic and social development has significantly improved. Public transit cities have increasingly become a crucial lever for implementing the strategy of prioritizing public transit development, serving as a vital platform for consensus on urban public transportation reform and development, and innovatively driving the development of urban transportation.

During the more than 10 years of the National Public Transit City Construction Demonstration Project, significant progress has been made in the construction of urban public transportation infrastructure and service systems. The assurance of urban public transportation infrastructure has continually strengthened, with the total length of nationwide public bus and trolleybus operating routes doubling, and the total length of urban rail transit operating routes and the number of operating vehicles quadrupling. Diversified public transit services have rapidly developed, with over 50 cities offering internet-based customized public transit services. Several cities have introduced customized, nighttime, community-based, tourism-specific, and other distinctive public transit services. The promotion and application of new energy vehicles have accelerated, with the number of new energy public buses and trolleybuses reaching 466,100 by the end of 2020, accounting for 66.2% of the total number of urban public buses and trolleybuses in operation. The payment methods for public transportation have shifted from physical cards and offline payments to a combination of physical cards and virtual cards, as well as online mobile payments using QR codes. In Beijing, for example, the system allows for automatic synchronization and verification of health code information when boarding by swiping a card or scanning a QR code. The exploration of intelligent urban public transportation has been ongoing. Cities such as Beijing, Shenzhen, and Jinan actively explore the application of autonomous driving in public transit, while some cities are also actively implementing Mobility as a Service (MaaS) application projects.

The construction of Public Transit Cities has further deepened the understanding of the government and various departments regarding the strategy of prioritizing public transportation development. This has led to a shift from departmental actions to government actions, fostering collaborative efforts in mechanism improvement, regulation formulation, and institutional reform to collectively drive implementation. In March 2022, the Ministry of Transport formulated the "Management Measures for the National Public Transit City Construction Demonstration Project," which constrains and standardizes the construction content, application for creation, organizational implementation, acceptance and naming, dynamic evaluation, and other aspects. This measure aims to promote the improvement and expansion of national public transit city development.

For public transportation to truly achieve prioritized development, it still faces various challenges in terms of conceptual understanding, resource allocation, technological innovation, management systems, and financial support. As China's urbanization has entered the "second half," with the construction of urban spatial and transportation systems at the regional level, public transportation will face more severe challenges and take on more significant functions. It is necessary to firmly implement the strategy of prioritizing public transportation development, integrate the development of rail transit and public buses and electric buses, seize the opportunity for prioritized development, and not forget the initial intention of solving traffic congestion issues, which is the essence of prioritizing public transportation.

Policies and Planning Guidance for Green Development of Urban Transportation

In 1983, the former State Science and Technology Commission, Planning Commission, Economic Commission, and Construction Commission jointly organized the "Development Direction of Urban Transportation in China" research, marking the first systematic and comprehensive analysis of China's urban transportation technology policy. This initiative laid the foundation for the research of urban transportation policies. The research results were incorporated into the 1985 Blue Book (No. 6) "Chinese Technology Policy—Urban and Rural Construction" by the State Science and Technology Commission. The document proposed policy directions such as scientifically determining the transportation structure, prioritizing public transportation in large cities, and coordinating the development of various modes of transportation. It also advocated controlling the development of bicycles in large cities to a certain extent, limiting the growth of social group and official vehicles, restricting the development of high-energy-consumption and highly-polluting motorcycles, and gradually establishing rapid rail transit in eligible large cities. Additionally, it outlined specific strategies for the active development of taxi services and other measures.

The inclusion of urban transportation planning as a specialized plan within the urban planning framework began in 1990. The "Urban Planning Law of the People's Republic of China," implemented in 1990, marked the first legal provision explicitly stating that the overall urban plan should incorporate the integrated urban transportation system.

In the years 1992–1995, the Ministry of Construction collaborated with the World Bank and the Asian Development Bank on the "China Urban Transport Industry Research Project," resulting in the influential "Beijing Declaration: China's Urban Transport Development Strategy." Its core content is summarized into five principles, four standards, and eight actions, forming the values and methodologies that urban transport policy and planning should adhere to. This laid the ideological foundation for the green and sustainable development of urban transportation.

In the initial period of urban transportation planning, the focus was on traditional road network planning. This approach was based on the urban layout, determining the structure, functions, and grade of road networks. This qualitative planning method persisted until the late 1970s, even after the introduction of traffic volume concepts in the 1950s and 1960s. Modern urban transportation planning, in the sense it is understood today, emerged in the late 1970s in China. It was influenced by lectures given by transportation engineering experts, including Mr. Zhang Qiu, a Chinese-American representing experts from the United States, Japan, the United Kingdom, Canada, and other countries [15]. These experts systematically introduced construction and management experiences in transportation planning, management, control, and safety from developed Western countries, contributing to the development of transportation engineering in China.

With the implementation of the reform and opening-up policies, foreign urban transportation planning concepts and technologies were introduced to China. This facilitated the shift in road network planning from qualitative to quantitative approaches, marking the exploration and construction periods of urban transportation planning. During this time, there was a fundamental change in planning ideology and technical methods. The focus expanded from road networks to integrated transportation networks, and the perspective shifted from road layout to a scientific understanding of traffic patterns and the rational organization of integrated transportation systems. The planning content expanded from spatial arrangements of transportation facilities to include traffic demand regulation, optimization of transportation supply structure, and the coordination of transportation and land use. This led to the development of comprehensive urban transportation planning and specialized planning for public transportation, static traffic, freight transportation, and others. Due to the expansion of urban land use and the rapid motorization, walking and cycling traffic diminished, receiving minimal representation in integrated urban transportation planning. There was also a lack of independent specialized planning for these modes. Affected by the expansion of urban land and rapid motorization, pedestrian and bicycle traffic has gradually diminished, with little representation in comprehensive urban transportation planning and almost no independent specialized planning.

Entering the 21st century, the planning of integrated urban transportation system has gradually improved and entered a period of transformation. In 2010, the Ministry of Housing and Urban–Rural Development issued the "Methods for the Compilation of Comprehensive Urban Transportation System Plans." Subsequently, guidelines for the compilation of comprehensive urban transportation system plans, traffic survey guidelines, and standards for comprehensive urban transportation system plans (GB/T 51,328—2018) were successively released. Information technology and digital technology have provided unprecedented conditions for urban transportation research. At the same time, comprehensive urban transportation system planning places more emphasis on a people-centered approach, adhering to principles of safety, environmental friendliness, fairness, efficiency, economic feasibility, and coordination to ensure the livability and sustainable development of cities. In the more than 40 years of research and practice since the reform and opening-up, urban transportation planning has continuously innovated and developed, drawing on theories and methods from various disciplines such as sociology, economics, and management. It has gradually formed a multi-disciplinary integrated approach to urban transportation planning and a multi-level professional planning framework tailored to practical needs.

In the year 2000, the Ministry of Public Security and the Ministry of Construction initiated the "Smooth Traffic Initiative" for urban road traffic management nationwide. This was a forward-looking initiative implemented at a crucial moment when motorization was entering a rapid development phase in China. It served as a powerful deployment of work to lead urban traffic management and even the healthy development of cities. The initiative became an innovative platform and carrier for advancing the scientific and modern processes of traffic management in China.

In August 2003, in order to promote the construction of urban traffic infrastructure and further advance the "Smooth Traffic Initiative" as well as strengthen urban ecological environmental protection, the Ministry of Construction and the Ministry of Public Security issued the "Notice on Carrying Out the Activity of Creating 'Green Transportation Demonstration Cities'" (Jian Cheng [2003] No. 169). This initiative was implemented nationwide within the jurisdiction of designated cities. Although this activity did not receive enthusiastic responses from cities, the established assessment standards for green transportation demonstration cities were ahead of their time and carried positive significance for the era. The assessment standards covered 66 indicators across five aspects: organizational management, planning and construction, public transportation, infrastructure, and traffic environment, providing a benchmark for the green development of urban transportation.

From 2010 to 2015, to promote the construction of urban pedestrian and bicycle transportation systems, the Ministry of Housing and Urban–Rural Development carried out the "Demonstration Program for Urban Pedestrian and Bicycle Transportation Systems" nationwide. This initiative identified over a hundred cities in three batches, including Hangzhou, Chongqing, Kunming, Jinan, Shenzhen, Xiamen, Beijing, and others. The main components of the demonstration program included formulating (or revising) professional plans for urban (or district) pedestrian and bicycle transportation systems, developing policies and measures to promote the planning and construction of urban pedestrian and bicycle transportation systems, and constructing demonstration areas with a certain scale of pedestrian and bicycle transportation facilities or improving public bicycle systems. Cities involved in the demonstration program implemented effective measures (Figs. 20.4, 20.5 and 20.6), not only improving conditions for pedestrian and bicycle travel but also significantly enhancing awareness of green commuting among different groups. During the progress of the demonstration program, the Ministry of Housing and Urban–Rural Development, together with the National Development and Reform Commission and the Ministry of Finance, issued the "Guiding Opinions on Strengthening the Construction of Urban Pedestrian and Bicycle Transportation Systems" (Jian Cheng [2012] No. 133) and the "Notice on Issuing Guidelines for Planning and Design of Urban Pedestrian and Bicycle Transportation Systems" (Jian Cheng [2013] No. 192). These laid the foundation for the subsequent development of guidelines for street design in many cities.

In 2011, the Ministry of Transport initiated the "Public Transit City Construction Demonstration Project" and subsequently carried out the construction of public transit cities in 87 cities nationwide in 2012, 2013, and 2016. This effort has propelled the enhancement of public transit capabilities, infrastructure, transportation equipment, and service quality to a new level. It has also achieved breakthroughs in the institutional reform, fiscal support, and the application of theories and engineering technologies for the prioritized development of urban public transit.

In August 2017, the Ministry of Public Security, the Office of the Central Committee for Spiritual Civilization, the Ministry of Housing and Urban–Rural Development, and the Ministry of Transport decided to further innovate urban road

Fig. 20.4 Hangzhou City promotes pedestrian and bicycle transportation development through road and river channel improvement (Sunken Pedestrian Walkway on Zhonghe Road, Hangzhou)

Fig. 20.5 Chongqing City enhances pedestrian space quality with mountain city walkways as a distinctive feature (Photography: Shi Honggang)

traffic management. Building on the implementation of the "Smooth Traffic Initiative" and the "Civilized Traffic Action Plan," they initiated the "Urban Road Traffic Civilization and Smoothness Enhancement Action Plan" nationwide starting from 2017. In November 2017, the National Expert Group for the Urban Road Traffic Civilization and Smoothness Enhancement Action Plan was established [20].

Fig. 20.6 Jinan City prioritizes road rights protection as a top task (Pedestrian and bicycle travel environment after the renovation of Yuxing Road in Jinan City)

Green Travel Initiatives in Chinese Cities

The green travel initiatives in Chinese cities are primarily influenced by events such as "World Car Free Day" and "European Mobility Week." The Car Free Day originated in Europe, with the idea of reclaiming city roads for residents to reduce the impact of automobiles on urban development being proposed by Eric Britton at the "Accessible Cities" conference held in Spain in 1994. In 1995, cities in Iceland, the UK, and France each hosted their first Car Free Day events. In 2001, the European Commission's Directorate-General for Environment decided to provide policy and financial support for Car Free Day, scheduling the event on September 22nd each year, which also became known as "World Car Free Day." Due to the attention and funding from the European Union, Car Free Day evolved into European Mobility Week in 2002, taking place annually from September 16th to 22nd.

China's green travel initiatives initially started as spontaneous efforts by some cities. On October 14, 2000, to coincide with the "21st Century International Conference on Urban Construction and the Environment," and to showcase Chengdu's commitment to ecological environmental protection in the western region, Chengdu took the lead in organizing and implementing a Car-Free Day event. From 9:00 to 19:00 on that day, all roads within Chengdu's inner ring road, excluding public transportation vehicles, special vehicles, designated vehicles, and non-motorized vehicles, were closed to all other vehicles, allowing residents to enjoy a day of less noise, reduced congestion, lower air pollution, and safer pedestrian conditions. Following

Chengdu's example, many other cities such as Beijing, Shanghai, and Wuhan also spontaneously initiated various Car-Free Day activities.

Taipei, China, holds its Car-Free Day on September 22. The Car-Free Day event in Taipei started its preparations in 2001, but it was forced to stop due to the impact of Typhoon Nari during the scheduled activities. Taipei officially organized the Car-Free Day event in 2002, focusing on activities such as cycling and carnivals. The content of the event gradually diversified in 2003, with different activities held over several days. Starting from 2004, the World Car-Free Day became the core of a series of events, and the duration of other activities was extended. For example, the "Walk to School Day" was extended to a "Walk to School Week," and the "Public Transportation Day" was extended to a "Public Transportation Week."

In September 2005, Mr. Wang Guangtao, the Minister of the Ministry of Construction, was invited to participate in the European Mobility Week held in London. He delivered a speech titled "Promoting the Strategy of Prioritizing the Development of Urban Public Transportation." In his speech, he pointed out that the European Mobility Week is highly instructive for China to conduct public transportation promotional activities. It is conducive to further raising awareness among various regions and relevant departments and guiding the public to choose public transportation as their primary means of travel. The Ministry of Construction will draw on the advanced concepts and measures of sustainable transportation in Europe and, at an appropriate time, launch China's Public Transportation Week. During Public Transportation Week, Car-Free Day activities can be organized. Car-Free Day is the most suitable time for citizens to observe the city, propose urban improvement suggestions, and provide feedback on public transportation services.

In December 2006, to implement the strategy of prioritizing public transportation development and promote sustainable urban transportation, the Ministry of Construction issued an initiative to all cities in the country, especially to the municipal governments of cities with a population of over 500,000, to carry out "China Urban Public Transportation Week and Car-Free Day" activities. The inaugural event received positive responses from 110 cities, including Beijing, Shanghai, Guangzhou, Shenzhen, Hangzhou, and others. It took place simultaneously nationwide from September 16 to 22, 2007, under the theme "Green Transportation and Health." In 2008, due to the impact of the restructuring of functions under the national large-scale system reform, the event was renamed "China Urban Car-Free Day," scheduled annually on September 22. Simultaneously, cities were encouraged to organize a week-long series of activities. In September 2016, the event was further adjusted to a green travel promotion activity. In September 2019, the "922 Green Travel" platform, which has long supported China's urban green travel activities, explored normalized operations. Green travel promotion transformed from short-term event-based campaigns to year-round dissemination of knowledge and cultivation of a green travel culture.

In June 2013, to strengthen the cultural development of public transportation, the Ministry of Transport decided to organize the "Public Transportation Promotion Week" annually in late September. The initiative aimed to mobilize urban public transportation enterprises and the general public to actively participate, collectively fostering a "low-carbon transportation, green travel" culture in urban public transportation.

In September 2013, the "Public Transportation Promotion Week" took place from September 16 to 22 with the theme "Public Transportation Priority, Convenient and Beneficial." In September 2018, to further implement the public transportation priority development strategy and promote green, safe, and civilized travel, the Ministry of Transport, in collaboration with the Ministry of Public Security, National Development and Reform Commission, and the All-China Federation of Trade Unions, organized the "2018 Green Travel Promotion Month and Public Transportation Promotion Week" initiative. The promotion period extended from "week" to "month." The theme for that year's activities was "Green Travel, Starting from Around Us," aiming to extensively promote the achievements and experiences in advancing public transportation priority, green travel, creating public transportation-oriented cities, and improving public transportation services over the 40 years of reform and opening up. The initiative actively advocated the concept and measures of public transportation priority and green travel, urging their integration into government agencies, enterprises, schools, communities, households, and public places to enhance public awareness of green travel.

Developing green transportation is the overarching theme of green travel activities, and prioritizing the development of urban public transportation is the initial goal of conducting green travel activities in China. Whether it's the "China Urban Public Transportation Week and Car-Free Day Activities" launched in 2007 or the "Public Transportation Promotion Week Activities" in 2013, the core objective is to implement the public transportation priority development strategy, thereby accelerating the construction of public transportation infrastructure, improving the level of public transportation services, and establishing a societal consensus on public transportation priority. The rapid urbanization and motorization have triggered widespread "urban diseases," and issues such as traffic congestion, environmental pollution, and noise pollution urgently need improvement. It is crucial to vigorously develop public transportation before people become heavily dependent on private cars.

Due to the interdependence and mutual promotion of public transportation with other green transportation modes, sustainable development in urban transportation is a global trend. Therefore, each year's green travel activities focus on a topic related to green transportation, aligning with current hot topics and major issues. Themes such as public transportation, walking and cycling, human-friendly streets, traffic safety, shared transportation, and smart transportation have successively become the focus of green travel activities, highlighting innovative and sustainable development with Chinese characteristics. The themes of green travel activities from 2007 to 2022 are detailed in Table 20.1.

Table 20.1 Green travel activities and themes from 2007 to 2022

Year	China car-free day (green travel promotion campaign)		Public transit promotion week	
	Theme	Interpretation	Theme	Interpretation
2007	Green transport and health	Emphasizes the implementation of the priority development of urban public transportation as an inevitable choice for sustainable urban development. It clarifies the responsibilities and obligations of the government, public transit enterprises, and individuals. The concept of green travel should form a social consensus	–	–
2008	Humanized streets	Emphasizes the redistribution of road space and points out that reducing the road space for cars is an effective solution for sustainable development and solving transportation problems. It advocates for the continuous increase of pedestrian areas, non-motorized vehicle lanes, and dedicated bus lanes	–	–
2009	Healthy and environmental-friendly bicycle and walking transportation	Emphasizes the creation of a safe, convenient, comfortable, and orderly environment for cycling and walking transportation. It highlights the significant role of bicycle and walking traffic in the comprehensive urban transportation system, suitable for short to medium-distance travel and integration with other modes. It also emphasizes their recreational and fitness functions	–	–
2010	Green transport and low-carbon living	Emphasizes that urban transportation is a crucial area for implementing the national energy-saving and emission reduction strategy. It clearly defines green travel as a healthy and positive lifestyle. Once travel habits are formed, changing them is challenging. The focus should be on implementing effective policy measures during this critical period of shaping transportation behavior		

(continued)

Table 20.1 (continued)

Year	China car-free day (green travel promotion campaign)		Public transit promotion week	
	Theme	Interpretation	Theme	Interpretation
2011	Green transportation, urban future	It emphasizes optimizing urban transportation structure through green transportation development strategies: integrating transportation planning with land use planning, implementing strategies for prioritizing public transportation development, equitably allocating road space, giving priority to pedestrian and bicycle traffic, and fostering public awareness and cooperation spirit		
2012	Caring for the city, green travel	It underscores the implementation of transportation policies to promote urban sustainable development, including rational pricing for on-street parking, encouraging "park-and-ride," exploring congestion charging policies, and guiding commuter traffic towards green transportation as part of demand management policies		
2013	Green transportation, fresh air	On one hand, it focuses on the impact of urban transportation activities on air quality, aiming to create healthier, happier, and more beautiful cities; on the other hand, it also reflects the powerful force of improving air quality through choosing sustainable travel modes	Public transit priority, convenience and benefit to the people	By integrating local realities, conducting diverse and flexible mass-themed promotional activities, promoting better understanding, care, support, and choice of public transportation throughout society
2014	Our streets, our choices	The emphasis is on the impact of transportation on urban quality of life, encouraging the redistribution and redesign of streets and public spaces to achieve a balance in allocating road space for various transportation modes	Opting for public transit, green travel	Through various forms of themed promotional activities and public cultural events, raising public awareness of participation in public transportation across all sectors of society, improving the external environment for the development of urban public transportation, and promoting the transformation of urban transportation development concepts, fostering a social trend of caring for, supporting, and choosing public transportation

(continued)

Table 20.1 (continued)

Year	China car-free day (green travel promotion campaign)		Public transit promotion week	
	Theme	Interpretation	Theme	Interpretation
2015	Green transportation—Choose · Transform · Integrate	It highlights the necessity of constructing various transportation modes, facilitating people's choice of public transportation, walking, and cycling as green transportation modes, changing the transportation development model overly reliant on private cars, and integrating green travel with social life	Opting for public transit, green travel	Guided by the construction of the "Four Transportation Services" and centered around the promotion of "prioritizing public transportation and advocating green travel," in conjunction with the activity themes, through various forms of urban public transportation experience, benefiting the people, and cultural promotional activities, enhancing the public's understanding, comprehension, and support for urban public transportation, fostering habits of preferring green travel among the public, and creating a favorable atmosphere for the whole society to support the prioritized development of urban public transportation
2016	Green transportation, smart travel	Emphasizing the wise selection of transportation modes to reduce the time, economic, health, and environmental costs of travel. The aim of the activity is to encourage individuals to consider both personal and societal transportation costs and to choose green transportation modes such as public transportation, walking, and cycling from both personal and societal perspectives	Preferential public transportation, green travel	By conducting diverse urban public transportation promotional activities, vigorously promoting the priority development policy of public transportation, nurturing a culture of "preferring public transportation, green travel" in urban public transportation, creating a favorable atmosphere in society to understand, care about, support, and choose public transportation, and actively guiding the public to participate in the development of urban public transportation

(continued)

Table 20.1 (continued)

Year	China car-free day (green travel promotion campaign)		Public transit promotion week	
	Theme	Interpretation	Theme	Interpretation
2017	Green transportation, shared travel	Advocating for shared travel solutions, emphasizing shared trips, transportation tools, and infrastructure to ensure that the personal and societal costs of travel are lower than those of using private cars, representing the sustainable future of urban transportation. The activity aims to encourage people to use shared transportation tools for both intra-city and inter-city travel	Preferential public transportation, green travel	Through organizing diverse urban public transportation promotional activities, actively promoting the priority development policy of public transportation, popularizing traffic safety knowledge, enhancing the awareness of service among urban public transportation personnel, passengers' awareness of traffic safety, and awareness of civilized travel, further cultivating a culture of "preferring public transportation, green travel" in urban public transportation, creating a favorable atmosphere in society to understand, care about, support, and choose public transportation, and actively guiding the public to participate in the governance of urban public transportation
2018	Green travel, plus one for safety	Advocating for green travel, focusing on ensuring the safety of city roads for people's travel and reducing the road traffic safety risks in urban areas	Green travel starts from around us	Vigorously promote the achievements and experiences of various places in advancing the priority development of public transportation and green travel, creating public transportation cities, and improving public transportation services over the past 40 years of reform and opening up. Actively promote the concept and measures of prioritizing public transportation and green travel, promote the integration of green travel into government agencies, enterprises, schools, communities, families, and public places, and enhance public awareness of green travel

(continued)

Table 20.1 (continued)

Year	China car-free day (green travel promotion campaign)		Public transit promotion week	
	Theme	Interpretation	Theme	Interpretation
2019	Safe walking and cycling	Highlighting the benefits of walking and cycling for health, environment, and economy, advocating for cities where walking and cycling are more attractive than cars, resulting in lower traffic congestion and higher quality of life	Public transportation priority, green travel	By organizing diverse and participatory promotional and experiential activities, deeply publicize the significant significance of green travel, comprehensively summarize the achievements of various places in advancing the priority development of public transportation, green travel, and improving urban transportation services since the 70th anniversary of the founding of the People's Republic of China, widely mobilize the whole society to actively participate in green travel actions, and accelerate the formation of a concept and behavioral guidelines for a green lifestyle in society
2020	Practicing green travel, building a beautiful China	The "922 Green Travel" platform proposes initiatives for cities to practice green travel: prioritizing the development of green transportation, advocating for reduced car usage frequency, improving the quality of public transportation services, enhancing walking and cycling infrastructure, enhancing social inclusivity, and ensuring road traffic safety	Practicing green travel, building a beautiful China	By various means such as concentrated promotion online and offline, widely mobilize the public to actively participate in green travel actions, and accelerate the formation of a green lifestyle
2021	Advocating green travel, promoting ecological civilization	The "922 Green Travel" platform suggests five measures to promote green travel through the "2021 Green Travel Activity City Handbook": encouraging walking and cycling, improving public transportation, reducing noise and air pollution, enhancing the safety of cyclists and pedestrians, and formulating sustainable transportation plans	Advocating green travel, promoting ecological civilization	By various means such as concentrated promotion online and offline, widely mobilize the public to actively participate in green travel actions, and accelerate the formation of a green lifestyle. Encourage enterprises to concentrate on publicity, innovate new formats of transportation services, focus on improving travel services for the elderly, carry out activities for civilized traffic safety travel, and carry out caring activities for transportation personnel, among others

(continued)

Table 20.1 (continued)

Year	China car-free day (green travel promotion campaign)		Public transit promotion week	
	Theme	Interpretation	Theme	Interpretation
2022	Green travel, let's go together	The "922 Green Travel" platform has authored the "2022 Green Travel City Handbook," proposing the theme "Green Travel, Let's Go Together," which reflects the core requirement of high-quality development centered around the people's aspiration for a better life and actively promotes a green lifestyle. The handbook focuses on age-friendly green transportation planning and construction, analyzes the travel needs characteristics of children, youth, the elderly, and women, identifies the development shortcomings of street spaces, addresses the impact of normalized online shopping on urban transportation, and discusses sustainable development strategies in human mobility and goods circulation through the formulation of sustainable urban transportation plans and strengthening road infrastructure construction	Green travel, let's go together	Vigorously publicize the achievements and experiences of various places in promoting the construction of national public transportation cities and the creation of green travel actions since the 18th National Congress of the Communist Party of China, widely mobilize all sectors of society to actively participate in green travel actions, accelerate the formation of a green lifestyle, and welcome the successful convening of the 20th National Congress of the Communist Party of China with practical actions

"Double Carbon" Goals and Resilient Transportation

Transportation activities are one of the main sources of global greenhouse gas emissions. There has been a growing social consensus on combating climate change through the development of green transportation. However, the task of carbon reduction still faces severe challenges. In September 2020, President Xi Jinping delivered an important speech at the 75th session of the United Nations General Assembly, announcing that China will adopt stronger policies and measures to strive to peak carbon dioxide emissions by 2030 and achieve carbon neutrality by 2060. As early as 2015, Beijing set a target to peak carbon emissions around 2020. Now, Beijing's carbon emissions have already peaked and are showing a stable downward trend, with achieving carbon neutrality by 2060 becoming an important goal for its future development [16]. Shanghai's urban passenger transportation carbon emissions may peak around 2030. While adhering to the improvement of the comprehensive transportation system, the proportion of electric vehicles in the city's fleet should be increased to 25–30% by 2030 to lay the foundation for subsequent carbon neutrality [17]. The "dual carbon" goals bring both opportunities and challenges to the green development of the transportation sector, promoting the low-carbon transformation of urban transportation.

On the road to low-carbon development, Chinese cities have already made many beneficial attempts. For example, promoting the electrification of transportation vehicles, especially small cars, public buses, and taxis; adopting various measures to encourage residents to reduce the use of private cars and shift to green transportation modes; implementing controls on the total number of private cars; promoting green freight transportation and implementing restrictions on freight vehicles in downtown areas, among others. However, achieving low-carbon transformation requires adopting more effective strategies and implementation pathways, not limited to the transportation sector but extending to urban spaces, lifestyles, and particularly the energy sector. At the same time, low-carbon transformation is a process, not an instant change, and in the future, there will be a need for more solid theoretical research and practical exploration.

Resilient transportation is an essential component of resilient cities. When facing external disruptions, a resilient urban transportation system is one that can adopt flexible response measures to maintain its development vitality, attract resource concentration, avoid potential losses, and address challenges and changes through autonomous regulation within the social system. Since the outbreak of the novel coronavirus pneumonia (COVID-19) in January 2020, the transportation systems of some cities have been stagnant, severely affecting the transportation of personnel and epidemic prevention materials. The unprecedented heavy rainfall in Henan Province in July 2021 also had varying degrees of impact on road transportation, with the Zhengzhou metro system being completely shut down and trains traveling to or passing through Zhengzhou being suspended nationwide. The transportation system is relatively vulnerable in the face of sudden public health emergencies and extreme weather events.

The World Bank's "Global Facility for Disaster Reduction and Recovery" (GFDRR) has been among the earliest and most comprehensive proponents of the concept of resilient transportation. A resilient transportation system can withstand disasters such as climate change by understanding the real risks faced by cities, planning alternative transportation modes and routes, utilizing new materials and standards to enhance emergency preparedness and response capabilities, and ensuring the reliability of resilient transportation systems [18]. Some successful practices in the field of urban transportation itself demonstrate strong resilience. For example, Transit-Oriented Development (TOD) models intensify and mix land use, increase the convenience of public transportation and pedestrian connections, thereby reducing travel demand and improving the accessibility of transportation systems. Policies such as creating car-free zones, promoting shared streets, increasing parking fees, or reducing parking spaces are implemented to reduce car usage. Developing various modes of transportation instead of relying excessively on motor vehicles is also emphasized. However, in the face of various risks and the diverse and complex urban environment, constructing resilient transportation systems requires a better understanding and evaluation of various risks and a comprehensive consideration of urban resilience as a whole, necessitating a shift in mindset.

References

1. National Bureau of Statistics. NBS: Number of Cities in China Reaches 672, Urbanization Rate of Permanent Residents Reaches 59.58% [EB/OL]. (2019-08-15) [2022-04-30]. https://www.fang.com/news/2019-08-15/33195493.htm.
2. Li Tianjian, Zhao Xuejun. The Evolution, Historical Contributions, and Implications of China's Urban Economic Policies Over the Past 70 Years [J]. Reform, 2019(8): 5–14.
3. State Council. State Council Endorses the Summary of the National Urban Planning Work Conference (Guo Fa [1980] No. 299) [A]. Beijing: State Council, 1980.
4. Wang Xianjun. The Thoughts and Practices of the Party and Government in Solving People's Livelihood Issues in the Early Years of New China [EB/OL]. CPC News Network. (2013-06-19) [2022-04-30]. http://theory.people.com.cn/n/2013/0619/c83867-21893604.html.
5. Pipipikaju. Detailed Planning and Design of Residential Areas (Learning Cases) [EB/OL]. (2019-04-25) [2022-04-30]. http://www.360doc.com/content/19/0425/08/63526814_831297891.shtml.
6. Zhang Nan, Luo Pinghan. Research on the Changes in Daily Life of Urban Residents Since the Founding of New China [D]. Beijing: Party School of the Central Committee of the CPC, 2020.
7. Cui Yong. Observing the Historical Changes of Urban Construction Through Beijing's Baiwanzhuang Residential Area [J]. Architecture, 2018(20): 49–51.
8. Xu Tao, Duan Wenjuan. The Century-Long Connection Between Bicycles and China - An Interview with an Assistant Researcher at the Institute of History, Shanghai Academy of Social Sciences [J]. The Earth, 2017(1): 34–36.
9. Tang Yong, Pan Haixiao. Research on the Development Characteristics and Roles of Public Bicycle Systems in Chinese Cities. Shanghai: Tongji University Press, 2015.
10. Liu Xuejie, Bai Tongzhou, Lin Xu, et al. Thoughts on Transportation Development During the 14th Five-Year Plan Period in Beijing [J]. Urban Transport of China, 2022(1): 19–24.

11. Liu Yishi. Research on the Cross-Sectional Design and Technical Standards of Urban Roads in Modern and Contemporary China - A Case Study Across the Threshold of 1949 [J]. New Architecture, 2016(5): 11–17.
12. Shu Shinan, Bian Yang, Li Shuang, et al. Development Positioning of Public Bicycles in Beijing [J]. Urban Transport of China, 2017(3): 58–62.
13. China Academy of Urban Planning & Design, China Urban Planning Society. 2017 922 Green Travel Activity Organization Manual [R]. Beijing: China Academy of Urban Planning & Design, 2017.
14. China Academy of Urban Planning & Design, China Urban Planning Society. 2021 Green Travel Activity Urban Toolkit [R]. Beijing: China Academy of Urban Planning & Design, 2021.
15. Xu Xunchu. Insights into the Development of Urban Transportation in China [J]. Urban Planning Forum, 2005(6): 15–19.
16. Song Xiaodong, Zhang Zhongxiu, Zheng Qiao, et al. Popularization Path of Electric Vehicle Battery Swapping Mode for Motor Vehicles in Beijing [J]. Urban Transport of China, 2021(5): 77–85.
17. Shao Dan, Li Han. Carbon Emission Reduction Benefits and Carbon Peaking Goals of Electrification in Urban Passenger Transport - A Case Study of Shanghai [J]. Urban Transport of China, 2021(5): 57–62.
18. Teng Wuxiao, Ma Lin, Pan Haixiao, et al. Epidemic Prevention and the Construction of Resilient Urban Transportation [J]. Urban Transport of China, 2020(4): 120–126.
19. Chen Xiaohong, Ye Jianhong, Yang Tao. Tracing the Dilemma and Exploring the Path of Prioritizing the Development of Urban Public Transport [J]. Urban Transport of China, 2013(2): 17–25.
20. Traffic Management Bureau of the Ministry of Public Security. Lu Huapu: From Smooth Traffic Project to Civilized and Smooth Traffic Improvement Action Plan - The Trajectory Towards Scientific and Modernized Traffic Management [EB/OL]. (2018-09-19) [2022-05-15]. https://baijiahao.baidu.com/s?id=1612048292656342414&wfr=spider&for=pc.

Chapter 21
Intelligent Empowerment: Enabling People to Enjoy Mobility, Facilitating Goods to Flow Freely

In the face of the rapidly changing and globally acclaimed achievements in the transportation development of our motherland, the author is filled with excitement and profound emotions. Summarizing the achievements of transportation development from the perspective of intelligence and looking forward to the future is the core theme of this chapter. To systematically and concisely introduce the development process of intelligent transportation, the chapter will first outline the current situation, followed by the development history and major advancements of intelligent transportation, then introduce the construction goals of "Transportation Power" and the main characteristics of the path to a transportation powerhouse with Chinese characteristics, and finally, provide insights into the future development of both human and material flows.

The Sacred Mission of Transportation

With the continuous and rapid development of China's economy and society, there have been significant changes in both the overall volume and structure of transportation demand. This is not only a sign of the country's prosperity but also presents new issues and challenges.

Since the reform and opening up, China's comprehensive transportation system has made significant progress. The road mileage has increased from 890,000 km in 1978 to 5.28 million kilometers in 2021, ranking second in the world. Among them, the mileage of expressways, starting from scratch, has reached 169,100 km in 2021, ranking first in the world. The railway mileage has increased from 51,700 km in 1978 to 150,000 km in 2021, ranking second in the world. The railway double-track rate is 59.5%, and the electrification rate is 73.3%. The mileage of high-speed railways has reached 40,000 km, ranking first in the world. There are 2,659 berths of 10,000 tons or above in ports, and the navigable mileage of inland waterways has reached

© China Communications Press Co., Ltd 2025

H. Lu (ed.), *A Brief History of Transportation in China*,
https://doi.org/10.1007/978-981-97-9851-3_21

127,700 km, ranking first in the world. Aviation transportation has developed rapidly, with the number of civil aviation transport airports increasing from 78 in 1978 to 248 in 2021, ranking second in the world [1, 2].

Furthermore, transportation services have also experienced significant development. By the end of 2021, the turnover volume of road freight was 6,908.77 billion ton-kilometers, road passenger turnover was 362.75 billion passenger-kilometers, railway passenger turnover was 9,567.8 billion passenger-kilometers, and waterway freight turnover was 11,557.75 billion ton-kilometers, all ranking first in the world. The railway freight turnover was 3,319.07 billion ton-kilometers, and civil aviation passenger turnover was 6,529.7 billion passenger-kilometers, both ranking second in the world [1, 2].

The achievements in urban transportation development are significant, and the comprehensive governance capacity, as well as the level of intelligence and integration in transportation, has reached new heights. However, the situation of traffic congestion remains severe, and it has become a bottleneck issue constraining urban development.

Since 1978, China's urbanization rate has increased from 17.92 to 64.72% in 2021, marking a growth of 3.6 times. During the same period, the Gross Domestic Product (GDP) has surged from 0.37 trillion yuan to 114.37 trillion yuan, a remarkable increase of 313 times. Additionally, the number of motor vehicles has risen from 1.59 million to 395 million, experiencing a 248-fold increase. The ownership of private cars (passenger vehicles) has soared from 259,000 to 262 million, an astounding growth of 1,011 times [1, 2]. With the rapid development of China's society, economy, urbanization, and motorization (Fig. 21.1), transportation infrastructure construction has been continuously implemented. The improvement of transportation equipment and the enhancement of transportation service guarantee capability have fundamentally improved the backwardness of transportation and the bottleneck constraints on economic and social development, achieving a historic leap forward.

From the initial bottleneck constraints in the early stages of reform and opening up to the initial alleviation by the end of the 20th century, and now to the current state of basic adaptation, transportation has provided strong support for the sustained and rapid development of the national economy. However, compared to leading

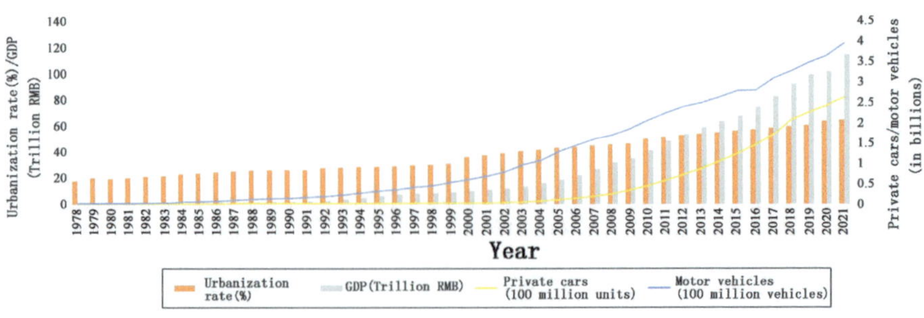

Fig. 21.1 Urbanization Rate, GDP Development, and Motorization Level in China since 1978

transportation powers worldwide, China still faces significant gaps in technology and equipment, service quality, efficiency, cost-effectiveness, safety standards, as well as international competitiveness and influence. In contrast to the requirements of high-quality development overall, there are prominent issues of imbalance and inadequacy in China's transportation development. It has yet to adequately meet the increasingly diverse, personalized, and quality-oriented travel demands of the people and fully adapt to the needs of building a socialist modernized country. While China is undoubtedly a major player in transportation, it has not yet reached the status of a transportation powerhouse.

In response to this situation, the Ministry of Transport keenly grasped the historical imperative for China, as a major player in transportation, to advance towards becoming a transportation powerhouse. In 2016, the ministry commissioned the Chinese Academy of Engineering to initiate a major consulting project titled "Strategic Research on Building a Transportation Powerhouse." The project was led by Academician Fu Zhihuan and Academician Sun Yongfu, with a research team consisting of over 100 experts, including 32 academicians from various research institutions. On January 18, 2017, the project officially commenced. After two years and three months of intensive work, the project team fulfilled its mission, and the research results, comprising three volumes titled "Strategic Research on Building a Transportation Powerhouse," were officially published. The research findings were adopted in the "Outline for Building a Transportation Powerhouse," issued by the Central Committee of the Communist Party of China and the State Council in September 2019. The construction of a transportation powerhouse was also incorporated into the report of the 19th National Congress of the Communist Party of China.

The construction goals of a "transport power" are as follows:

By 2035, to basically establish a transport power. A modernized comprehensive transportation system is basically formed, with a significant increase in people's satisfaction and a significant enhancement in the ability to support national modernization construction; having developed a well-connected express network, a comprehensive trunk network, and an extensive basic network, achieving a new level of coordinated development of urban and rural regional transportation; basically forming the "National 123 Travel Transportation Circle" (1-h commuting in urban areas, 2-h access in urban agglomerations, and 3-h coverage of major cities nationwide) and the "Global 123 Express Logistics Circle" (1-day delivery domestically, 2-day delivery to surrounding countries, and 3-day delivery to major cities globally), with convenient and smooth intermodal transportation for passengers and efficient and economical multimodal transportation for goods; significant improvement in the development level of intelligent, safe, green, and shared transportation, basic alleviation of urban traffic congestion, and basic improvement of the barrier-free travel service system; basic establishment of a transportation technology innovation system, advanced and safe transportation key equipment, excellent talent team, and favorable market environment; basic realization of modernization of transportation governance system and governance capacity; significant enhancement of international competitiveness and influence of transportation.

By the middle of the twenty-first century, to comprehensively build a transport power that satisfies the people, guarantees effectively, and ranks among the top in the world. The scale and quality of infrastructure, technological equipment, technological innovation capability, level of intelligence and greening rank among the top in the world, while the level of transportation safety, governance capacity, civilization, international competitiveness, and influence reach the international advanced level, comprehensively serving and guaranteeing the construction of a socialist modernized strong country, and providing the people with excellent transportation services.

Building a Strong Transportation Nation: China's Path

What kind of strong transportation nation to build and how to build it are the first questions that transportation professionals in the new era need to consider and answer. The project team of the Chinese Academy of Engineering and the Ministry of Transport recognize that building a strong transportation nation requires both a global perspective and strategic thinking, as well as consideration of national conditions and practical integration.

Developed countries and regions have established robust transportation systems, and their successful experiences are worth learning from and emulating. However, the choice of transportation mode for a country depends on its own resource endowment, environmental conditions, characteristics of transportation demand, and stage of development. Due to significant differences between China's national conditions and those of many developed countries, it is not feasible to simply copy the development models of other countries. It is necessary to fully consider China's population, resources, environment, industry, urbanization, and the strategic needs to achieve the second centenary goal, and to follow the path of building a transportation power with Chinese characteristics.

China's total population is large, and there are significant differences in the ecological environment and population density between the eastern and western regions. Therefore, a one-size-fits-all approach to transportation development is not feasible. The "Hu Huanyong Line" notably divides China into densely populated and sparsely populated areas: the southeastern side occupies 43.18% of the national territory, hosting 93.77% of the population and contributing to 95.70% of the GDP, with a population density of 311.7 people per square kilometer. On the northwest side, which accounts for 56.82% of the national territory, the total population represents only 6.23% of the country's population, with a population density of only 15.7 people per square kilometer [2] (see Fig. 17.2). In densely populated areas, the total demand for transportation is high, and it is relatively concentrated in terms of spatial and temporal distribution. This requires the adoption of high-capacity and efficient intensive transportation methods. Particularly as China has entered a new stage of urbanization development dominated by urban agglomerations, many large cities are transitioning towards the development pattern of urban clusters and city clusters. Currently, motor vehicle traffic volume on main routes in Chinese cities and urban clusters is rapidly increasing, and traffic flow is approaching saturation continuously, which is unsustainable. Therefore, urban clusters and large cities in densely populated areas should rely on rail transit as the backbone to build comprehensive transportation systems, which differs significantly from the approach in Western developed countries.

China's energy structure is characterized by a high reliance on coal and a relatively low consumption of oil. As a developing country, China cannot follow the path of developed countries, especially the model of "cars plus airplanes" seen in the United States. Instead, China should vigorously promote green transportation and shared mobility. With the rapid growth of the Chinese economy, total energy consumption has also been increasing rapidly, making China the world's largest energy consumer. The country's external dependence on oil has reached 71%, highlighting significant energy security concerns. The transportation, storage, and postal sectors alone account for 40% of total oil consumption. High oil prices and the non-renewable nature of petroleum dictate that a development model centered around automobiles, powered mainly by oil, does not align with the requirements of sustainable development. Therefore, there is an urgent need to promote the clean and energy-efficient transformation of transportation vehicles.

Additionally, China's land resources are limited. To meet the demand for car travel, the expansion of urban roads and the construction of facilities serving automobiles have extensively occupied the limited land resources in cities. This is particularly evident in many rapidly developing economic cities in China, which are often located in areas where arable land is concentrated. The indiscriminate expansion of cities to improve urban traffic conditions will reduce China's limited and valuable arable land area, leading to severe and irreversible consequences. Therefore, China's urban transportation development strategy and corresponding policies should encourage the development of travel modes with low land occupancy rates to avoid excessive use of limited arable land.

China's urban areas not only face the challenge of achieving energy conservation and emission reduction goals but also must tackle the problem of urban traffic congestion. Therefore, the main characteristics of the road to building a traffic-strong nation with Chinese characteristics are [2, 3]:

People-oriented: Putting the people at the center is the fundamental characteristic and important approach to realizing China's distinctive road to building a traffic-strong nation.

Regional differentiation: There is a significant difference in ecological environment and population density between the eastern and western regions. Different population densities determine distinct traffic demand characteristics, which in turn determine different traffic development models and structures.

Traffic integration: Utilizing the "window period" of transformation and upgrading to achieve the integration and high-quality development of traffic infrastructure, while promoting the integrated development of various transportation modes and services.

Green transportation: Given China's basic national conditions of large population density, scarce land resources, and energy shortages, the essence of China's characteristic green transportation lies in the predominance of intensive and low-carbon transportation modes. The greening of urban transportation in China needs to achieve the dual goals of energy conservation and emission reduction while ensuring smooth traffic flow.

Intelligent transportation: Intelligent transportation is one of the important breakthroughs in building a traffic-strong nation and is also an important means to alleviate traffic congestion, improve traffic efficiency, enhance safety levels, and provide high-quality services. The current economic development and challenges faced by transportation provide opportunities and possibilities for the rapid development of intelligent transportation in China.

Shared transportation: Sharing is a feasible approach to addressing supply shortages, and there is great potential for its application in the transportation sector. Transport vehicles, parking facilities, as well as traffic data and information, can all play a greater role through sharing models.

Equalization of basic public transportation services: Equalization of basic public transportation services needs to be provided to all citizens, ensuring that every citizen has equal access to transportation rights. This is a characteristic and advantage of China's system.

"Modern transportation+": "Modern transportation+" refers to the deep integration and development of transportation and the economy, which is the most promising development area and is expected to become a new engine of development.

Intelligent Transportation

Intelligent transportation is a vast field that encompasses urban transportation as well as integrated transportation. Overall, China's intelligent transportation sector started in the late twentieth century, with a relatively late initial stage but rapid progress in the mid-term. Currently, it has achieved significant advancements and is leading in some areas.

Remarkable Achievements in Transportation Informatization

The level of transportation informatization in China has significantly improved, with breakthroughs achieved in core technologies such as comprehensive monitoring of transportation status and electronic toll collection, which have been widely applied.

In 2004, in accordance with the requirements of the Ministry of Public Security for "uniform standards, uniform processes, and uniform software," the Traffic Management Institute of the Ministry of Public Security completed the development and nationwide promotion of core business application software for motor vehicle registration, driver management, traffic violation processing, and traffic accident handling.

On June 12, 2011, the nationwide railway online ticketing system was put into operation. Starting with ticket sales for the Beijing-Tianjin line and extending to all trains across the country by December 23 of the same year, this system greatly facilitated railway passenger travel [4].

In 2014, the Ministry of Transport promoted the nationwide application of the "Two Passenger and One Dangerous" vehicle intelligent supervision platform, gradually achieving dynamic monitoring and management of the operational status of key vehicles, thus improving the safety level of their operations.

Based on the widespread use of mobile terminals, the acquisition and integrated application of status information are widely applied. Technologies such as comprehensive traffic information collection, processing, and collaborative services are widely used in major cities nationwide, showing four trends: massive unstructured traffic big data provides more knowledge information; high-dimensional heterogeneous traffic big data contains more regularities; cross-analysis of multi-source traffic big data identifies more features; and intelligent connected vehicle traffic big data supports more applications.

By the end of September 2015, the nationwide Electronic Toll Collection (ETC) system for highways achieved full interconnection [5]. ETC interconnection for highways was implemented in all 31 provinces, autonomous regions, and municipalities, essentially enabling nationwide ETC passage. Users could obtain an ETC pass card in any province and travel smoothly across the country. As of the end of 2020, a total of 24,588 sets of ETC gantry systems were installed nationwide, along with the completion of modifications for 48,211 ETC lanes and 11,401 sets of highway non-stop weighing detection systems. Starting from January 1, 2020, all 487 provincial toll stations at highway borders in 29 interconnected provinces were abolished, significantly improving the travel experience for the people and contributing to energy conservation, emission reduction, and cost-effectiveness. As of 2021, there were approximately 257 million ETC users nationwide, accounting for about 85% of the total vehicle ownership (3.02 billion by the end of 2021), making it the world's largest ETC operation network with a networked mileage of 142,600 km.

In 2016, in accordance with the notice issued by the General Office of the Ministry of Transport regarding the implementation of the inter-provincial networking of road passenger transportation ticketing systems, the construction of provincial and ministerial-level road passenger transportation ticketing systems was systematically advanced, following the steps of first providing intra-provincial services and then achieving nationwide interconnection for long-distance highway passenger transportation. This approach ensured the interconnected operation of the provincial and ministerial-level systems, guaranteeing the overall operation of the nationwide road passenger transportation ticketing system. It reduced the hidden costs for the public and operational costs for enterprises, facilitated public travel, and enhanced the level of road transportation services [6].

The Significantly Enhanced Level of Integrated Application of Intelligent Technologies in Integrated Transportation.

Intelligent transportation has played a prominent role in ensuring transportation during major international events. Intelligent transportation technology was widely integrated and applied during the transportation support for the Beijing Olympic Games, Shanghai World Expo, and Guangzhou Asian Games.

The intelligent transportation integrated system for the Beijing Olympics included ten major intelligent transportation management systems, ensuring smooth traffic flow for the 29th Olympic Games. These systems comprised modern traffic command and dispatch systems, automatic detection and alarm systems for traffic incidents, automatic recognition of odd–even license plate numbers, comprehensive monitoring systems with digital high-definition capabilities for the Olympic central area, closed-loop management for digital traffic law enforcement systems, intelligent regional traffic signal systems, flexible control systems for expressway traffic, traffic signal control systems prioritizing public transportation, continuous guidance from large roadside variable message signs, and real-time traffic condition prediction and forecasting systems. These ten intelligent transportation management systems greatly improved the level of scientific traffic management, providing strong technical support and guarantees for ensuring road traffic safety, orderliness, and smoothness, and achieving a safe and successful Olympic Games.

The intelligent transportation technology integrated system for the Shanghai World Expo was based on the strategic principles of "time–space diversion, impact reduction; regional coordination, balanced demand; multi-level diversion, intensive conversion; dedicated line priority, rational control". It established a comprehensive integrated application and service system for intelligent transportation, including travel services, traffic monitoring, smart public transportation, decision support, electronic toll collection, emergency rescue, and other functions tailored to the specific needs of the World Expo users. The system realized real-time interconnection among the Expo Priority Traffic Coordination Management System, urban transportation hub traffic information service system, and comprehensive traffic information platform. It also established key technology demonstration projects such as the Expo Comprehensive Traffic Emergency Event Management System, providing technical support for realizing priority traffic management during the World Expo.

The intelligent transportation system for the Guangzhou Asian Games was primarily based on the informationization core of Asian Games security, with computer communication networks and intelligent command and control management as the foundation. It established an integrated intelligent road traffic management system incorporating advanced technologies, aiming to improve the operational order of the road network, enhance the effective utilization rate and traffic capacity of roads, reduce traffic congestion, accident frequency, exhaust emissions, and travel delays caused by congestion and accidents. The system achieved the intelligence of road traffic management by expanding the capacity of command centers and sub-control centers monitoring systems, extending the storage time of video signals, optimizing system functions, and ensuring that remote management, hierarchical management, and multi-user management functions are advanced, reliable, clear, complete, and efficient.

The quality of travel services continues to improve. As of 2021, electronic ticketing services for road passenger transport have been implemented in over 1600 bus stations at or above the county level nationwide. Additionally, 318 prefecture-level cities have achieved interconnection and interoperability of transportation smart cards [7].

In 2011, Beijing established the first comprehensive Traffic Operation Coordination Center (TOCC) in the country. Subsequently, cities such as Shanghai, Guangzhou, Shenzhen, and Chengdu have successively built TOCCs. These centers enable intelligent monitoring, real-time coordination, and command dispatch of comprehensive traffic operations. They facilitate integrated coordination and management of urban road traffic, rail transit, civil aviation, as well as urban parking and public transportation services.

Significant progress has been made in the field of air traffic management and civil aviation intelligence. China began planning the development of a new generation air traffic management system in 2007, with the goal of completing it by 2025 [8]. Substantial technological breakthroughs have been achieved in the new generation air traffic management system, including the establishment of core technical frameworks for high-precision aviation navigation, collaborative aviation surveillance, air traffic control operations, and civil aviation air traffic management information service platforms. These breakthroughs have reached the international advanced level for the same period. The new generation air traffic service platform has been successfully applied in air traffic control and airline departments, with more than 100 airports adopting Air-Ground Collaborative Decision Making (A-CDM). Airports such as Shanghai Hongqiao International Airport and Beijing Daxing International Airport have comprehensively promoted intelligent management and services, constructing a modern three-dimensional transportation system. At Beijing Daxing International Airport, intelligent technology covers terminals, airfields, and supports 19 platforms and 68 systems, providing comprehensive intelligent services for airport operations and passenger travel [9].

Significant progress has been made in intelligent shipping, smart ports, and ship-shore coordination. In 2020, China's inland waterway mileage reached 127,700 km, with 22,142 berths in ports dedicated to production, and a total of 126,800 water transport vessels nationwide [10]. The smartification of waterway transportation is an integral part of China's intelligent transportation development. Innovations in green intelligent technology for inland waterway vessels, key technologies and applications for high-grade inland waterway navigation systems, and fully automated terminals such as the Yangshan Port are among the innovative achievements in China's water transportation smartification field. In recent years, there has been increased attention to the construction of smart waterways and ports, with emphasis on enhancing shore-based monitoring and support facilities and focusing on ship intelligence navigation and control based on shore-based coordination [9].

The intelligent transportation hub is the key to improving travel efficiency and transportation service levels, and achieving convenient and efficient travel. From the national to the local level, there is a great emphasis on the intelligent construction of integrated transportation hubs. Beijing Daxing International Airport represents the intelligent integrated transportation hub in China. The Ministry of Transport explicitly proposes in the "Action Plan for the Construction of New Infrastructure in the Transportation Sector (2021–2025)" [11] to promote the intelligent construction of integrated passenger transportation hubs and freight transportation hubs (logistics parks). This includes establishing intelligent integrated passenger transportation hubs dominated by aviation in cities such as Jinan and Changsha, realizing functions such as traffic operation monitoring, intelligent analysis decision-making, traffic information services, and integrated transportation coordination. Additionally, there will be the creation of intelligent integrated passenger transportation hubs dominated by railways in cities such as Chongqing, achieving functions such as station linkage, passenger flow control, one-code access, intelligent parking, and smart customer service. Furthermore, efforts will be made to promote the construction of intelligent freight transportation hubs (logistics parks) in cities such as Beijing and Hebei, advancing the intelligent upgrading of facilities and equipment for warehousing, transportation, and distribution, to achieve services such as vehicle-cargo matching, intelligent scheduling, collaborative distribution, and multimodal transportation.

Intelligent Transportation Provides Convenient Services and Safety Guarantees for Public Travel.

Inter-city Public Transit Card has been implemented across more than 100 cities, with the peak daily ticketing service users reaching 13 million on 12,306, the national railway ticketing network. Inter-provincial long-distance passenger transport ticketing and nationwide highway ETC network have provided convenient and efficient travel services to the public. The scale development of mobile terminals has led to the widespread use of nationwide mobile and vehicle navigation, with approximately 900 million users as of 2021. Various transportation information and ticketing service apps have been widely adopted.

The rapid development of new models of shared transportation services has been witnessed. In 2020, there were 400 companies operating shared new energy vehicles in 60 cities nationwide, with a fleet of over 70,000 vehicles. The registered users of shared bicycles exceeded 400 million [7].

The intelligent connected vehicles (ICVs) and vehicle–road cooperative (VRC) technologies have attracted widespread attention and emphasis. Through a series of efforts, a system of intelligent VRC technology based on the Chinese context has been established and improved. Automated driving technologies based on VRC have been successfully applied in large-scale mines and logistics parks. The Ministry of Industry and Information Technology, the Ministry of Public Security, and the Jiangsu Provincial People's Government have jointly established the "National Intelligent Transportation Comprehensive Test Base". Additionally, the Ministry of Industry and Information Technology has promoted the construction of six "demonstration bases for intelligent vehicles and smart transportation applications based on broadband mobile Internet". As of May 2021, the Ministry of Transport has identified seven closed-loop test bases for automated driving, and more than 60 intelligent connected vehicle and automated driving test bases have been established nationwide. The integration and development of intelligent vehicles and intelligent transportation continue to expand the industry and market [9, 12].

The "National Road Traffic Safety Science and Technology Action Plan" is jointly led by the Ministry of Science and Technology, the Ministry of Public Security, and the Ministry of Transport. It was launched in 2008 with the first-phase technology support projects, mainly including seven demonstration projects such as information sharing platforms, safety assurance for mountainous roads, safety control for expressways, safe operation of commercial vehicles, enhancement of traffic safety behaviors for the general public, road network safety situational monitoring, and traffic safety law enforcement. The second phase primarily focuses on driving safety on expressways and lower-grade roads, featuring six demonstration projects including the improvement of expressway driving conditions, safety networking and control for expressways, safety prevention and control for lower-grade roads, intelligent monitoring and safety service guarantee for the operation status of expressway networks, integrated monitoring of traffic behaviors and informatized law enforcement for expressways, and emergency response and rescue for major emergencies on expressways. These demonstration projects have played a significant role in promoting the development of intelligent transportation in China [9].

The Rapid Development of Urban Intelligent Transportation

The rapid development of urban intelligent transportation management systems.

The development history of intelligent urban traffic management in China can be traced back to the signal control system construction in Beijing in the 1970s. Before that, signal control was mainly manual, relying on empirical control schemes and requiring a large amount of manual labor, which lagged far behind the international advanced level. With the development of communication technology, control technology, and the accumulation of international experience, the field of urban traffic signal control has initially met the conditions for automation and intelligence. In 1976, the Beijing Municipal Traffic Management Bureau introduced the SCOOT (Split Cycle Offset Optimizing Technique) system from the United Kingdom for the first time, as part of a World Bank project. Since then, the field of intelligent traffic management has continued to develop rapidly, from signal control optimization to the integration of the "situation awareness and dynamic dispatching" intelligent police system, the visualization of traffic command, the convenience of traffic management public services and information guidance, and the preliminary development of intelligent analysis and judgment. The development speed is fast, the innovation is strong, and it has reached the world's advanced level.

In response to China's actual situation, the field of urban intelligent transportation has initially formed a system of intelligent traffic management with Chinese characteristics, primarily manifested in the following aspects:

Significant breakthroughs have been achieved in the analysis and integration of multi-source heterogeneous data in urban transportation, distributed heterogeneous multi-system integration technology, and integrated command and dispatch technology based on Geographic Information Systems (GIS). A framework of intelligent traffic management system has been established, centered around core components such as centers, platforms, and application systems. These systems are highly integrated with application subsystems including video surveillance, individual positioning, 122 emergency response, Global Positioning System (GPS) vehicle positioning, signal control, and cluster communication. This integration enhances the operational capabilities of intelligent traffic management. Moreover, modernized traffic command and control centers have been established, equipped with functions such as command dispatch, traffic control, comprehensive monitoring, and information services [13].

With the development of informatization and intelligence, urban traffic management models have gradually shifted from surface patrols to digital police stations and online patrols. Based on this foundation, new modes of precise governance and rapid implementation of road management and traffic command have been established. In this new policing system, the analysis and judgment of traffic conditions, as well as the automatic generation of countermeasure plans, have become core objectives and tasks. Hangzhou city began setting up digital police stations as early as 1998.

While achieving intelligent control and visualized command, traffic management departments have comprehensively expanded smart-based convenience services. Shanghai, focusing on addressing people's livelihood demands and shortcomings, has made significant achievements in promoting the "vehicle management at home" traffic management service model, the "police-medical-post" service model, integration and sharing of government information resources, establishing comprehensive monitoring platforms for in-process and post-process supervision, establishing a one-stop "Internet" public service platform, and intelligent illegal evidence collection. Since 2018, Wuhan has advocated the concept of "letting data travel more and citizens travel less," focusing on the establishment of a comprehensive intelligent traffic management system integrating traffic information collection, big data analysis and utilization, precise service management, convenient traffic information services, and traffic signal control around the big data center. Using smart transportation as a lever, Wuhan has integrated traffic data resources from various departments, established the Wuhan Smart Transportation "big data" center, and leveraged the "Internet Plus" approach to comprehensively promote the construction of "smart traffic management" from the perspectives of smart governance, smart travel, smart emergency response, smart supervision, and smart decision-making. This transformation leads traffic management from the traditional government-led model to a combination of social governance and smart modernization, achieving a transition from manual to intelligent management approaches [14].

Improving the automation and precision of analysis and judgment, as well as the efficiency and real-time nature of on-site management, are crucial challenges facing traffic management. Since 2016, Shenzhen has been exploring and establishing a new police management system with "the most powerful brain" and "the fastest legs" throughout the entire process, characterized by intelligence and high efficiency. They have achieved significant results by standardizing data collection, practicalizing data analysis, simulating signal control, precise management of duty, and providing convenient traffic services, tailored to the city's needs, creating a city traffic brain. Leveraging big data for precise duty and "Internet Plus" for information services have been notably effective. They pioneered the use of facial recognition systems for law enforcement nationwide. Through continuous research and development of various new technologies, they have created "one police officer, multiple capabilities" law enforcement devices, expanding the scope of capturing violations such as running red lights and speeding. Additionally, police-enterprise collaboration has driven technological innovation, establishing joint innovation laboratories for signal control, video analysis, traffic simulation, and more, propelling Shenzhen to become a leader in the field of traffic management [14].

The means of detection are diverse, and determining the most effective detection method and how to achieve intelligent traffic management tailored to the data characteristics of a city are valuable topics. In this regard, Chongqing has made significant explorations. They have utilized Radio Frequency Identification (RFID) technology as a fundamental means of information collection, constructing an intelligent traffic management system that includes various components such as traffic information collection systems, traffic incident detection systems, traffic violation monitoring and recording systems, traffic checkpoint systems, traffic surveillance video systems, traffic signal intelligent control systems, traffic information dissemination systems, and communication network systems. They have been at the forefront nationwide in the application of RFID with electronic license plates as information carriers.

Overall, China's intelligent traffic management systems are still in the process of development and refinement. While there have been initial strides in areas such as signal optimization control, visualized traffic command, integrated analysis and judgment, and automated convenience services, there is still a long way to go.

The rapid development of smart urban public transportation.

Several cities in China, including Beijing and Shanghai, initiated the construction and implementation of intelligent public transportation systems since the 1990s. The focus was on the implementation of location systems for buses and the management of bus dispatching. This enabled satellite positioning of operating vehicles and intelligent scheduling for departure. Intelligent dispatch command centers could monitor the real-time operation of vehicles on various routes, while electronic signs at bus stops could simultaneously display arrival information for buses.

From 2004 to 2015, the national government successively introduced a series of policies to promote the development of priority public transportation, which propelled the advancement of intelligent public transportation in various cities. In 2012, the Ministry of Transport initiated the "Urban Public Transportation Intelligent Application Demonstration Project" in the first batch of 10 cities. This project established a technical system, including equipment in bus factories, communication protocols for transmission, application systems, etc. It also established relevant standard systems and facilitated the conversion of industry standards and national standards.

Today, many cities across the country have established intelligent dispatch command centers for their public transportation systems, enabling intelligent dispatching management and services for buses. Cities like Beijing, Guangzhou, and Shanghai have tailored their approaches to develop high-capacity Bus Rapid

Transit (BRT) systems, with support from national science and technology initiatives facilitating the widespread application of BRT system equipment and technology. Initiatives such as green buses and responsive public transit are actively promoted, and the development of intelligent technologies based on Artificial Intelligence (AI) and big data for passenger services and ensuring safe operations is rapidly advancing.

Since 2016, with the development of new technologies, intelligent public transportation has been rapidly advancing. Many cities across the country have initiated the construction of integrated intelligent public transportation platforms, enabling various convenient payment methods and statistical settlement, office automation, intelligent bus dispatching, real-time information services, application of mobile smart terminals for services and convenient query systems, station video surveillance, and more. The development of intelligent public transportation is moving towards electrification, efficiency, and cleanliness of the power system, as well as system integration, networking, and coordination. Moreover, it reflects trends towards experiential, shared, and comprehensive user consumption demands, and quantification, precision, and scientific decision-making in management.

Continuously Evolving Vehicle–Road Collaboration and Autonomous Driving Technology.

During the "Five-Year Plan" and "Eleventh Five-Year Plan" periods, researchers conducted extensive research in areas such as automotive assisted driving, in-vehicle navigation devices, driver state recognition, vehicle operation safety status monitoring and early warning, traffic information collection, and vehicle self-organizing networks. They have essentially mastered core technologies such as common intelligent vehicle technologies, vehicle operational status identification, high-precision navigation and map matching, and high-reliability information collection and interaction. Some universities and research institutions have conducted special technical research focusing on vehicle–road collaboration, including the National Key Research and Development Program "Intelligent Highway Technology Tracking" and the National 863 Program project "Research on Intelligent Road System Information Structure and Environmental Perception and Reconstruction Technology" and "Research on Road Intelligent Identification and Perception Technology Based on Vehicle–Road Coordination," achieving phased results.

During the "Twelfth Five-Year Plan" period, the Ministry of Science and Technology initiated the "Key Technologies for Intelligent Vehicle–Road Cooperative Systems" project, leading to the formation of the basic framework for intelligent vehicle–road cooperative systems in China. In the "Thirteenth Five-Year Plan" period, the National Key R&D Program provided extensive support for research in the fields of Internet of Vehicles (IoV), vehicle–road cooperation, and autonomous driving. This support resulted in a wealth of research outcomes, promoting the construction of IoV and vehicle–road cooperation systems, as well as the demonstration, application, and industrialization of autonomous driving technologies.

In recent years, China has successively launched the construction of demonstration zones and testing areas for autonomous driving applications. An overview of the construction of some demonstration zones is shown in Table 21.1.

Table 21.1 Overview of the construction of some autonomous driving application demonstration zones in China [15]

Demonstration zone	City	Description
Wuxi national intelligent transportation comprehensive test base and IoV application demonstration area	Wuxi	Constructed jointly by the Traffic Management Science Research Institute of the Ministry of Public Security, enterprises, and local governments. It is planned in three stages: open road testing research, urban-scale demonstration applications, and the establishment of a connected IoV industry base. It covers multiple areas around the comprehensive test base and deploys new business applications based on LTE-V2X
National Intelligent Connected vehicle (Shanghai) pilot demonstration zone	Shanghai	Constructed by the Shanghai International Automobile City with the goal of serving intelligent vehicles and V2X communication. In the second phase of construction, it has deployed GPS differential base stations, LTE-V2X communication base stations, roadside units, intelligent traffic lights, various cameras, and newly built 13 LTE-V2X base stations. It has established a complete set of C-V2X Server data center platform, providing research and testing support for C-V2X vehicle–road communication applications for vehicle and component companies
National intelligent vehicles and smart transportation (Beijing–Hebei)	Beijing	Led by the Beijing Innovation Center for Mobility Intelligent (BICMI). In September 2017, the demonstration zone officially launched intelligent connected vehicle tidal test road services. It has an open road length of about 12 km and has completed various roadside traffic facility transformations, realizing applications such as pedestrian collision warning. In February 2018, the demonstration zone's closed test site for autonomous vehicles—Haidian Base—was officially put into use. The test site covers urban and rural complex road environments in the Beijing-Tianjin-Hebei region, supports the construction of hundreds of static and dynamic typical traffic scenarios, and has deployed V2X equipment and systems to support networked driving research and testing
National Intelligent Connected Vehicle Application (Northern) Demonstration Base	Changchun	Led by Qiming Information Technology Co., Ltd., the first phase of the project has been completed, with 11 major scenario and 233 minor scenario testing and demonstration functions. It has implemented V2X applications such as information prompts and safety warnings for FAW's independent brand intelligent connected vehicles and has been designated as a China-Russia V2X joint testing and application base

(continued)

Table 21.1 (continued)

Demonstration zone	City	Description
Zhejiang intelligent vehicles and smart transportation demonstration zone	Wuzhen	Established in the core areas of Yunqi Town in Hangzhou and Wuzhen in Tongxiang City, it integrates intelligent vehicles, smart transportation, and broadband mobile Internet. Yunqi Town has deployed 34 LTE-V2X roadside stations and built various interactive scenarios. The Tongxiang area has constructed an integrated traffic big data and information service model, established a connected vehicle comprehensive operation platform, and completed research and testing on various advanced driving assistance and autonomous driving technologies
Chongqing (i-VISTA) intelligent vehicle integrated system test demonstration area	Chongqing	Led by China Automotive Engineering Research Institute Co., Ltd., highlighting the unique mountain-city road traffic and communication characteristics. The demonstration area has completed the transformation of 9.6 km of open roads to V2X connectivity and the installation of V2X onboard equipment, achieving V2X test-related monitoring, statistical services, and possessing V2X testing capabilities. The second phase of the project will complete the construction and transformation of the comprehensive test area of Chongqing Western Automobile Test Field to address V2X-related system and technology testing issues in highway and rural road environments. The third phase will demonstrate large-scale intelligent vehicles and smart transportation applications in complex open traffic scenarios
Hubei intelligent vehicles and smart transportation demonstration zone	Wuhan	Constructed by Wuhan Economic and Technological Development Zone, focusing on building "autonomous driving" smart towns, with plans to transition from a closed test area to a semi-open demonstration application area to an urban traffic open environment in three phases over five years, conducting multiple application demonstrations such as intelligent vehicle testing evaluation, intelligent vehicle autonomous driving, smart transportation, and smart towns

(continued)

Table 21.1 (continued)

Demonstration zone	City	Description
Enterprise or local self-built/commercial operation projects	–	Domestic industry organizations such as automobile manufacturers, equipment vendors, telecommunications operators, and universities rely on their own advantages to collaborate with local governments to promote the construction of V2X demonstration roads: Chery Automobile has completed the construction of V2X demonstration roads in Anhui Province; Jianghuai Automobile has achieved the transformation to vehicle networking in its Hefei park; Changan University has established an autonomous driving test base in its comprehensive performance test field

The Intelligentization of Integrated Transportation [13, 14]

The development focus of intelligentization in integrated transportation is reflected in the intelligentization of various transportation modes and the construction of intelligent service, management, and decision support systems across transportation modes.

The intelligentization construction of highways in China is led by expressways and is reflected in the intelligentization of facilities, services, decision-making, and management, driving the overall intelligentization development of highways. Currently, the basic database of the highway transportation industry has been established, and the data collection rate of important traffic infrastructure and key transportation equipment operation status has steadily increased. Over 40% of the key sections of the national and provincial trunk highway network, as well as major bridges and tunnels, have basically achieved dynamic monitoring of operational status. The informationization system for infrastructure operation management has been basically completed, basic information resources are gradually improving, information sharing is comprehensively promoted, public information services are fully launched, and the conditions for informationization development have significantly improved.

With the rapid development of high-speed railways and the deepening of railway informatization construction, China's railways are advancing the construction of modern comprehensive passenger transportation hubs based on the concepts of efficient transfers, intelligent convenience, multi-dimensional development, and integration with urban areas. According to these requirements, railway stations are being developed as central hubs, seamlessly integrated with other modes of transportation, to achieve convenient transfers and integrated transportation services. The level of "one-stop" service has significantly improved, with the establishment of the China Railway Customer Service Center website and the China Railway 12,306 Internet ticketing system. The proportion of online ticket sales has reached 70%, and various ticket sales channels such as Internet/mobile, telephone, station and agency windows, and automatic ticket machines have been established. Smart navigation services have been piloted at major stations such as Beijing South Station, Jinan West Station, and Nanjing South Station. In terms of freight transportation, the China Railway 95,306 website has been launched and operated, and the construction of the freight e-commerce system has been accelerated, with over 70% of freight business conducted online. A railway electronic payment platform has been established to enable online payment for railway ticketing and freight services. The integrated transport information platforms at both the national and bureau levels have been completed, facilitating resource sharing among different information systems of the railway corporation, railway bureaus, and stations. Meanwhile, unmanned driving has been implemented on some metro lines in cities like Beijing.

In terms of water transportation, China has prioritized the construction of electronic ports, the development of port container multimodal transportation information service systems, and the promotion of ship networking technology applications. The technological level in areas such as ship traffic management and port electronic data exchange has been aligned with international standards. Major port enterprises have established intelligent production scheduling and command systems, with the level of informationization in coastal large container ports reaching world-class standards. New information technologies such as digital waterways, waterborne Electronic Toll Collection (ETC), "sensory waterways," and electronic supervision of key vessels have been vigorously promoted. The digital waterway of the Yangtze River main channel has provided public services to society. In Jiangsu Province, a smart lock system for inland waterway vessels (waterborne ETC) has been implemented, focusing on areas with numerous ship locks. The construction of "sensory waterways" has been carried out in the Wuxi section of the Beijing-Hangzhou Grand Canal, enabling the all-weather, all-area, and all-process monitoring of pilot waterways. Leveraging the unified management system for ports, waterways, maritime affairs, and ship inspections, Zhejiang Province has made efforts to advance electronic supervision of key vessels and has begun promoting electronic visa services for certain non-stop vessels. In the upper reaches of the Yangtze River in Chongqing and the middle reaches in Wuhan, water transportation centers have been developed rapidly with the support of information technology. This has effectively promoted the further concentration of shipping service elements, providing more professional and efficient freight transactions, ship transactions, talent exchanges, freight rate information, and other services for the inland water transportation market, significantly enhancing shipping service capabilities.

China's civil aviation business information system is at the forefront of the world. Technologies such as RFID and GPS are widely used in civil aviation transportation and logistics distribution. Advanced avionics technologies such as low-altitude data link technology, ground network interconnection protocols (IP), airborne navigation, airborne collision avoidance, airport map and traffic information display are given priority development. However, China has yet to develop systematic solutions for runway performance perception, foreign objects on runways, runway incursions, and flight area collision prevention. There is a lack of comprehensive application and demonstration platforms for intelligent flight area facility management.

Traffic operation supervision and coordination require an intelligent traffic management system that involves cross-departmental collaboration. With the rapid development of urban clusters, the integrated transportation systems of these clusters have become vast and complex networks that are constantly evolving. Current research is transitioning from static analysis to dynamic identification. By leveraging real-time dynamic analysis techniques, combined with online data and situational inference systems in intelligent traffic environments, real-time intelligent analysis of the operational situation of integrated transportation systems in urban clusters can be achieved.

New Formats of Intelligent Traffic Services

With the development of new technologies, new models and formats of intelligent traffic services continue to emerge.

Intelligent shared bikes, distinct from traditional public bicycles, feature characteristics such as no need for inflation, puncture resistance, and satellite positioning. Users do not need to obtain a card; they simply download the app on their mobile phones, register an account, and pay a deposit. They can then search for and unlock bikes by scanning a QR code, and return them without the need for fixed parking stations. Currently, next-generation Internet of Things (IoT), cloud computing, and big data technologies connect shared bikes with user apps on smartphones, enabling precise location tracking of vehicles and real-time monitoring of their status, thus facilitating fine-grained management of each bike. In 2020, there were approximately 20 million shared bikes in China, with around 1.5 million of them being electric bikes.

In recent years, the development of Mobility as a Service (MaaS) has been rapid. Cities like Beijing and Shanghai have launched one-stop public transportation service platforms, allowing travelers to access comprehensive public transportation services through these platforms. It is anticipated that in the future, MaaS will enable personalized travel planning based on users' travel needs. Utilizing data, operators can improve services, allowing users to freely plan personalized travel from any origin to any destination, using combinations of different transportation modes. Ultimately, this will result in the optimal experience, planning, and cost for users.

Currently, some enterprises and organizations are dedicated to providing integrated solutions for travelers, including accommodation, transportation (solutions, information, and transactions), sightseeing (tickets), entertainment, shopping, and dining services. They integrate services such as booking flights, train tickets, bus tickets, ferry tickets, and arranging for scheduled taxis, ride-sharing, chartering, and car rentals, offering customers one-stop services. In addition to these services, facial recognition big data can be utilized to collaborate with scenic spots to introduce "facial recognition" turnstiles, enabling customers to enter the park quickly without waiting in line.

Beidou System Facilitates the Development of Transportation and Logistics

The BeiDou Satellite Navigation System (referred to as the "BeiDou System") is a global satellite navigation system independently developed and operated by China to meet the needs of national security and economic and social development. It is a crucial national spatial–temporal information infrastructure that provides all-weather, all-time, high-precision positioning, navigation, and timing services to users worldwide. The BeiDou System began construction in the late twentieth century, following a "three-step" development roadmap of regional first, global later, active first, passive later implementation, forming a distinctive development path with prominent regional coverage and global orientation. Especially since the 18th National Congress of the Communist Party of China, within two and a half years, 18 launches with 30 satellites have achieved a high-density network, setting a record of success with all 55 satellites launched into orbit since the first launch in 2000.

Today, the BeiDou system provides positioning, navigation, timing, short message communication, international search and rescue, and other services globally. In addition to these services, it also offers satellite-based augmentation, ground-based augmentation, precise point positioning, regional short message communication, and other services in China and surrounding regions. It has become deeply integrated into human life, comprehensively integrated into economic and social development, and will continue to expand its influence.

As an essential component of social livelihoods, transportation is not only the largest civil industry for BeiDou system applications but also the vanguard of its implementation. It has laid the foundation for pioneering efforts, accumulated experience in large-scale applications, and demonstrated its effectiveness. Leveraging the functionalities of the BeiDou system such as positioning, navigation, and timing, the transportation industry continues to improve its information infrastructure, accelerate the construction of integrated transportation networks, promote high-quality development, and expedite the construction of a transportation powerhouse.

The United States' Global Positioning System (GPS), Russia's Global Navigation Satellite System (GLONASS), the European Union's Galileo satellite navigation system, along with China's BeiDou system, constitute the four major satellite navigation systems in the world today. BeiDou, the latest to be developed among them, has made significant technological strides despite its later inception. With the rapid growth of China's scientific and technological capabilities and economic prowess, the BeiDou system has transitioned from an experimental system to a regional one and finally completed its global system deployment in 2020. Presently, the positioning performance of the BeiDou system is on par with that of GPS.

The transportation industry, as the "vanguard" of national economic development, has enormous demand for positioning services across various sectors. Deeply integrating the BeiDou system with various sectors of the industry is an important means to promote the informatization and modernization of transportation, and it holds significant significance for establishing a modern transportation system that is smooth, efficient, safe, and environmentally friendly. The transportation industry has always been at the forefront of the construction and application of the BeiDou system. By combining industry-specific needs with the development of the BeiDou system, it has initiated a series of application demonstration projects and promoted the internationalization of the BeiDou system using its industry advantages. Through industry efforts to support the construction and application of the BeiDou system, the spirit of the Central Committee of the Communist Party of China is implemented in practical actions. According to data released by the Ministry of Transport, as of the second quarter of 2021, there are over 7 million key operational vehicles nationwide equipped with BeiDou system terminals, nearly 40,000 vehicles in postal and express delivery trunk lines, over 400 general aviation aircraft, and more than 1,800 government vessels in the transportation system. Additionally, there are over 13,000 waterway navigation and aids to navigation facilities, and more than 200 BeiDou ground augmentation stations have been constructed along the coastal areas and the Yangtze River.

During the "BeiDou-1" experimental system phase, the Ministry of Transport actively participated in the construction of the BeiDou system. It collaborated closely with the BeiDou system construction units to promote testing and verification of the system's service performance across various sectors of the industry. With the initiation of the "BeiDou-2" system construction, the Ministry of Transport participated in the "BeiDou-2 Civil Market Development and Industrialization Project" organized by relevant units. This initiative aimed to explore and validate civil applications of the BeiDou system, making the transportation industry one of the earliest civilian sectors to initiate practical applications of the BeiDou system.

During the "12th Five-Year Plan" period, the transportation industry took the lead in applying the BeiDou system by constructing BeiDou demonstration projects, providing an experiential foundation for early adoption. In 2011, the Ministry of Transport, in conjunction with the former General Equipment Department, initiated the first civil BeiDou application demonstration project, the "Key Transportation Process Monitoring and Management Service Demonstration System Project." This project was the first to commence and the first to be accepted, earning the reputation of being a "demonstration within demonstrations." In 2013, the Ministry of Transport participated in the construction of the national BeiDou ground-based augmentation system, establishing an industry-specific BeiDou augmentation data processing system and a batch of BeiDou augmentation reference stations within the industry. Simultaneously, effective validation of BeiDou's high-precision positioning and navigation services within the industry was conducted. In 2014, the Ministry of Transport, in collaboration with the former General Equipment Department, launched the transportation industry's second BeiDou application demonstration project, the "China Maritime Search and Rescue Information System Demonstration Project Based on BeiDou," marking the expansion of BeiDou system applications from the road transportation sector to the maritime domain.

During the "13th Five-Year Plan" period, which was a crucial period for the global networking construction of the BeiDou system, it was also an opportune time for the transportation industry to undergo transformation and development.

Entering the "13th Five-Year Plan" period, the application of the BeiDou system in the transportation industry entered a new stage. During this period, the organizational structure of the BeiDou system leadership group of the Ministry of Transport was further improved, and progress was made in the internationalization of the BeiDou system. The Ministry of Transport issued policy documents such as the "Guiding Opinions on Promoting the Application of the BeiDou Satellite Navigation System in the Industry," and the ministry's leaders issued directives to "achieve full coverage, substitutability, and security of the BeiDou system in the transportation industry first," establishing the overall direction for the application of the BeiDou system in the industry. During this period, the transportation sector continued to expand the scale of industry applications through demonstration projects while gradually building the industry's BeiDou service support capabilities.

The "14th Five-Year Plan" period is an important strategic opportunity, during which the major project of "BeiDou Industrialization" will be implemented.

The Future Passenger Transport System: Human-Centric Mobility

Based on intelligent transportation technology, establish efficient, convenient, one-stop intelligent passenger transportation services (Fig. 21.2), significantly improve the level of passenger transportation services, achieve intelligence in various aspects such as public transportation services, hub transfers, terminal travel, and responsive customized services, thereby realizing door-to-door one-stop efficient and convenient passenger transportation services.

One-Stop Travel Service Platform

A one-stop travel service platform, featuring functions such as travel route planning, mode of transportation selection, travel support services, point-to-point travel services, and on-demand travel services, provides intensive, personalized, diversified, end-to-end, and appointment-based intelligent services. Before the trip, users are provided with comprehensive travel planning services, offering travel-related content including hotel services based on customer needs, providing a "one-stop" solution; during the trip, real-time information is available to dynamically optimize and arrange the itinerary; after the trip, users can easily handle reimbursement and other follow-up procedures.

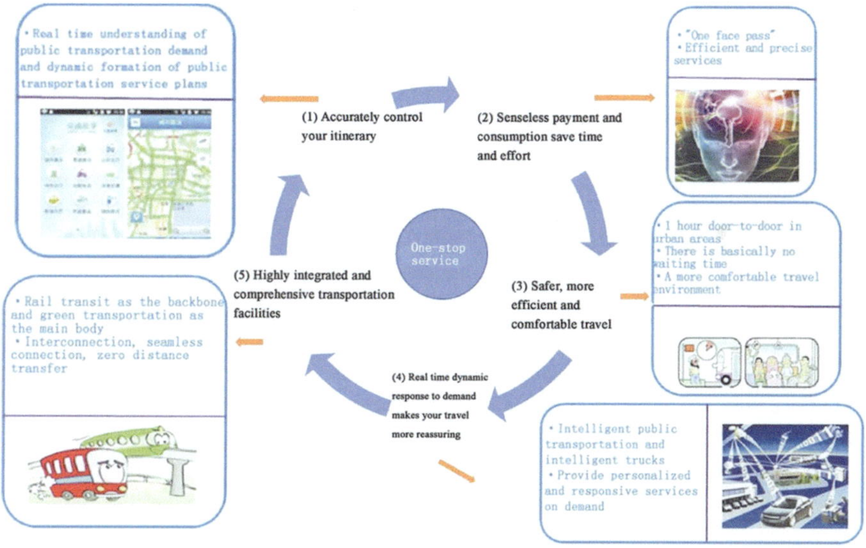

Fig. 21.2 One-stop intelligent passenger transportation service

Intelligent Public Transit System

Building an advanced intelligent public transit system involves gradually improving functions such as intelligent dispatching, public transit operation management, safety management, information services, performance management, as well as features like public transit priority control, intelligent demand-responsive transit services, and diversified and personalized public transit services.

Promoting diversified intelligent public transit services includes implementing intelligent payment for boarding and alighting, transitioning from card-based and mobile payments to contactless payments. Additionally, it involves providing precise predictions of vehicle arrival and stop times based on mobile terminals and accurate forecasts of vehicle locations, along with the implementation of intelligent virtual reality (VR) onboard systems.

Promoting intelligent public transit infrastructure and vehicles involves deploying intelligent interactive bus stops and VR human-vehicle interaction systems. With the advancement of autonomous driving technology, certain scenarios allow for the provision of autonomous driving services by buses.

Intelligent Parking System

Establishing an intelligent parking system to achieve smart guidance, smart parking, contactless payment, intelligent comprehensive management, and services.

Setting up an intelligent parking management platform to achieve unified management and information sharing of parking lots and parking spaces within a designated area. Based on parking resources, real-time accurate information on available parking spaces is disseminated, along with intelligent forecasting and guidance.

Promoting automated parking systems to enhance parking efficiency. Researching and applying new technologies to ultimately achieve automated parking.

Utilizing new technologies to automate parking fee collection, promoting contactless payment and electronic barrier technologies to save time entering and exiting parking lots. Rapidly transitioning from manual payment to RFID and mobile payments, ultimately achieving contactless payment.

Gradually implementing graded guidance for urban parking and parking space allocation in parking lots.

Intelligent Hub System

Transport hubs are crucial points for convenient and efficient travel. Establishing intelligent transport hubs is key to improving travel efficiency and service levels.

Setting up a comprehensive operation monitoring system. Information acquisition forms the foundation for all functionalities and services, enabling real-time monitoring of traffic flow and pedestrian flow in the surrounding environment round-the-clock and comprehensively. This facilitates rapid acquisition of transportation demand information and integrates data analysis to support informed decision-making.

Developing an intelligent analysis and decision-making platform. Building the "hub brain" enables intelligent analysis and decision-making, achieving demand-based monitoring and prediction, implementing intelligent management, and facilitating coordination between different modes of transportation and schedules.

Provide efficient and convenient public services. To achieve efficient evacuation, enhance the convenience at all stages, and avoid the formation of bottlenecks. For example, utilize smart and non-invasive security check technology and promote its widespread use, as well as ensure full coverage of self-service ticketing, ticket collection, and recharge services. Precisely predict travel times and provide real-time updates to travelers, transforming waiting stations into pass-through stations. For intermodal transfers, coordinate schedules and provide intelligent guidance within hubs to minimize passenger waiting times.

Establishing a stable safety control and emergency rescue system. By leveraging intelligent technology, comprehensively integrate various transportation modes into a platform that monitors daily operations, analyzes key performance indicators, forecasts warnings, and handles emergency incidents. Achieve a dynamic warning accuracy rate of $\geq 99\%$ for major events and automatically implement measures to mitigate harm.

Smart Slow Mobility System

To achieve green and sustainable development, cities must prioritize the development of intensive public transportation. In the urban public transportation chain, travel time is a crucial factor affecting the service level of public transportation, with last-mile transportation having a significant impact on travel efficiency. It is urgent to

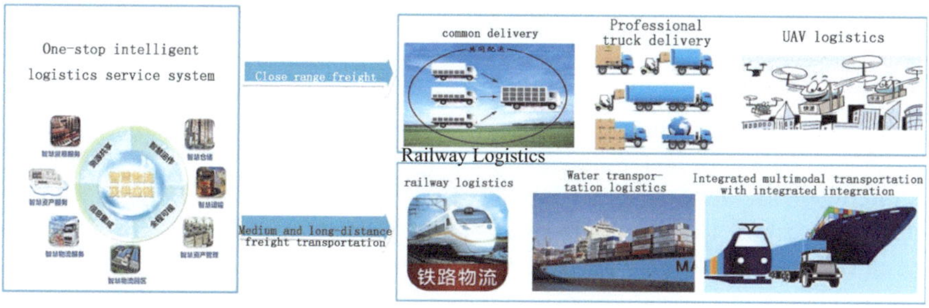

Fig. 21.3 The construction of a one-stop logistics system

establish an intelligent, convenient, and continuous slow mobility system, focusing on the following two aspects:

Intelligent guidance near public transportation stops. When travelers are cycling or walking, it can be inconvenient to stop and check maps or their phones. In such situations, having intelligent guidance facilities on the street or wireless smart voice guidance devices would be greatly beneficial.

Smart bike-sharing. Operating companies utilize big data acquisition and analysis technology to match supply with demand, thereby formulating scientifically sound dispatching plans. They also employ electronic fences and electronic credit systems to address issues such as bike parking and damage.

The Future Freight Transport System: Ensuring Smooth Flow of Goods

Utilizing advanced information technologies such as the Internet, to reshape enterprise freight business processes, innovate freight organization methods, promote online and offline integration, establish intelligent service modes of "one-time consignment" throughout the entire process, "single bill to the end" for waybills, and "one-time collection" for settlements. This aims to improve the efficiency and safety levels of warehousing, transportation, distribution, and other freight transport links, providing door-to-door one-stop intelligent freight services and achieving a unified bill system.

In short-distance freight transport, collaborative delivery, professional truck delivery, and drone logistics will be widely utilized. For long-distance freight transport, the proportion of green transportation methods such as railways and waterways will be increased. Unified and integrated intelligent logistics service systems will become more widespread, and the proportion of multimodal transportation will continue to increase, resulting in smoother connections between various modes of transportation (Fig. 21.3).

The establishment of a comprehensive intelligent freight transport infrastructure network: involving various market entities such as freight backbone enterprises, industry associations, public service institutions, etc., in the construction of intelligent freight transport infrastructure including cloud (cloud computing), network (broadband network), and terminal (various terminals).

Intelligent Freight Transport Cloud Platform: Achieving end-to-end visibility and intelligent control to enhance the level of intelligent freight transport. The integrated information and transport services across the entire logistics chain enable new freight organizational models such as "zero inventory", meeting the evolving market demands for segmentation, decentralization, and refinement. Focused on user experience, industrial upgrading, and consumption upgrading needs, it enables agile responses to various adjustments and changes in the supply chain, significantly improving freight efficiency.

Advanced multimodal transportation and collaborative delivery: Utilizing electronic freight manifests extensively, establishing electronic tags containing essential information, forming unique encoding and electronic identity, promoting mutual recognition and traceability throughout the process, accelerating the development of multimodal transportation "single-document" services; establishing an intelligent dispatch linkage system for multimodal transportation of goods.

Advanced freight technology research and application: Focused on key areas such as product traceability, online dispatch management, intelligent transportation, and distribution, based on IoT technology, conduct research and application of key technologies such as cargo tracking and positioning, RFID identification, visualization, mobile information services, integrated navigation systems; gradually realize automated driving of freight vehicles.

Smart City

The Essence and Core Features of Smart City

Smart city is a new concept and modern urban development model that involves the comprehensive application of modern science and technology, integration of information resources, and coordination of business application systems to enhance the intelligence of urban planning, construction, management, services, and governance. Specifically, it entails the comprehensive integration of tools and methods that utilize next-generation information technologies and information resources, such as the Internet of Things, cloud computing, mobile Internet, and big data. It blends advanced concepts of urban management and services to elevate the modernization level of urban governance capabilities, thereby realizing new paths, models, and forms for sustainable urban development. It also serves as a systematic project to implement the national new urbanization development plan, enhance people's happiness and satisfaction, and promote the transformation and upgrading of urban development methodologies [16].

The core characteristic of a smart city lies in its "intelligence," and the realization of intelligence relies on the construction of a widely covered information network, the formation of a deeply interconnected and shared information system, the establishment of a high-level intelligent development mechanism with business linkage, and the achievement of system self-evolution and deep application of artificial intelligence.

The Current Development Status of Smart Cities in China

Overall, the development of smart cities in China has gone through three stages:

Nascent Stage (Before the Concept of Smart Cities Emerged): Emphasized the concept of "digital cities," focusing more on data acquisition and transmission.

Pilot Exploration Stage (2010–2015): The smart city construction boom began, with participation from various departments such as the Ministry of Housing and Urban–Rural Development and the National Development and Reform Commission. Significant progress was made in data processing, application, decision support, and interconnection.

Systematic Innovation Stage (2016-Present): The concept of new smart cities emerged, while the idea of a smart society also became a new development direction. The goal is to concentrate data resources and achieve precise and meticulous urban governance through efficient and unified urban operation scheduling.

There are nearly 600 smart city pilot projects nationwide, with over 300 cities proposing smart city planning. Among them, 100% of the sub-provincial-level and above cities, 89% of the prefecture-level and above cities, and 47% of the county-level and above cities have proposed the construction of smart cities. As China's smart city development enters the stage of scale expansion, cities such as Beijing, Shanghai, and Shenzhen have achieved a series of successes in smart city construction, providing valuable experience for the construction of other cities. However, overall, China's smart city construction is still in its initial stage [16].

Future Smart City Functional Requirements and Development Prospects

The development of smart cities should revolve around urban development needs, aiming to address urban congestion, enhance traffic safety, improve governance efficiency, promote green economy, enhance service levels, and elevate the quality of life for residents (Fig. 21.4).

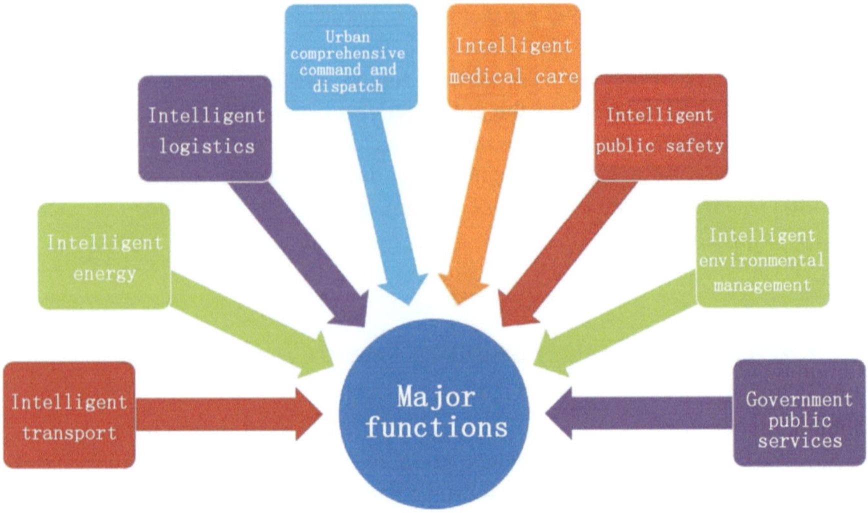

Fig. 21.4 Functional requirements of a smart city

To achieve such goals, it is necessary to rely on high-level top-level design, leverage powerful "urban brain" and robust basic units of smart cities, and possess capabilities for in-depth analysis and judgment, precise scientific decision-making, efficient business coordination, and accumulation of self-evolutionary genes. To acquire such capabilities, intelligent comprehensive sensing technology, integration of new technologies such as big data, AI, and IoT, 5G information rapid transmission technology, next-generation network security technology, cloud-edge-end integrated and supercomputing technology, real-time closed-loop feedback self-optimization technology, multi-level optimization global optimization technology, as well as integrated institutional mechanisms and policy regulations to ensure the advancement of this advanced system are needed. The development goals and strategies of smart cities are shown in Fig. 21.5 [16].

The architecture of the smart city system is divided into five layers on top of urban communication infrastructure resources: the perception layer, communication layer, platform layer, application layer, and user layer. Building upon the "urban brain," it achieves functions such as intelligent transportation, smart urban governance, intelligent livelihood services, smart industrial economy, and smart civilization and livability. Among these, intelligent transportation serves as the "forerunner" and breakthrough in smart city construction.

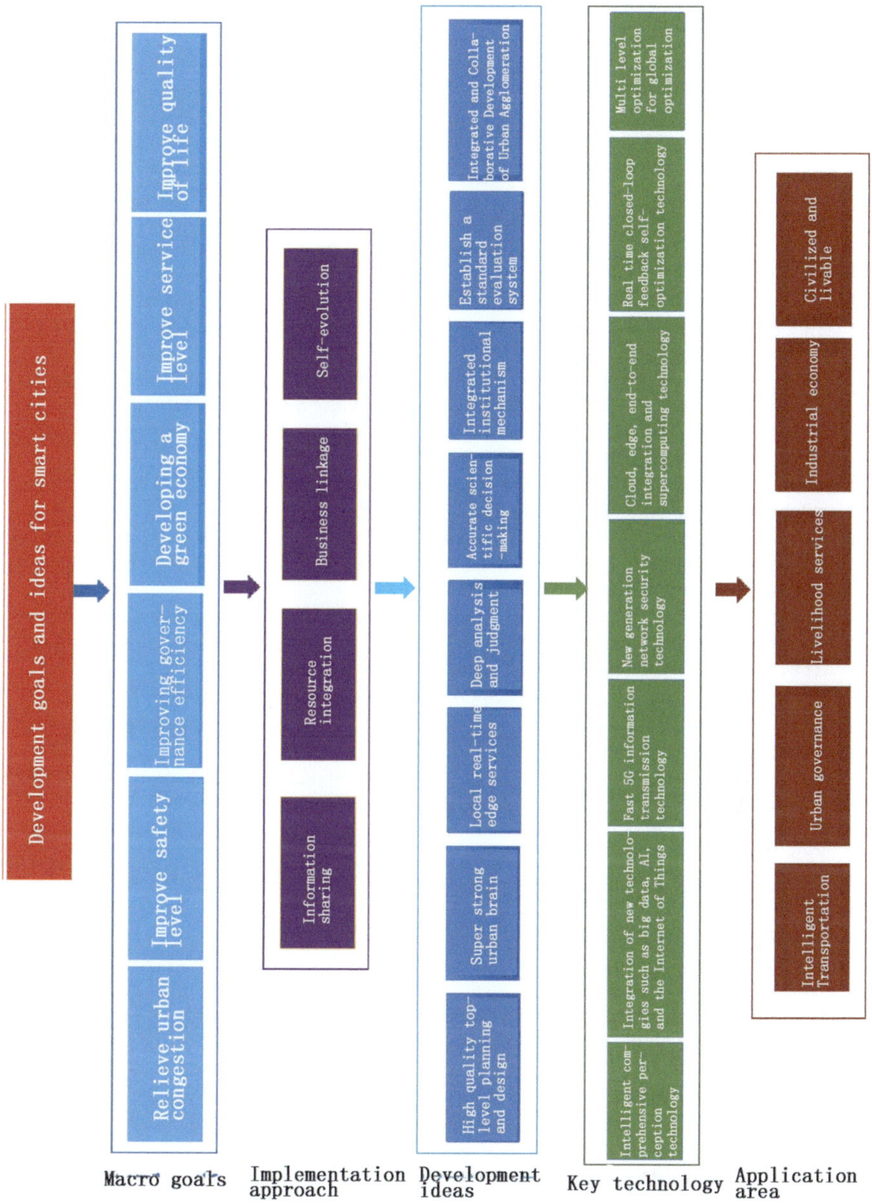

Fig. 21.5 The development goals and strategies of smart cities

References

1. National Bureau of Statistics. Statistical Communique on the National Economic and Social Development of the People's Republic of China in 2021 [EB/OL]. [2022-05-15]. http://www.gov.cn/xinwen/2022-02/28/content_5676015.htm.
2. Institute of Transportation, Tsinghua University. Research Report on the Strategy of Building a Powerful Transportation Nation (Phase II) [R]. Beijing: Institute of Transportation, Tsinghua University, 2021.
3. Lu Huapu. Requirements and Key Countermeasures for Urban Transportation Development under the Strategy of Building a Powerful Transportation Nation [J]. Urban Transport of China, 2020, 18(6): 9.
4. Wang Yang. Ministry of Railways: Online Ticket Sales for Beijing-Tianjin Intercity Railway to Begin Trial on June 12 [EB/OL]. (2011-06-12) [2022-05-15]. http://www.gov.cn/jrzg/2011-06/12/content_1882159.htm.
5. Lu Qian. Electronic Toll Collection System for Expressways Achieves Nationwide Interconnection in China [EB/OL]. (2015-09-28) [2022-05-15]. http://www.gov.cn/xinwen/2015-09/28/content_2939851.htm.
6. Ministry of Transport. Notice of the General Office of the Ministry of Transport on the Implementation of the Road Passenger Transport Online Ticketing System with Provincial Interconnection [EB/OL]. (2016-07-08) [2022-05-15]. http://xxgk.mot.gov.cn/2020/jigou/ysfws/202006/t20200623_3315275.html.
7. Guan Jizhen. Achievements and Future Prospects of Intelligent Transportation Innovation and Development in China [J]. Sustainable Development Review, 2021.
8. Xinhua News Agency. China Achieves Phased Results in the Construction of a New Generation of Air Traffic Management System [EB/OL]. (2017-09-04) [2022-05-15]. http://www.gov.cn/xinwen/2017-09/04/content_5222589.htm.
9. Transportation Research Society. Guan Jizhen: The Development and Innovation of Intelligent Transportation in China [EB/OL]. (2020-10-29) [2022-05-15]. https://baijiahao.baidu.com/s?id=1681878251734020939&wfr=spider&for=pc.
10. Ministry of Transport. Statistical Communique on the Development of the Transportation Industry in 2020 [EB/OL]. (2021-05-19) [2022-05-15]. http://www.gov.cn/xinwen/2021-05/19/content_5608523.htm.
11. Ministry of Transport. Notice of the Ministry of Transport on the Issuance of the Action Plan for the Construction of New Infrastructure in the Transportation Sector (2021–2025) [EB/OL]. (2021-09-23) [2022-05-15]. http://www.gov.cn/zhengce/zhengceku/2021-9/29/content_5639987.htm.
12. Lu Huapu. Prospects for the Development of Transportation Science and Technology . Beijing: China Communications Press Co., Ltd., 2022.
13. Institute of Transportation, Tsinghua University. Research on the Goals, Green, and Intelligent Transportation Strategies of the Strategy of Building a Powerful Transportation Nation [R]. Beijing: Institute of Transportation, Tsinghua University, 2019.
14. Institute of Transportation, Tsinghua University. Research Report on Intelligent Transportation Management Systems in Chinese Cities [R]. Beijing: Institute of Transportation, Tsinghua University, 2018.
15. Institute of Transportation, Tsinghua University. Research on the Current Status of Intelligent and Connected Vehicles [R]. Beijing: Institute of Transportation, Tsinghua University, 2021.
16. Lu Yang, Li Ping, Zhou Qinghua. Basic Unit of Smart Cities: Functional Positioning and Deep Application of Edge Servers [J]. Science & Technology Review, 2020, 38(9): 7.

Chapter 22
Glory Decade: Filling Gaps, Creating Miracles, Transport Development Reaches New Heights

Standing at the historical juncture of 2022, looking back at the past decade of transportation development, we cannot help but feel deeply moved and reflect on the journey. Since the 18th National Congress of the Communist Party of China, under the strong leadership of the Party Central Committee with Comrade Xi Jinping at its core, the Chinese people, with the spirit of fearlessness in the face of difficulties, have forged ahead, overcoming obstacles and bridging gaps. The transportation industry has achieved remarkable development, making historic leaps, filling one transportation gap after another, and creating one human miracle after another. The development of transportation has transformed the economic landscape, turning obstacles into smooth roads, leading the people toward prosperity, ushering in a new era, and changing the lives of hundreds of millions of Chinese people.

The Lalin Railway Has Opened a New Chapter of Happiness in the Lives of the People in Xizang

From the forests and snow-covered plains to the water towns of the Jiangnan region, from the deserts and steppes to the mountain peaks, China's railways traverse mighty rivers, traverse towering mountains, and connect in all directions.

The Lalin Railway (see Fig. 22.1) is located in the southeastern part of the Xizang Autonomous Region, starting from Lhasa City in the west, passing through Gonggar County, Zhanang County, Naidong District, Sangri County, Gyaca County in Shannan Prefecture, and finally reaching Lang County and Milin County in Linzhi City. The line traverses the southern valleys between the Gangdise Mountains, the Nyenchen Tanglha Mountains, and the Himalayas, with the highest elevation reaching 3,700 m. It branches out from the Xierong Station on the Lhasa-Nyingchi Railway, crosses the Gangdise Mountains to the south into the Yarlung Zangbo River Valley, then heads eastward through Zhanang County, Naidong District, Sangri

© China Communications Press Co., Ltd 2025

H. Lu (ed.), *A Brief History of Transportation in China*,
https://doi.org/10.1007/978-981-97-9851-3_22

Fig. 22.1 Lalin Railway

County, Gyaca County, Lang County, and Milin County, and finally reaches the downtown area of Linzhi City in the Xizang Autonomous Region along the Nyang River. The main line has a total length of 403.144 km [1]. The Lalin Railway crosses the Yarlung Zangbo River 16 times and overcomes multiple challenges of plateau railways. The average altitude of the stations exceeds 3,500 m, with bridges and tunnels accounting for 75% of the total length.

Building a railway on the Qinghai-Xizang Plateau faces three major world-class engineering challenges: permafrost, high-altitude, and fragile ecology. The construction of the Lalin Railway was extremely difficult due to these challenges. From the start of construction on December 19, 2014, to its opening for operation on June 25, 2021, the construction of the Lalin Railway took a total of seven years. China's transportation professionals overcame one obstacle after another in terms of terrain, geology, and ecology. Terrain challenges: The Lalin Railway is located in the valley between the Gangdise Mountains and the Himalayas on the Qinghai-Xizang Plateau. Over 90% of the route is at an altitude exceeding 3,000 m, with extremely complex terrain, requiring the line to pass through several natural barriers, including massive peaks reaching about 5,600 m above sea level. Geological challenges: Major geological challenges encountered during construction include collapses, landslides, loose rock formations, mudslides, sand slopes, water damage, ice damage, wind and sand erosion, high-intensity earthquakes and active faults, thermal damage, high-altitude stress, steep rocky coastlines during the growth period, radioactive and harmful gases, weathered rock, expansive rock, and soft soil. Ecological challenges: The Yarlung Zangbo River Valley, through which the Lalin Railway passes, is the "river source" and "ecological source" of South Asia and Southeast Asia. The protection of pristine vegetation and wildlife posed challenges to railway construction [2].

In the past seven years, 17 railway stations have sprung up along the banks of the Yarlung Zangbo River, with an average altitude of over 3,500 m. These stations serve as hubs for the Dali-Xizang (Kunming-Lhasa), Sichuan-Xizang (Chengdu-Lhasa), and Gannan-Xizang (Lanzhou-Lhasa) railways, with all three lines converging on the Lalin Railway. As the first electrified railway in the Xizang Autonomous Region, the completion and opening of the Lalin Railway have put an end to the history of no railways in southeast Xizang, allowing the Fuxing high-speed trains to achieve full coverage of all 31 provinces, autonomous regions, and municipalities in China [2]. The Fuxing high-altitude electrified multiple-unit trains, independently developed and manufactured in China, have ushered in the "era of high-speed trains" in the Xizang Autonomous Region, reaching the "roof of the world."

The Lalin Railway is not only the lifeline of the people in Xizang but also the lifeline of the Chinese nation, playing a significant role in improving economic development, enhancing regional connectivity, and consolidating national defense. The construction of the Lalin Railway has improved and optimized the layout and planning of the railway network in Xizang, while also contributing significantly to the promotion of the development of the western regions of China, enhancing ethnic unity, and safeguarding national security. It has propelled Xizang's transportation system to a new level [3].

The opening and operation of the Lalin Railway resemble a new pathway traversing the snowy plateau, injecting fresh vitality into local prosperity and the development of Xizang. The Lalin Railway provides convenient and efficient passenger services. As of June 2022, there are four high-speed trains and two regular express trains running daily between Lhasa and Nyingchi, as well as two high-speed trains between Shigatse and Nyingchi. With a peak daily passenger volume exceeding 4,300, it significantly facilitates travel for passengers along the route [4]. Additionally, the Lalin Railway effectively ensures freight transportation, having transported a total of 47,800 tons of goods within its first year of operation, effectively serving local economic development and meeting the needs of residents.

The Lalin Railway has reduced the travel time from Lhasa to Nyingchi to just 3 h, connecting the essence of the scenic landscapes in southeast Xizang, earning it the title of "China's Most Beautiful Sky Road." Along the railway, Xizang reveals its beauty with its serene waters and vast white clouds. The railway's capacity and operational speed no longer pose obstacles to Tibet's economic development. This is the new "Sky Road" on the Roof of the World, the "Unity Line" on the Qinghai-Xizang Plateau. The Lalin Railway has opened a new chapter of happiness for the people of Xizang.

The Highways Running and the Railways Soaring Have Realized the Life Dreams of Mountainous People for Thousands of years

"Roads Built Through Mountains and Bridges Erected Over Water" in Guizhou's Transportation Construction

Guizhou is the only province in China without plains. "It's not that Yelang [an ancient kingdom in Guizhou] was truly arrogant, but because there was no road leading to the Central Plains" reflects the past helplessness and bitterness of the people of Guizhou [5]. Unique geological conditions and backward transportation have long hindered the development of Guizhou. Poverty and backwardness seem to be labels firmly attached to the land of Guizhou. With the kind care and strong support of the Party and the country, the Guizhou Provincial Party Committee and Provincial Government have led the people of the whole province with the ambition and determination to overcome difficulties and work hard, continuously composing magnificent chapters of Guizhou's great transformation in transportation, and promoting the "Golden Decade" of catching up and surpassing in transportation development in Guizhou.

In 2012, Guizhou launched the "Three-Year Battle for Expressway Construction" campaign. By the end of 2015, Guizhou became one of the few provinces in China to achieve "county-wide access to expressways". In 2017, Guizhou took the lead in the western region to achieve the "Double Hundred" goals of 100% accessibility and passenger transportation rate for every administrative village. In the same year, Guizhou Province was the first in the country to propose and implement the "village-wide access" project for rural areas. By 2019, 78,700 km of village roads had been hardened, bridging the "last mile" on the road to poverty alleviation. This benefited nearly 40,000 villages and 12 million rural residents, including 1.83 million registered impoverished individuals. Guizhou's experience in transportation poverty alleviation has contributed to the remarkable achievements in poverty reduction in China, and the masses are now enjoying the convenience of "hardened roads and accessible transportation", breaking through geographical barriers, freeing themselves from poverty, and embarking on a new journey of rural revitalization.

Nearly 20,000 bridges of various types have transformed Guizhou, known for its karst topography and lack of flat land, into a "high-speed plain" [6]. Guizhou became the first province in the western region where "every city has access to high-speed rail, and every county has access to expressways". Eighteen world-class bridges have been constructed, with 50 of them ranking in the top 100 and 5 in the top 10 globally, marking a magnificent transformation from a land of uneven terrain to a "bridge museum". With over 7,600 km of expressways, more than 2,100 tunnels, and over 12,000 bridges, a "high-speed plain" covering nearly 500 square kilometers has been created, turning impassable barriers into accessible routes.

At the end of 2014, the Guizhou-Guangzhou High-Speed Railway (referred to as "high-speed rail") commenced operations, marking Guizhou's entry into the "high-speed rail era". With the completion of the Shanghai-Kunming High-Speed Railway, the Chongqing-Guiyang High-Speed Railway, and other lines, a Guizhou high-speed rail "big cross" centered around Guiyang was formed, creating a rapid transit corridor linking the Yangtze River Delta, the Pearl River Delta, the Beijing-Tianjin-Hebei region, and the Sichuan-Chongqing-Guizhou-Yunnan region.

"The road ahead is as solid as iron; let's forge ahead with renewed determination." Guizhou will continue to carry forward the spirit of "enduring hardships, daring to innovate, forging paths through mountains, and building bridges over water" in transportation development. We will reorganize and set out again, pressing forward with renewed vigor and further enhancements.

The Construction of "Multi-dimensional Transportation" in Chongqin

Chongqing, located in the southwestern part of China, sits at the upper reaches of the Yangtze River, surrounded by layers of mountains. Additionally, with its position on the edge of a basin where the two rivers meet, it is shrouded in mist year-round, earning it the nicknames "Mountain City" and "Fog Capital." Over the past decade, there has been a dramatic transformation in transportation in Chongqing. In this complex and ever-changing cityscape, the people of Chongqing have built an integrated comprehensive transportation system.

Over the years, Chongqing has seen an unprecedented number of highway construction projects, with an unprecedented length of completed roads and unprecedented scale of ongoing projects [7]. After the completion of the planned "Second Ring and Eight Radials" highway network framework in Chongqing, the city initiated the construction of a new 1000-km highway network called the "Third Ring and Ten Radials Plus Multiple Interconnected Lines". In July 2017, Chongqing launched the "Three-Year Action Plan for Transportation Construction". Over the course of three years, the city has completed 62,600 km of "Four Good Rural Roads". Villages with suitable conditions have achieved 100% road connectivity, with 92% having hardened roads. Townships and administrative villages with suitable conditions have achieved 100% road hardening and 100% bus access. In 18 deeply impoverished townships, 100% of village groups with suitable conditions have road access.

In 2015, the Chengdu-Chongqing High-Speed Railway was officially opened, marking the beginning of the high-speed rail era in Chongqing and ushering in a new era of high-speed rail construction in the city. At the end of November 2016, the first high-speed railway in the Three Gorges Reservoir Area of the Yangtze River—the Chongqing-Wanzhou High-Speed Railway—was officially completed and opened to traffic, reducing the travel time from Wanzhou to Chongqing and Chengdu from 4 and 7 h, respectively, to just over 1 and 3.5 h. Now, taking high-speed rail from

Chongqing to surrounding major cities such as Chengdu, Guiyang, Kunming, and Xi'an only takes about 1, 2, 4, and 5 h, respectively [8].

The history of public transportation in Chongqing dates back to the 1930s. In 1933, Chongqing had only one diesel bus operating between Zengjiayan and Qixinggang. In 1938, the first ferry route in Chongqing, the Chumen-Haitangxi line, was opened. Before the founding of the People's Republic of China, Chongqing had only 11 public buses, 15 ferries, and 1 cable car. However, today, Chongqing's urban transportation is considered a marvel of the world. Known as the "8D City," Chongqing is renowned for its unique "vertical transportation," with the Liziba Station on Metro Line 2 and the Crown Escalator being typical examples.

The Liziba Station on Chongqing Metro Line 2 is the first in China to be integrated with commercial and residential buildings, making it a unique attraction known as the "skyscraper train passing through" (Fig. 22.2). The station and the commercial/residential buildings were designed, constructed, and put into use simultaneously, employing a "station-bridge separation" structural form to effectively address the transmission and vibration issues between the two structures [9].

The Crown Escalator (see Fig. 22.3) is a large-scale escalator located at an intersection in Chongqing, connecting the Lianglukou and Chongqing stations. It is one of the distinctive features of Chongqing's transportation system. With a total length of 112 m and a width of 1.3 m, it elevates passengers by a height of 52.7 m, making it the second longest first-tier inclined escalator in Asia. In 2021, it served 2.46 million passengers throughout the year, with the highest daily ridership reaching 29,000 passengers.

Fig. 22.2 Chongqing light rail through the city [10]

Fig. 22.3 Crown Escalator [11]

The Construction of the Taklamakan Desert Railway Has Filled a Gap in World Transportation History

In the past decade, China has advanced into transportation voids with unprecedented determination and effort, rapidly expanding its transportation network and infrastructure. Across the vast expanse of the desert, the world's first desert railway loop, stretching 2712 km, encircles the Taklamakan Desert, eliminating the barrier of shifting sands and filling a void in world transportation history.

On June 16, 2022, the Heruo Railway joined hands with the Geku Railway and the Southern Xinjiang Railway, marking the official completion of the circular route of the Taklamakan Desert Railway Loop. This also signifies the world's first railway loop around a desert, spanning a length of 2712 km [12].

The Heruo Railway officially commenced construction at the end of 2018. Stretching from Hetian City in the southern part of Xinjiang Uygur Autonomous Region to Ruoqiang County in Bayin'guoleng Mongolian Autonomous Prefecture, it spans a total length of 825 km. The designed operating speed is 120 km per hour, with 22 stations including Hotan, Luopu, Cele, and Yutian.

Before construction began, the survey and design process took more than 8 years. The Taklimakan Desert is China's largest desert and the world's second-largest mobile desert. About three-quarters of the Heruo Railway passes through the shifting sands of the Taklimakan Desert. Preventing wind erosion and sand fixation, and effectively addressing the hazards of wind and sand, are global challenges in railway construction. The Heruo Railway is located on the southern edge of the world's second-largest mobile desert, the Taklimakan Desert, with 534 km passing through wind and sand areas, accounting for 65% of the total length of the railway. The wind and sand, as well as the high temperatures of the Taklimakan Desert, pose significant challenges to railway engineering construction: here, the main windy season can last up to 7 months annually, with as many as 90 days of dust storms each year; from May to September, the highest noon temperature can reach 40 °C, and the desert surface temperature at construction sites can reach 80 °C.

The builders adopted a "sand-blocking barrier + grass checkerboard" approach, establishing a multi-layered sand-blocking system using reeds, straw, and other materials in a combination of high and low structures. Each layer of sand-blocking barrier has appropriate gaps to prevent the formation of "wind-blocking walls" due to overcrowding, making it difficult to withstand strong winds. Between the sand-blocking barriers and the grass checkerboards, plants such as Haloxylon, Tamarix, and Populus euphratica, which are easy to grow in the desert, were planted. In the future, these plants will form permanent windproof and sand-fixing forest belts on both sides of the railway (see Fig. 22.4). Over more than 1,000 days and nights of construction, builders have constructed nearly 50 million square meters of grass checkerboards along the Heruo Railway and planted nearly 13 million bushes and trees such as Haloxylon, Tamarix, and Hippophae, covering an area of over 700,000 square kilometers. This 300-km-long "green corridor" has become a barrier against wind and sand.

In areas particularly prone to sandstorms, the Heruo Railway adopts a design solution of "bridges instead of roads", constructing a total of five sand-overpass bridges spanning 49.7 km in length. This allows sandstorms to pass underneath the bridges, reducing their impact on the railway tracks and train operations.

Fig. 22.4 The combination of sand barriers, grass checkerboards, and drought-resistant plants along the Heruo Railway forms a sand prevention "trio" [53]

Along the route, the 11 passenger stations, each reflecting local characteristics, are ingeniously designed to integrate local history and cultural heritage, creating unique "scenery at each station". For example, the Luopu Railway Station draws inspiration from the local national-level intangible cultural heritage, the Adelis Silk, with its vibrant colors adding a touch of liveliness to the small station [13].

This national Class I single-track railway has put an end to the historical lack of train service in multiple areas such as Luopu, Cele, Yutian, Minfeng, and Bachu County in Hotan Prefecture, greatly facilitating travel and freight transport for people of various ethnic groups along the route. Its inauguration marks the formation of the world's first desert railway loop, ending the history of millions of people and numerous towns and villages surrounding the Taklamakan Desert without railway access. It has shortened the outbound distance for people in southern Xinjiang by nearly 1000 km, allowing residents of remote towns to easily traverse vast stretches of shifting sands and access various supplies and support materials from the mainland more conveniently [14].

The 'Fuxing' High-Speed Train, Embodiment of Chinese Wisdom, Opens a New Era in Railway Transportation

High-speed rail is an important symbol of modern transportation and a significant reflection of a country's industrialization level. The name "Fuxing" represents the pursuit and deep affection of Chinese railway workers for the great rejuvenation of the Chinese nation.

In May 2018, General Secretary Xi Jinping stated at the 19th Academician Conference of the Chinese Academy of Sciences, "The Fuxing high-speed trains have taken a crucial step from catching up to leading."[1] By 2021, the Fuxing high-speed trains have developed into a "big family" of high-speed trains capable of accommodating various operational environments such as plateaus, high-altitude regions, humid and hot areas, and windy and sandy conditions. This family includes products with different speed grades of 350 km per hour, 250 km per hour, and 160 km per hour, as well as different formations of 8-car sets, 16-car sets, and 17-car sets. It incorporates different power traction modes of centralized and distributed power, catering to the needs of high-speed rail lines such as the Beijing-Shanghai, Beijing-Tianjin, Beijing-Zhangjiakou, and Chengdu-Chongqing. The Fuxing high-speed trains (Fig. 22.5) have set a national record as the only ones in the world to achieve commercial operation at a speed of 350 km per hour on high-speed rail lines, establishing a benchmark for the commercial operation of high-speed rail worldwide [15]. They vividly demonstrate "China's speed" to the world and have created a Chinese high-speed rail brand with world-class operational quality.

---

[1] Refer to the "Speech at the 19th Academician Conference of the Chinese Academy of Sciences and the 14th Academician Conference of the Chinese Academy of Engineering" by Xi Jinping, published in People's Daily, May 29, 2018, Page 02.

Although China's high-speed rail development started more than 40 years later than developed countries, it has achieved a historic transformation from nothing to something, from catching up to keeping pace, and then leading the way, thanks to the leadership of the Communist Party and the advantages of the new type of nationwide system. This progress is the result of generations of railway workers' continuous efforts.

In 2012, the research and development work on China's standard high-speed train began. The railway department strengthened its role as an innovation leader and utilized its enormous market advantages by integrating industry, academia, research, and application. Comprehensive efforts were made in key areas such as bogies, traction systems, braking, network control, wheelsets, and overall vehicle design. Extensive research experiments were conducted based on fundamental theories and technological innovations, achieving breakthroughs in key technologies and components. The formulation of overall technical conditions and the design review process were completed. After the prototype was officially produced, it underwent operational testing under various conditions covering a distance of 600,000 km.

On June 26, 2017, two Fuxing bullet train sets departed simultaneously from Beijing South Station and Shanghai Hongqiao Station, marking the beginning of a new era in China's railway technology and equipment. On September 21 of the same year, Fuxing bullet trains achieved commercial operation at a speed of 350 km per hour on the Beijing-Shanghai high-speed railway, setting a new benchmark for high-speed rail commercial operations worldwide, with same-day round trips within a 1000-km range.

Fig. 22.5 On April 21, 2022, two Fuxing bullet trains successfully met in Xunxian County, Henan Province. The trains were traveling at a speed of 435 km per hour during operation, resulting in a relative meeting speed of 870 km per hour. The time taken for the trains to meet was 0.86 s [16]

On December 30, 2019, the intelligent Beijing-Zhangjiakou high-speed railway commenced operation, marking a significant milestone for China’s high-speed rail development. This achievement propelled China’s high-speed rail technology from ground zero to the world’s most advanced level. It marked the world’s first implementation of automatic driving function at a speed of 350 km per hour and the inaugural use of China’s independently developed Beidou satellite navigation system for safeguarding… [17]. The century-old Beijing-Zhangjiakou line bears witness to the remarkable leap in China’s comprehensive national strength.

On January 19, 2021, General Secretary Xi Jinping emphasized during his inspection at the Taizicheng Station aboard the Beijing-Zhangjiakou high-speed railway that a successful example of China’s independent innovation is the high-speed rail. From inception to leadership, China has progressed from introduction, digestion, absorption, and innovation to independent innovation, now leading the world in this field. It is crucial to summarize experiences, continue efforts, and strive for greater development during the “14th Five-Year Plan” period. Implementing the important instructions of General Secretary Xi Jinping, summarizing and applying China’s experience in independent innovation in high-speed rail are of great significance. This will fully leverage the advantages of the new type of all-round national system and promote greater development of China’s high-speed rail during the “14th Five-Year Plan” period, better serving the comprehensive construction of a socialist modern country and benefiting the vast masses of the people.

As of 2021, the nationwide deployment of the Fuxing series high-speed trains has exceeded 1,200 sets, with a cumulative safe operation distance of over 13.58 billion kilometers and the transportation of 1.37 billion passengers. Represented by the Fuxing series, China’s high-speed rail has become a shining national emblem. Moreover, the development of the Fuxing series has effectively driven the optimization and upgrade of the industrial structure, playing a significant role in promoting the transformation of the economic development mode. The Fuxing high-speed train, as an integration of modern high-tech, comprises over 100,000 components and more than 260 independent technical systems. The core enterprises involved in the design and production of train components exceed 100, with over 500 closely linked companies spanning across more than 20 provinces and municipalities. The high-speed rail equipment manufacturing industry has thus become a strategic emerging industry in China with international competitive advantages throughout the entire industrial chain [16].

On April 21, 2022, China’s independently developed world-leading new Fuxing high-speed comprehensive inspection train set a world record with a relative meeting speed of 870 km per hour on the clear track.

The Fuxing high-speed trains, running across the vast land of our motherland, represent the fastest and highest-quality high-speed rail service in the world today. They stand as a shining emblem and a developmental marvel of China’s high-speed rail. The inauguration of their operation has ushered in a new era in global railway transportation.

Innovative Development of Tunnel Boring Machines Creates a World Miracle

Unimaginable achievements have become a reality over the past decade. At an altitude of 4,700 m in the Sejila Mountains, snowfall covers the area for more than half of the year. In this challenging terrain, the domestically developed hard-rock tunnel boring machine, known as the 'Snowland Pioneer' (see Fig. 22.6), has been operational for over six months. Operating under extreme conditions of high altitude and cold temperatures, this project represents one of the most challenging railway engineering endeavors humanity has ever faced.

Over the past decade, advanced equipment like the "Snowland Pioneer" has been developed and deployed into similarly arduous tasks. In the central and western regions, the number of tunnel boring machines has increased nearly eightfold compared to 2014. With the intelligent positioning systems installed on each device, we can pinpoint the locations of these equipment on the map. Mountain ranges like the Hengduan, Nyainqentanglha, and Qilian, which were once considered unconquerable obstacles and inaccessible gaps in our transportation network, have now become integrated into our comprehensive transportation network system.

In the past decade, China's tunnel boring machine (TBM) technology has experienced rapid development. TBMs are primarily used in infrastructure projects such as railways, highways, subways, and water conservancy tunnels, earning them the title of "king of construction machinery" and the nickname "underground dragons." The research and development of TBMs involve multiple disciplines including geology,

Fig. 22.6 'Snowland Pioneer' shield tunnel boring machine [18]

civil engineering, mechanical engineering, mechanics, hydraulics, electrical engineering, control systems, and surveying. Moreover, TBMs require customized design and manufacturing tailored to different geological conditions, resulting in a high technological threshold. In just over a decade, China has transitioned from trailing behind to leading the development of TBM technology, showcasing another remarkable achievement in Chinese manufacturing following the success of high-speed railways.

On April 25, 2008, China's first domestically developed compound shield tunnel boring machine, known as "China Railway No. 1," officially rolled off the production line, marking a historic leap forward for China's tunnel boring machine industry. Despite being a latecomer, China has continuously broken records in the global tunneling field. These records include the world's largest rectangular shield tunnel boring machine, the world's first horseshoe-shaped shield tunnel boring machine, the world's largest cross-section hard rock tunnel boring machine, the world's smallest diameter hard rock TBM (full-face tunnel boring machine), the world's first mining-oriented small-radius TBM, the world's first inclined shaft dual-mode TBM, and the world's first full-face hard rock vertical shaft boring machine, among others, all being developed in China [19].

In December 2019, China's first undersea metro shield tunnel, part of the Xiamen Rail Transit Line 2, commenced operation. The section between Haicang Avenue Station and Dongdu Road Station marked the country's inaugural shield tunnel crossing beneath the sea. This project faced significant construction challenges and required advanced technical expertise, as there was no prior experience to draw upon. The geological conditions in this section were complex, featuring a combination of soft soil, hard rock, variable soil conditions, water seepage from the top, and protruding bedrock at the bottom, earning it the nickname "Geological Museum." During the excavation phase, the construction team conducted a total of 3,475 pressurized advances and manually cleared over 1,000 cubic meters of isolated rocks. This project enabled the mastery of tunneling techniques for uneven geological conditions beneath the seabed and techniques for handling isolated rock formations. Qian Qihu, a recipient of China's highest scientific and technological award and an academician of the Chinese Academy of Engineering, stated that the integrated innovation demonstrated in the Xiamen Metro Line 2 undersea tunnel signifies that China's integrated innovation in undersea metro shield tunnel technology has reached an international leading level [18].

The shield tunneling machine, with its complex structure and diverse applications, has completely revolutionized traditional manual drilling and blasting methods, greatly enhancing safety and efficiency. However, due to significant geological variations across different cities in China, a single type of shield tunneling machine cannot fully meet practical demands. Urban railway projects present challenges such as complex underground environments and numerous surface risks. For example, in the case of the control project for the Beijing section of the Beijing-Shenyang High-Speed Railway, a tunneling section using a 10.9-m diameter shield tunneling machine requires continuous excavation of approximately 3,680 m, traversing through various structures including the South Gaoli Road and box culvert, the Capital Airport auxiliary road, the Capital Airport expressway, airport rail transit, as well as multiple residential areas and tall buildings. In intercity projects such as the South-to-North Water Diversion Project, Beijing Daxing International Airport, and the Guangdong-Hong Kong-Macao Greater Bay Area transportation hub, the presence of "Made in China" shield tunneling machines is ubiquitous. On August 23, 2021, the domestically produced "Shenjiang No. 1" slurry balance shield tunneling machine was unveiled in Changsha. With a total length of 130 m and a weight of 3,300 tons, it boasts a excavation diameter of 13.42 m. This machine has been deployed for the major transportation infrastructure project in the Greater Bay Area—the construction of the Zhujiang River Estuary Tunnel for the new Shenjiang (Shenzhen to Jiangmen) high-speed rail line, marking the beginning of its underwater journey. The Zhujiang River Estuary Tunnel, situated between Dongguan and Guangzhou at the mouth of the Pearl River, spans a total length of 13.69 km, with approximately 11.05 km submerged underwater. The maximum depth of the tunneling section is 106 m, and it faces a maximum water pressure of 1.06 megapascals. As the deepest and highest water pressure underwater tunnel in China at the time, and a key control project for the Shenjiang High-Speed Rail, it presents extremely complex geological conditions and numerous sensitive environmental factors, demanding the utmost precision in shield tunneling technology.

Over the past decade, China's shield tunneling industry has surged forward, creating one miracle after another. In 2021, Chinese-made shield tunneling machines led the world in terms of annual production, ownership, and construction mileage of shield tunneling projects. Domestic shield tunneling machines accounted for 90% of the domestic market share and two-thirds of the global market share. With the ability to shape the main structure of tunnels in a single formation, shield tunneling machines have provided strong support for the development of China's high-speed rail network. Not only have shield tunneling machines ventured beyond China's borders, but Chinese shield tunneling technology has also made its mark on the international stage, demonstrating its prowess in the Belt and Road Initiative. Serving as a hallmark of "Made in China," it has become a global phenomenon [18].

"The Kunlun" Track Laying Integrated Machine Shows Its Strength, Creating a Miracle in China's Railway Construction

"The Kunlun" Thousand-Ton Track Laying Integrated Machine integrates lifting, transporting, and placing functions into one, with a high degree of intelligence. It is the high-speed railway box girder construction equipment with the most functions, widest application range, and highest technological content in China. It is also the first equipment in the country that can be used for the construction of double-track box girders on high-speed railways under all road conditions.

Fig. 22.7 Aerial view of the construction site of the Meizhou Bay Cross-Sea Bridge for the Fuzhou-Xiamen High-Speed Railway [20]

"The Kunlun" embodies the diligent efforts of numerous researchers at Hanjiang Heavy Industries. Its R&D team has successively overcome more than 20 key technologies, including automatic synchronous control technology and automatic centering driving technology for the integrated track laying machine, ultra-low position beam transporting technology and rapid span-changing technology for the split-type machine, and high-precision automatic correction technology for the girder lifter, with many of these technologies reaching the international leading level.

In early June 2020, the thousand-ton-level integrated track laying machine arrived at the construction site of the Meizhou Bay Cross-sea Bridge. The R&D team named the bridge erecting machine "Kunlun" symbolizing its ability to "move mountains and pave the way in rugged terrain," as depicted in Fig. 22.7.

Once assembled, the "Kunlun" machine is 116 m long, equivalent to the length of four high-speed train carriages. It stands at 9.3 m tall, akin to the height of a three-story building. With 42 pairs of giant wheels, it can transport bridge segments measuring 40 m in length and weighing 1000 tons under the control of the operator.

The "Kunlun" is equipped with ball joints with a diameter of 1.4 m, serving as the balancing pivot for the machine. These ball joints consist of two spherical surfaces, allowing directional adjustment through sliding motion. They can withstand tens of tons of tension while smoothly rotating under pressures of thousands of tons. This design is the secret to its stability and balance.

In addition, the "Kunlun" makes it possible to install 40-m-high railway box girders. Compared to traditional 32-m box girders, using 40-m box girders can reduce the number of piers by six per kilometer. It also effectively reduces the impact of train resonance, making high-speed rail operation smoother. This provides technical support and equipment assurance for advancing China's high-speed rail to speeds of up to 400 km per hour.

The "Kunlun" has also been optimized for offshore construction conditions. The Muzhou Bay Cross-Sea Bridge has a total length of 14.7 km, with 10 km in the sea area. It faces adverse weather conditions such as rainy seasons, strong winds, and typhoons throughout the year. This model is capable of operating in temperatures ranging from minus 20 °C to plus 50 °C and can withstand wind speeds of up to 7 on the Beaufort scale during operation and up to 11 when not in operation.

As a versatile bridge erection machine, the "Kunlun" is capable of simultaneously accommodating the construction of high-speed railway box girders with spans of 24, 32, and 40 m, replacing the need for three separate bridge erection machines. Taking the Chongqing section of the Zhengzhou-Wanzhou high-speed railway, where the bridge-to-tunnel ratio exceeds 90%, as an example, constructing high-speed railways through mountainous terrain requires a combination of tunnels and bridges. The "Kunlun" can accomplish erection tasks under conditions of 30‰ steep slopes and 2000-m small curves, with a lighter overall weight compared to sectional machines. It holds advantages in transporting girders through tunnels and across small curved sections, particularly capable of achieving zero-meter girder erection at tunnel entrances and exits, as well as inside tunnels. This flexibility and efficiency make it particularly suitable for scenarios involving frequent transitions between bridges and tunnels.

On November 13, 2021, the Fuzhou-Xiamen High-Speed Railway's Meizhou Bay Cross-Sea Bridge successfully completed closure, marking a significant milestone in the construction of this sea-crossing bridge, which is a crucial component of the Fuzhou-Xiamen High-Speed Railway. With a total length of 14.7 km, including a 10-km stretch over the sea, the bridge underwent a year-long process of construction on the sea surface. During this period, the "Kunlun" high-speed bridge erection machine played a crucial role, ensuring the smooth completion of the high-speed railway box girder erection tasks (see Fig. 22.8).

The advent and successful deployment of the "Kunlun" have made the layout and geological alignment of high-speed railway bridges more flexible, effectively reducing the impact on river navigation and construction land, and providing technical support for the construction of higher-speed railways. It holds epoch-making significance for the future development of high-speed railways in China, marking a significant advancement in technology and equipment for high-speed railway construction and adding another powerful tool to China's high-speed railway construction endeavors.

In 2021, the "Kunlun" integrated bridge erection machine was named one of the "Top Ten National Heavyweights of 2021" among central enterprises. It shares the list with well-known super equipment such as the "Tianwen-1" probe, the "Hualong One" nuclear power unit, and the "Deep Sea No. 1" energy station.

On January 17, 2022, the upgraded version of the "Kunlun" bridge erection machine, known as "Kunlun II," was put into operation on the Chihuang High-Speed Railway at the foot of Huangshan Mountain. The new equipment is lighter, more energy-efficient, and more efficient. The "Kunlun" series of bridge erection machines has also become a new symbol of China's infrastructure construction.

Fig. 22.8 "Kunlun" conducting girder erection operations on the Meizhou Bay Cross-Sea Bridge of the Fuzhou-Xiamen High-Speed Railway (Photo by Yang Guang) [21]

"Island-Making Wonder" Dredger

In 1966, China imported a self-propelled dredger from the Netherlands. It was not the most advanced ship at the time, but the Tianjin Navigation Bureau of the Ministry of Communications paid the equivalent of 4 tons of gold for it [22].

Until 2000, dredger technology had been a major gap in China's capabilities, with no independent design and manufacturing capacity for such specialized vessels. Even for dredging the Yangtze River channel, entire ships needed to be imported from abroad. However, Chinese engineers and technicians, driven by a strong sense of responsibility for the great rejuvenation of the Chinese nation, the diligence and wisdom of the Chinese people, the indomitable spirit of perseverance, and a scientific attitude of relentless exploration, finally succeeded in independently developing the first dredger, the "Tianshi" in 2006. Subsequently, with the continuous development of production technology and productivity, China has successively developed the "Tianjing" and "Tiankun" dredgers.

The dredging depth of the "Tianjing" dredger (see Fig. 22.9) can reach 30 m, with an average dredging capacity of 4500 cubic meters per hour, making it the top dredging vessel in Asia.

When working, the huge cutterhead of the 'Tianjing' constantly rotates, instantly turning ordinary sediment or underwater reefs into crushed sand and stones. The dredged sediment is continuously discharged from the rear of the vessel, and if connected to a pipeline pump, it can be transported up to 6 km away. In addition, the 'Tianjing' has unlimited range and can navigate safely in any ocean worldwide.

The main capability of the 'Tianjing' lies in artificial island construction. A single 'Tianjing' can blow out 4,500 cubic meters of sand and gravel per hour. In 2013,

Fig. 22.9 "Dredger Tianjing" in operation

during 200 days of operation in the South China Sea, the 'Tianjing' ejected 10 million cubic meters of sand and gravel, consecutively creating eight small islands including the Meiji Island, Zhubi Island, Yongshu Island, Huayang Island, Nanxun Island, Dongmen Island, Chigua Island, and Yongxing Island, significantly enhancing China's guardianship capabilities in the South China Sea (Fig. 22.10).

On the international stage, the Tianjing dredger spent 3 years and 7 months, successfully completing the construction tasks proposed by the Belt and Road Initiative. Along the azure African coast, the Chinese giant whale turned the tide, showcasing China's speed and strength to the world [23].

On October 18, 2017, the Tianjing dredger set off for Ghana in West Africa, preparing to participate in the construction of the dredging project at the Tema Port's new container terminal, the largest port in Ghana. After nearly 50 days and 12,000 nautical miles of sailing, the Tianjing dredger successfully arrived at the construction site of the Tema Port's new container terminal project in Ghana on December 10, as shown in Fig. 22.11.

The Tema Port's new container terminal project, in which the Tianjing dredger is involved, is Ghana's largest hydraulic engineering project as of 2021. It is also the largest expansion project in the history of the Tema Port. The project involves the construction of four new 150,000-ton container berths, a 3,848-m breakwater, and the reclamation of nearly 130 ha of land, with a duration of 51 months. The expanded capacity of the Tema Port's container handling facilities will significantly enhance trade relations in West Africa, improve the standard of living for the people of Ghana, and promote socio-economic development in the region.

Fig. 22.10 Islands created by the Tianjing dredger in the South China Sea [22]

Fig. 22.11 Construction site of the Tema Port's new container terminal project involving the Tianjing dredger [23]

Furthermore, the dredging and reclamation project at Lamu Port in Kenya officially commenced in September 2016. The vessels "Tianji" and "Tianjing" successively participated in the project construction, which includes dredging the port basin, channel, and temporary cofferdams for reclamation areas. The total area of the port basin is 2.352 million cubic meters, the channel extends over a length of 1.89 km, and the dredging and reclamation volume amounts to 14.67 million cubic meters. The project faces challenging marine and meteorological conditions, with a short window for effective vessel operation due to adverse weather. Moreover, the construction area features complex soil conditions, with a mix of clay, coral rocks, and other soil types, posing significant excavation challenges.

In December 2018, the "Tianjing" officially joined the construction project at Lamu Port. During the construction period, the "Tianjing" fully demonstrated its role as a flagship of the nation, capable of tackling challenging tasks with its formidable excavation capabilities. Leveraging its powerful digging capacity, it crushed coral reefs embedded on the seabed, consuming a total of 12,987 cutting teeth of various types. The vessel completed a total volume of 10.6965 million cubic meters of work. Additionally, addressing the project's unique challenges, several process reforms and technological innovations were implemented. These efforts resulted in setting new productivity records under the complex soil conditions comprising fine sand, coarse sand, clay, and coral rock mixtures, ultimately ensuring the smooth completion of the project.

Located within the Manda Bay in Lamu County, Kenya, Lamu Port is part of the larger Lamu Port-South Sudan-Ethiopia-Transport (LAPSSET) Corridor Project, a key initiative under Kenya's Vision 2030 development plan and an integral component of the Belt and Road Initiative. Once completed, the port will serve as a vital maritime gateway for South Sudan, Ethiopia, Uganda, and the eastern Democratic Republic of the Congo. The successful completion of the Lamu Port dredging and reclamation project not only provides ample land for the subsequent construction of port terminals but also lays a solid foundation for the entire port area's navigation and operations. This achievement holds significant importance in promoting global trade, driving regional economic development, and expanding the influence of companies in the East African region.

China's dredgers are continuously upgrading, reshaping the world's perception of China. In addition to the "Tianjing" dredger, the technological prowess demonstrated by the "Tiankun" dredger is also remarkable.

As a pioneer in the construction of the Belt and Road Initiative, the "Tiankun" dredger (see Fig. 22.12) was put into operation and immediately deployed to the Far East region for construction. Faced with a new and unknown market, it showcased China's dredging capabilities. As a newly built vessel, the "Tiankun" overcame challenges such as equipment adaptation, personnel integration, and cultural assimilation, achieving perfect project delivery and demonstrating China's dredging quality. By crushing hard rocks with a hardness of 70 megapascals in the construction area, it successfully challenged the design standards for rock excavation vessels, showcasing China's dredging strength [23].

Fig. 22.12 "Tiankun" is equipped with four different types of cutter heads [23]

On January 10, 2020, the project "Independent Research and Industrialization of Large-scale Cutter Suction Dredgers at Sea," represented by the "Tiankun," was awarded the Special Prize of the National Science and Technology Progress Award. This marked the first time that China's dredging industry received such an honor. These dredgers serve national strategic tasks and major projects, deeply implement the national "Belt and Road" initiative, and have successfully implemented major projects such as the construction of remote island reefs and the Hong Kong-Zhuhai-Macao Bridge. They have demonstrated the "Chinese speed" that has shocked the world in the historical process of building a maritime power and safeguarding national territory and maritime sovereignty, significantly enhancing China's discourse power and influence on the global dredging stage, and proving with practical actions the remarkable achievements of the industrialization of this project. The technology of Chinese cutter suction dredgers has reached a leading level worldwide, with the "Tiankun" ranking third globally and first in Asia among specialized vessels. The value of domestically produced dredgers lies in their independent research and development, as well as mass production. Not only do they match or surpass similar foreign products in performance, but they are also priced significantly lower.

From Mongolia Railway to Jakarta-Bandung High-Speed Railway: The Dream of Shared Prosperity is Transitioning from Ideal to Reality

Mombasa-Nairobi Standard Gauge Railway (SGR), also known as the Mombasa-Nairobi Standard Gauge Railway, is a railway connecting Mombasa and Nairobi within the Republic of Kenya. Constructed with the assistance of China according to Chinese National Railway Class I standards, it forms part of the East African railway network. It is the largest infrastructure project undertaken in Kenya since independence and serves as a flagship project for Kenya's Vision 2030 national development agenda (Fig. 22.13). The railway stretches from Mombasa West Station in the east to Nairobi South Station in the west, with construction commencing on December 12, 2014, and opening for operation on May 31, 2017. Over the past five years, the SGR has not only improved the lives of Kenyan people and propelled socioeconomic development but has also facilitated the transfer of Chinese railway technology and reshaped transportation and commercial landscapes across East Africa. With trains roaring and whistles blowing, it tells compelling stories of development, cooperation, and friendship [24] (Fig. 22.13).

The Mombasa-Nairobi Standard Gauge Railway (SGR) has become a catalyst for Kenya's economic and social development, as well as for improving people's livelihoods. In the five years since its operation began, the SGR has consistently achieved a ridership rate of over 90%, making it highly popular among the Kenyan people. Preliminary estimates suggest that the SGR has contributed over 2% to Kenya's Gross Domestic Product (GDP). Moreover, the SGR has created nearly 50,000 job opportunities in Kenya, with over 80% of these jobs being localized. Additionally, the project has helped train more than 1,700 high-quality railway professionals and managers in Kenya. The Mombasa-Nairobi SGR was recognized by the UK's Telegraph as one

Fig. 22.13 Mombasa-Nairobi Standard Gauge Railway [25]

of the 13 most popular railways in the world. It was also ranked fifth in the list of the top 20 things to do in Kenya recommended by the CNN Travel Channel to American tourists. As Kenyan President Kenyatta remarked on the day of the SGR's official opening on May 31, 2017, "This is the beginning of transformation that will create jobs, hope, opportunities, and prosperity for all Kenyans."

The Jakarta-Bandung High-Speed Railway (JBHSR), scheduled to commence operation in June 2023, is a high-speed rail connecting the Greater Jakarta metropolitan area in the Republic of Indonesia with West Java Province. It marks Southeast Asia's first high-speed railway and represents the first full-system, comprehensive, and entire industrial chain deployment of Chinese high-speed rail technology overseas [26]. Stretching 142 km, the JBHSR is a flagship project under the Belt and Road Initiative and a testament to practical cooperation between China and Indonesia. Upon completion, travel time between Jakarta and Bandung will be reduced from over three hours to just 40 min. The JBHSR spans from the Halim Station in Jakarta to the Karawang Station in Bandung, with a total of four stations along its 142.3-km line. Designed for a maximum speed of 350 km per hour, the operational speed will be 300 km per hour. Representing China's first export of high-speed trainsets, the JBHSR trainsets are another milestone in the global outreach of Chinese high-end equipment. Leveraging the technology platform of China's "Fuxing" high-speed trainsets, renowned for the world's fastest commercial operation speeds, the JBHSR trainsets are customized for Indonesia, integrating local culture and tailored for the local operating environment (Fig. 22.14).

In recent years, China's railway expansion overseas has accelerated, spanning across Asia, Europe, North America, and Africa. It has become a shining example of the Belt and Road Initiative and international capacity cooperation, contributing significantly to global development. Transitioning from establishing itself as a "national symbol" to leading "international standards," China's railway expansion abroad has made remarkable strides. By addressing critical core technologies in

Fig. 22.14 Jakarta-Bandung high-speed railway [27]

the high-speed rail sector, China's standards have reached international levels, even leading the way in some areas. The International Union of Railways has implemented several high-speed rail standards developed by China, incorporating China's advantageous technologies and key parameters into international standards. This standardization framework provides the foundation for the comprehensive deployment of China's railway system, covering all aspects and elements of the industry, on the global stage.

From the TAZARA Railway, Mombasa-Nairobi Standard Gauge Railway, and China-Laos Railway, to the Hungary-Serbia Railway and the Jakarta-Bandung High-Speed Railway, railways have not only brought Chinese products, technology, and services but also created more development opportunities for the countries involved. Moreover, they have fostered friendly relations between China and the relevant countries. The stories of Chinese railways taking root and thriving around the world demonstrate that only by jointly creating a development pattern characterized by universal benefits, coordination and inclusiveness, cooperation and mutual benefit, and common prosperity can we truly promote global development [28].

From "Calling Card" to Standard: International Railway Union Releases Three Chinese High-Speed Rail Standards

In July 2022, the International Railway Union (UIC) released and implemented the standards "High-Speed Railway Design—Infrastructure" (IRS 60,680:2022) and "High-Speed Railway Design—Power Supply" (IRS 60,682:2022), both of which were the first international railway standards in their respective fields to be led by China [29]. In February 2022, the "Iron Alliance" had previously issued the standard "High-Speed Railway Design—Communication Signaling" (IRS 60,681:2021), making the "Infrastructure" and "Power Supply" standards the second and third international standards in the series of "High-Speed Railway Design" [30]. These two standards are the first international railway standards in their respective fields, earning praise and recommendation from experts from multiple countries, thus adding a pair of "wings" named Chinese wisdom and Chinese solutions to the construction and operation of high-speed railways worldwide.

The International Union of Railways (UIC) was established in 1922 as a non-governmental international railway organization. It consists primarily of four departments responsible for corresponding technical activities: Passenger, Freight, Rail System, and Fundamental Values. The organization also established a standardization agency to coordinate standardization work across departments. Currently, the UIC has 208 member entities, including two formal members and eight affiliate members from China. The UIC has a total of 599 IRS standards and Leaflet standards. Since 2011, the UIC has been organizing the conversion of Leaflet standards into IRS standards, and no new Leaflet standards have been issued. In recent years, the China National Railway Administration has actively promoted the participation of railway industry-related entities in UIC standardization work. The number of standards led or participated in by China has continued to increase. As of June 2022, China has led the development of 31 UIC standards (14 of which have been implemented) and participated in the development of 30 UIC standards. The quantity and quality of standards led or participated in by China have significantly improved.

The Chinese railway system, full of creativity and influence, not only facilitates people's travel but also drives the economic development of many regions, reflecting the prosperity of our homeland. Over a hundred years ago, the "Father of Chinese Railways," Zhan Tianyou, expressed such sentiments: "How can the vast land of China remain poor and weak? It will be rich and strong, relying on industry and learning." A century has passed, and from having no railways to having the world's longest high-speed rail network, we have independently built remarkable achievements such as the Qinghai-Xizang Railway, the Daqin Railway with the heaviest load, and the intelligent and environmentally friendly Beijing-Zhangjiakou High-Speed Railway. Each of these achievements represents unprecedented milestones in "Chinese standards" and is a true reflection of the Chinese nation's resilience and determination in overcoming challenges.

Looking back on the development of China's railway, there is a consistent core principle of "railways for the people." It has continuously progressed through dialogue with the times. As one of the important carriers leading the world with "Chinese standards," the Chinese railway undoubtedly embodies Chinese wisdom. It demonstrates the country's commitment to building a community with a shared future for mankind and reflects China's lofty ideals and relentless pursuit of constructing a better world. It is something that every Chinese person can be proud of.

From the "Four Vertical and Four Horizontal" to the "Eight Vertical and Eight Horizontal" of China's Railway: A Decade of Monumental Changes

Over the past decade, the comprehensive construction of the "Four Vertical and Four Horizontal" high-speed rail network has been completed, and the densification of the "Eight Vertical and Eight Horizontal" high-speed rail network has taken shape, providing strong support for major national strategies. The rapid construction and development of the high-speed rail network have effectively supported poverty alleviation, rural revitalization, regional development, and the creation of "urban clusters on rails." At the same time, the service scope of the high-speed rail network has enhanced China's international influence. The high-quality construction of the Mombasa-Nairobi, Jakarta-Bandung, and China-Laos railways has effectively promoted the development of the New Western Land-Sea Corridor, contributing to the joint construction of the "Belt and Road Initiative." From the "Four Vertical and Four Horizontal" to the "Eight Vertical and Eight Horizontal," the monumental changes in high-speed rail over the past decade have witnessed the continuous enhancement of China's international influence.

The "Four Vertical and Four Horizontal" refers to the national medium and long-term railway layout plan proposed in the "Medium and Long-term Railway Network Plan (Adjusted in 2008)" by the Ministry of Railways of the People's Republic of China: Railway development focuses on building a rapid passenger transportation network with a speed of over 200 km per hour for "Four Vertical and Four Horizontal" passenger dedicated lines as the backbone. The "Four Vertical" refers to: Beijing-Shanghai Passenger Dedicated Line, Beijing-Wuhan-Guangzhou-Shenzhen (Hong Kong) Passenger Dedicated Line, Beijing-Shenyang-Harbin (Dalian) Passenger Dedicated Line, Hangzhou-Ningbo-Fuzhou-Shenzhen Passenger Dedicated Line. The "Four Horizontal" refers to: Xuzhou-Zhengzhou-Lanzhou Passenger Dedicated Line, Shanghai-Hangzhou-Nanchang-Changsha-Kunming Passenger Dedicated Line, Qingdao-Shijiazhuang-Taiyuan Passenger Dedicated Line, Shanghai-Nanjing-Wuhan-Chongqing-Chengdu Passenger Dedicated Line. The completion of the "Four Vertical and Four Horizontal" high-speed rail network, serving as the backbone of the entire railway and the main artery of the country, marks a significant milestone in the history of railway development, and is also an integral part of the "Chinese Dream" of national prosperity, rejuvenation, and people's well-being. The high-speed rail network of "Four Vertical and Four Horizontal" not only connects major cities but also drives the economic pulse of the entire country. With the gradual networkization of high-speed rail, the radiation effect of high-speed rail has become increasingly evident. The travel time between regions has been greatly reduced after the construction of high-speed rail. The high-speed economic circles of 1–2 h between adjacent provincial capitals and 0.5–1 h between urban agglomerations within provinces have gradually formed. People have completely broken away from the outdated concept of geographical restrictions, and high-speed rail is gradually bridging the distance between cities, making it possible for resources to be shared among regions and easing the contradiction between driving economic growth and alleviating "urban ills" [31, 32].

From a comprehensive perspective, after the basic completion of the construction of the "Four Vertical and Four Horizontal" high-speed rail network, the tight capacity situation of China's railways has been basically alleviated, bottleneck constraints have been basically eliminated, and it has basically adapted to the needs of economic and social development. However, compared with the new situation and requirements of development, there are still shortcomings such as incomplete network layout, the need to improve operational efficiency, and prominent structural contradictions.

In July 2016, the National Development and Reform Commission, the Ministry of Transport, and China Railway Corporation jointly issued the "Medium and Long-Term Railway Network Plan", outlining the grand blueprint of the "Eight Vertical and Eight Horizontal" high-speed rail network in the new era. Among them, the "Eight Vertical" channels include coastal channels, Beijing-Shanghai channels, Beijing-Hong Kong (Taiwan) channels, Beijing-Harbin-Beijing-Hong Kong-Macao channels, Huhai-Nan channels, Beijing-Kunming channels, Bao (Yin)hai channels,

Lan (West)guang channels; the "Eight Horizontal" channels include Suiman channels, Beijing-Lanzhou channels, Qingyin channels, land bridge channels, along the Yangtze River channels, Shanghai-Kunming channels, Xiamen-Chongqing channels, Guangzhou-Kunming channels. From "connecting" to "networking", the "Eight Vertical and Eight Horizontal" injects new vitality into economic and social development; from "being able to travel" to "traveling well", the "Eight Vertical and Eight Horizontal" opens up the road to prosperity; from "following" to "leading", the achievements of the "Eight Vertical and Eight Horizontal" are attracting worldwide attention [33].

"The 'Heart-to-Heart' Road Across Six Provinces and Autonomous Regions—The Beijing-Xinjiang Expressway Creates the Miracle of a High-Grade Desert Highway".

One hundred years ago, Dr. Sun Yat-sen outlined a grand vision in his work "Strategies for National Reconstruction": "To establish a second major transportation corridor leading into Xinjiang, starting from Beiping (now Beijing), passing through Alashan, and extending to Dihua (now Urumqi in Xinjiang)."

This dream finally came true in 2021. On June 30, 2021, China Railway 22nd Bureau Group announced that the Beijing-Urumqi Expressway (referred to as "Jingxin Expressway"), funded by China Railway Construction Investment Group and constructed by China Railway 22nd Bureau Group and others, was completed and opened to traffic from Wutong Daquan to Mulei section. This highway, spanning the world's longest desert and Gobi, is also China's second major expressway corridor into and out of Xinjiang—the full line of the Beijing-Urumqi Expressway (G7) is now open (Fig. 22.15).

The Beijing-Urumqi Expressway, also known as the Jingxin Expressway, is the 7th radial artery in the national expressway network plan. It is also a vital transportation

Fig. 22.15 Jingxin Expressway [34]

artery for the western development of China, serving as a "connecting" road that spans six provinces and autonomous regions. The route of the Jingxin Expressway runs from Beijing to Urumqi, passing through Beijing, Hebei, Shanxi, Inner Mongolia, Gansu, and Xinjiang. With a total length of 2,822 km, it is the fastest highway corridor connecting Northeast, North China, and Northwest regions. Compared to the Lianyungang-Horgos Expressway (G30), the Jingxin Expressway shortens the distance from Urumqi to Beijing by more than 1,300 km, creating the most efficient Eurasian land bridge from Tianjin Port to Rotterdam Port in the Netherlands.

"Expanding the sky is our road, standing tall is our monument." The Jingxin Expressway traverses vast deserts and desolate wilderness, presenting immense construction challenges. It is another emblematic project of China characterized by its typical hardships, following the Qinghai-Xizang Railway. Situated in the western part of Inner Mongolia, the Lianbai section of the Jingxin Expressway faces harsh natural conditions. The Badain Jaran, Tengger, and Ulan Buh deserts span the region, and it is said that "when the wind rises in Ejina, the sand falls in Beijing." Confronted with frequent sandstorms and large temperature differences between day and night, the construction team developed unique techniques such as laying geotextile membranes to overcome the challenge of constructing 30.5 km of saline soil embankments. Additionally, considering the unique climate of Xinjiang, they intelligently utilized the terrain by installing snow and wind barriers at necessary locations, effectively reducing snow accumulation on the road surface and ensuring vehicle safety. In this area devoid of water, electricity, communication facilities, and human settlements, almost cut off from modern civilization, China has built a high-grade desert highway, a testament to the country's engineering prowess and hailed as a miracle of China.

The design and construction of the Jingxin Expressway also take into account ecological conservation and local economic development. In order to facilitate the migration of wildlife in the area, the construction team designed a total of 62 wildlife passages. Among them are two instances of three-tiered viaducts, where railway bridges are positioned on the top level, wildlife passages in the middle, and the expressway at the bottom (refer to Fig. 22.16).

The Jingxin Expressway is also known as the "Miraculous China Route 7." Along this expressway, people can appreciate a variety of landscapes including grasslands, rivers, forests, deserts, plateaus, lakes, glaciers, mountains, villages, and cities. It offers views of ten different sceneries: grassland green, desert gray, desert yellow, lake blue, sunset red, poplar gold, glacier white, forest green, and iridescent lake pink.

The Jingxin Expressway serves as the fastest maritime gateway from the Khorgos Port to the northern part of Tianjin Port, playing a crucial role in advancing the Belt and Road Initiative and establishing Xinjiang as a core area along the Belt and Road. Its full completion is also of great significance in driving the development of the economy along the northern slope of the Tianshan Mountains, ending the history of no expressways in Barkol County and Yiwu County. With the realization of the long-awaited prosperity road, fortune road, and dream road for the ordinary people living at the foot of the eastern Tianshan Mountains, the Silk Road is revitalized, injecting new vitality into the region. With the opening of the Wutongdaquan to

Fig. 22.16 Three-tiered viaducts of Jingxin Expressway [35]

Mulei section of the Jingxin Expressway, the total length of highways in Xinjiang exceeds 6,000 km, making the road network smoother and clearer in achieving the goal of improving internal connectivity and facilitating entry and exit to and from Xinjiang. Xinjiang's position as a core area of the Belt and Road Initiative and a pioneer in opening up to the west will be further highlighted [36, 37].

The Hong Kong-Zhuhai-Macao Bridge, One of the "Seven Wonders of the New World," Serves as the "Spine" of the Guangdong-Hong Kong-Macao Greater Bay Area

In the vast expanse of the South China Sea, a magnificent "Bridge Across the Sea" spans the waters. This is the Hong Kong-Zhuhai-Macao Bridge (HZMB), a super project integrating bridges, islands, and tunnels, connecting the Hong Kong Special Administrative Region, Zhuhai in Guangdong Province, and the Macao Special Administrative Region (see Fig. 22.17). As the world's longest sea-crossing bridge, it has been hailed by The Guardian as one of the "Seven Wonders of the New World." The HZMB is an integral part of the national highway network plan in the Pearl River Delta region and a crucial project spanning the Lingdingyang area, providing a new road transport link between the eastern and western banks of the Pearl River.

The HZMB is a mega-scale sea-crossing project jointly constructed for the first time by Guangdong, Hong Kong, and Macao under the framework of "One Country, Two Systems." It spans a total length of 55 km and is designed to have a service life of 120 years, with a total investment of approximately 120 billion RMB. The preliminary work for the bridge commenced in August 2003, construction started in

Fig. 22.17 Hong Kong-Zhuhai-Macao Bridge [38]

December 2009, and after a lengthy preparation and construction period of 15 years, it was opened to traffic in October 2018.

The main works of the bridge is managed by the Hong Kong-Zhuhai-Macao Bridge Authority, jointly established by Guangdong, Hong Kong, and Macao, responsible for construction, operation, management, and maintenance. The ports and connecting lines in each region are constructed and operated separately by local governments. The main project adopts a combination of bridge, island, and tunnel structures, with a total length of approximately 29.6 km. About 6.7 km of the route passes through the Lingding Channel and the Tonggu Channel as tunnels, with artificial islands (Dolphin Island and White Dolphin Island) set at both ends, resembling sparkling "twin pearls." The remaining section spans approximately 22.9 km of bridges, featuring three cable-stayed bridges: the "Chinese Knot" Qingzhou Bridge symbolizing unity among the three regions, the "Dolphin Tower" Jianghai Bridge symbolizing harmonious coexistence between humans and nature, and the "Sail Tower" Jiuzhou Bridge representing embarking on a new journey.

The artificial island at the Zhuhai-Macau Port has a total area of 208.87 ha, divided into three zones: the Zhuhai Port Management Area (107.33 ha), the Macau Port Management Area (71.61 ha), and the Bridge Management Area (29.93 ha). Each zone is independently managed by the respective authorities. The Zhuhai Connection Line, spanning 13.4 km, connects the Zhuhai Port to the western coastal highway, extending from the Moon Ring Road to the Nanping Branch Extension Line, thereby integrating the bridge into the national highway network. The Macau Connection Line accesses the Macau Reclamation Zone from the Macau Port via bridge [39].

The most challenging aspect of the Hong Kong-Zhuhai-Macao Bridge project is the immersed tube tunnel construction. The undersea tunnel of the bridge spans 5.6 km and consists of 33 giant immersed tubes, each measuring 180 m in length and 38 m in width. The working depths of E9-E27 range from 45 to 50 m, with a foundation groove depth of 35–40 m. Each immersed tube weighs 80,000 tons, equivalent to placing 33 aircraft carriers underwater, for which there is no prior construction experience at such depths with similar grooves worldwide. The project imposes stringent requirements on meteorological and hydrological conditions. The core processes of the Hong Kong-Zhuhai-Macao undersea tunnel construction are floating transportation and immersion, with the former facing significant challenges due to meteorological and oceanic factors. The massive resistance generated by the 80,000-ton immersed tubes, coupled with water currents of 10 cm per second per meter, presents considerable towing control difficulties. Construction can only proceed under suitable weather, tidal, wave, and current conditions, requiring precise coordination of multiple factors to identify brief operational "windows," which occur only twice a month. Therefore, construction necessitates extremely accurate meteorological and hydrological forecasts. While a 70% to 80% accuracy rate may suffice for general forecasts, for the Hong Kong-Zhuhai-Macao Bridge undersea tunnel project, even a minor increase of 0.1 m in wave height or a flow velocity increase of 0.1–0.3 cm per second can determine success or failure. Such margins of error are unacceptable, necessitating a forecast accuracy rate of 100%.

In terms of construction standards, the Hong Kong-Zhuhai-Macao Bridge is jointly constructed by Guangdong, Hong Kong, and Macao, and is subject to joint supervision by the three parties. However, the design and construction quality standards proposed by the three parties differ: Hong Kong follows British standards, Macao adopts European standards, and the Mainland adheres to Chinese standards. Therefore, the bridge's design adheres to the highest standard to meet the requirements of all parties, following the principle of "choosing the highest standard."

Behind the 'super project' lies 'super innovation.' The construction of the Hong Kong-Zhuhai-Macao Bridge posed immense challenges, leading to the emergence of new materials, new techniques, new equipment, and new technologies. With as many as 400 patents, it filled numerous gaps in various fields, resulting in the creation of the world's most difficult, longest, and deepest underwater highway immersed-tube tunnel. Several world records were set, including the construction of the largest-scale steel bridge sections, the production and installation of the world's longest underwater tunnel, and the rapid formation of two artificial islands [40].

The opening of the Hong Kong-Zhuhai-Macao Bridge has significantly enhanced the flow of people, goods, capital, and information within the Guangdong-Hong Kong-Macao Greater Bay Area. Previously, the logistics industry faced development bottlenecks due to congestion at the Humen Bridge. With the diversion brought about by the bridge, these bottlenecks have been alleviated, greatly promoting integrated development within the region.

The Hong Kong-Zhuhai-Macao Bridge and the transportation corridor it establishes have enhanced the economic radiance and driving force of Hong Kong and the eastern bank of the Pearl River, fully tapping into the development potential of the western bank of the Pearl River. The smooth operation of the bridge has reduced the travel time from Hong Kong to Zhuhai from 3 h to just 30 min, facilitating closer economic and social ties between the western bank of the Pearl River and Hong Kong. This promotes coordinated economic and social development across the Pearl River Delta, enhances the overall competitiveness of the region, and maintains the sustained prosperity and stability of Hong Kong and Macao, all of which hold epoch-making significance [41, 55].

The World's Highest Bridge: Beipanjiang First Bridge Integrates the Guizhou-Sichuan-Yunnan Border Region into the National Expressway Network

The Beipanjiang First Bridge, jointly constructed by Guizhou and Yunnan provinces, originally known as the Ni Zhu River Bridge or Beipanjiang Bridge, is a mega-bridge located within China, connecting Puli Township in Xuanwei City, Qujing, Yunnan Province, and Duoge Town in Shuicheng District, Liupanshui, Guizhou Province. It spans the Ni Zhu River and is part of the Hangrui Expressway. Renowned both domestically and internationally for surpassing the Four-Dan River Bridge in relative height and setting a new record as the world's highest bridge (see Fig. 22.18) [42].

The bridge commenced construction in 2013 and achieved closure of the main span on September 10, 2016. It officially opened to traffic on December 29 of the same year. Stretching from Duoge Town in the north, spanning the Ni Zhu River Grand Canyon, to Lalong Village in the south, it has a total length of 1,341.4 m. The distance from the bridge deck to the river surface is 565.4 m. It is designed according to the standard of a two-way, four-lane expressway, with a design speed of 80 km per hour. Constructed through cooperation between Yunnan and Guizhou provinces, the total investment in the bridge amounted to 1.028 billion yuan. Additionally, the heights of the main bridge piers on the eastern and western banks are 269 and 247 m, respectively. With a main span of 720 m, it ranks second in the world in terms of the span length among similar bridge types.

Fig. 22.18 Beipanjiang first bridge [43]

As a world-class bridge, since the commencement of construction in 2012, the project site has been characterized by rugged mountains, crisscrossing valleys, complex geology, and harsh climate conditions. During construction, the project faced five major challenges in the mountainous region: temperature control for large-volume concrete foundations, high-performance concrete pumping for ultra-high cable towers, integral hoisting of heavy steel anchor beams in mountainous areas, top pushing of steel truss girders with no horizontal force on the high piers, and the final closure of the large-span steel truss girder cable-stayed bridge. To overcome these challenges and ensure the smooth completion of the bridge, the designers adopted "smart" concrete and cloud computing technologies, ensuring both safety and successful project execution. The "smart" concrete developed for the bridge is known as "self-compacting mechanism sand concrete," which possesses high fluidity and excellent resistance to segregation and bleeding. It does not require vibration and can be uniformly compacted and molded solely by its own gravity, meeting the modern structural complexity and dense reinforcement requirements of concrete construction. With the successful opening of the bridge, all these challenges have been effectively overcome by China's technology and engineering teams.

After the bridge is opened to traffic, the journey from Xuanwei City in Yunnan to Liupanshui in Guizhou will be reduced from about 5 h to just over 1 h. The highway connected by the bridge, the Bidu Expressway, extends northeast to the Zunbi Expressway to Zunyi, north to the Bisong Expressway to Luzhou, northwest to the Biwei Expressway to Weining, west to the Puxuan Expressway to Qujing in Yunnan, east to the Zhina Expressway to Guiyang, and south to the Shuipan Expressway to Xingyi. It can quickly integrate the border areas of Guizhou, Sichuan, and Yunnan into the national highway network, providing modern, high-quality, and efficient transportation guarantee for the leapfrog development of Bijie and Liupanshui, promoting the comprehensive urban–rural reform pilot in Liupanshui, and improving the modern highway transportation skeleton system in Guizhou and even the whole country. It is of great significance to realize the strategy of "accelerated development, faster transformation, and promotion of leapfrogging" in Guizhou and respond to the national "Belt and Road" initiative [44].

"The First Tunnel of Sichuan-Xizang Highway": The New Erlangshan Tunnel on the Yakan Expressway Aids in Constructing the Sichuan-Xizang "Economic Corridor"

In 1998, the first phase of the Erlangshan Tunnel, with a total length of 4,176 m, was successfully opened to traffic. Nineteen years later, the new Erlangshan Tunnel on the Yakan Expressway, the second tunnel traversing Erlangshan, was also successfully opened to traffic (see Fig. 22.19). With a total length of 13,459 m, it is the fourth longest highway tunnel in the country and the longest in high-altitude regions of China. Serving as the first long tunnel on a highway entering Ganzi Prefecture from the Chengdu Plain, it connects Tianquan County in Ya'an City with Luding County in Ganzi Prefecture and is acclaimed as the "First Tunnel of Sichuan-Xizang Highway." The project officially achieved breakthrough on September 26, 2017, and commenced trial operation on December 31, 2017.

Erlangshan is the first high mountain on the Sichuan-Xizang Highway from the Chengdu Plain to the Qinghai-Xizang Plateau, known as the "thousand-mile Sichuan-Xizang Highway, with Erlangshan as the natural barrier." Three-quarters of the year experience rainy or snowy weather, coupled with the harsh climate and complex

Fig. 22.19 The New Erlangshan tunnel on the Yakang Expressway [56]

terrain. The Erlangshan road features numerous bends, steep slopes, and narrow paths, often necessitating one-way traffic control. The New Erlangshan Tunnel, located in the VIII-degree seismic intensity zone, traverses 13 regional fault zones. Its construction has reduced the time required to cross Erlangshan from over an hour to just 15 min. Notably, the project has achieved two "firsts": the longest single-end excavation of 7,333.6 m, ranking first among all highways tunnels in China, and the longest stretch of asphalt concrete pavement of 13.4 km, ranking first among all highways in areas involving Xizang.

In terms of engineering innovation, the tunnel has achieved five "firsts": it is the first to incorporate a double-lane large-section traffic conversion channel within the tunnel, enhancing the tunnel's disaster prevention and rescue capabilities. It is also the first to adopt an earthquake-resistant design concept beyond expectations, with the tunnel section inside the active fault zone expanded by 40 cm to provide space for post-earthquake reinforcement while ensuring that the service level of the tunnel is not compromised. Moreover, it is the first to completely implement reverse excavation within inclined shafts, effectively protecting the ecological environment of the giant panda conservation area. Additionally, it is the first to install a self-flowing high-level firefighting water pool within the tunnel, eliminating the need for pumping equipment, thereby improving firefighting reliability. It successfully utilizes the height difference in the inclined shafts to generate electricity from firefighting residual water, and takes advantage of the pressure difference formed by the different climatic conditions at both ends of the tunnel to establish a natural air passage, resulting in approximately 15% energy savings. Lastly, it is the first to employ a visual dynamic lighting system to create a conducive driving environment within the tunnel, alleviating psychological pressure on drivers and reducing driving risks.

On December 6, 2021, the China Construction Enterprise Management Association announced the list of projects selected for the "National Quality Engineering Award" for the years 2020–2021, with the New Erlangshan Tunnel of the Yakang Expressway included. This marks another national-level honor for the Yakang Expressway following the "Li Chun Award" [45].

The New Erlangshan Tunnel of the Yakang Expressway, as a critical segment along the Sichuan-Xizang Highway, serves as the vital link between past and future routes. By bypassing the lengthy and steep slopes as well as the icy sections from Xingou to G318 Erlangshan Tunnel, it ensures year-round accessibility from Ya'an to Luding and shortens the journey by nearly two-thirds. This contributes significantly to the establishment of the Sichuan-Xizang Economic Corridor and promotes the openness and development of the Ganzi and even Xizang regions, holding profound political, economic, and cultural significance [46].

The 12.5-Meter Deepwater Channel Project Below Nanjing on the Yangtze River Provides Strong Support for the Development of the Yangtze River Economic Belt.

The 12.5-Meter Deepwater Channel Project Below Nanjing on the Yangtze River is a major project explicitly outlined in the development plan for the Yangtze River Economic Belt, as well as in the "Twelfth Five-Year Plan" and "Thirteenth Five-Year Plan" for national economic and social development. It is the largest inland waterway

Fig. 22.20 Full completion of the 12.5-meter deepwater channel project below Nanjing on the Yangtze River [47]

investment project in China, with the most complex technical and construction environment. The project covers the section of the Yangtze River main channel from Nanjing to Taicang, with a total length of approximately 283 km. Following the construction philosophy of "overall planning, phased implementation, bottom-up approach, and ensuring passage before smooth flow," the project is being implemented in stages. Located at the intersection of the "Belt and Road Initiative," this river section boasts the best navigational conditions, highest ship traffic volume, and most significant economic benefits along the Yangtze River main channel. Despite representing only 1/7 of the total navigable length of the Yangtze River, it handles 70% of the total cargo traffic, with an annual volume exceeding 1.6 billion tons (Fig. 22.20).

In 2014, the State Council issued the "Guiding Opinions on Promoting the Development of the Yangtze River Economic Belt by Leveraging the Golden Waterway," which clearly stated the "Key Implementation of the 12.5-Meter Deepwater Channel Project below Nanjing in the downstream." To implement the decisions and arrangements of the Party Central Committee, the Ministry of Transport and the Jiangsu Provincial People's Government jointly established a leadership group and command headquarters for the Deepwater Channel Project to promote the construction of the project. After 8 years of efforts, the 12.5-m deepwater channel achieved full-line connectivity in May 2018, providing stronger support for the development of the Yangtze River Economic Belt, the integration strategy of the Yangtze River Delta, and the implementation of the Belt and Road Initiative.

The project is located in the lower reaches of the Yangtze River, from the alluvial plain river section to the estuary section. It is affected by both runoff and tides, with numerous shallow river sections branching out, making the bars and channels prone to frequent and complex changes in water and sediment exchange. Some sections of the river have rapid currents and deep waters, posing significant construction challenges. There are many external constraints to channel regulation, requiring comprehensive consideration of multi-channel navigation. There are few large-scale tidal river channel regulation experiences available for reference domestically and internationally.

The project is located in the most developed area downstream of the Yangtze River, with numerous industries along the banks and dense port facilities. It is the busiest section of the Yangtze River navigation, characterized by high navigational density and complex vessel traffic. The construction period of the project is long, covering a wide range. Moreover, simultaneous channel improvement projects are being implemented, requiring a large number of vessels and construction personnel, leading to significant challenges in on-site safety management. Ensuring both navigation and construction safety is not only a requirement for promoting the smooth progress of the project but also essential for ensuring the unimpeded flow of the Golden Waterway artery.

The construction scope of the project spans a long distance, involves multiple river sections, and deals with complex environments. The project area harbors diverse aquatic species and abundant fisheries resources. It also encompasses several natural reserves, aquatic germplasm resource protection areas, drinking water source protection areas, and ecological red line control zones. Given the high density of environmentally sensitive areas and the high social sensitivity, there are stringent requirements for environmental protection work. During the project construction process, the competent authorities attach great importance to and dialectically grasp the relationship between ecological protection and construction. They have established a comprehensive system and mechanism for ecological environmental protection construction management, innovatively designed ecological structures such as eco-bottoms, eco-banks, eco-shores, and embankments suitable for aquatic organisms to grow and inhabit. Experimental facilities such as eco-fish nests and floating islands have been constructed in the Yangtze River waters. For the first time in similar projects, work stoppages have been implemented during the fish breeding period. Environmental monitoring during construction has been strengthened, and relevant measures have been implemented. Targeted ecological compensation measures such as enhancement stocking and release have been implemented. While achieving the goal of river regulation, the project has effectively protected the ecological environment, yielding theoretical and practical achievements in the field of ecological waterway construction.

Under the correct leadership of the Ministry of Transport and the People's Government of Jiangsu Province, the Command Headquarters and all participating units of the Yangtze River deepwater navigation channel construction project have diligently implemented General Secretary Xi Jinping's important directives on the Yangtze River Economic Belt, emphasizing "ensuring overall conservation while refraining from large-scale development" and advocating for "ecological priority and green development." They actively practiced the new development philosophy, completed the construction task six months ahead of schedule, achieved the full-line connection of the deepwater navigation channel in May 2018, and attained a 100% qualification rate for project quality. Furthermore, they realized zero accidents in safety production and zero complaints related to Party conduct and government integrity.

In 2021, the Phase II project of the 12.5-m deepwater navigation channel below Nanjing on the Yangtze River was awarded the "National Quality Engineering Gold Award" in the National Quality Engineering Awards for the years 2020–2021. This accolade follows its previous recognition in 2020, where it received the "China Construction Engineering Luban Award," marking another national-level quality award for the project. The Phase I project also received accolades such as the National Quality Engineering Award and the Quality Engineering Award for Water Transport. Grounded in practicality, the project achieved a series of innovative results in tidal estuary channel regulation theory, design methods, new structures, process technology equipment, acceptance technical standards, and engineering construction management, overcoming significant technical challenges in construction under deepwater and high-flow conditions. It has garnered a total of 30 awards at provincial and ministerial levels, obtained 85 approved patents, 13 construction methods, and 10 software copyrights.

The comprehensive increase in the navigable depth to 12.5 m in the project has significantly enhanced the loading capacity of large ocean vessels, further promoting the optimization and adjustment of ship transportation organization methods, thereby reducing logistics costs. The large dock berthing capacity along the Jiangsu River ports has been fully utilized, forming a hub for cargo transportation. Additionally, it has further elevated the level of logistics services along the river, facilitating the development of industrial scale aggregation along the river and creating conditions for the transformation and upgrading of industries along the river.

The Deepwater Navigation Project is a significant water transportation project serving the implementation of major national strategies. The construction of the project has notably improved the navigation conditions along the Yangtze River below Nanjing, enhancing its navigational capacity. This improvement is conducive to further enhancing the economic value of the Yangtze River Golden Waterway, thereby promoting structural reforms in water transportation supply, building a comprehensive and multi-dimensional transportation corridor along the Yangtze River Economic Belt, and driving economic and social development in the Yangtze River Basin [48, 49].

Yangshan Deep Water Port: "China Core" Breaks Foreign Technological Monopoly and Achieves Surpassing

Shanghai Yangshan Deep Water Port (referred to as "Yangshan Port") is located on the rugged islands outside the mouth of Hangzhou Bay. It consists of Xiaoyangshan Island area, East China Sea Bridge, and Yangshan Free Trade Port Zone. It was opened on December 10, 2005, and operationally falls under the Shanghai Port Area, while administratively it belongs to Shengsi County, Zhoushan, Zhejiang Province. The planned total area of Yangshan Port area exceeds 25 square kilometers, including four port areas: East, West, South, and North. Following the principle of "one-time planning, phased implementation," construction was carried out in four phases from 2002 to 2020.

In 1995, Shanghai officially proposed the construction of a deep-water port. In 1996, following the establishment of the Shanghai International Shipping Center Shanghai Area Leading Group Office, the preliminary selection, assessment and project processing work began in September of the same year, in accordance with the requirements of the Ministry of Transport's notice. By March 2002, after six years, the Yangshan Port construction project was approved by the State Council. In December 2005, the Phase I project of Yangshan Port (including the East China Sea Bridge) was completed and put into operation. By the end of 2017, the Phase IV automated terminal of Yangshan Port will begin trial operation, becoming the largest fully automated terminal in the world. On August 6, 2019, the State Council issued the Overall Plan for the China (Shanghai) Pilot Free Trade Zone Lingang New Area, and Xiaoyangshan Island was included in the scope of the Lingang New Area (Fig. 22.21).

Construction of Phase I of the Yangshan Port project began in June 2002, and by December 2005, Phase I was completed and put into operation, with a total of 16 large container deep-water berths built across three phases by 2008. Due to the rapid development of business since the opening of Yangshan Port, the Shanghai International Shipping Center required a new container terminal to undertake more challenging tasks. Therefore, the Yangshan Port Phase IV Automated Terminal Project (referred to as "Yangshan Port Phase IV Project") was initiated.

The Yangshan Port Phase IV Project, built on the reclaimed land of the Kezhushan Island and the Da and Xiao Wugui Islands, covers a total land area of 2.23 million

Fig. 22.21 Yangshan Deep Water Port in Shanghai [50]

square meters, with a quay length of 2350 m. It comprises 7 deep-water container berths and 61 automated container yards. The entire project commenced construction in December 2014 and was completed and put into operation in December 2018, becoming the world's largest intelligent container terminal. On November 30, 2019, it was selected for the Second Excellent Marine Engineering List.

From 2.01 million TEUs in 2018 to 3.27 million TEUs in 2019, to 4.2 million TEUs in 2020, and further to 5.7 million TEUs in 2021, the unmanned terminal of the Yangshan Port Phase IV Project has demonstrated immense technological prowess. It even won the Shanghai Science and Technology Progress Special Award in 2020, showcasing its significant technological advancements.

As the youngest terminal in Shanghai Port, the automated terminal of Yangshan Port Phase IV (referred to as "Yangshan IV") boasts a smaller coastline and land area compared to the previous three phases, yet it is more eye-catching and carries greater expectations. It is one of the world's most highly automated container terminals and is also the largest automated container terminal completed and put into operation at one time globally, earning it the title of "masterpiece of consolidation." Its completion and operation signify a milestone leap and significant transformation in the operational model and technological application of China's port industry. Moreover, it solidifies Shanghai Port's position as the world's leading container throughput hub and accelerates its progress towards becoming a top-tier global shipping center, providing fresh impetus for its advancement.

The data shows that in 2021, the container throughput of Yangshan Deep Water Port reached 22.813 million standard containers, a year-on-year increase of 12.8%. Among them, the annual throughput of Yangshan Phase IV exceeded 5.7 million standard containers for the first time, with a year-on-year growth of 35.7%. It is estimated that the throughput will reach 6.3 million standard containers in 2022.

With minimal space and personnel, achieving maximum efficiency is made possible by the key term "intelligence." Thanks to its high level of intelligence, the loading and unloading efficiency of Yangshan Phase IV has increased by nearly 30% compared to traditional terminals, while labor costs have been reduced by almost 70%. Moreover, the working environment for crane operators has been significantly improved, transitioning from "blue-collar" to "white-collar" jobs. Additionally, the application of large-scale electrically driven equipment, automatic guided vehicles (AGVs), paperless operations throughout the process, and solar-assisted heating technologies have made Yangshan Phase IV a model of a green port.

If the rise of the Yangshan Port area has propelled the Port of Shanghai towards deep blue waters, earning it the title of "Oriental Giant Port" and initiating a streak of championships in maintaining the world's top container throughput, then the birth and development of the automated terminal at Yangshan Phase IV has enabled the Port of Shanghai to complete its transformation from an "Oriental Giant Port" to a "World-Class Port," leading the way in technology and operational models and spearheading a "port revolution" on a global scale.

As this "Chinese core" breaks foreign technological monopolies and achieves a leading position, its replication overseas becomes inevitable. On September 1, 2021, the Shanghai International Port Group officially opened the Haifa New Port in

Fig. 22.22 Haifa New Port in Israel [51]

Israel, replicating the technological highlights of the Yangshan Phase IV terminal (see Fig. 22.22). This marks the first time that China has exported advanced "smart port" technology and management experience to developed countries. From 2012 to 2020, Shanghai has selected eight innovative achievements with special economic and social value for the "economic main battlefield" (Science and Technology Progress Award), with Yangshan Phase IV winning this prestigious honor. It is evident that in the pursuit of becoming world-class, the automated terminal at Yangshan Phase IV has become another remarkable innovation hub in Shanghai.

Yangtze River Ecological Waterway: A Faithful Practitioner of "Pursuing Large-Scale Conservation and Avoiding Large-Scale Development"

Flowing through the magnificent land of China, the Yangtze River gathers countless tributaries and has consistently ranked as the world's top inland waterway for cargo transportation for many years. Compared to other modes of transportation, Yangtze River shipping boasts advantages such as minimal land occupation, low costs, low energy consumption, light pollution, high capacity, and high efficiency. The 18th National Congress of the Communist Party of China clearly stated the goal of "building a beautiful China and achieving sustainable development for the Chinese nation." This requires the implementation of scientific management and moderate development of the Yangtze River waterway, making reasonable use of its natural depth for water transportation. This approach can effectively save land resources, reduce energy consumption, and mitigate environmental pollution, thus alleviating pressures on resources, environment, and ecology, and promoting the construction of a beautiful China (Fig. 22.23). Following the strategy of "deepening the downstream, smoothing the midstream, extending the upstream, and connecting the tributaries" in construction and maintenance, the "gold content" of the Yangtze River's golden waterway has been increasing year by year. The construction of green ecological waterways has been progressing vigorously. The Yangtze Eco-Navigation is becoming a faithful practitioner of "pursuing large-scale conservation and avoiding large-scale development."

In September 2013, the Jingjiang River Channel Improvement Project (referred to as the "Jingjiang Project") in the middle reaches of the Yangtze River from Changmen Creek to Xiongjiazhou officially commenced construction and was completed in April 2017. The Jingjiang Project has enhanced the navigational capacity of the high-grade transportation main channel and fully leveraged the benefits of the golden waterway navigation, providing strong support for the economic and social development of the regions along the Xiang and E Rivers, as well as promoting economic development and external openness in the upper reaches of the Yangtze River. The Jingjiang Project attaches great importance to ecological conservation and actively promotes the construction of an ecological waterway. It integrates ecological conservation concepts throughout the entire process of channel planning, design, construction, and maintenance. In September 2014, it was approved as an ecological conservation demonstration project by the Ministry of Transport.

Four years after the completion of the Jingjiang Project, an ecological waterway has been established in the Yangtze River basin. The populations of critically endangered species such as the finless porpoise and the Reeves's muntjac have significantly

Fig. 22.23 Yangtze River ecological waterway [52]

increased within the three national-level protected areas along this stretch of the river. Walking along the banks of Jingjiang River in Jiangling County, lush greenery lines both sides of the embankment, and the slopes are covered with grass thanks to the vegetated steel wire mesh slope protection structures. This type of wire mesh boasts exceptional corrosion resistance with a lifespan of up to 120 years. What was once a desolate sandbar devoid of vegetation has now transformed into a vast oasis, where reeds sway and wild ducks find refuge.

The Jingjiang Project allocated over 200 million yuan specifically for ecological construction, including ecological revetment structures and ecological embankment reinforcement, as well as the restoration of vegetation on the banks and the rehabilitation of terrestrial and aquatic environments spanning 2.18 million square meters. During the restoration process, a large number of underwater "artificial reefs" were deployed. These ecological permeable frameworks, made from new materials and techniques, have become new habitats for animals such as the finless porpoise, which prefer to live at the head and tail of sandbars.

In 2018, the 6-m deep waterway improvement project (referred to as the "Wuhan-Anqing Project") along the Wuhan to Anqing section of the Yangtze River main channel officially commenced construction. Following the Jingjiang Project, this marked another systematic improvement project for a long river section. The project area spans across 3 provinces and 8 cities, including Hubei, Jiangxi, and Anhui, with numerous points, extensive lines, and complex terrain. After 29 months of intensive construction, by the end of 2020, the Wuhan-Anqing section project had entered its final stages. The water depth of the navigation channel increased from 4.5 m to 6 m, allowing large vessels of tens of thousands of tons to sail directly from the sea to Wuhan, adding a significant enhancement to the magnificent Yangtze River navigation. On May 20, 2022, the Wuhan-Anqing section project successfully passed the completion environmental protection acceptance. Serving as a benchmark for the implementation of the ecological navigation channel construction concept in the long river section improvement, the project has formulated a replicable and promotable ecological construction "Yangtze Plan". It has significant demonstration effects in the Yangtze River and even nationwide inland river management, with remarkable achievements in ecological construction.

For the constructors of the Yangtze River navigation channel, the ecological and environmental protection of the Wuhan-Anqing section project poses a sensitive and complex "must-answer question". The project spans across Hubei, Jiangxi, and Anhui provinces, passing through seven cities including Wuhan, Huanggang, Jiujiang, and Anqing. Along the way, there are seven protected areas, two wetland reserves, and 32 water intake points, presenting significant ecological, flood control, and safety pressures.

With a commitment to "ecological priority and green development" as its guiding principle, the project seizes the opportunity of being included in the pilot construction of a transportation power country for the Wuhan-Anqing section to create a model project for the construction of a green channel on the Yangtze River mainline and for the beauty of the Yangtze River. In the project's "top-level design," an investment of 450 million yuan was allocated for ecological protection, with ecological protection implemented in all seven river sections, making it the project with the highest ecological protection investment, the widest ecological construction scope, and the most ecological structures and technology applications among all national inland river channel improvement projects. In terms of project site selection, full consideration was given to ecological sensitivity factors of the river, avoiding encroachment on irreplaceable habitats due to the layout of improvement projects. While ensuring the effectiveness of the project's improvement, ecological and environmental protection needs were comprehensively considered.

Looking at the process of green ecological development along the Yangtze River, it is not difficult to see that channel improvement and environmental protection are closely integrated. By optimizing engineering schemes, adopting ecological structures, and implementing environmental protection measures, the impact of engineering on the environment can be minimized to the greatest extent, achieving coordinated development between engineering construction and environmental protection. This allows the Yangtze River, which carries the soul of the nation and the benefits of development, to become more magnificent and poetic, unfolding a green ecological picture of harmony between people and water.

"In order to maintain historical patience and strategic determination, we must persist in carrying out our blueprint to the end, tackling one task after another, and ensuring that the clear waters of the Yangtze River will benefit future generations and bring blessings to the people."[2] In November 2021, at the symposium on comprehensively promoting the development of the Yangtze River Economic Belt, General Secretary Xi Jinping's resolute words echoed through history, sounding the clarion call of the times and charting the course for the high-quality development of the Yangtze River Economic Belt.

In the past decade, over 50,000 administrative villages have been connected by public transportation. In the same period, the railway mileage in western regions has surpassed 60,000 km, accounting for 40% of the national total. Ten years ago, migrant workers returning home for the Spring Festival formed massive motorcades, but today, this scene has become history as safe and rapid high-speed railways link construction sites to their hometowns in China.

In the past decade, the Chinese people have never ceased to overcome challenges. Among the mountains and rivers, in remote villages and deserts, they have created one human miracle after another, connecting more and more previously isolated areas to the transportation network. Countless Chinese have found warmth and happiness, pursuing their ambitions and dreams through mobility.

[2] "Xi Jinping emphasized at the Symposium on Promoting the Comprehensive Development of the Yangtze River Economic Belt to implement the spirit of the Fifth Plenary Session of the 19th CPC Central Committee and promote the high-quality development of the Yangtze River Economic Belt."—People's Daily, November 16, 2020, Page 01.

References

1. Baidu Baike. Lhasa-Linzhi Railway [EB/OL]. [2022-08-19]. https://baike.baidu.com/item/拉林铁路/1764731?fr=aladdin.
2. National Railway Administration. One-Year Anniversary of the Opening and Operation of Lhasa-Linzhi Railway [EB/OL]. (2022-06-27) [2022-08-19]. http://www.nra.gov.cn/xwzx/tpsp/tpxx/202206/t20220627_327480.shtml.
3. China Economic Net. Lhasa-Linzhi Railway: New Heavenly Road Writes a New Height on the Roof of the World [EB/OL]. (2021-05-10) [2022-08-19]. http://www.ce.cn/cysc/jtys/tielu/202105/10/t20210510_36545830.shtml.
4. People's Political Consultative Conference News. One-Year Anniversary of the Opening and Operation of Lhasa-Linzhi Railway [EB/OL]. (2022-06-27) [2022-08-19]. (Note: Original link is repeated with [1], assuming a different but related link here)
5. Guizhou Political Research Network. To Completely Tear Off the Poverty Label of Thousands of Years [EB/OL]. [2022-08-19]. http://www.gzswzys.gov.cn/mobile/content.aspx?id=116a9f7b-44ba-4f2a-a931-aab67ba4822b.
6. Lu Linyu, Ren Lingjie. Traversing Mountains and Valleys, Driving the High-Quality Development of Colorful Guizhou [EB/OL]. (2022-08-16) [2022-08-19]. http://comment.gog.cn/system/2022/08/16/018204268.shtml.
7. China Government Network. 15 Years of Chongqing Municipality: From Natural Barriers to Thoroughfares [EB/OL]. [2022-08-19]. https://news.sina.com.cn/o/2012-09-03/170925091608.shtml.
8. Chongqing News. How Much Have Chongqing's Transportation Changed in These Five Years? [EB/OL]. [2022-08-19]. https://www.sohu.com/a/138018377_449808.
9. Chen Zhu. 2021 Chongqing Transportation Annual Report ③|Coexistence of Magic and Beauty! These Unique Transportation Facilities in Chongqing are Worth Visiting - Upstream News • Gathering Upward Momentum (cqcb.com) [EB/OL]. [2022-08-19]. (Note: Original link is assumed as it was not provided)
10. Lao Su. A Marvel in Chongqing - The Light Rail Passes Through a Building [EB/OL]. (2021-05-28) [2022-08-19]. https://www.clzg.cn/article/277671.html.
11. Ninglexi Travel Notes. Top 10 Filming Locations in Chongqing from the Movie Better Days, Including the Crown Escalator and Tongyuanju Subway Station [EB/OL]. (2019-10–30) [2022-08-19]. https://ninglexi.com/shaoniandeni.html.
12. Haibao News. The World's First Desert Loop Railway Opens! China's Railway Miracles Go Beyond This Strongest Closed Loop [EB/OL]. (2022-06-17) [2022-08-19]. https://baijiahao.baidu.com/s?id=1735869829578992818&wfr=spider&for=pc.
13. Southeast Express. It's Here! Opening Today! The World's First Desert Loop Railway! How Spectacular It Is [EB/OL]. (2022-06-16) [2022-08-19]. https://baijiahao.baidu.com/s?id=1735793956254726443&wfr=spider&for=pc.
14. CCTV.com. The Heruo Railway in Xinjiang Opens, Completing the World's First Desert Railway Loop [EB/OL]. (2022-06-17) [2022-08-19]. https://baijiahao.baidu.com/s?id=1735843732687870356&wfr=spider&for=pc.
15. Guangming Net. From Catching Up to Leading the Way: China's Overall Railway Technology Advances to the World's Advanced Level [EB/OL]. (2022-08-16) [2022-08-19]. https://m.gmw.cn/baijia/2022-08/16/1303094614.html.
16. China Railway. How Did Fuxing Create the World Record for Intersection Speed? All Your Concerns Are Here [EB/OL]. (2022-05-18) [2022-08-19]. https://mp.weixin.qq.com/s/8x9MejFNKbOWVBVMa3piWg.
17. China Youth Net. From Catching Up to Leading the Way, China's High-Speed Rail Amazes the World! [EB/OL]. (2022-08-16) [2022-08-19]. https://baijiahao.baidu.com/s?id=1741298599036838885&wfr=spider&for=pc.

18. China Construction Magazine. Large-Diameter TBM for Plateau and Alpine Railways - Pioneer of the Snowy Region Successfully Initiated with Remote Control [EB/OL]. (2021-09-15) [2022-08-19]. https://baijiahao.baidu.com/s?id=1710980521729621149&wfr=spider&for=pc.
19. Dayoo News. Great Power, Great Machines: China's Tunnel Boring Machines Win Big in Infrastructure Construction [EB/OL]. (2022-06-23) [2022-08-19]. https://news.dayoo.com/china/202206/23/139997_54294579.htm.
20. China Railway Construction Corporation. With Nearly 80% Market Share, This Set of High-Speed Rail Equipment Becomes the New Business Card of China Railway Construction Corporation [EB/OL]. (2021-05-13) [2022-08-19]. https://www.thepaper.cn/newsDetail_forward_12663332.
21. Big Wuhan. The Valiant Iron Army Paves Roads, Builds Bridges, Constructs Tunnels, and Makes Heavy Equipment, China Railway 11th Bureau Group Contributes to the Construction of a Powerful Country in Transportation [EB/OL]. (2021-04-12) [2022-08-19]. http://www.app.dawuhanapp.com/p/270197.html.
22. IT Home. Digging Out 3,000 Square Kilometers of Land for China, Tianjing: Impressive! [EB/OL]. (2020–01-10) [2022-08-19]. https://mp.weixin.qq.com/s/PaXuXIwdl0Fj5dmubahRw.
23. China Discipline Inspection and Supervision News. Tiankun Sets Sail on the Waves [EB/OL]. (2022-05-07) [2022-08-19]. https://mp.weixin.qq.com/s/SR6esKMPhWeqD9Nx30oghg.
24. 360 Baike. Mombasa-Nairobi Railway [EB/OL]. [2022-08-19]. https://upimg.baike.so.com/doc/7911188-8185283.html.
25. Zilong Observer. What Has the Mombasa-Nairobi Railway, Built by China for Kenya, Brought to the Local Area After 5 Years of Operation? [EB/OL]. (2022-04-21) [2022-08-19]. https://baijiahao.baidu.com/s?id=1730705279084927200&wfr=spider&for=pc.
26. Baidu Baike. Jakarta-Bandung High-Speed Railway [EB/OL]. [2022-08-19]. https://baike.baidu.com/item/%E9%9B%85%E4%B8%87%E9%AB%98%E9%80%9F%E9%93%81%E8%B7%AF/23774765?fr=aladdin.
27. Guangming Net. Designed and Manufactured Using Chinese Standards! Significant Progress Made in the Construction of the Jakarta-Bandung High-Speed Railway [EB/OL]. (2022-08-05) [2022-08-19]. https://m.gmw.cn/baijia/2022-08/05/1303076553.html.
28. High-Speed Rail Network. China's Railway Adds Color to the Belt and Road Initiative [EB/OL]. (2022-08-15) [2022-08-19]. http://news.gaotie.cn/jianshe/2022-08-15/618770.html.
29. National Railway Administration. China-led International Union of Railways Standard High-Speed Railway Design - Infrastructure Officially Released and Implemented [EB/OL]. (2022-06-30) [2022-08-19]. http://www.nra.gov.cn/xwzx/xwxx/xwlb/202206/t20220630_328241.shtml.
30. National Railway Administration. China-led International Union of Railways Standard High-Speed Railway Design - Communication Signals Officially Released and Implemented [EB/OL]. (2022-02-10) [2022-08-19]. http://www.nra.gov.cn/xwzx/xwxx/xwlb/202204/t20220405_280580.shtml.
31. National Development and Reform Commission of the People's Republic of China. Medium- and Long-Term Railway Network Plan (Adjusted in 2008) [EB/OL]. (2008-10–08) [2022-08-19]. https://www.ndrc.gov.cn/xxgk/zcfb/qt/200906/t20090605_967738.html?code=&state=123.
32. National Railway Administration. Notice on the Issuance of the Medium- and Long-Term Railway Network Plan [EB/OL]. (2016-07-21) [2022-08-19]. http://www.nra.gov.cn/xwzx/xwxx/xwlb/202204/t20220405_279864.shtml.
33. Guangming Daily. Eight Verticals and Eight Horizontals: A Bold Continuation of the Dream of Building a Powerful Railway Nation [EB/OL]. (2016-07-24) [2022-08-19]. http://www.xinhuanet.com/politics/2016-07/24/c_129172386.htm.
34. Gao Jianping. The Paper: Exploring China Along the Highways|G7 Beijing-Xinjiang Expressway: Crossing the Desert and Gobi, Known as the Magical China's No. 7 Heavenly Road [EB/OL]. (2021-05-04) [2022-09-16]. http://m.thepaper.cn/quickApp_jump.jsp?contid=12476954.

35. Yang Zhi. Xinjiang News Online: A National First! Three-Layer Stacked Interchange on Expressway Ensures Safe Migration of Wild Animals [EB/OL]. (2021-06-30) [2022-09-16]. http://www.xjtvs.com.cn/hy/xw/tpxw/29316899.shtml.
36. China News Service. Beijing-Xinjiang Expressway Fully Open to Traffic, Adding Another Expressway Access to and from Xinjiang [EB/OL]. (2021-06-30) [2022-09-14]. https://www.chinanews.com.cn/cj/2021/06-30/9510571.shtml.
37. China Ethnic News. Transforming Deserts into Thoroughfares: A Chronicle of the Construction of the Beijing-Xinjiang Expressway [EB/OL]. (2018-11-26) [2022-09-14]. http://www.minwang.com.cn/mzwhzyk/674771/682200/682203/739795/index.html.
38. Hong Kong-Zhuhai-Macao Bridge Authority. Hong Kong-Zhuhai-Macao Bridge [EB/OL]. (2018-10–01) [2022-09-16]. https://www.hzmb.org/Home/Images/Museum/cate_id/20.
39. Hong Kong-Zhuhai-Macao Bridge Authority. Introduction to the Hong Kong-Zhuhai-Macao Bridge Project [EB/OL]. (2018-10–01) [2022-09-14]. https://www.hzmb.org/Home/Enter/Enter/cate_id/19.html.
40. State-owned Assets Supervision and Administration Commission of the State Council. The Hong Kong-Zhuhai-Macao Bridge: A 56-Kilometer Long Challenge in Construction [EB/OL]. (2014-11-13) [2022-09-14]. http://www.sasac.gov.cn/n2588025/n2588139/c2822915/content.html.
41. China News Service. Hong Kong Political Figures Speak Highly of the Significance of the Opening of the Hong Kong-Zhuhai-Macao Bridge [EB/OL]. (2018-10–23) [2022-09-14]. https://baijiahao.baidu.com/s?id=1615083138124533164&wfr=spider&for=pc.
42. CCTV.com. Beipanjiang Bridge, the World's Highest Bridge, Opens to Traffic on the 29th [EB/OL]. (2016-12-30) [2022-09-14]. http://news.cctv.com/2016/12/30/ARTIJWBkw0BcWNLOLcFFBRlo161230.shtml.
43. China Engineering Science and Technology Knowledge Center. Beipan River Bridge [EB/OL]. (2016-09-10) [2022-09-16]. http://cbe.ckcest.cn/de/1052.
44. Xuanwei Municipal People's Government Portal. Beipan River Bridge (Nizhu Bridge) Opens to Traffic on the 28th of This Month [EB/OL]. (2016-12-21) [2022-09-14]. https://baike.baidu.com/reference/23354551/d04f0JacBLEnWJK9NEO8JQwnUv2XobYdzgAIy5ucDbjdPfZslbuAM5c3rSR5VZnxKEhbYaiAU0T-HNP0ZNPEzouHHQZaDvfnyDk54kdpVPyu_oGI.
45. Ganzi Tibetan Autonomous Prefecture People's Government. Ya'an-Kangding Expressway Erlang Mountain Tunnel Wins National Quality Engineering Award [EB/OL]. (2021-12-09) [2022-09-14]. http://www.gzz.gov.cn/gzzrmzf/c100053/202112/c070581bcf064e9a996a5a832b2e7254.shtml.
46. Tianquan County Transportation Bureau. Significance of the Construction of the Ya'an-Kangding Expressway Erlang Mountain Tunnel [EB/OL]. (2016-03-14) [2022-09-14]. http://www.tqx.gov.cn/gongkai/show/20160314182724-059292-00-000.html.
47. China Harbor Dredging. Congratulations|The Second Phase of the 12.5-Meter Deepwater Channel Project Below Nanjing on the Yangtze River Wins China Communications Construction Quality Engineering Award [EB/OL]. (2019-11-15) [2022-09-16]. https://www.sohu.com/a/354000417_99903819.
48. Yangtze River Deepwater Channel Command Below Nanjing. Introduction to the 12.5-Meter Deepwater Channel Project Below Nanjing on the Yangtze River [EB/OL]. (2019-05-18) [2022-09-14]. https://www.sohu.com/a/314829767_114731.
49. Sohu.com. Another National Quality Award! The Second Phase of the 12.5-Meter Deepwater Channel Project Below Nanjing on the Yangtze River Wins the National Quality Engineering Gold Award [EB/OL]. (2021-12-14) [2022-09-14]. http://news.sohu.com/a/508262399_120099883.
50. Xinhua Net. An Overview of the New Coordinate of Shanghai International Shipping Center - Yangshan Deepwater Port Area [EB/OL]. (2018-07-27) [2022-09-16]. http://m.cnr.cn/jdt/20180727/t20180727_524313380.shtml.
51. Wang Li. Yangshan Phase IV: Breaking Foreign Technology Monopoly, Leading the Global Dock Revolution [EB/OL]. (2022-07-01) [2022-09-16]. https://www.shobserver.com/news/detail.do?id=503062.

52. Xiong Feng. CNR News: [Green Yangtze River, Walking into the Golden Waterway] Harmonious Coexistence between Man and Water: How the Yangtze River Channel Regulation Gives Way to Fish [EB/OL]. (2019-09-11) [2022-09-16]. https://www.cnr.cn/hubei/yuanchuang/20190911/t20190911_524773138.shtml.
53. Haibao News. How Was the World's First Desert Loop Railway Forged? [EB/OL]. (2022-06-17) [2022-08-23]. https://baijiahao.baidu.com/s?id=1735810610407067084&wfr=spider&for=pc.
54. Lu Dongfu. Forging a Beautiful Business Card for China's High-Speed Rail [J]. The Belt and Road Initiative Report (Chinese and English), 2021(05): 93–97.
55. Xinhua Net. China's Comprehensive Technology for Undersea Subway Tunnel Boring Reaches the International Leading Level [EB/OL]. (2022-07-02) [2022-08-19]. https://baijiahao.baidu.com/s?id=1637928206769034119&wfr=spider&for=pc.
56. Xinhua News Agency. China's Decade • China Story|Sichuan: Transportation Construction Helps Overcome the Difficulty of Shu Road [EB/OL]. (2022-09-07) [2022-09-16]. http://www.qhbtv.com/new_index/gjgn/folder2638/2022-09-07/1806675.html.

Chapter 23
Carrying on the Past and Opening up the Future: Comprehensive Transportation Enters the Era of Integrated Development

Transportation possesses foundational, service-oriented, guiding, and strategic importance. Its development directly serves various travel needs of the people, including production, circulation, consumption, commuting, business, and daily life, which are closely related to comprehensive national strength, competitiveness, national security, and geopolitical considerations. It serves as the vanguard for building a modern socialist country with Chinese characteristics.

The development strategy of transportation is inseparable from the country's development goals and governance philosophy, representing the manifestation of national strategy in the transportation sector. China, as a country with a long history of civilization, has seen the propagation and transmission of Chinese national ideology and culture, as well as the philosophical propositions of ancient sages, in the contemporary realization of the Chinese Dream of national rejuvenation. This has formed a development path and governance philosophy with Chinese characteristics, including the scientific outlook on development, ecological prioritization, green development, and a profound understanding of the relationship between transportation, economic development, and people's livelihoods. It is in this historical context that the 19th National Congress of the Communist Party of China explicitly proposed the construction of a transportation powerhouse. Since then, China's transportation development has entered a new era, not only contributing to the realization of the Chinese Dream of national rejuvenation but also opening up broader prospects for sustainable development worldwide and the construction of a community with a shared future for humanity.

In September 2019, the Communist Party of China Central Committee and the State Council issued the "Outline of Building a Transportation Powerhouse," which clearly stated that to achieve the grand goal of building a transportation powerhouse, it is necessary to promote the transformation of transportation development from pursuing speed and scale to focusing more on quality and efficiency. This involves shifting from the relatively independent development of various transportation modes to a more integrated and coordinated development, and transitioning

© China Communications Press Co., Ltd 2025
H. Lu (ed.), *A Brief History of Transportation in China*,
https://doi.org/10.1007/978-981-97-9851-3_23

from reliance on traditional factors to emphasizing innovation-driven development. The aim is to construct a modern comprehensive transportation system that is safe, convenient, efficient, green, and economical, and to create first-class facilities, technologies, management, and services. The goal is to build a transportation powerhouse that meets the satisfaction of the people, ensures strong support, and ranks among the world's best, providing robust support for the comprehensive construction of a socialist modernized country and the realization of the Chinese Dream of national rejuvenation.

Entering a new era, the integrated and coordinated development of various transportation modes is the fundamental premise and essential guarantee for achieving the goal of building a transportation powerhouse, and it serves as the foundation for achieving all objectives. This chapter focuses on the integrated development of comprehensive transportation, providing an outlook on the promising future prospects of transportation and logistics.

The Philosophical Thoughts of Ancient Sages

The rich and diverse treasure trove of ancient Chinese thought provides profound insights into the development of transportation, which cannot be fully explored within the limited scope of this book. Here, we will briefly draw upon the brilliance of Chinese civilization to illustrate how it supports the realization of "interconnectedness" and "integration," nourishing the past and future of Chinese transportation.

From the day of its birth, the great Chinese nation has embarked on a journey of exploring the laws of the universe, pondering the philosophy of life, fostering harmony among all things, and pursuing a better life, endowed with its qualities of diligence, kindness, tolerance, resilience, and wisdom.

According to the summary by Mr. Feng Youlan, Western philosophy is characterized by an exoteric approach, delving into metaphysical ideas and philosophical concepts. In contrast, Chinese philosophy follows an esoteric approach, reflecting on and exploring the direct connection with the governance system, harmonious coexistence with the environment, understanding of the universe, and guiding principles for life [1].

The essence of classical Chinese philosophy, including Daoism's emphasis on aligning with the natural order and the principle of "doing good without contention," Confucianism's teachings of loyalty, benevolence, and impartiality, as well as the advocacy for pursuing common prosperity and world harmony, has had a profound and lasting impact. At the core of these philosophical tenets lies the importance of living in harmony with nature, exercising moderation in all endeavors, and striving for shared prosperity among humanity. Therefore, it can be said that the Chinese nation has long been a creator of a good life and a promoter of harmonious coexistence among all beings. These philosophical ideas have also deeply influenced transportation development throughout China's history.

"Follow the Natural Way"

The phrase "道法自然" originates from Chapter 25 of the Dao De Jing (also known as Tao Te Ching or Laozi): "人法地,地法天,天法道,道法自然." It means that humans follow the earth, the earth follows the heavens, the heavens follow the Dao (Way), and the Dao follows nature. "Dao" refers to the approach and method of solving problems. Based on profound observation and contemplation of the deep laws of nature, earth, humans, and even the entire universe, Laozi realized the key to harmonious coexistence between humans and nature. He proposed the most fundamental behavioral principles for human activities, emphasizing that all things in the universe should follow the law of "naturalness" spontaneously and naturally.

The concept of "Dao follows nature" is also expressed as "The Dao constantly acts without action yet accomplishes everything" [2]. Acting without action means following the natural course, as it is inherently so. Only by doing so can the Dao naturally bring about all things in the development and changes of phenomena.

One common misunderstanding of the term "ziran" (自然) in Laozi's philosophy is that it is interpreted as a higher, more abstract entity than the Dao. Another misconception is to equate "ziran" with the natural world corresponding to human society. However, in reality, the "ziran" in Laozi's teachings is neither an abstract entity above the Dao nor an external objective entity apart from human beings; rather, it is inherent and spontaneous.

The philosophical concept of "Dao follows nature" has permeated various aspects of Chinese society. For example, people choose to build cities in "auspicious" locations near water sources and facing mountains, and to construct residences in places with good ventilation and sun exposure, all following the principle of respecting and harmonizing with nature. The flood control strategies employed by Yu the Great, combining dredging and blocking, also reflect the idea of respecting the natural laws of water flow and adapting to its inherent properties.

"the Virtue of Water" and "the Way of Loyalty and Forgiveness"

"The Virtue of Water" originates from Chapter Eight of the Tao Te Ching: "The highest virtue is like water. Water benefits all things without contention, flowing to where others disdain to dwell. Hence, it is close to the way. In dwelling, be close to the land; in meditation, go deep in the heart; in dealing with others, be gentle and kind; in speech, be true; in ruling, be just; in business, be competent; in action, choose the right timing. Because one does not contend, there is no blame." [2]. This phrase actually speaks to the method of conducting oneself as a human being, suggesting that one should emulate the characteristics of water. Water nourishes all things but does not strive or contend with them; embodying such qualities brings one closest to the "Way."

At this point, the Confucian principle of "The Way of Loyalty and Forgiveness" also serves as a behavioral guideline for human interaction. "The Way of Loyalty and Forgiveness" is one of the core ideas in Confucian philosophy. Confucius regarded "The Way of Loyalty and Forgiveness" as a principle for handling interpersonal relationships. It represents the highest realm of human interaction. By embodying loyalty and forgiveness, one not only dispels misunderstandings others may have towards oneself, but also alleviates interpersonal conflicts, stabilizes societal order, and fosters mutual affection among people. This principle promotes harmony and can be considered a condition for advancing towards the ideal of global unity.

Another expression of "The Way of Loyalty and Forgiveness" is "Do not do unto others what you would not want done unto yourself," which comes from Confucius in the Analects of Confucius during the Spring and Autumn Period. In the passage "Wei Ling Gong," Confucius said, "Zigong asked: 'Is there any one word that could guide a person throughout life?' The Master replied: 'How about 'reciprocity'? Never impose on others what you would not choose for yourself.'" [3].

Between individuals and between objects, there exist both homogeneity and differences. Regarding homogeneity, "Do not do unto others what you would not want done unto yourself" is correct; concerning differences, one should also add "Do not do unto others what they would not want done unto themselves." The Guanzi, in the "Xiao Wen" section, states: "Not doing unto others what they would not want done unto themselves is benevolence." [4]. Take eating habits as an example: some people dislike spicy food, while others enjoy it. One should not impose their own preferences onto others just because they personally enjoy it. Zhuangzi provides a concise and clear explanation for this. He said: "Fish thrive in water, while humans perish in it. They differ in their likings and dislikings. Therefore, the sages do not impose uniformity in ability or uniformity in tasks." [5].

The ancient sages not only emphasized respecting nature when dealing with the environment but also stressed the importance of respecting others in interpersonal relationships. This represents the highest realm of human interaction. This mindset is reflected in Chinese diplomacy, where regardless of a nation's size or ethnicity, the principles of the Five Principles of Peaceful Coexistence should be adhered to as a common standard.

The Pursuit of Global Unity

The concept of "Great Unity" was first proposed by Confucius. The Great Unity ideology is in harmony with the highest moral character, and the Confucian ultimate social ideal is the Great Unity of the world. This Great Unity world is not purely conceptual but concrete. The Book of Rites describes the social scene of the Great Unity world as follows: "When the Great Way prevails, the world is shared by all. The worthy are chosen for public office, honesty and integrity are practiced, and harmony is cultivated. Therefore, people do not only show affection for their own relatives, nor do they only raise their own children. The elderly are provided for until the end of their lives, the able-bodied are employed according to their capabilities, and the young are educated and nurtured. Even the widowed, orphaned, lonely, impoverished, and disabled are all cared for. Men have their roles, and women have their places. When something of value is abandoned on the ground, there is no need to hoard it for oneself. When strength is needed, it does not have to come from one's own body. Therefore, schemes for robbery and theft do not arise, and disorderly elements do not emerge. Thus, the gates remain open without being shut, and this is what is meant by Great Unity." [6].

Based on the concept of "Great Unity" advocating for common prosperity, development, and protection for all humanity, China has put forward initiatives such as the "Belt and Road" initiative and the idea of building a community with a shared future for mankind. This not only represents the fundamental path towards achieving sustainable development for humanity but also serves as an inevitable choice beneficial to the development of all humanity.

Transportation and Modernization in China

The philosophical doctrines of ancient sages have been integrated into China's governance principles and external development strategies, profoundly influencing the country's transportation construction from ancient times to the present. Transportation, characterized by its foundational, service-oriented, leading, and strategic attributes, serves as a pioneer in opening new paths, meeting the needs of people's travel and cultural exchanges, driving economic and social development, shaping the current transportation landscape in China, and demonstrating China's responsibilities. In the early stages of economic and social development, the foundational and service-oriented aspects of transportation were widely recognized. As China enters a new era of high-quality development, the leading and strategic roles of transportation have become its primary functional orientations.

As early as the early 20th century, Mr. Sun Yat-sen highly valued the role of transportation in his book "The Principle of Building the Nation" and proposed the following transportation construction plan: the construction of approximately 160,000 km of railways to connect China's coastal areas, inland regions, and border areas; the construction of 1.6 million kilometers of highways to form a nationwide road network, including access to the Qinghai-Xizang Plateau; the excavation and renovation of national waterways and canals, construction of the Three Gorges Dam, and development of inland waterway transportation, as well as water conservancy, electric power industries; and the construction of world-class seaports along the northern, central, and southern coasts of China [7].

Since the founding of the People's Republic of China, especially after the implementation of reform and opening-up policies, China has adhered to the principle of giving priority to transportation in economic and social development. Significant changes have occurred in the transportation sector, with accelerated development of transportation infrastructure networks and improved transport service capabilities. Transportation has played a leading role in the development of the national economy.

Transportation is the cornerstone of people's well-being in modern times. The report of the 19th National Congress of the Communist Party of China pointed out that "after long-term efforts, socialism with Chinese characteristics has entered a new era, which represents a new historical direction for China's development," and "the principal contradiction in Chinese society has evolved into the contradiction between the ever-growing needs of the people for a better life and the unbalanced and inadequate development." Among the necessities of life such as clothing, food, shelter, and transportation, the needs for clothing, food, and shelter have been largely met, and people's access to transportation has become a key indicator and benchmark of a better life, serving as the primary element for constructing a better life.

Transportation is key to enhancing competitiveness. As Mr. Sun Yat-sen once said, "Transportation is the mother of industry, and railways are the mother of transportation." One of the important factors in measuring the economic competitiveness of a city, region, or country is transportation. Developed transportation is the first condition for attracting investment, it is the primary element for reducing costs and enhancing product competitiveness, thus it is crucial for the success or failure of development. Superior transportation conditions are also crucial for attracting people and talent. History has proven that where talent gathers, regions thrive, and where talent is lost, regions decline.

"Transportation+" represents the most promising development direction and the highest realm of transportation construction. Deep integration of transportation with land use is essential to achieve goals such as intensive and efficient land use, convenient and efficient travel, environmental protection and emission reduction, and providing world-class services. Deep integration of transportation with tourism is necessary to enable "enjoying the journey" throughout the entire travel experience, allowing people to make the most of every moment and fully enjoy the pleasures of life. Deep integration of transportation with industry is essential for precise transportation and efficient logistics, leading to zero inventory and low costs, thereby enhancing the competitiveness of "Made in China". Combining transportation with high-end resources and the environment allows for the comprehensive utilization and widespread sharing of high-end services and scarce scenic resources to the fullest extent possible.

The transportation and logistics safety system serves as an effective support for implementing the national security strategy. National key transportation channels and critical hub facilities are the lifelines of economic and social operation, as well as the guarantee for the smooth and secure transportation of important commodities such as energy, minerals, and food, and the security of critical transportation channels and key hubs. A comprehensive emergency support system and capability for transportation and logistics will become a strong guarantee for safeguarding the safety of people's lives and property as well as national security overall. Furthermore, when combined with the national security assurance system, transportation can provide the fastest, most reliable, and most efficient transportation capabilities at the lowest cost, thus forging an impregnable defense for national security.

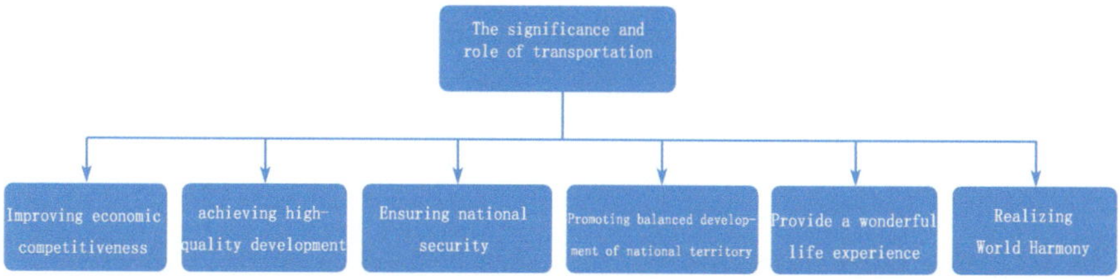

Fig. 23.1 The significance and role of transportation in China's modernization

From this, it is evident that transportation is essential for national prosperity, the foundation of a strong nation, the path to wealth, the vessel of ideals, and the basis of people's needs. Figure 23.1 summarizes the significance and role of transportation in China's modernization, including enhancing economic competitiveness, achieving high-quality development, safeguarding national security, promoting balanced regional development, providing a quality living experience, and advancing towards global unity.

Sustainable Transportation Driving Sustainable Development

Since the founding of the People's Republic of China, successive generations have upheld the principle of prioritizing transportation, forging paths through mountains and building bridges over water, transforming China into a major transportation power. Currently, efforts are being accelerated to construct a strong transportation nation. It is important to recognize that as transportation demands continue to grow, associated issues such as resources, energy, and the environment are also becoming increasingly prominent. Achieving sustainable development in transportation is of significant importance, both for the industry itself and for the high-quality development of the economy and society as a whole.

From October 14th to 16th, 2021, the Second United Nations Global Sustainable Transport Conference was held in Beijing. President Xi Jinping delivered a keynote speech titled "Engaging with the World, Conforming to the Times, and Striding Forward on the Path of Sustainable Development." In his speech, he proposed five principles: upholding openness and connectivity to promote interconnection; adhering to common development to promote fairness and inclusiveness; driving innovation to enhance development momentum; prioritizing ecology to achieve green and low-carbon development; and upholding multilateralism to improve global governance.[1] The conference adopted the "Beijing Declaration," emphasizing the

[1] Please refer to "Striding Forward on the Path of Sustainable Development: Keynote Speech at the Opening Ceremony of the Second United Nations Global Sustainable Transport Conference" delivered on October 14, 2021, published in the People's Daily, October 15, 2021, page 02.

importance of sustainable transportation and elucidating that accelerating the transition to sustainable transportation is an important aspect of promoting the construction of a community with a shared future for mankind.

The Relationship Between Sustainable Transportation, Green Transportation, and Low-Carbon Transportation

A sustainable transportation system is a comprehensive transportation system that minimizes resource input and environmental costs while maximizing the satisfaction of reasonable transportation needs. Its optimization goals include resource conservation, environmental protection, and meeting demands. In this system, various modes of transportation are reasonably divided and coordinated, with green transportation as the mainstay. Different transportation users have equal rights to choose various modes of transportation. The system matches the characteristics of urban transportation demand and is coordinated with the land use nature of the surrounding area. It is characterized by safety, convenience, smoothness, comfort, environmental friendliness, energy conservation, high efficiency, high accessibility, and people-oriented principles [8].

Green transportation is a concept for urban comprehensive transportation system planning and construction aimed at reducing traffic congestion, lowering energy consumption, promoting environmental friendliness, and saving construction and maintenance costs. The narrow concept of green transportation emphasizes the environmental friendliness of the transportation system, advocating for attention to environmental protection and the quality of living environments in the planning, construction, and operation management of urban transportation systems. The broader concept of green transportation includes promoting the priority development of public transportation, encouraging people to choose cycling and walking for short-distance travel, saving energy, protecting the environment, and establishing urban comprehensive transportation systems led by public transportation [8].

The essence of green transportation is to establish and maintain a transportation system that supports sustainable urban development, meeting people's transportation needs while emphasizing resource conservation, environmental protection, and social equity. Green transportation has clear strategic goals for sustainable development, achieving the highest transportation efficiency with minimal social costs. It is coordinated with the urban environment, adapts to urban land use patterns, and fosters the harmonious coexistence and complementary advantages of various transportation modes.

Low-carbon transportation further emphasizes the reduction of greenhouse gas emissions, a global issue crucial to the future of human society. It focuses on taking various measures to reduce the CO2 emissions generated by transportation.

In a broad sense, green transportation is equivalent to urban sustainable transportation. In a narrow sense, sustainable transportation encompasses green transportation.

The Pathway to Achieving Sustainable Transportation

Sustainable transportation is both a fundamental requirement and an essential pathway to achieving the goal of building a strong transportation nation. It is also one of the important characteristics and key breakthrough points of a strong transportation nation. Sustainable transportation serves as both a development concept and a guiding principle for the planning and construction of comprehensive transportation systems. A transportation system constructed based on the concept of sustainable development aims to enhance transportation efficiency, reduce resource consumption, promote environmental friendliness, and save construction and maintenance costs. Therefore, the development goals of sustainable transportation are essentially the same as the goals and pursuits of building a strong transportation nation [9].

The key breakthrough points in the construction of a strong transportation nation lie in the integration, greening, intelligence, and sharing of transportation technologies. This represents the mainstream trend in global transportation development and will trigger profound changes in transportation infrastructure, energy sources, technological equipment, operational management, and service models. Seizing these key breakthrough points is akin to grasping the "bull's nose" of building a strong transportation nation, and it is also the pathway to achieving sustainable transportation [10].

Therefore, based on the concept of sustainable development, to embark on the path of building a strong transportation nation with Chinese characteristics, it is imperative to firmly grasp the following key points.

Based on the concept of sustainable transportation, it is essential to intensively and efficiently utilize resources, vigorously promote the integrated development of comprehensive transportation, and achieve the green and sustainable development of infrastructure from planning and construction to operation and management. By employing a "one-map" planning approach, efforts should be made to establish a comprehensive three-dimensional transportation network system characterized by high-level construction capability, comprehensive layout, interconnectivity, green intelligence, scientific maintenance, and durability. This includes creating a multi-level comprehensive transportation hub system with "zero transfers" and "zero transitions," as well as constructing highly intelligent and integrated systems for passenger and freight transportation services and organizational management.

Based on the concept of sustainable transportation, the goal is to achieve green transportation as the dominant mode, significantly increasing the share of intensive and green transportation modes such as public transit, walking, and cycling. This aims to prioritize ecological considerations and promote green development in the transportation sector. To accomplish this, it is crucial to undertake top-level design and comprehensively implement the construction of integrated transportation systems under the supervision of green transportation dominance. Furthermore, policies supporting green transportation modes such as railways and waterways should be strengthened. In densely populated areas such as urban agglomerations and metropolitan areas, the focus should be on building comprehensive urban transportation systems with rail transit as the backbone, extensive coverage by public bus transit, convenient and efficient last-mile microcirculation through walking and cycling, and private cars serving as a supplementary mode of transportation.

Based on the concept of sustainable transportation, enhancing the efficiency of transportation systems and vigorously developing intelligent transportation are key priorities. Leveraging intelligent transportation methods to promote integrated transportation development, significantly improve transportation safety levels and service quality, and intensify efforts to achieve sustainable transportation represent invaluable opportunities. Utilizing a comprehensive transportation data platform as a foundation, guided by systematic scientific thinking, traffic engineering principles, and the laws of transportation development, the goal is to address traffic congestion, enhance transportation safety, efficiency, and service quality. This entails constructing an intelligent comprehensive transportation system characterized by holistic perception, information sharing, deep analysis, scientific decision-making, precise management, accurate services, and proactive safety measures. This will contribute to realizing a comprehensive transportation system characterized by safety, convenience, efficiency, greenness, and economy.

Based on the concept of sustainable transportation and innovative management principles, efforts are made to enhance resource utilization efficiency and encourage and support the development of shared transportation. The application space for shared models in the transportation sector is extensive, encompassing transportation tools, parking facilities, as well as transportation data and information, which can all play a greater role through shared models. The scientific and reasonable development of shared transportation will also provide important support for the improvement and quality enhancement of last-mile transportation, constituting a crucial element in achieving sustainable transportation. Additionally, it not only facilitates the maximization of various existing transportation resources but also reduces transportation and logistics costs. By avoiding the path of "one car per household" seen in developed countries, it helps alleviate the contradiction between limited resources and the growing demand for personalized transportation options.

Future City and Transportation

As described in Chapter Fifteen, the ideal city of the future should possess characteristics such as greenness, intelligence, humanity, livability, innovation, and resilience.

The urban transportation system plays a pivotal role in the development of cities, serving as both a backbone and a guiding system. On one hand, it supports the city's structure, land use patterns, and urban spatial development strategies. On the other hand, urban transportation exerts a significant influence and guiding role on the city's development. The experience of urban development worldwide indicates that, unlike other urban subsystems, the urban transportation system should lead the city's development. Considering the needs of future cities, clarifying the development goals of urban transportation systems, and promoting the development of integrated transportation that is green, intelligent, and humanistic are crucial for establishing future cities characterized by greenness, intelligence, humanity, livability, innovation, and resilience. The main features of future cities and the development goals of their transportation systems are illustrated in Fig. 23.2.

Safe and Reliable. Safety and reliability are prerequisites for transportation systems, reflected in high levels of traffic safety, strong emergency response capabilities, and effective military-civilian integration.

Convenient and Efficient. Convenience and efficiency are not only basic requirements for urban transportation systems but also the highest standard, representing important aspects of an attractive city and a happy life. The key to achieving this goal lies in integration, including the integration of transportation and land use, seamless connections and zero-distance transfers between different modes of transportation and transportation hubs, integrated operation management, integrated transportation services, and institutional mechanisms.

Green and Low-carbon. Green and low-carbon direction emphasizes the construction of a comprehensive urban transportation system led by green transportation, increasing the share of green transportation, and promoting green development throughout all aspects of transportation. This includes green planning guidance, green travel leadership, green transportation as the main component, green facilities support, and green management assurance.

Smart Sharing. Smart transportation is a key lever for improving transportation efficiency, safety, and service levels. By establishing a comprehensive transportation system that deeply applies intelligent technologies such as holographic perception, information sharing, in-depth analysis, scientific decision-making, refined management, precise service, active safety, and vehicle–road coordination, and along the trajectory of resource sharing, system integration, business linkage, and self-evolution, overall intelligence and modernization levels of transportation can be comprehensively enhanced. This helps address transportation challenges, achieve higher transportation efficiency, and provide more convenient transportation services. Shared transportation is a powerful measure to improve the efficiency of transportation resource utilization and address the sharp contradictions between transportation supply and demand when transportation supply is insufficient. This includes sharing transportation facilities (such as parking and charging facilities), transport tools (such as non-motorized vehicles, cars, and trucks), transportation information and data, and transportation services to enhance resource utilization efficiency.

Civilized Order. Civilized order is a fundamental characteristic of a highly civilized society, and orderly traffic primarily manifests in the good quality of traffic participants, traffic culture, and traffic ethics. It involves abiding by laws and regulations while traveling, actively choosing green transportation methods, thereby enhancing the experience, sense of achievement, and happiness.

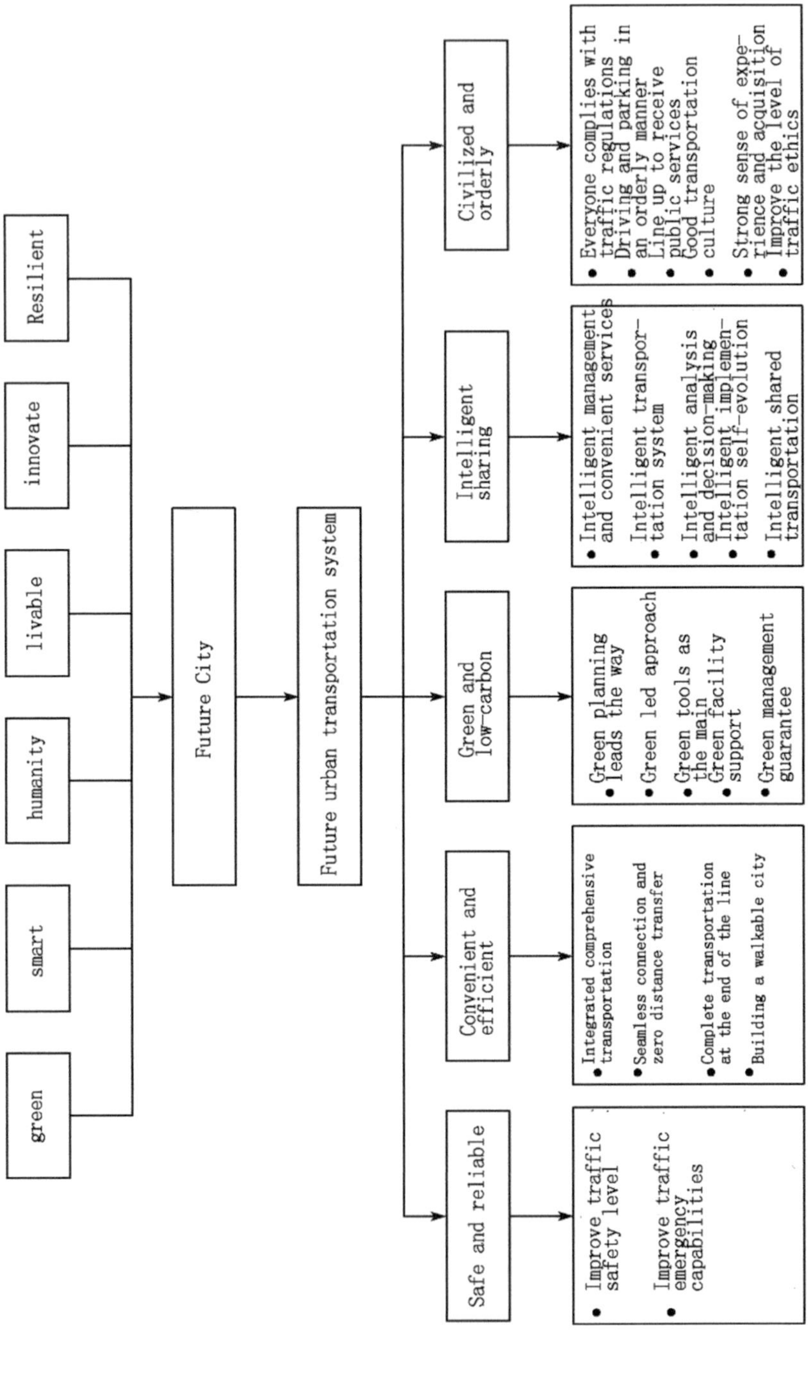

Fig. 23.2 Main characteristics of future cities and development goals of their transportation systems

The Beautiful Blueprint of Chinese Transportation

"The Outline of Building a Strong Transportation Nation" and "The Outline of National Comprehensive Three-dimensional Transportation Network Plan" have outlined the magnificent blueprint to comprehensively build a strong transportation nation and establish a modern, high-quality national comprehensive three-dimensional transportation network by the middle of the twenty-first century. They serve as the "overall strategy" and "master plan" for the development of comprehensive transportation in the new era. These plans require continuous implementation over several five-year plans and will contribute to the realization of the Chinese Dream of the great rejuvenation of the Chinese nation.

The Shared Vision for Building a Strong Transportation Nation

In October 2017, the 19th National Congress of the Communist Party of China clearly proposed the construction of a strong transportation nation, marking the beginning of a new journey for transportation professionals. In September 2019, the Central Committee of the Communist Party of China and the State Council issued the "Outline for Building a Strong Transportation Nation" (referred to as the "Construction Outline"), which serves as the "general strategy" for the construction of a strong transportation nation in China. It coordinates the advancement of the construction of a strong transportation nation and serves as a common action plan for the transportation industry [11].

The "Outline" outlines a comprehensive vision for building a strong transportation nation, summarized as "one, two, three, four, five, nine, three". It encompasses an overarching goal of achieving people's satisfaction, providing a strong guarantee, and attaining world-class status, with two-stage objectives targeting a basic achievement of this goal by 2035 and a comprehensive realization by 2050. This vision entails three transformations, shifting from speed and scale to quality and efficiency, integrating various transportation modes, and transitioning from traditional to innovation-driven approaches. It emphasizes four first-class elements: infrastructure, technology, management, and service, characterized by safety, convenience, efficiency, greenness, and economy. Nine key tasks encompass infrastructure development, technological innovation, talent cultivation, and governance enhancement, with the Party's leadership, financial support, and implementation management as three vital pillars.

In March 2021, the "Outline of the Fourteenth Five-Year Plan for National Economic and Social Development and the Long-Range Objectives Through the Year 2035" was issued, issuing a call to accelerate the construction of a strong transportation nation. It provided clear directives for transportation and logistics across various domains, including the development of modern infrastructure systems, the advancement of digital economy initiatives, and the promotion of regional coordination and development.

The Overall Plan for Constructing a Comprehensive Three-Dimensional National Transportation Network.

In February 2021, the Communist Party Central Committee and the State Council issued the "Outline of the National Comprehensive Three-dimensional Transportation Network Plan" (referred to as the "Plan Outline"), which serves as the "master plan" guiding the layout of the comprehensive transportation network and implementing relevant tasks outlined in the "Outline of Construction". This plan provides robust support and technical guidance for the implementation of the construction of a transportation powerhouse at the national, provincial, and municipal levels [12].

The main contents of the "Outline of the National Comprehensive Three-dimensional Transportation Network Plan" can be summarized as "One, Two, Three, Four, Five, Twelve". This includes one overarching goal: to construct a modern and high-quality national comprehensive three-dimensional transportation network. It also entails two stage objectives: to achieve basic completion of the modern and high-quality national comprehensive three-dimensional transportation network by 2035, and to comprehensively complete it by 2050. The plan outlines three main lines of work: optimizing the layout of the national comprehensive three-dimensional transportation network, promoting integrated development of comprehensive transportation, and advancing high-quality development of comprehensive transportation. It further includes four principles and requirements: serving the overall situation and the people, based on national conditions and reform and opening up, optimizing the structure and coordinating integration, and innovating intelligence and ensuring safety and greenness. Additionally, it emphasizes five areas for enhancement and assurance: the leadership of the Party, organizational coordination, resource support, funding assurance, and implementation management. Finally, the plan delineates twelve key tasks: the comprehensive network, channels, hubs, and transport networks; coordinating various transportation modes, integrating the four networks, regional coordination, and industrial integration; safety, intelligence, green humanities, and governance.

The New Journey of Constructing a Modern Integrated Transportation System

In January 2022, the State Council issued the "14th Five-Year Plan for the Development of Modern Integrated Transportation System" (referred to as the "14th Five-Year Plan for Integrated Transportation"). Aligned with China's "14th Five-Year Plan," the "Construction Outline," and the "Planning Outline," this plan outlines the goals and tasks for transportation development during the "14th Five-Year Plan" period, aiming to accelerate the construction of a transportation powerhouse and establish a modern integrated transportation system.

The "14th Five-Year Plan for Integrated Transportation" sets the goal of achieving basic integration and development in comprehensive transportation by 2025, with substantial breakthroughs in intelligence and green initiatives. It aims to significantly enhance comprehensive capabilities, service quality, operational efficiency, and overall benefits, propelling transportation development towards world-class standards. To achieve these objectives during the "14th Five-Year Plan" period, efforts will focus on ten key areas: constructing a high-quality comprehensive three-dimensional transportation network; consolidating the foundational support for coordinated urban–rural regional development; advancing the modernization of transportation in urban agglomerations and metropolitan areas; expanding the supply of high-quality transportation services; accelerating the widespread application of intelligent technologies; comprehensively promoting green and low-carbon transformation; enhancing safety and emergency preparedness capabilities; driving high-level international cooperation and opening-up; strengthening the construction of modern governance capabilities; and ensuring effective implementation through robust planning support [13].

To Contribute to the Realization of the Chinese Dream of National Rejuvenation

Since General Secretary Xi Jinping first proposed the concept of the Chinese Dream during his visit to the "Road to Rejuvenation" exhibition on November 29, 2012, he has elaborated on the essence and pathways to achieve the Chinese Dream in a series of important speeches. This succinctly encapsulates the aspiration for the great rejuvenation of the Chinese nation. In the report of the 19th National Congress of the Communist Party of China on October 18, 2017, it was emphasized that realizing the great rejuvenation of the Chinese nation is the greatest dream of the Chinese people since modern times. In December 2018, General Secretary Xi Jinping stated at the ceremony commemorating the 40th anniversary of reform and opening-up that the establishment of the Communist Party of China, the founding of the People's Republic of China, and the promotion of reform, opening-up, and the socialist cause with Chinese characteristics represent the three historic events since the May Fourth Movement, marking three milestones in the modern realization of the great rejuvenation of the Chinese nation.[2]

The basic connotations of the Chinese Dream encompass three dimensions: national, ethnic, and individual. From the national perspective, the Chinese Dream

[2] Refer to Xi Jinping's speech at the 40th anniversary celebration of reform and opening up, People's Daily, December 19, 2018, edition 02.

signifies the aspiration for a strong country. China aims to become a powerful modernized nation, gaining global recognition and assuming a leading role in global development. From the ethnic viewpoint, the Chinese Dream represents the rejuvenation of the Chinese nation. The Chinese people seek to make greater, more significant contributions to human development. From the individual perspective, the Chinese Dream pertains to the aspirations of every Chinese citizen, striving to continuously satisfy the people's aspirations for a better life. These three dimensions unite the nation, the ethnicity, and the individual into a community of shared destiny, thereby closely linking national interests, ethnic interests, and the specific interests of each individual.

The Chinese Dream aligns with the ancient Chinese vision of a world of great harmony. It belongs not only to China but also to the world. The Chinese Dream resonates with the ideal pursuit of peace and development by humanity, embodying openness, inclusiveness, and shared prosperity. The concept of a community with a shared future for mankind and the Belt and Road Initiative serve as international expressions of the Chinese Dream. Within the strategy of building a strong transportation nation, the domestic "123 travel circle" and the international "123 fast freight logistics circle" represent concrete manifestations of the Chinese Dream in the field of transportation and logistics.

"The Chinese nation is a peace-loving nation. Eliminating wars and achieving peace have been the most urgent and profound aspirations of the Chinese people since modern times. Pursuing the path of peaceful development is the inheritance and development of the fine cultural traditions of the Chinese nation, as well as the inevitable conclusion drawn by the Chinese people from the sufferings they have experienced since modern times." [14].

"We also hope that all countries in the world will pursue the path of peaceful development, engage in equal exchanges and mutual learning among different civilizations, and work together to promote the construction of a harmonious world of lasting peace and common prosperity." [15].

The Chinese Dream, rooted in the great wisdom of the nation, embodies the common ideals and aspirations of all Chinese people. It represents the collective hope of the Chinese people for national stability, prosperity, and rejuvenation, as well as their ideal pursuit in closely engaging with and mutually depending on the international community. Its goal is to build a modernized country that is prosperous, democratic, civilized, and harmonious, dedicated to constructing a harmonious society domestically and a harmonious world internationally. The realization of the Chinese Dream is not only a process that benefits the Chinese people but also a process that continuously brings new impetus for development and contributes positive energy to the world.

Promoting the Construction of a Community with a Shared Future for Mankind

China adheres to an open strategy of mutual benefit and win–win cooperation, deepens cooperation with other countries in the field of transportation, actively promotes global connectivity, actively participates in global transportation governance, conscientiously fulfills international responsibilities and obligations for transportation development, and strives to achieve win–win cooperation and common development in more areas and at higher levels. This aims to promote the construction of a global transportation community and contribute to the establishment of a community with a shared future for mankind.

In 2012, the 18th National Congress of the Communist Party of China clearly advocated for the awareness of a community with a shared future for mankind, emphasizing the consideration of other countries' legitimate concerns while pursuing national interests. In January 2017, President Xi Jinping delivered a keynote speech titled "Building a Community with a Shared Future for Mankind" at the United Nations Office at Geneva. In October 2017, the report of the 19th National Congress of the Communist Party of China reiterated the commitment to the path of peaceful development, promoting the construction of a community with a shared future for mankind, and advancing the reform of the global governance system.

Today, the world is experiencing unprecedented changes of a century. People increasingly recognize the importance and urgency of a shared human destiny. Where is the world heading? What is the path of development? China's voice and proposals will serve as an important reference point in global history. With the promotion of building a community with a shared future for mankind as its "root" and "backbone," China has successively proposed the construction of a new type of international relations featuring mutual respect, fairness, justice, and win–win cooperation, as well as the innovation-driven, coordinated, green, open, and shared development philosophy. Additionally, China has advocated for a global governance concept of consultation, contribution, and shared benefits, a new security concept featuring common, comprehensive, cooperative, and sustainable security, and a civilized view of equality, mutual learning, dialogue, and inclusiveness. These series of Chinese wisdom and propositions reflect the diplomatic principles of the Chinese nation, which aim for harmony among nations and the lofty ideal of a world with shared prosperity, enriching the treasury of human civilization continuously.

The concept of a community with a shared future for mankind is a set of values that reflects China's profound wisdom in understanding the trends of global development and the direction of human destiny. General Secretary Xi Jinping remarked, "A community with a shared future for mankind, as the name suggests, means that the future and destiny of every nation and every country are closely interconnected. We should weather the storms together, share both honor and disgrace, and strive to build this planet, where we were born and raised, into a harmonious global family, turning the aspirations of people around the world for a better life into reality."[3]The rich connotations of building a community with a shared future for mankind can be understood from five aspects: political, security, economic, cultural, and ecological.

In politics, mutual respect, equality, and consultation should be upheld, firmly rejecting Cold War mentality and power politics, and embracing a new type of state-to-state relations featuring dialogue without confrontation and partnership without alliance. Regarding security, disputes should be resolved through dialogue and differences through consultation, while addressing both traditional and non-traditional security threats and combating all forms of terrorism. Economically, solidarity should be fostered to promote trade and investment liberalization and facilitation, advancing economic globalization towards greater openness, inclusiveness, universality, balance, and win–win outcomes. Culturally, diversity of world civilizations should be respected, promoting cultural exchanges to transcend cultural barriers and achieve mutual learning beyond cultural superiority. Ecologically, environmental friendliness should be upheld, cooperating to tackle climate change and safeguarding the Earth, our common home, on which humanity depends for survival.

In doing so, China upholds the principle of consultation, joint construction, and shared benefits, accelerating infrastructure connectivity cooperation with relevant countries, jointly building the Belt and Road Initiative, and jointly creating a path of high-quality development that is open, inclusive, and mutually beneficial, as well as a road of peace, cooperation, and happiness. China actively promotes the reform of the global transportation governance system, working with other developing countries to promote global governance reforms for sustainable development in transportation, creating a favorable international environment for the development of developing countries, and actively assuming international responsibilities in line with its own development stage and national conditions. In the field of transportation, China actively responds to climate change, contributing to global ecological civilization construction and sustainable development. Following the principles of equality, mutual benefit, and win–win cooperation, China deepens exchanges and cooperation with countries in the field of transportation, constantly expanding breadth and depth, and promoting the construction of an open world economy.

Engaging with the world and keeping pace with the times, we adhere to the path of safety, convenience, efficiency, environmental friendliness, and economy. We persist in promoting connectivity and advancing the sustainable development of transportation. By doing so, we are certain to contribute China's strength to the joint construction of a better world.

---

[3] Refer to "Xi Jinping's keynote speech at the Dialogue between the CPC and World Political Parties High-Level Meeting", published in the People's Daily on December 2, 2017, page 2.

References

1. Feng Youlan. Collected Works of Feng Youlan. Changchun: Changchun Publishing House, 2017.
2. Tao Te Ching. Nanjing: Jiangsu People's Publishing House, 2016.
3. Yang Bojun. The Analects of Confucius: Annotated and Explained. Beijing: Ancient Books Publishing House, 1958.
4. Guanzi. Annotated by Fang Xuanling, Supplementary Annotation by Liu Ji, Proofread by Liu Xiaoyi. Shanghai: Shanghai Ancient Books Publishing House, 2015.
5. Zhuangzi. Shanghai: Shanghai Lexicographical Publishing House, 2003.
6. Dai Sheng. Book of Rites. Beijing: Zhonghua Book Company, 2017.
7. Sun Yat-sen. The Fundamental Principles of National Reconstruction. Beijing: SDX Joint Publishing Company, 2014.
8. Lu Huapu. Approaches to Realizing Urban Green Transportation [J]. Urban Transport of China, 2009, 7(6): 5.
9. Lu Huapu, Zhang Yongbo. Achievements, Changes, and Prospects of China's Powerful Transportation Nation Construction from the Perspective of Sustainable Development [J]. Journal of Sustainable Development Economics, 2021(6): 4.
10. Fu Zhihuan, Sun Yongfu, Weng Mengyong, et al. Strategic Research on Building a Powerful Transportation Nation. Beijing: China Communications Press Co., Ltd., 2019.
11. The State Council of the People's Republic of China. Outline for Building a Powerful Transportation Nation [R]. Beijing: The State Council of the People's Republic of China, 2019.
12. The State Council of the People's Republic of China. Outline of the National Comprehensive Three-Dimensional Transportation Network Plan [R]. Beijing: The State Council of the People's Republic of China, 2021.
13. The State Council of the People's Republic of China. 14th Five-Year Plan for the Development of a Modern Comprehensive Transportation System [R]. Beijing: The State Council of the People's Republic of China, 2022.
14. Xi Jinping Stressed at the Third Collective Study Session of the Political Bureau of the CPC Central Committee to Better Coordinate Domestic and International Situations and Strengthen the Foundation for Peaceful Development [N]. People's Daily, 2013-01-30(1).
15. Lan Hongguang. Xi Jinping Stressed in a Joint Interview with BRICS Media on Unswervingly Following the Path of Peaceful Development and Unswervingly Promoting World Peace and Development [N]. People's Daily, 2013-03-20(1).

MIX
Papier aus verantwortungsvollen Quellen
Paper from responsible sources
FSC® C105338
FSC
www.fsc.org

If you have any concerns about our products,
you can contact us on
ProductSafety@springernature.com

In case Publisher is established outside the EU,
the EU authorized representative is:
Springer Nature Customer Service Center GmbH
Europaplatz 3, 69115 Heidelberg, Germany

Printed by Libri Plureos GmbH
in Hamburg, Germany